W9-CJE-416

Power Supplies, Switching Regulators, Inverters, and Converters

2nd Edition

Irving M. Gottlieb

TAB Books
Division of McGraw-Hill, Inc.
Blue Ridge Summit, PA 17294-0850

SECOND EDITION
FIRST PRINTING

Library of Congress Cataloging-in-Publication Data

Gottlieb, Irving M.
Power supplies, switching regulators, inverters, and converters / by Irving M. Gottlieb. — 2nd ed.
p. cm.
Rev. ed. of: Power supplies, switching regulators, inverters & converters, 1st ed, c1984.
Includes index.
ISBN 0-8306-4405-9 (H) ISBN 0-8306-4404-0 (pbk.)
1. Electronic apparatus and appliances—Power supply. 2. Electric power supplies to apparatus. I. Gottlieb, Irving M. Power supplies, switching regulators, inverters & converters. II. Title.
TK7868.P6G66 1993
621.381′044—dc20 93-1029
CIP

Acquisitions editor: Roland S. Phelps
Book editor: Andrew Yoder
Book designer: Jaclyn J. Boone
Cover design: Denny Bond, East Petersburg, Pa.
EL1
4394

Contents

Preface

MOST ELECTRONIC CIRCUITS AND SYSTEMS REQUIRE A SOURCE OF ELECTRICAL energy for their operation and such sources are collectively known as *power supplies.* These can be classified in more detail by identifying them as the utility mains, rectifiers, regulators, inverters, converters, batteries, solar power supplies, and alternators and generators. This book focuses primarily on those dc and ac sources delineated in the title; a little reflection will reveal that these are "kissing cousins," exhibiting many common features.

Perhaps the most important mutual trait of power supplies, switching regulators, inverters, and converters is their use of solid-state devices. Also, ever increasingly, IC "chips" are found as the "brains" of the more-advanced power sources. Interesting, too, it is seen that advanced systems and sophisticated designs no longer automatically imply greater complexity of implementation—often a single control chip readily performs the myriad of circuit functions that required dozens of discrete stages in erstwhile practice.

At the same time, superceded techniques tend to retain much usefulness, rather than being relegated into obsolescence. This is because many applications can be implemented with relatively simply power sources; tight regulation and the host of "bells and whistles" often provided by control ICs are not always needed. It is for this reason that the technical coverage of this book embraces both the "bare-bones" power sources and the sophisticated power systems that have evolved from primitive beginnings.

Truly, the processing of dc and ac operating power has become a specialized branch of electronic technology and the author is grateful for the assistance rendered by the various firms. Their cooperation reinforces the hope that the contents of this book will interest and benefit designers, constructors, hobbyists, and experimenters in their activities with applied electronics.

Irving M. Gottlieb
Redwood City, California

Acknowledgments

- Steve Bakos: Sales Engineer, Linear Technology Corporation
- John P. Breickner: Manager, Electronics Industry Sales; Indiana General Electronic Products
- W. C. Caldwell: Distributor Sales Administrator; Delco Electronics
- Neil Cleere: Manager, Marketing Services; International Rectifier Corp.
- Walter B. Dennen: Manager, News and Information; RCA Solid State Division
- Mr. Robert Dobkin, Director, Advanced Circuit Development—National Semiconductor Corporation
- J. David Fuchs: Marketing Manager, Magnetics Division; Allen-Bradley Co.
- Forrest B. Golden: Customer Engineering; General Electric Semiconductor Products Dept.
- R. W. Kassiotis: Sales Manager, Germanium Power Devices Corp.
- Alan Koblinski: Engineering Applications Manager, Micro Linear Corp.
- R. A. Mammano: Manager, Advanced Development; Silicon General, Inc.
- Lee Shaeffer: Analog Applications Manager; Siliconix Inc.
- Ken Smith: Manager, Business Development, GENNUM Corporation
- Lothar Stern: Manager, Technical Information Center; Motorola Semiconductor Products Inc.

Introduction

THE INTERESTED READER WILL FIND THIS BOOK TO BE INORDINATELY FLEXIBLE as a practical guide in working with the topics embraced in its title. Some, with patience and a thirst for knowledge, will wish to study it from beginning to end. Others, probably those with considerable experience, might derive optimum benefit from homing in on the specific discussions bearing direct relevance to a project at hand, or to a present problem. Finally, many will, no doubt, be best described as occupying a position between these extremes. This third group of users of this book can be expected to peruse through its pages in random fashion, and at random times.

The author welcomes any or all of these approaches; to a large extent, each topic has been presented on a stand-alone basis. The editorial format will be found to have closer resemblance to a compilation of engineering application notes, rather than to a college textbook in which grasp of later material is very dependent upon mastering earlier concepts.

This being the case, it should not be too far-fetched to hope that the user of this book will gain many useful insights, and that such activity will be imbued with pleasantness, if not downright enjoyment.

1 PART

Inverters and converters

Solid-state inverters and converters take their places alongside inductors, capacitors, discrete semiconductors, and integrated circuit modules. Indeed, a whole new branch of electronic technology is involved. As has generally been the case in such situations, it should prove rewarding to become acquainted with these new techniques for manipulating electrical energy. Over the next five chapters we will examine the theory, design, and application of these devices.

1
CHAPTER

An overview

INVERTERS AND CONVERTERS HAVE BEEN WITH US IN VARIOUS FORMS FOR decades, but in recent years they have become key building blocks in almost endless variety of circuits and systems. Whatever your involvement in electronic or electrical technology might be, it is becoming increasingly difficult to function effectively without a good working knowledge of inverters and converters.

A few examples of practical applications for inverters are shown in Fig. 1-1. Figure 1-1A shows a method whereby 75 percent overall efficiency in converting electrical energy into light can be attained with high frequencies. Figure 1-1B illustrates a method for controlling the speed of an ac motor. With appropriate feedback techniques, extremely constant speed can be obtained. Figure 1-1C is a diagram of a regulated power supply in which the need for magnetic cores at 60 Hz is eliminated; this results in important savings in size, weight, and cost. The electronic ignition system of Fig. 1-1D offers such advantages as precise timing, high spark voltage at high engine speeds, and the ability to fire-fouled plugs.

Other applications include battery chargers, clocks, electric vehicles, frequency converters, induction heating, phonographs, radar, sonar, and TV. An interesting and significant implementation of inversion is presently being developed. It happens that there are decided benefits to be had from the use of dc for transmission of electrical energy over long distances. Giant solid-state inverters appear promising for use at the receiving end of such a transmission line. And, in view of the numerous and diverse applications already cited, it should come as no surprise that computer and microprocessor control of inverters has given further impetue to the art.

Definition of an inverter

An *inverter* is a device, circuit, or system that delivers ac power when energized from a source of dc power. Stated another way, inversion is the reciprocal function of *rectifi-*

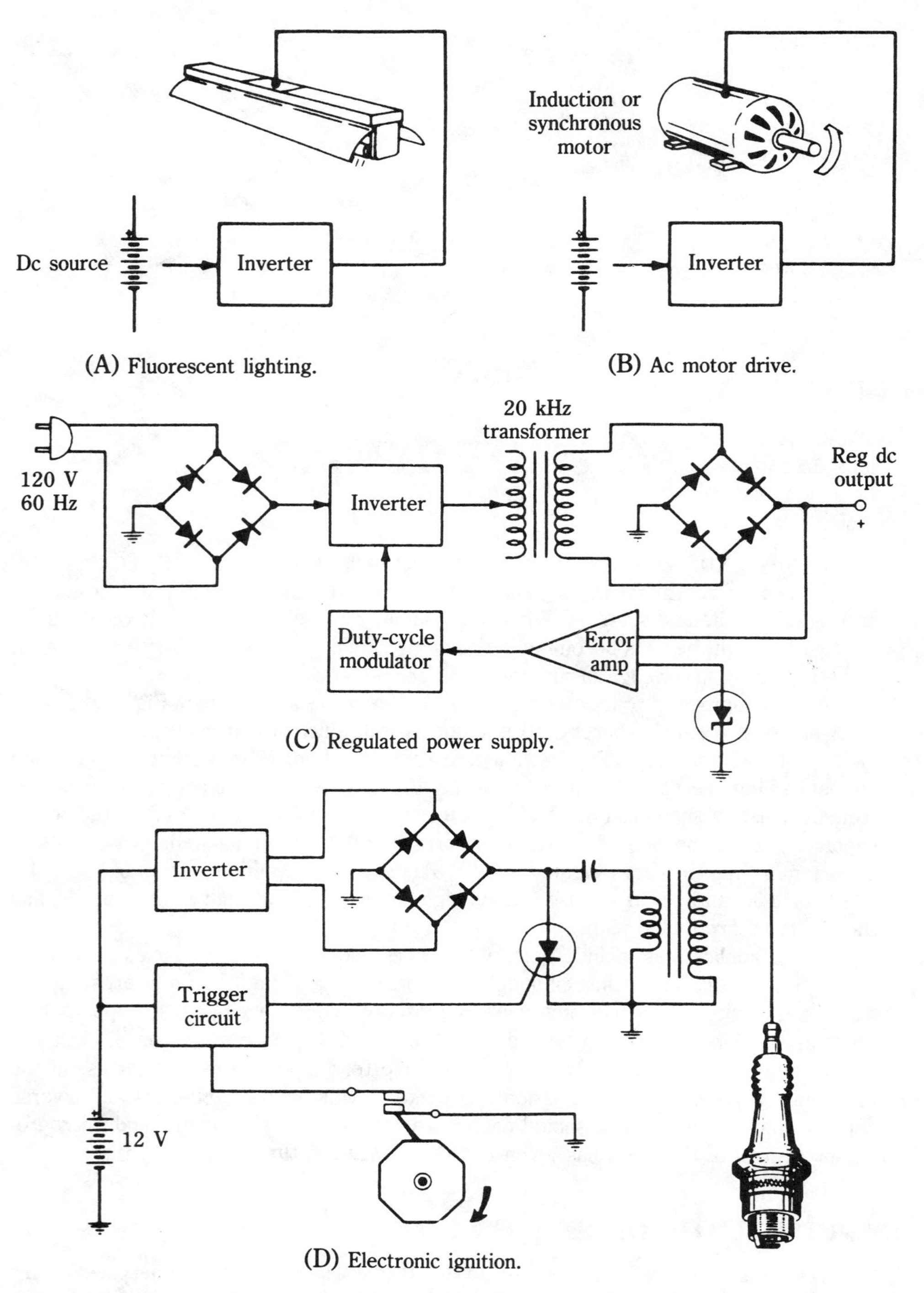

1-1 Typical applications of inverters.

cation. Rectifiers change ac into dc, whereas inverters behave in the converse fashion —they turn dc into ac.

Inverters are not rare entities at all. Under other names, they appear in myriad applications. Choppers, feedback oscillators, and relaxation oscillators certainly appear to qualify as inverters. Do they not change dc into ac? Actually, the use of the terms "inverter" and "oscillator" is somewhat arbitrary. An inverter can be an oscillator, and an oscillator can be used as an inverter. General usage used to favor the term "inverter" when the operating frequency was less than about 100 kHz, and when the implemented function was to provide ac power for some other circuit or equipment. Modern inverters are no longer limited in frequency.

Because there is no sharp line of demarcation between inverters and oscillators, it can be said that many, but not all, inverters are special types of oscillators. Some inverters, however, are more in the nature of driven amplifiers or actuated switches. The choice of terms really boils down to a matter of emphasis. A circuit for producing radio-frequency energy with relatively tight frequency stability has traditionally been called an *oscillator.* A circuit in which the emphasis is on such parameters as efficiency, regulation, and overload capability, and which can operate at audio or low supersonic frequencies, can qualify as an inverter even though its operating principle is similar to that of an oscillator.

Fortunately, in actual practice, the distinguishing features of so-called inverters and oscillators become clear enough when we consider the end use of the circuit. Second nature will tell us immediately whether it is more appropriate to speak of an oscillator as such or to label it an inverter. Usually, it will be found that the inverter is, in essence, a power supply.

Definition of a converter

I have stated that an inverter is powered from a dc source and delivers ac, and a rectifier is powered from an ac source and delivers dc. There is a third possibility, a circuit or system that operates from a dc source and provides dc power to a load. A *converter* does this, but not all circuit arrangements that are powered by dc and deliver dc are referred to as converters. For example, potentiometers, voltage dividers, and attenuators do, indeed, "convert" from one dc level to another. But these generally would not be referred to as converters. The missing element is an inverter, chopper, or oscillator in the middle of the process. In other words, the sequence of processes in the true converter is dc, ac, dc. A usable definition of a converter, then, is as follows: a circuit or system that both receives and provides dc power, in which ac is generated as an intermediate process in the flow of energy (the descriptive phrase *dc-to-dc converter* is sometimes used).

The practical significance of the converter definition is that the converter functions essentially as a *dc transformer.* This property makes possible the manipulation of dc voltage and current levels in the same way that we use transformers in ac systems. Moreover, like the transformer, the converter provides isolation between the input and output circuits. This promotes electrical safety and greatly simplifies a number of system design problems.

Consider a converter with an additional step added. Suppose that the overall sequence of events is ac, dc, ac, dc; that is, the device receives power from an ac line, rectifies it, inverts it to ac, and rectifies it again. This is the basic format for many power supplies. Does it appear to be needlessly redundant? It is not, for the inversion process usually generates ac at a much higher frequency than that of the power line. This makes possible the elimination of a massive and costly 60-Hz power-line transformer. The inverter transformer (often a 20 kHz to several-MHz design) can be very small, yet provide complete isolation.

Mechanical inverters and converters

Most of this book deals with inverters and converters that utilize modern solid-state devices. However, it is necessary to review briefly some of the older methods of inversion and conversion. Many of these schemes are still with us and continue to require service and maintenance. It would be wrong to assume that they are automatically obsolete because they involve tubes or mechanical motion. In industry, many inverters and converters operate at high power levels. Although we tend to think of the life spans of semiconductors as indefinite, if not infinite, large transistors and thyristors, unfortunately, are subject to both aging and failure mechanisms.

If we are to do justice to the solid-state approach, we must be fully aware that an electronic replacement for mechanical actuation is not necessarily better because it seems to be a more sophisticated or elegant solution. The applications engineer must look at many factors besides scientific novelty. In industry, ruggedness and immunity to hostile environments are vital considerations. As a case in point, it has been quite difficult to produce thyristor welders as "tough" as the motor-generator type. A thyristor converter appears to be a beautifully precise and strikingly efficient means of controlling welder current. But only recently have devices and designs become available that are able to cope with the extremely high electrical and thermal stresses encountered in welding.

The vibrator power supply

The predecessor of the modern solid-state converter was the *vibrator power supply.* These supplies were designed primarily for operating automobile radios. However, amateur radio operators often modified them for greater output in order to power their mobile communications equipment. Vibrator power supplies generally had efficiencies in the vicinity of 70 percent. The vibrators themselves had limited life spans, so they were designed as plug-in units that could be replaced as easily as electron tubes. Some types had exotic metal contacts and were hermetically sealed. Special units were evacuated. The extensive research and development poured into the production of these vibrators brought them to a high state of technical evolution before solid-state techniques rendered them unnecessary. Although their use has dwindled, they still are potentially useful to the hobbyist and the experimenter. They are still used occasionally in certain instrumentation fields, such as Geiger counters and portable ultraviolet lamps.

In any event, a brief review of the vibrator supply will help gain an insight into the nature of more sophisticated inverters and converters. The vibrator type is similar to the solid-state types in that both deliver ***chopped dc*** to a transformer. Both, therefore, generate ac power having a square waveform.

The circuit shown in Fig. 1-2 is an inverter, although it was usually associated with an output rectifier and filter for the purpose of radio operation. The contact-carrying armature occupies a "neutral" position when S1 is open. When S1 is closed, the magnetic flux in the solenoid attracts the iron armature, bringing the upper vibrator contacts together. Thus, current passes through the upper half of the center-tapped primary winding of transformer T1. But, this action also shorts the solenoid, thereby allowing the armature to spring back. Momentum then causes the armature to overshoot its neutral position, bringing the lower contacts together. Current now passes through the lower half of the transformer primary. However, the armature cannot remain in this position because of its springiness and because of the pull again exerted by the magnetic field of the solenoid. As a result, the armature bounces back and closes the upper contacts again. The sequence of events continues to repeat itself, and the alternate contact closures impress a "push-pull" waveform on the primary winding of the transformer.

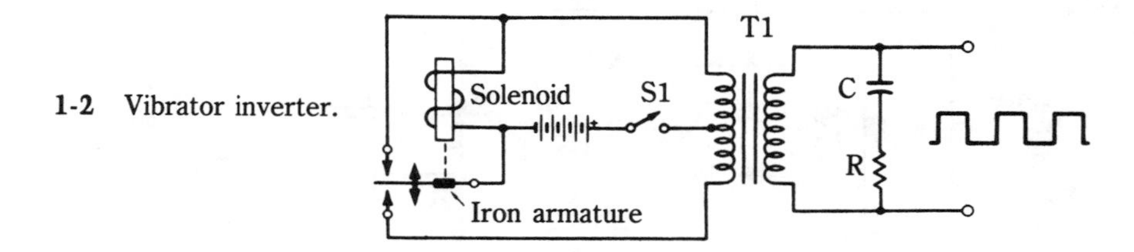

1-2 Vibrator inverter.

By appropriate design, these vibrators can be made to produce symmetrical square-wave voltage waveforms in the secondary of the transformer. In most instances, T1 is a step-up transformer because the overall objective of these supplies is to derive 200 volts or so from the low-voltage vehicle battery. Vibration frequencies are anywhere from about 50 to several hundred hertz, depending on the application and the state of the battery. (The automobile-radio types generally "buzzed" at about 120 Hz. The vibrator units were frequently packed with felt or other sound-absorbing material to attenuate their audible noise output.)

The RC network across the secondary winding of T1 is called a ***buffer circuit.*** Proper selection of R and C results in minimal sparking of the contacts and greatly reduces the spikes in the generated waveform. This reduces the "hash" output of the supply. However, additional RF filtering in the form of chokes and bypass capacitors in the input and output leads of the supply is generally necessary in order to obtain reasonably clean radio operation. Nominal values for R and C are 1000 Ω and 0.01 μF, respectively. Higher C and lower R will do the same job in the primary, assuming T1 is a step-up transformer.

The design of the transformer T1 is "conventional" in the sense that deep magnetic saturation is avoided. Electrostatic shields between the primary and secondary windings have sometimes been used, but grounding the core gives results that are nearly as good.

Although not shown in conjunction with Fig. 1-2, various full-wave rectifiers were used in auto-radio supplies. These included cold-cathode gaseous tubes, thermionic rectifier tubes, and selenium rectifiers. A modern version of such a supply would use silicon diodes, probably in a bridge configuration. Such combinations have indeed been used in recently manufactured electronic flash units for photographers. Half-wave and voltage-multiplier circuits have also been used in such equipment.

The synchronous vibrator The vibrator-transformer arrangement shown in Fig. 1-3 is a dc-to-dc converter that does not need a thermionic, gaseous, or semiconductor rectifier. This is because rectification is accomplished by additional contacts that are driven in synchronism with the chopped primary current. Such mechanical rectification had compelling advantages prior to the availability of inexpensive germanium and silicon rectifying diodes. From the vantage point of present technology, the introduction of the extra contacts would raise questions regarding the reliability and life span of the vibrator. It would seem wiser to invest more in the integrity of the primary contacts and use the simpler inverter of Fig. 1-2 in conjunction with a silicon full-wave rectifier—if you decided to use a vibrator in the first place.

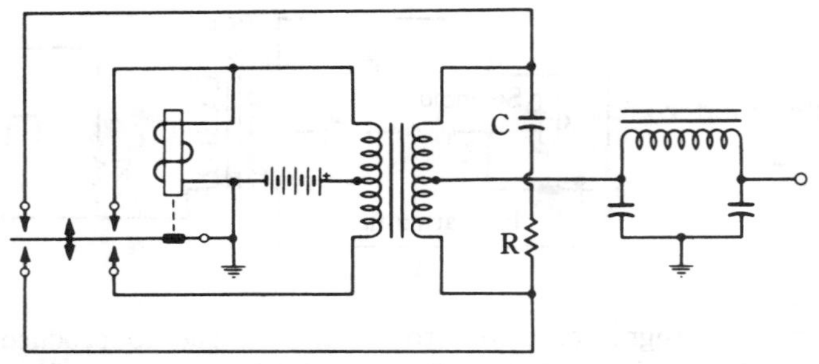

1-3 Synchronous vibrator converter.

The vibrator ignition system Vibrator ignition coils were well known as part of the Model T Ford automobile, and they were also used on other early-vintage cars. Although there might be a tendency to think of such a method as crude and primitive, it is interesting to note that one goal of modern solid-state ignition designs has been to recapture the sustained firing feature of the vibrator ignition system. The reason for recalling it here is that its objective was diametrically opposite to an ever-present problem in modern inverters; whereas we try to damp or suppress transients in inverters, the vibrator ignition coil made optimal use of transient generation. In Fig. 1-4, it can be seen that the ignition coil is not unlike the vibrator power supply described previously, except that the vibrator is actuated by the magnetism of the transformer core itself. Some vibrator power supplies also used leakage flux from the transformer core in this fashion.

The central idea of the vibrator ignition coil was to achieve a very high rate of current change in the primary winding. By the law of self-induction, this must cause a high induced voltage (counter EMF) to be developed across the primary when the contact points open. Nature opposes the abrupt cessation of current in an inductive

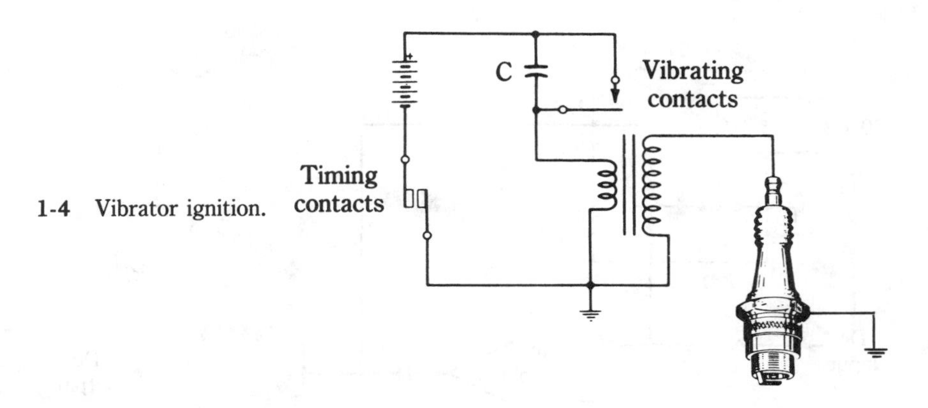

1-4 Vibrator ignition.

circuit, with the result that arcing tends to occur between the contacts as they open. But, if a capacitor (C in Fig. 1-4) is placed across the contacts, the voltage that produces the arc is absorbed as electrostatic energy. Accordingly, the air between the contacts is not ionized, and the primary current is interrupted abruptly. The counter EMF developed across the primary is therefore very much higher than the battery voltage. A high number of secondary turns can then further raise this voltage for the spark plug.

The vibrator operates at an audio-frequency rate, but the open-core construction of the step-up transformer is not designed for reproduction of the fundamental interruption frequency. Rather, high-frequency components contained in the steep decay rate of the current interruption contribute most of the useful energy to the spark plug. Shock excitation of stray inductance and capacitance produces damped radio-frequency wave trains, which occur at the chopping rate of the vibrating contacts. The transient and spikes that tend to accompany the operation of modern solid-state inverters originate in much the same way. Such transients, unless adequately suppressed, are potentially destructive to a modern inverter because of the vulnerability of solid-state devices to excessive voltage levels and to excessively fast current transitions. Also, faulty operation often results from high rates of change of voltage. Even if the active devices are not endangered or the circuit operation is not made erratic, transients could still cause objectionable interference in sensitive circuits, such as those in communications equipment.

The chopper amplifier

The arrangement illustrated in Fig. 1-5 might be considered a converter because it can accept dc at one voltage-current combination and deliver it at another. Furthermore, the sequence of events is dc, ac, dc (thus, an ordinary dc amplifier would not qualify as a converter). The function of the chopper is to pulse the dc impressed at the input of the amplifier and simultaneously to short out one polarity of the pulse train at the output of the amplifier. The second process constitutes synchronous rectification similar to the process in the device of Fig. 1-3. The arrangement in Fig. 1-5 eliminates the problem of drift when the amplification is high. Another attribute of this converter is that the level of the dc output power can be much greater than the input. Depending on the application, this might or might not be significant, for the major portion of the output

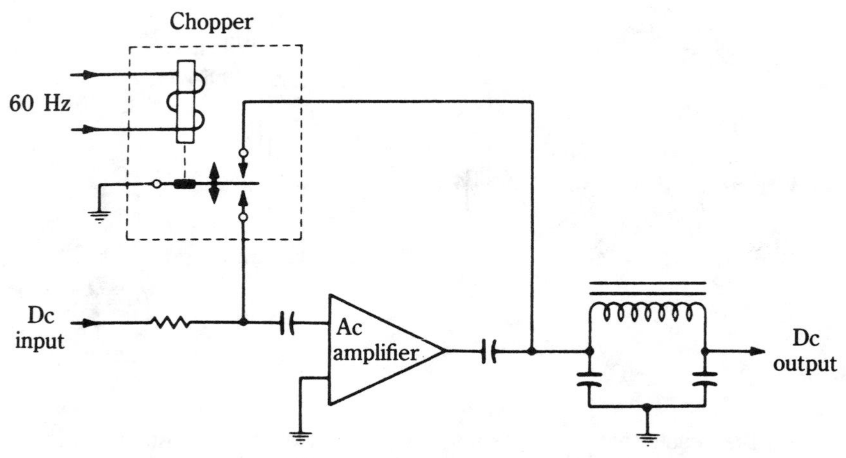

1-5 Chopper amplifier or converter.

power comes from the dc supply for the ac amplifier (not shown). In more recent designs, the electromagnetic chopper has often been replaced by solid-state switching circuits.

Rotary machinery For many years, the functions of inversion, conversion, rectification, level shifting, and isolation have been performed by ***rotating machinery.*** If is often possible to substitute solid-state equipment for such large and heavy machines, but it is not necessarily certain that static semiconductor systems will completely displace these mechanical "monsters." There are those who strongly feel that, despite the need for considerable maintenance, rotating equipment has high reliability, can withstand even severe abuse, and naturally provides various degrees of operational flexibility. In addition, advocates of rotating machinery point out that there are many applications where size, weight, vibration, and noise are not necessarily negative features.

The motor-generator A straightforward combination of rotating machines can be made from a squirrel-cage induction motor and a permanent-magnet dc motor. If power from the three-phase ac line is fed to the induction motor, the dc motor will behave as a dc generator. The system will then simulate the function of rectification with the advantage that the dc load will be electrically isolated from the ac power line. If, instead of an induction motor, a synchronous (or synchronous-induction) motor is utilized, essentially similar operation can be had. However, the converse sequence is also attainable. That is, if dc is fed to the permanent-magnet motor, the driven synchronous motor will then behave as an alternator. (A squirrel-cage motor can also operate as an alternator, but only when it is overdriven to supply additional power to an already-energized ac power line.) The system will therefore perform the function of an inverter. Such a combination of electrical machines is shown in Fig. 1-6. For added flexibility, the field current of the alternator can be externally controlled, and the dc machine can also have field windings, rather than a permanent magnet. With these machine parameters available for adjustment, it is not difficult to control the regulation, frequency, and power factor of the ac output power.

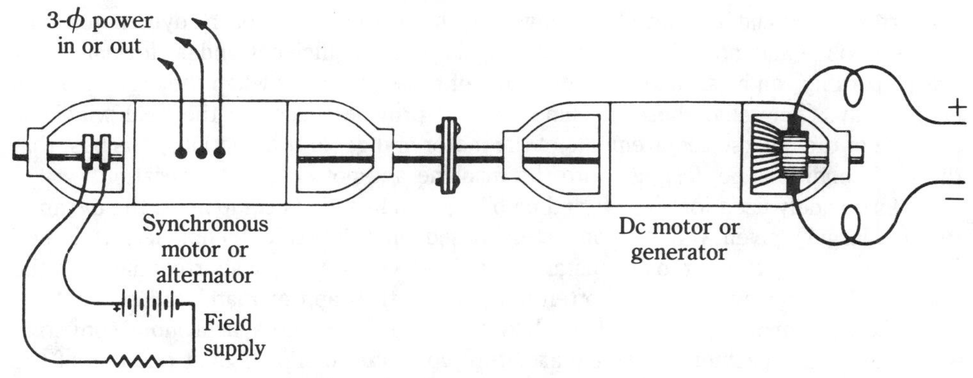

1-6 A motor generator system.

An advantage of the synchronous motor or alternator is that the heavy ac currents are conducted into and out of the stator winding by direct connection. The slip-ring assembly in this case handles the relatively small current needed by the rotating field.

The dynamotor The machine illustrated in Fig. 1-7 is known as the *dynamotor,* or *genemotor.* It is a true converter. If dc is applied to one set of brushes, dc power will then be available from the other set of brushes, but generally at a different voltage and current level. Although obviously not a static device, the dynamotor is suggestive of a "dc transformer." Typically, the "motor" part of the machine will be designed for a voltage that is associated with batteries (such as 6, 12, or 24 volts). The "generator" part is generally designed to deliver several hundred volts for vacuum-tube plate power. The machine is essentially a double dc dynamo in which the armature carries two separate and mutually insulated windings that share a common field.

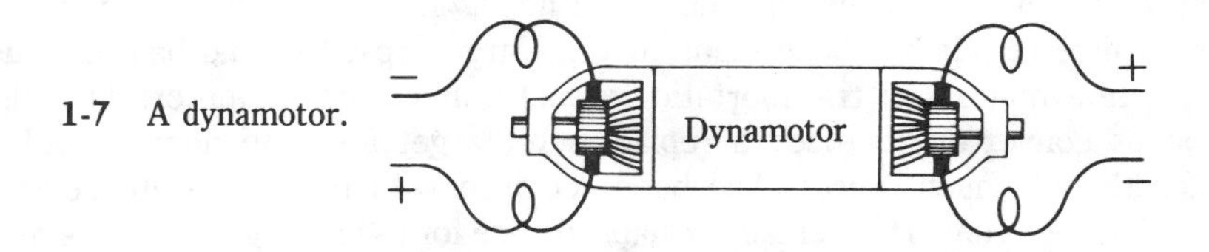

1-7 A dynamotor.

It should be noted that this machine qualifies as a true converter, inasmuch as the sequence of change in the format of electrical energy is dc, ac, dc. This is true because the armature windings of all dc commutator-type machines carry ac. The commutator performs as a synchronous switch during motor action in order to produce continual rotation, rather than a stationary "latched up" condition. During generator action, the commutator behaves as a synchronous rectifier, continually delivering the same polarity to each brush.

From the preceding description of the dynamotor, it follows that its operational mode can be reversed. That is, if dc at several hundred volts is applied to the "generator" end, a low-voltage, high-current output will be available from the "motor" end. Such interchangeability is not unique to the dynamotor, but applies to ordinary dc motors and generators; such mode reversal is not generally attainable from solid-state converters. It should be pointed out, however, that the operation of the dynamotor in its reversed mode may not necessarily occur with optimum efficiency and performance, for brush sparking can be anticipated from most of these machines when they are operated in this way. Generally, these machines do not provide access to the field windings. Because of the diverse requirements of the motor and generator sections, it is best that the field conditions be designed into the machine and not subject to variation.

Once widely used for aircraft and mobile communications equipment, the dynamotor has largely given way to converters based on solid-state techniques. It is still frequently encountered in older installations, however. Its major shortcoming was the low overall efficiency attained—often not greater than approximately 45 percent.

The synchronous converter Like the dynamotor, the synchronous converter is a double-duty machine. Its single armature connects to slip rings at one end of the machine and to a commutator at the opposite end (Fig. 1-8). The machine is driven by applying ac to the slip rings, and dc power is then available from the commutator brushes. The machine also can be operated in the reverse mode, whereupon it functions as an inverter—that is, it can provide ac power when it is driven from a source of dc power. Inasmuch as a single armature winding is used, rather than two separate ones as in the dynamotor, there is no isolation between the input and output terminals.

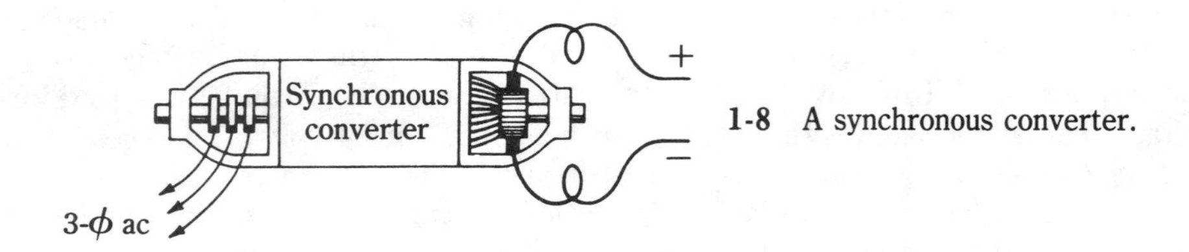

1-8 A synchronous converter.

In the past, this machine has been used more often as a substitute for rectifiers than as an inverter. It has been made in very large capacities and has been used to power traction motors on transportation vehicles and huge dc motors in mills. The synchronous converter has been a replacement target for proponents of solid-state control. Although much success has been achieved with the solid-state devices, the large synchronous converter remains a viable device for use in large power installations.

When the synchronous converter is operated as an inverter, the output frequency can be varied by the simple expedient of adjusting the speed of the dc-motor portion of the machine. One of the several possible output formats is an array of polyphase sine waves. Single-phase versions of this machine have been rare because of starting difficulties when they are used to supply dc from an ac power line. However, such a single-phase design could function satisfactorily as an inverter. Its efficiency would not be as good as that of the polyphase machine, however. An interesting feature of the synchronous converter is that it can provide both dc and ac power when driven by another prime mover. In such operation, the available power simply divides between the ac and dc loads.

The ac part of the synchronous converter is a synchronous motor with the poles wound in the stator. The rotor, therefore, conveys and receives heavy currents through its slip-ring assembly. When the machine is used as an inverter, the synchronous motor becomes an alternator. The stationary poles also contain damper windings. These serve two purposes: (1) they serve as a stationary "squirrel cage" to enable the synchronous motor to start as a polyphase induction motor. When the motor attains, for example, 85 percent of synchronous speed, it pulls into synchronism and remains locked in synchronism with the power-line frequency. (2) The damper windings also prevent hunting and transient departures from synchronous speed.

Feedback oscillators

It has been pointed out that inverters and oscillators are closely related. It often does not introduce either error or confusion to use the two terms interchangeably. Moreover, some practical inverters actually use circuits that have traditionally been called *oscillators.* In any event, a brief review of the salient features of oscillators is in order. Most of these oscillators are feedback types. Although most types of oscillators probably would give good performance if proper consideration was given to design parameters, some circuits are more convenient for certain applications than others. The Hartley circuit is probably the one most often encountered in inverters. Sometimes, because a tapped inductor is not required, the Colpitts circuit is also used. Four feedback oscillators that are often used as inverters are shown in Fig. 1-9.

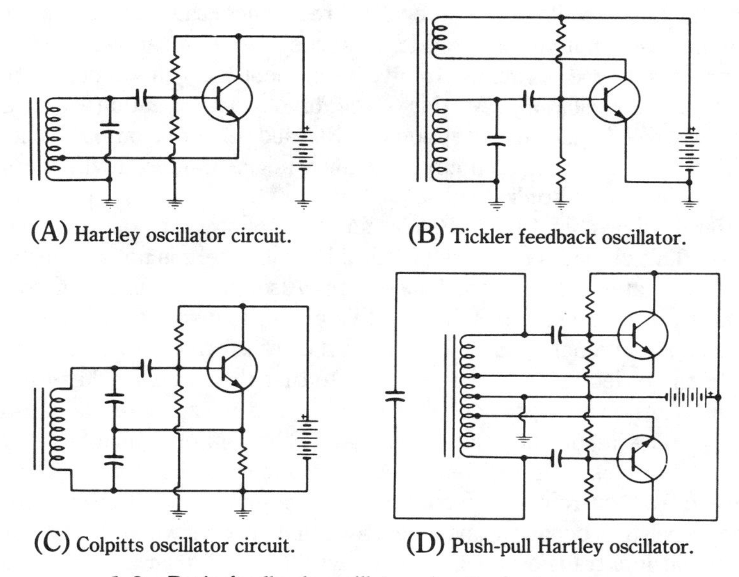

(A) Hartley oscillator circuit.

(B) Tickler feedback oscillator.

(C) Colpitts oscillator circuit.

(D) Push-pull Hartley oscillator.

1-9 Basic feedback oscillator circuits for inverters.

An important feature of feedback oscillators is that a near sine wave can be generated, which greatly reduced the amount of RF interference produced. However, in the interest of efficiency, such feedback oscillators are often allowed to operate in an overexcited condition. This, of course, results in a definitely nonsinusoidal output. Nevertheless, the harmonic output tends to be much less than in the case of conventional square-wave inverters, which involve very short rise and fall times, to say nothing of spikes and switching transients. Generally, feedback oscillators are utilized in inverters with low power capability, such as those associated with Geiger counters, small strobe lights, and some photoflash units. Although ferrous-core inductors are usually used in the resonant tank circuits, operation is confined to the essentially linear portion of the B-H curve.

The basic feedback circuits can be used in either the single-ended or push-pull form. The active device(s) can operate in class A, B, or C or anywhere between. Very useful inverters can be designed around push-pull feedback oscillators operating in class B, or slightly into class C, and utilizing tuned tank circuits. Such inverters produce reasonably good sine-wave outputs and can attain efficiencies above 70 percent.

RC and L/R relaxation oscillators

Relaxation oscillators can be defined in several ways. Sometimes the emphasis is placed on the generation of nonsinusoidal waveform, such as sawteeth, pulses, or square waves. Saturated operation of the active device is sometimes cited as an important aspect of such oscillators. Yet another viewpoint draws attention to the cyclic storage and discharge of energy in inductive and capacitive networks (but not in circuits with resonant LC tanks). Practical relaxation oscillators generally assume the form of RC multivibrators. Both feedback and negative resistance can be associated with RC networks to produce multivibrator action. It is true, also, that saturated-core oscillators (the most important type pertaining to inverters) meet the general descriptive criteria defining relaxation oscillators. At this stage of our investigation of inverters and converters, we will focus our attention on RC and L/R relaxation oscillators. The objective is to point out that such circuits should not be confused with saturated-core oscillators, which will be studied subsequently.

The circuit shown in Fig. 1-10A is an L/R counterpart of a conventional RC multivibrator. This circuit is not very practical because there is not a wide choice in the selection of the ratio L/R in actual inductors. Also, the requirement for two core components is a drawback. Nevertheless, this circuit is instructional (output could be derived from an additional winding on one of the transformers, or through capacitors connected to the collectors of transistors). Note that this circuit makes use of positive feedback; the transformers provide cross coupling as the capacitors do in ordinary RC multivibrators. The important point is that the mechanism of oscillation does not entail saturation of the cores. Rather, the collector windings of the transformers are alternately charged from current drawn from the power supply and discharged through the diodes. The transistors behave as switches. Switching occurs when the rate of change of collector current in a transformer associated with an "on" transistor approaches zero. Insufficient secondary voltage is then developed in the base winding to keep the alternate transistor in its "off" state. The transistors alternate their conductive states regeneratively as in the more familiar RC multivibrators. The resistance involved in the L/R time constant is that of the collector windings of the transformers.

A more practical form of the L/R multivibrator is shown in Fig. 1-10B. This circuit illustrates the basically different oscillation modes of L/R multivibrators and saturable-

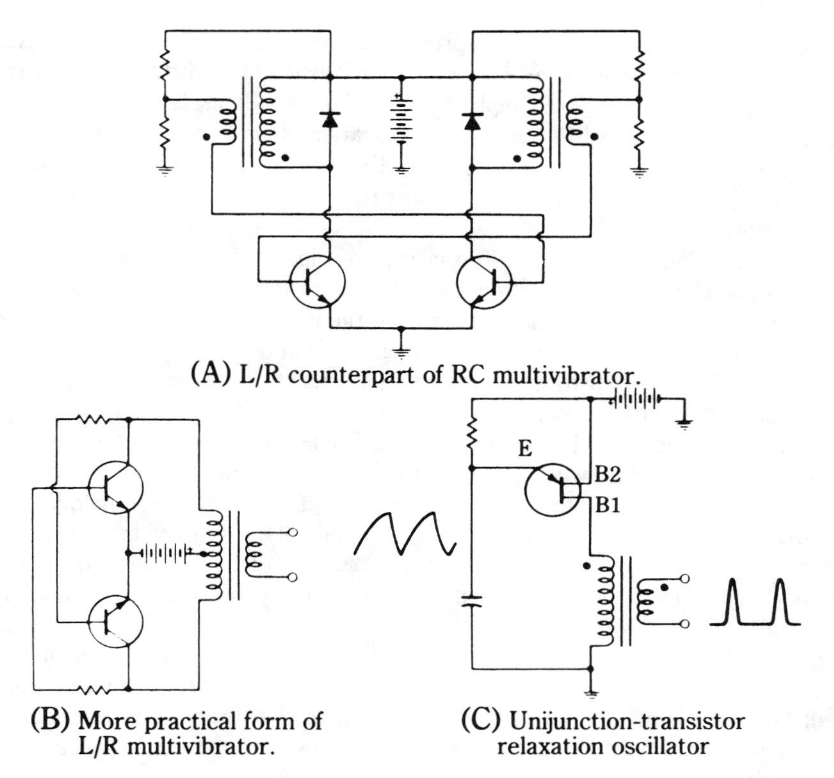

(A) L/R counterpart of RC multivibrator.

(B) More practical form of L/R multivibrator.

(C) Unijunction-transistor relaxation oscillator

1-10 Relaxation oscillators that do not depend on core saturation.

core inverters. If the base resistances are kept at relatively high values, this circuit functions in a manner similar to that of Fig. 1-10A. The oscillation frequency is a function of the wire resistance of the center-tapped collector winding. Notice that there is mutual inductance between the two halves of this winding. Suppose the lower transistor is turning on. This action is reinforced by base voltage induced in the upper half of the winding. But this action can be sustained only while the rate of change of current in the lower half of the winding remains high. Ultimately, the rate of change approaches zero, whereupon the lower transistor no longer receives turn on bias from the upper half of the winding. At the same time, the upper transistor receives sufficient forward bias to turn on. The transistors then switch their conductive states. The process is both regenerative and repetitive.

If the base resistances are lowered sufficiently, the circuit goes into a different mode of oscillation. Although it remains a relaxation oscillator according to the usual definitions, its frequency is no longer strongly dependent on the ohmic resistance of the collector winding. In this operational mode, the core of the transformer is driven into saturation. The oscillation frequency now is a function of the flux density at which magnetic saturation occurs. In practice, this circuit is often operated in an intermediate condition such that its operation involves some aspects of both oscillatory modes. With this type of operation, it is difficult to predict the performance of the circuit. Also, the dissipation in the base resistances sometimes seriously impairs its efficiency. However, the circuit has the advantage of simplicity, and it has proven useful for some low-power applications in which precision and efficiency are not vital considerations. It would appear desirable to design the transformer so that the oscillation frequencies of the two modes would be close to one another. However, basic incompatibilities normally make this impossible.

The unijunction-transistor (UJT) oscillator of Fig. 1-10C merits attention because it utilizes a transformer, yet is not dependent on core saturation for its operation. This circuit is basically an RC relaxation oscillator—the inductance of the transformer has a relatively minor effect on frequency or waveform. The primary purpose of the transformer in this circuit is to provide a convenient means of extracting the output. This is a very practical inverter for low-power applications, and it is often used to produce high voltage at low current. Good efficiency is attainable because negligible current is drawn from the power supply between pulses. The pulses are generated when the capacitor charges to the emitter triggering point. The resultant action is analogous to the ionization that occurs between the anode and cathode of a thyratron tube when the grid potential is raised to a certain level. When the emitter, E, is thus triggered, the previously high impedance between E and B1 abruptly becomes very low, thereby discharging the capacitor. But once the voltage of the capacitor is substantially depleted, the action ceases, and the capacitor repeats its charge cycle. For all practical purposes, the emitter-base-1 circuit functions very much like a simple neon bulb. Indeed, the waveform monitored at the junction of the charging resistance and the capacitor is very nearly the same exponential sawtooth produced by a neon-bulb relaxation oscillator.

Figure 1-11 is the current-vs.-voltage characteristic of the emitter-base-1 circuit. Notice that the voltages below the value of the triggering point, B, the current is quite small. Once point B is reached, the device becomes a negative resistance, and the current rapidly accelerates to high values. If the source of this current is a capacitor, as in the case of the relaxation oscillator, the voltage falls below the value at point C, and the action can no longer be sustained. In essence, the propagation of charge carries inside the device becomes extinguished, and the circuit current abruptly returns to point A. The capacitor charging cycle then repeats. The action is analogous to the deionization of a gaseous tube. The time interval between points A and B is determined by the time constant of the resistance-capacitance network and the voltage triggering level of the device.

The UJT oscillator excels the neon-bulb circuit for production and fast pulses. The load and firing circuit are isolated, and, in addition, the frequency is largely unaffected by changes in the supply voltage (the triggering voltage changes in proportion to the supply voltage). The UJT oscillator is widely used as a triggering device for SCRs and triacs in high-power inverters.

The thyratron inverter

From the standpoint of circuit applications, the thyratron, a gas tube, is the ancestor of the SCR, a solid-state device. Although they are rapidly being rendered obsolete by

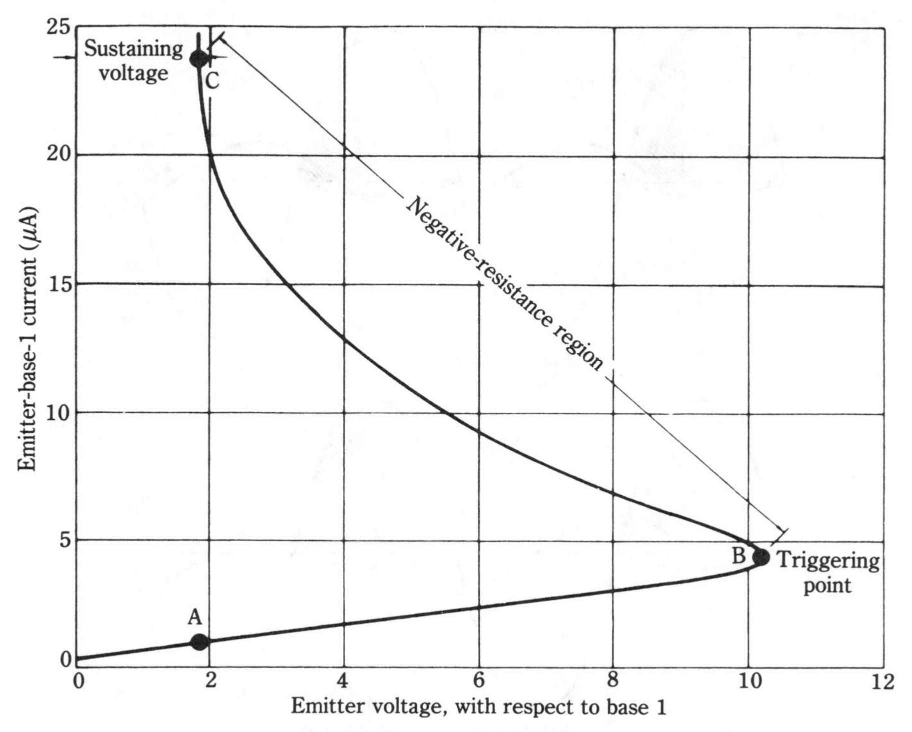

1-11 Current vs. voltage in the emitter-base 1 circuit of a unijunction transistor.

modern semiconductor devices, thyratrons have been extensively used in industrial applications, as well as for modulator service in radars, and it is probable that they will be encountered in older equipment for some years to come. From a servicing and maintenance viewpoint, it is important to be acquainted with these devices. Although an in-depth study of thyratrons will not be undertaken in this book, a fairly comprehensive knowledge of thyratron technology is recognized as an asset to those working in power electronics. If for no other reason, this is true because thyristor technology is very much a takeover of techniques previously developed for thyratrons.

Figure 1-12 shows how the conduction angle, and therefore the average output voltage, of a thyratron can be controlled by varying the phase of an ac voltage applied to its control grid. Once triggered, the conductive path between cathode and plate in the thyratron can no longer be controlled by the grid voltage. However, it can be seen in Fig. 1-12 that conduction is extinguished every time the plate-cathode voltage passes through zero. Accordingly, if you wish to turn the tube off cyclically in an inverter circuit, you must find some way to depress the plate voltage momentarily to zero or to a negative value, with respect to the cathode. In such inverter operation, the thyratrons are turned on by a grid signal, but are turned off by a transient that reduces the plate voltage sufficiently to enable the tube to deionize. This turn-off technique is known as *commutation.* Note that whereas turnoff can be automatic in ac-powered circuits, an inverter, which is dc-powered, needs a special commutation technique.

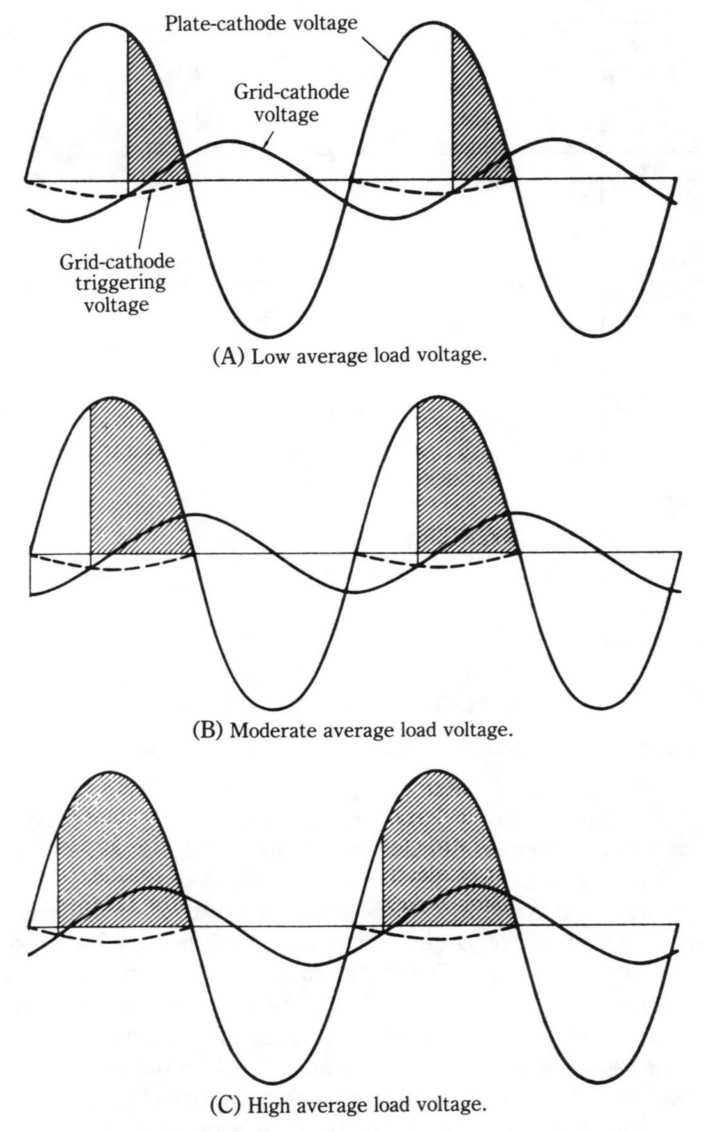

(A) Low average load voltage.

(B) Moderate average load voltage.

(C) High average load voltage.

1-12 Phase control of a thyratron.

A basic thyratron inverter is shown in Fig. 1-13. The resemblance to a multivibrator circuit is obvious. Not obvious from inspection is the fact that transformer saturation plays no part in the switching process. Also, the notion that C_x is used in conjunction with the transformer to form a resonant tank circuit would prove misleading. It is often instructive to analyze the operation of a circuit by assuming it to be in operation and trying to account for its continued operation. Apply this technique to the thyratron

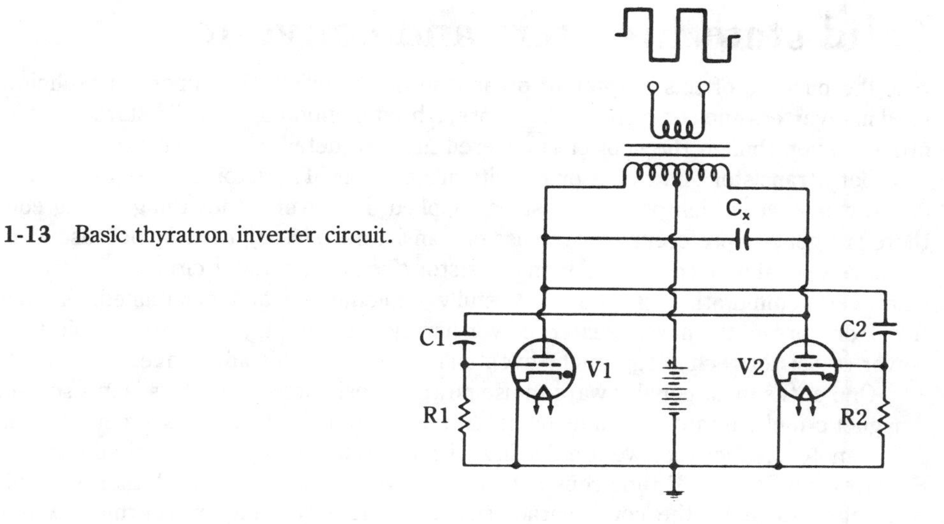

1-13 Basic thyratron inverter circuit.

inverter. Suppose that the inverter of Fig. 1-13 is in a steady-state oscillatory condition and try to see why such oscillation is its natural behavior.

When you initially monitor the performances of this circuit, you find that tube V1 has just turned on. In doing so, it causes negative-going pulses to be transmitted through the two capacitors connected to its plate. One of these pulses is coupled through C_x to the plate of tube V2, thereby turning V2 off. This pulse is dissipated quickly because of the power required by the deionization process. This is perfectly acceptable inasmuch as a prolongation of this pulse could conceivably interfere with the subsequent turn on of tube V2.

Now, focus your attention on the turn on of V2. After the grid of V2 is driven negative by the negative-going pulse received through C2, this negative potential immediately begins a relatively slow discharge toward cathode potential. The rate of this discharge is governed by the time constant of the grid network, R2/C2. Ultimately, the grid is no longer sufficiently negative to hold off ionization, and tube V2 turns on. In so doing, it causes a similar sequence of events to affect V1, first turning V1 off and subsequently turning it on again. Thus, the two tubes alternately exchange conductive states, which means that oscillation persists indefinitely.

In practical inverters, switching is not instantaneous. This is partially because a step-function transition is not attainable in practice. The situation is further aggravated because of the ionization and deionization times required. Unfortunately, deionization time is considerably greater than ionization time. As a result, an interval of simultaneous conduction can exist. This degrades efficiency and is not good for the tubes. Simultaneous conduction assumes greater importance at higher oscillation frequencies, where the mutually on interval becomes comparable to half period of the switching cycle. A choke is often placed in series with a power-supply lead to absorb such spike energy.

If this inverter delivers power to an inductive load, the commutating action of C_x will be reduced. Within limits, this can be remedied by making C_x larger.

Solid-state inverters and converters

It is the purpose of this chapter to present in quick review the important techniques used in inverters and converters. Therefore, a brief mention of the solid-state types is in order—even though the subject is covered in more detail in later chapters.

Both transistor and thyristor circuits are popular. Thyristor circuits takes precedence where very high powers must be supplied. However, other things being equal, there is a general preference for the use of transistors. This is primarily because you do not have to deal with commutation in transistor circuits. Thyristor circuits perform well wherever commutation can be successfully implemented and maintained. Although thyristors presently have greater power-handling capability, the rapid evolution of power-transistor technology might eventually overcome this advantage.

One of the most popular ways to use power transistors as inverters is in a so-called "magnetic multivibrator" arrangement. Although somewhat of a misnomer, the term does denote the fact that we are dealing with a relaxation-type, self-excited oscillator. However, no RC or L/R time constants are involved in its operation. Rather, switching is brought about by the core characteristics of the associated transformer. A more appropriate name for this important solid-state inverter is ***saturable-core oscillator.*** The basic circuit, shown in Fig. 1-14, appears simple enough. Its uninvolved configuration should not, however, obscure the fact that it, together with various modifications, constitutes the prototype for many inverter and converter applications encountered in electronics. It is only in heavy power engineering that other basic formats, generally thyristor circuits, assume importance.

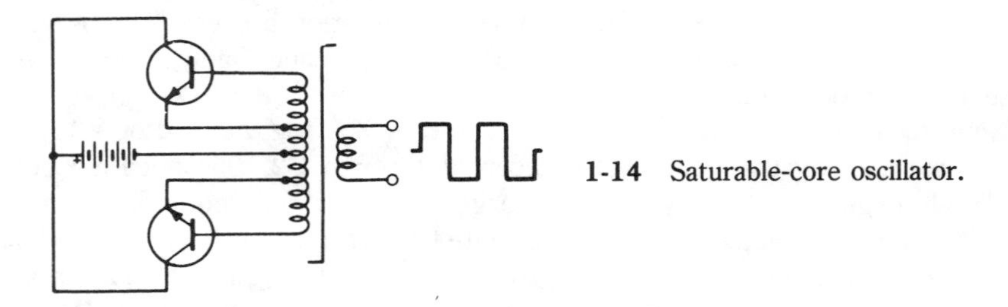

1-14 Saturable-core oscillator.

A different class of inverters is represented by Fig. 1-15. Here, the power transistors are not self-oscillatory, but are driven. Usually the transistors are driven from a square-wave source, but in some instances, they can be driven from a sine-wave source. Driven inverters lend themselves well to duty-cycle control and to pulse-width modulation.

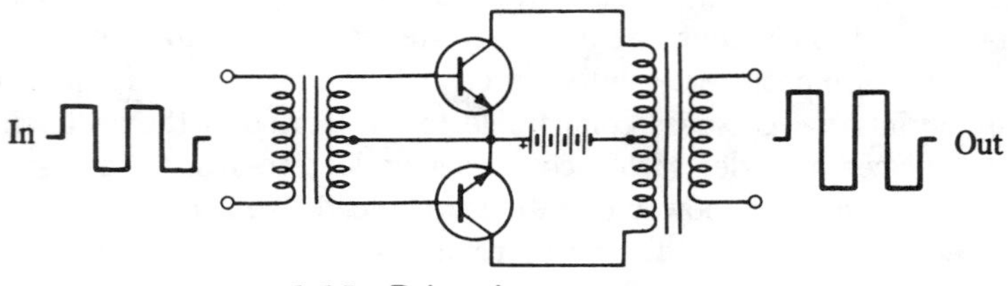

1-15 Driven inverter system.

The circuit of Fig. 1-16 is that of a parallel-type SCR inverter. It is somewhat similar in concept to the thyratron inverter of Fig. 1-13. However, the SCR version is not self-oscillatory; rather, gate signals are required from an appropriate logic circuit. This, however, does not qualify this inverter as a driven type in the generally accepted sense, for the output is not even an approximate replica of the sharp trigger pulses fed to the SCR gates. This and most other SCR and triac arrangements used to produce inversion or conversion can best be described as *triggered circuits.* They perform in the manner of triggered flip-flops, but with high power-handling capability.

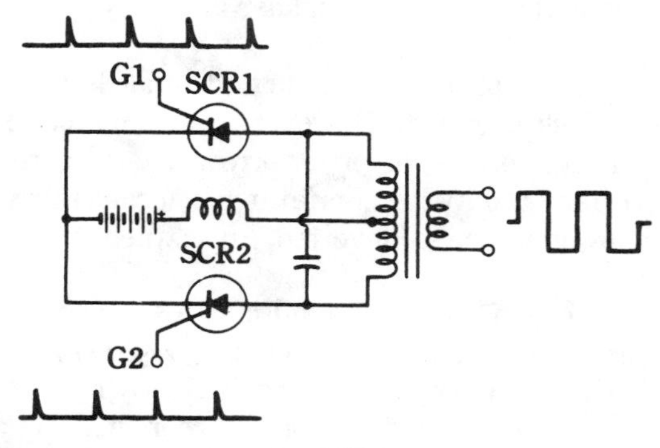

1-16 An SCR inverter.

Notice that the logic circuitry for producing the gate trigger signals is not shown. This is in keeping with the objective of this chapter, reviewing inverters and converters, according to their salient features, rather than in detail. However, it is also the general practice in technical literature to depict SCR and triac inverter and converter circuits in this way. Sometimes the logic circuitry is shown separately. Frequently, however, it is not shown at all. The philosophy underlying this practice is that the circuit will operate if supplied with appropriate gate signals. Inasmuch as different applications will have different requirements for frequency, control, isolation, regulation, etc., it has become fairly standard practice to omit the gate logic on the schematic diagram of the power system. Whether or not this is justified has been disputed. However, the inclusion of a gate logic circuit would result in an extremely cluttered diagram for some of the more complex systems.

This dilemma has not been confined to the graphic arts; it has been a stumbling block in the implementation of hardware as well. In the past, the need for dozens of discrete devices and the electrical problems involved in synchronizing numerous interdependent circuits made good reliability difficult to obtain. This problem is being overcome by the introduction of IC modules that are especially designed for actuating inverters, converters, choppers, and switching regulators. Not only does this new breed of IC control logic make inverter and converter design more of a "cut-and-dried" procedure, but programming functions, start and stop techniques, isolation, and automatic protection are much more readily obtainable than they were from discrete logic circuits.

Related circuits and techniques

Inverters and converters use many circuits and techniques in addition to saturable-core oscillators. Other types of oscillators, as well as amplifiers, choppers, "switchers," and logic-gated power circuits are, by their very nature, well-suited to certain driven-inverter uses, particularly when a sine wave output is desired. An audio amplifier can be used as such an inverter merely by replacing the loudspeaker with the desired load. An output transformer can be used if isolation or voltage transformation is required. This is not to infer that minor changes, such as in the feedback network, might not be needed—the basic idea is that an audio amplifier is readily adaptable to service as a driven inverter.

Switching-type power supplies often contain identifiable inverters or converters within their functional block diagrams. However, even the "series-switcher" of Fig. 1-17, which has no saturable-core oscillators or driven inverters, relates very closely to our definition of an inverter or converter. For example, it can readily be seen that the dc from the unregulated power source is "inverted" to ac, then "converted" again to a dc output.

The cycloconverter concept, illustrated in Fig. 1-18, is another "gray-area" application of switching techniques. It is, in essence, a *frequency converter.* Frequency conversion, especially when switching is involved, is closely related to the processes of inversion, conversion, rectification, and chopping. Accordingly, relevant material pertaining to the cycloconverter will be found in later chapters. The cycloconverter, being a frequency reducer, has proved to be an efficient and reliable method of converting 400-Hz aircraft power to 60 Hz. It is presently being used in experimental vehicle drive systems that depend on induction motors for traction (output frequency, and therefore motor speed, are controlled by selection of modulating frequency).

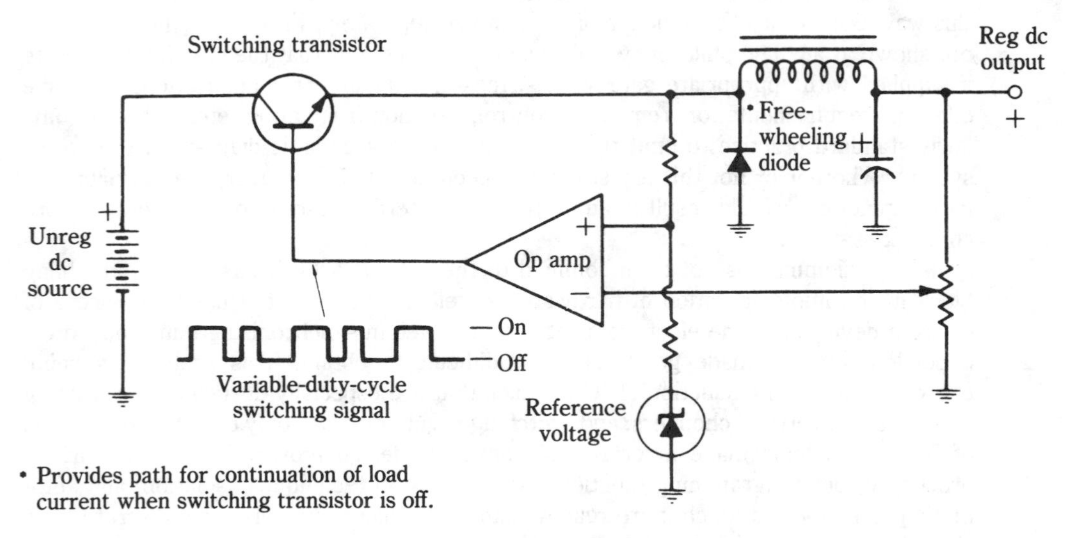

1-17 A self-oscillating switching-type regulator.

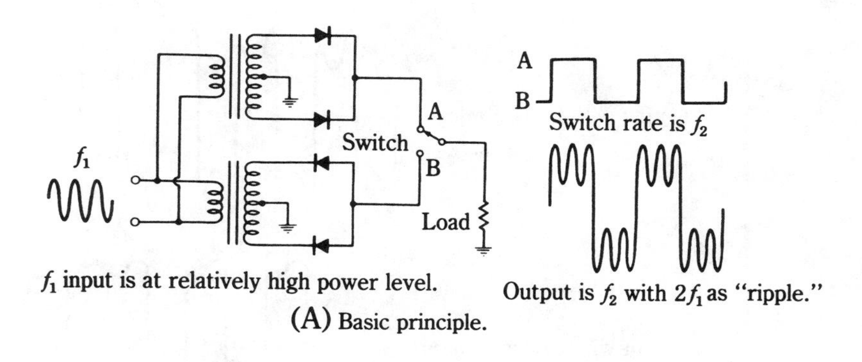

(A) Basic principle.

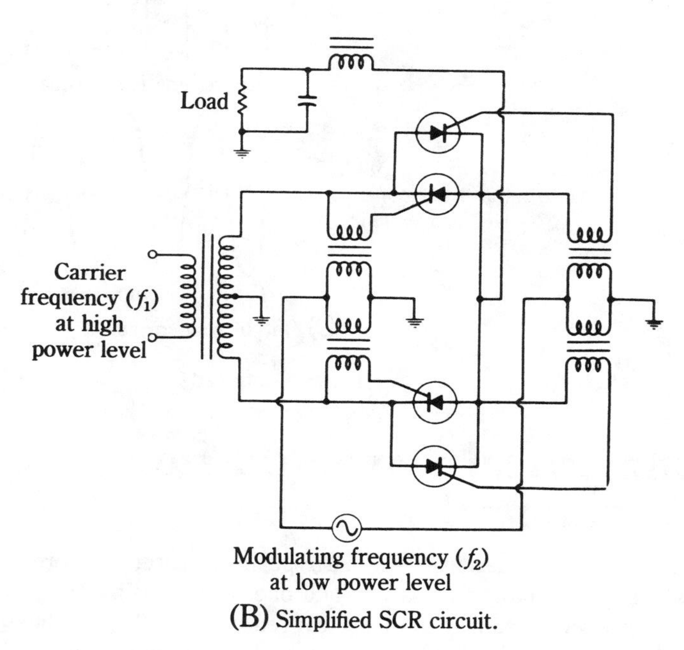

(B) Simplified SCR circuit.

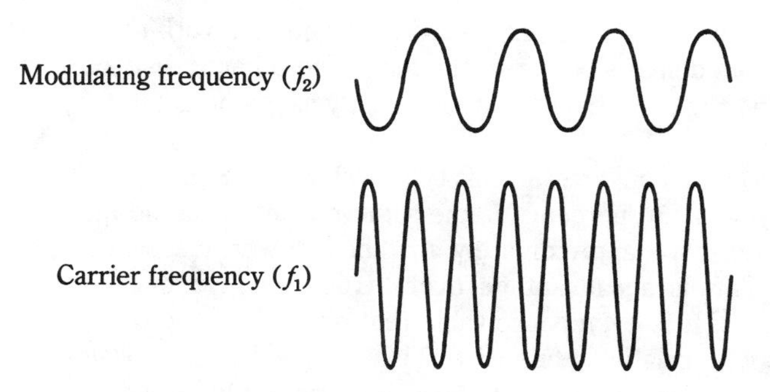

1-18 Principle, circuit, and waveforms of a cycloconverter.

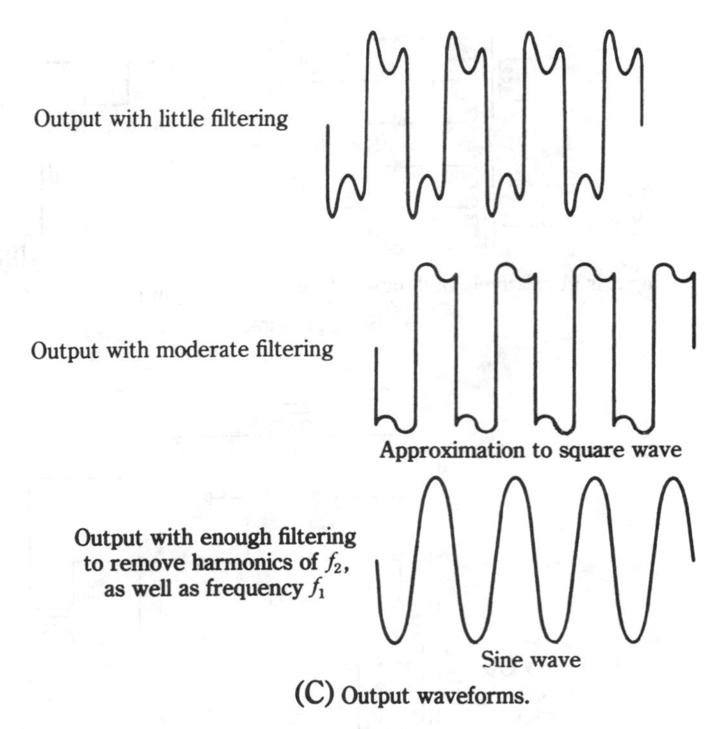

(C) Output waveforms.

1-18 Continued

Symmetrical waveform from an asymmetrical inverter

An interesting modification of the basic push-pull saturable-core oscillator or inverter makes use of one power transistor and one much smaller transistor—otherwise the circuit remains very similar to those covered thus far. At first though, such an arrangement of push-pull transistors with great disparity in power-handling capability appears absurd—what useful purpose can be served by such asymmetry? It turns out, however, that if a half-wave rectifier is used in conjunction with the load winding, only one of the transistors delivers power to the load. The other transistor then serves the function of resetting the core magnetization, exactly as it would if it, too, was delivering power into the load.

A typical converter of this type is shown in Fig. 1-19. This converter maintains the duty cycle at 50 percent for the same reasons that the more conventional inverter or converter with approximately matched power transistors generates a symmetrical wave. This is an advantage over "true" single-transistor circuits, such as blocking oscillators. It is even conceivable that one-half of the tapped primary winding could be wound with relatively fine wire. This is a good circuit for developing high dc voltage—especially because many of the voltage multipliers operate as half-wave rectification circuits. When placing this type of converter in operation, it must be ascertained that

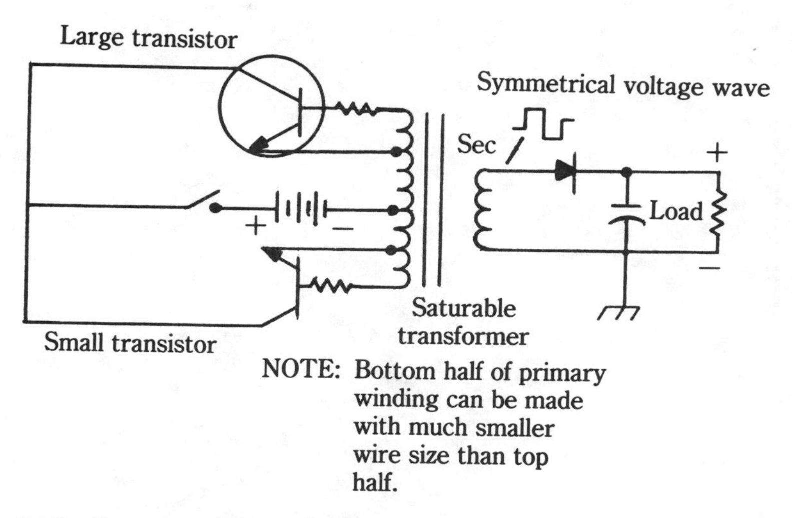

1-19 Converter with a single large transistor to supply load power. The small transistor merely resets the core. Proper operation depends upon the phasing of the secondary winding.

the larger power transistor is associated with the appropriate half of the primary winding so that it can deliver the requisite load current through the half-wave rectifier. If this is not the case, the situation can be remedied by reversing the connections of the secondary or load winding (wrong polarization will be evident because rated load current will not be forthcoming and the smaller transistor will quickly heat up and possibly be destroyed).

2 CHAPTER

Theory

YOU HAVE SEEN INVERTERS USING VIBRATORS OR THYRATRONS IN THEIR associated transformers in an essentially linear mode of operation. That is, magnetic saturation of the core was avoided. The reason for this was simple: operation in the saturation region greatly increases the hysteresis component of the "iron" losses, thereby impairing efficiency. Indeed, to the classical designer of transformers, the cyclic generation of a large hysteresis loop was to be carefully avoided, and a square loop (had the magnetic materials been available) surely would have been double undesirable. Yet, today, some of the best inverters are designed around cores purposely driven into deep magnetic saturation.

The difference between the erstwhile "linear" operation of cores and the magnetically saturated operational mode is much more than a matter of degree. Indeed, if the only advance over previous inverter technology was the simple transistor saturable-core oscillator, an appreciable portion of modern design would remain intact. Saturated-core operation and power transistors form a hand-in-glove relationship. This was not so with thyratrons, ignitrons, or vacuum tubes. With those devices, saturable-core oscillators received little developmental impetus.

It remains true that a transformer driven into magnetic saturation becomes a source of power dissipation today, just as was the case in yesterday's transformers. However, the combination of such a hard-driven transformer with power transistors (also hard-driven) can produce an inverter with very high efficiency, as well as with other compelling performance features. Core saturation is the mechanism responsible for switching and timing in these unique oscillators.

Theory of operation of the saturable-core oscillator

The manner in which oscillation occurs in saturable-core inverter circuits is interesting. The following are the salient features of these oscillating circuits:

1. Switching transitions stem from the mutual characteristics of the magnetic saturation in the core and the behavior of the power transistors. Core saturation alone cannot account for the operation; neither can the transfer characteristics of the power transistors alone.
2. Despite positive feedback, the oscillation is not of the LC variety.
3. Despite classification as a relaxation oscillator, oscillation is not governed by L/R or RC combinations.
4. Ideally, the frequency of oscillation of such an inverter is independent of transistor parameters and temperature. The ideal circuit also exhibits zero regulation. These attributes can be closely approximated through appropriate design.
5. The ideal saturable-core oscillator generates a frequency that is directly proportional to the power-supply voltage. This can be a useful function for certain purposes.
6. The ideal waveshape is a symmetrical rectangular or square wave with a 50 percent duty cycle.

Here is a trace of the sequence of events in this inverter. Actually, there are a variety of circuits, all embracing the same basic principles. One such circuit was depicted in Fig. 1-14. Another, the one to be analyzed now, is shown in Fig. 2-1. Others will be shown subsequently. Each variety possesses unique attributes—the choice is generally influenced by the application intended. However, these two-transistor, saturable-core circuits all operate in essentially the same way.

Suppose that in the circuit of Fig. 2-1 the voltage from the dc power source has just been applied. You might suspect that both transistors would be placed simultaneously in their active regions and that the circuit would quickly assume a latched- up condition as before, that Q1 begins to turn on, the transition from x to d embraces the regenerative sequence of events culminating with Q1 in its fully saturated state. This conduction state is maintained until point a is approached. Now the core begins to saturate in the negative direction, and the electromagnetic induction responsible for the regenerative turn on of Q1 can no longer prevail—the rate of change of flux linkages between the windings falls off rapidly as the slope of the B-H curve approaches the horizontal. Then, Q1 comes out of collector-current saturation, and this further reduces its induced forward bias. As a consequence of this rapidly accelerating degenerative action, Q1 begins switching off approximately when point a is reached.

The cessation of conduction in transistor Q1 in the vicinity of point a is followed by the transition on the hysteresis loop from point a to point b. This occurs very rapidly inasmuch as the rate of change of flux linkages is minimal. This is another way of saying that there is negligible inductance left in the core windings to impose appreciable time

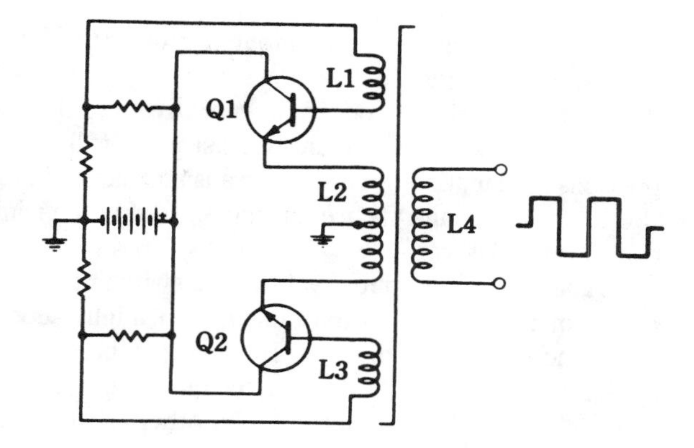

2-1 Saturable-core inventor with two transistors and single transformer.

delay. The sequence proceeds from a to b because of the nature of a hysteresis loop—if there were no hysteresis effect in the core material, path a-o would be traced as transistor Q1 turns off (Fig. 2-2).

As point b is approached, the core goes out of saturation, regaining high permeability and allowing strong electromagnetic induction to occur in the windings. But, this time, the voltages induced in feedback windings L1 and L3 have polarities opposite to

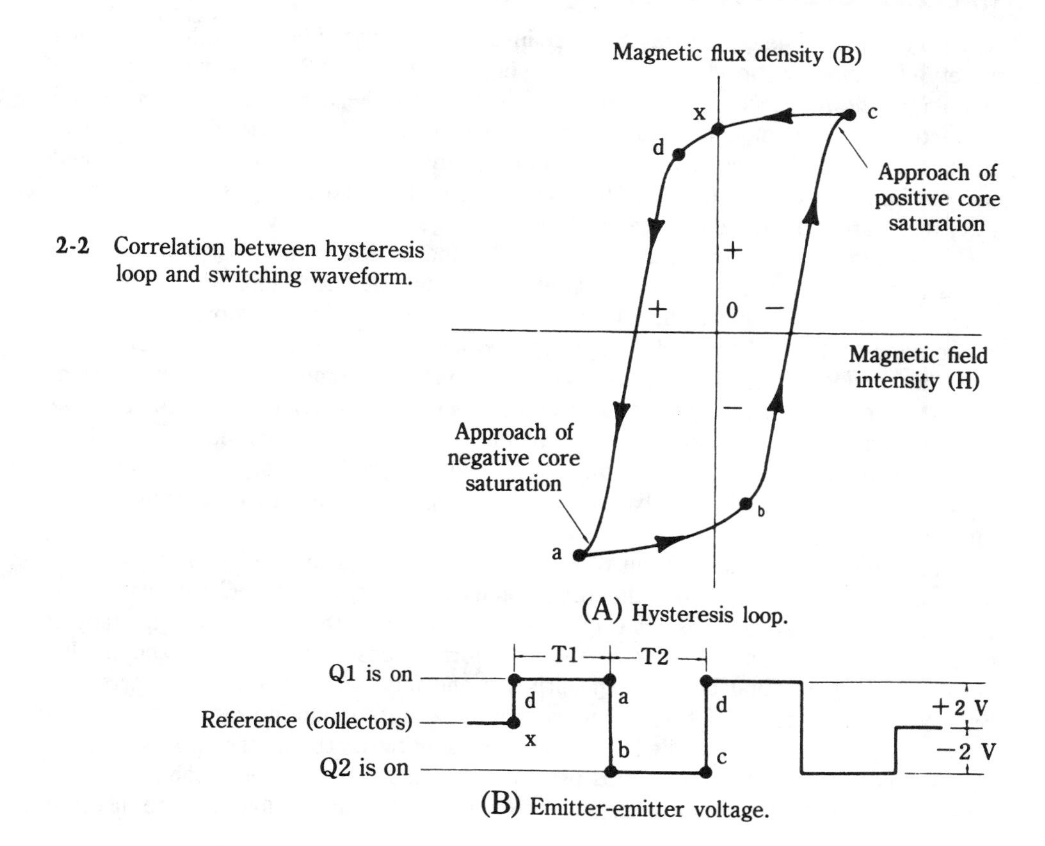

2-2 Correlation between hysteresis loop and switching waveform.

those that correspond to the initial turn on of Q1. Therefore, Q1 now receives ***turn-on bias,*** whereas transistor Q2 receives *forward,* or ***turn-on, bias.*** Moreover, these two switching transitions proceed regeneratively so that in the vicinity of point b, transistor Q2 will be fully turned on and transistor Q1 will be deprived of forward base bias. The previous explanation of core and transistor actions still applies, except that the conduction states of Q1 and Q2 are interchanged. In the vicinity of point c, the entire switching cycle begins anew.

The 50 percent duty cycle of the switching oscillations stems from the basic symmetry of the hysteresis loop itself. You might suppose that unlike transistors would upset the equal time division. However, in practice this effect is negligibly small with modern devices. By the time the inequality in transistor parameters was great enough to affect the duty cycle appreciably, other performance characteristics of the inverter would be seriously degraded.

You can see that the hysteresis phenomenon is not self-oscillatory—the control feature of active devices is needed to complete the feedback system. By the same token, transistors associated with an air-core transformer cannot generate this unique type of oscillation.

Other circuit configurations

As has been mentioned, the saturable-core inverter dealt with in Fig. 2-1 is but one of a variety of circuits. Four additional ones are depicted in Fig. 2-3. They differ from one another in the feedback arrangements and in the way the transistors are connected (common base, common emitter, or common collector). Secondary differences can involve the type of transistor (npn or pnp) and the ways in which base biasing and current limiting are implemented. Although such circuits can often be used interchangeably, there are definite advantages and disadvantages associated with each.

The common-emitter circuits of Fig. 2-3A and 2-3B are the least demanding in terms of base driving power. This is an important factor when emphasis is placed on overall operating efficiency. When high base drive is needed, the transformer efficiency is degraded because of the additional copper losses in the feedback winding(s).

The circuit of Fig. 2-3C utilizes the transistors in a common-base configuration. With this arrangement, there tends to be reduced vulnerability to transistor damage from switching spikes. Also, given transistors are likely to exhibit lower switching losses—especially at higher oscillation frequencies. These benefits, however, are usually offset by the relatively high base drive that is required to maintain the transistors in saturation.

The common-collector circuit of Fig. 2-3D is very popular. This popularity is not necessarily justified by purely theoretical considerations, but the circuit has some compelling practical features. One of these is the use of the single-winding, tapped transformer. This generally makes the transformer easier to produce, and it also promotes tight coupling between the primary and feedback sections of the winding. But the salient feature of this circuit is that it permits mounting of the transistors directly on heatsinks so that the thermal resistance between the transistor case and the heatsink is minimal. Because of this practical expedient, the power capability of this arrangement often exceeds that of other circuits in which insulation must be inserted

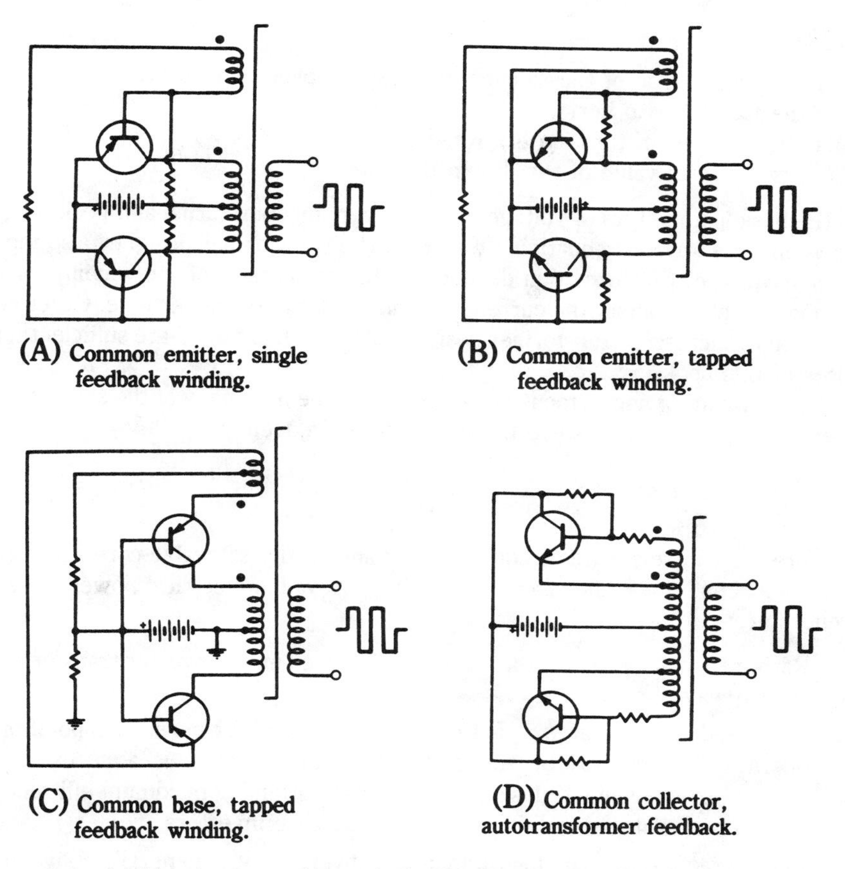

2-3 Additional two-transistor, single-transformer saturable-core inverters.

between transistor and heatsink. For this circuit, the base drive power is intermediate between the requirements of the common-emitter and common-base circuits. And, as with the common-emitter circuit, it is important to ascertain that the switching speed of the transistors is sufficient for the desired oscillation frequency. It should be noted that the phasing of the primary and feedback windings is not the same in all of the circuits shown in Fig. 2-3.

The oscillation frequency of the saturable-core inverter

Much insight into the behavior of the saturable-core inverter can be gained from the basic equation defining its frequency of oscillation. From classical transformer theory, we have the following relationship:

$$E = (4.44 \times 10^{-8}) f N \phi_{max}$$

Where:

E is the peak value of the sine wave voltage applied or induced,
f is the frequency in hertz,
N is the number of turns in the winding,
ϕ_{max} is the peak value of the flux in the core.

Thus, the basic parameters of voltage, frequency, number of turns, and peak magnetic flux are connected by this equation. Of course, the transformer designer also must be aware that voltage transformation depends on the turns ratio of the windings, and he must know something about the current-carrying capacity of copper wire. Other details are involved in successful transformer design, but those mentioned are sufficient for the purposes of this book.

Because we are going to focus our attention on the frequency of the saturated-core inverter, it is convenient to solve the equation for frequency:

$$f = \frac{E \times 10^8}{4.44 \times N \times \phi_{max}}$$

Before we can relate this frequency equation to the saturable-core inverter, we must ask how conditions differ from those which prevail in electrical power work. The differences are:

1. Rather than a sine wave, the waveform involved in the inverter transformer is a square wave.
2. We already know that ϕ_{max} is the saturation flux of the core. This point of core operation is shunned by the classical designer, but it is a necessary, and fortunately convenient, design parameter for attaining maximum efficiency and optimum operating conditions for saturable-core inverters.
3. In the practical design of saturable-core inverters, it is generally convenient to use the relationship:

$$\phi_{max} = B_{sat} \times A$$

Where:

B_{sat} is the saturation flux density of the core material,
A is the cross-sectional area of the core.

The convenience stems from the fact that core manufacturers supply this information in their literature.

You can now derive an expression for the frequency of the saturated-core inverter. First, you must make the appropriate modification for the square voltage wave. The ratio of peak-to-average value in a sine wave is 1.11. The square wave is unique in that the peak, average, and effective values of the wave are all equal to each other. Therefore, to modify the frequency equation for use with a square wave instead of a sine wave, you must divide 4.44 by 1.11.

The equation then becomes:

$$f = \frac{E = 10^8}{4 \times N \times B_{sat} \times A}$$

Where:

f is the frequency in hertz,
E is the voltage (see next paragraph),
N is the number of turns in one-half the primary winding,
B_{sat} is the saturation flux density of the core material,
A is the cross-sectional area of the core.

In this equation, E is the peak voltage developed across one-half of the primary winding, that is, from center tap to an emitter-connected or collector-connected termination in any of the arrangements shown in Figs. 2-1 and 2-3. However, in most practical situations, it is simpler and more convenient to use the close approximation that E is equal to the supply voltage from the dc source.

Any units of area can be used for B_{sat} and A, as long as they are both the same. Thus, if B_{sat} is expressed in lines per square centimeter, A must be expressed in square centimeters, etc.

Significance of the frequency equation

Although the factors that make up the frequency equation are informative, it is interesting to consider what is not included. For example, you can see no evidence of capacitance or a resonant tank circuit. Resistance does not appear, nor does the number of turns in the feedback windings. Although inductance exerts an indirect effect through its connection with the number of primary turns, do not be concerned with the permeability of the core or the area of the hysteresis loop. None of these factors, including transistor parameters, govern frequency, except for their possible secondary effects on E (the same can be said with regard to temperature, except for its influence on B_{sat} and E).

It can be readily seen that the frequency stability of the saturable-core inverter can be made quite good. It is not difficult to devise a dc source with regulated output voltage, and core materials are available in which B_{sat} is nearly constant over a wide temperature range. The only variables left are the voltage drop across the transistors and the voltage drop in the primary winding. Variations in these two voltage drops can influence the value of E, and therefore the frequency. By using suitable wire size, by supplying a relatively high dc voltage, and by providing sufficient base drive to the transistors to saturate them, the frequency can be made more stable than is generally required in inverter applications. Frequency stability, per se, is not a high priority design problem. It is, however, reassuring to be able to design inverters with predictable performance.

Forms of the frequency equation

It is profitable to solve the basic frequency equation for the other quantities. This makes it convenient to decide on the voltage of the dc supply, to specify the core material, to determine the physical size of the transformer, and to compute the number of turns needed. Furthermore, the various forms of the equation provide added insights into the relationships among the parameters.

Repeating the form previously given:

$$f = \frac{E \times 10^8}{4 \times N \times B_{sat} \times A}$$

Rearranging to solve for E yields:

$$E = f \times 4 \times N \times B_{sat} \times A \times 10^{-8}$$

Recall that E is approximately equal to the dc supply voltage.
Solving for N gives:

$$N = \frac{E \times 10^8}{4 \times f \times B_{sat} \times A}$$

Recall that N represents one-half the total number of turns in the primary winding:
Solving for B_{sat} gives:

$$B_{sat} = \frac{E \times 10^8}{4 \times f \times N \times A}$$

Different magnetic core materials are available with a wide range of values for B_{sat}.
Solving for A gives:

$$A = \frac{E \times 10^8}{4 \times f \times N \times B_{sat}}$$

Notice that nothing is stated with regard to the length of the flux path in the core. In practice, the core length is automatically taken care of by ascertaining that there is sufficient space for the windings.

A closer look at the oscillation waveform

The oscillation theory developed thus far constitutes an adequate explanation of the essential operating behavior of the saturable-core inverter. There is, however, a peculiar side effect pertaining to practical circuits. This involves the generation of *voltage spikes* that are superimposed on the voltage waveform. These spikes can be quite narrow on a 60- or 400-Hz square wave and can even be overlooked if an oscilloscope with poor high-frequency response is used for observing the waveform. However, spikes are often the explanation for the destruction of transistors. The argument is sometimes advanced that the energy content of these spikes must be negligible when the case temperature of the transistors is low. There is much evidence, however, to suggest that even if such spikes do not cause immediate damage, they exert a deteriorating effect and are likely to bring about eventual destruction of the transistors. The fact is that spikes can inflict damage without causing any appreciable rise in the case temperature of the transistors. Also, spikes are an added source of radio-frequency interference. It is highly desirable to make deliberate design attempts to limit their generation and to suppress what spike energy is generated.

Spikes are the result of high rates of change of current in inductive circuits. If the transformers used in the inverters were perfect, and if switching was instantaneous, the transistors would not "see" any inductive effects when turning off. This is because the opposite half of the transformer would, at that instant, be "clamped" to the other transistor. In practical situations, the off-switching transistor interrupts its current through the equivalent ***leakage inductance*** inherent in all practical transformers. In Fig. 2-4A, the added inductances, *L*, represent the effects of the transformer leakage inductance. Leakage inductance is the result of less than perfect electromagnetic coupling between windings, or between the turns of a single winding. In addition to the leakage inductance, another factor contributes to the generation of spikes. This factor is the increase in collector-emitter current prior to turn off.

The main reason inverters are plagued with voltage spikes is frequently misunderstood. It is well known that the abrupt cessation of current in an inductor is accompanied by a self-induced voltage, or counter EMF. Such a voltage transient attains a magnitude directly proportional to the speed at which the current can be turned off and the maximum inductance "seen" during the transition. Thus, if a battery, a coil, and a knife switch are connected in a simple series circuit, it is possible to produce a visible arc between the blades of the switch as it is opened. (This "voltage spike" is put to good use in ignition systems.) The same process must be operative in saturable-core inverters. Yet, spike production would be relatively benign if this were the sole explanation. The induction of a voltage spike by this process is limited by the "clamping" action of the alternate transistor. For example, in Fig. 2-4, when Q1 turns off, it tends to generate a voltage spike. But the turn on of transistor Q2 constitutes a short circuit on the lower half of the center tapped winding. This short circuit is electromagnetically coupled to the upper half of the winding, thereby holding down the amplitude of the developing spike.

This explanation assumes that switching is instantaneous, that the turned-on transistor is a perfect short circuit, and that perfect coupling exists between the two sections of the center tapped winding. Even after all allowances are made for the nonideal conditions, however, it turns out that further insight is needed to explain the strength of the spikes. Empirical observation shows that a transformer with high leakage inductance definitely promotes spike production. From this, you might conclude that the leakage inductance inhibits the clamping effect of the transistor that is turning on; this surely would tend to aggravate spike generation. Another empirical observation leads to the real explanation, however. A saturable-core inverter must have sufficient feedback to maintain the transistors in saturation during their "on" periods. It can be shown easily that the spikes become worse with increased base drive. Thus, leakage inductance in conjunction with an effect related to the base drive of the transistors bears much responsibility for the phenomenon being tracked down.

Look at the collector current. In Fig. 2-4B, the collector waveform is a ramp with a sudden rise near its termination. (These collector-current waves correspond to unloaded operation of the inverter. It is convenient for this purpose to deal with this mode of operation. However, the explanation being developed is not materially effected when the inverter supplies power to a load.) The abrupt change in slope of the collector current ramp coincides with the attainment of magnetic saturation in the core. This is to be expected, because the rate of rise of current in an inductor is governed by the inductance presented to the voltage source. In an inductor with a ferromagnetic core, the inductance is greatly influenced by the magnetic permeability of the core. Specifi-

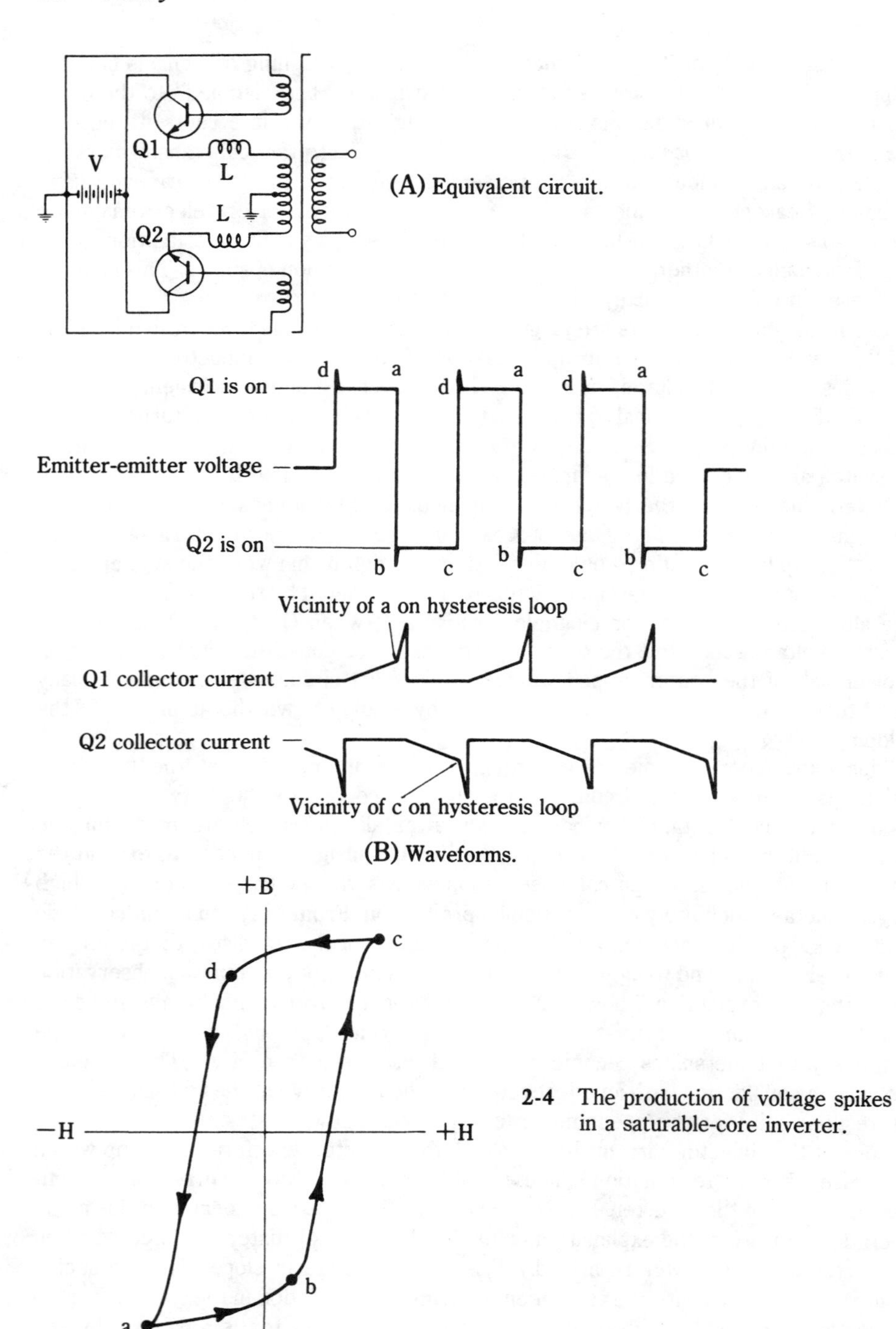

(A) Equivalent circuit.

(B) Waveforms.

(C) Hysteresis loop.

2-4 The production of voltage spikes in a saturable-core inverter.

cally, the permeability is high in the "linear" range of the B-H curve, or hysteresis loop (Fig. 2-4C). When saturation sets in, the permeability drops to a relatively low value. Therefore, the voltage source, which had initially been presented with a high-inductance device, suddenly "sees" a much lower inductance. This enables the rate of rise of the current to increase abruptly.

This situation is nicely explained by the equation $E = L\ d_i/d_t$. In this form, the equation is usually brought to bear on the generation of a voltage as a consequence of a time-changing current in an inductance. However, it also shows that if E represents a constant voltage impressed on an inductance (L), the resultant current will have a constant rate of change. That is, current plotted with respect to time will have the form of a ramp with a constant slope. By the same reasoning, if L should suddenly decrease (as happens at magnetic saturation), the rate of change of the current must increase in order to satisfy the equality indicated by the equation. This accounts for the double-sloped ramp of collector current shown in Fig. 2-4.

At first glance, the foregoing phenomenon might seem unimportant. After all, peak collector-current capability has not been identified as any problem. And, with reasonably high-gain transistors, it should not be difficult or detrimental simply to supply a little more base drive so that the "current hunger" of the transformer could be accommodated. This is true enough. Unfortunately, however, the sudden burst of current passes through the equivalent leakage inductance of the transformer (L in Fig. 2-4A). Using the equation $E = L\ d_i/d_t$ again, we can see clearly how a voltage transient is produced. To be sure, L, the leakage inductance, might be quite small compared to the inductance of the transformer primary prior to saturation. Offsetting this, however, the new rate of change of current, d_i/d_t, can be very high.

In practical inverters, it is not a trivial matter to hold leakage inductance down to manageable levels. Best results are usually obtained with the use of toroidal cores and bifilar windings. Generally, after the best transformer possible within the constraints of economic and production considerations has been designed, the base drive of the transistors is chosen to be sufficient for rated-load operation, but not much beyond that. By this technique, the entire current demand of the saturated transformer will not be met. This can take much of the "sting" out of the voltage spikes, for they will then be reduced both in peak magnitude and in energy content. Often, when everything feasible has been done with transformer design and with base drive, the spikes might still endanger the transistors. Sometimes, you can select transistors with voltage ratings that provide a good safety factor. But, such an expedient can prove costly. It must not be forgotten that transistor parameters are highly interdependent. Emphasizing one can degrade another. So, if you try to select transistors with very high voltage ratings and with the requisite switching speed, current gain, etc., the economics of such a "brute force" approach can become prohibitive. In any event, the concern is often with the reaction of the load to these spikes. It is best to try to reduce them.

Various *damper* and *snubber networks* can be added to the basic inverter circuit to attenuate the residual spikes that prevail despite appropriate considerations given to the transformer and feedback, or base drive. These will be described in the next chapter. Suffice it to say here that they are not generally suitable as compensation for sloppy design. If the transformer is not good and/or the base drive is excessive, the application of such networks can cause degradation of other operating characteristics.

Other solutions to the spike problem involve the use of different types of inverters than the two-transistor, single-transformer types considered thus far. From what you have seen about the simple saturable-core inverter, you are led to the conjecture that the spike problem could be alleviated if the output transformer could be operated in its linear, rather than its saturable, mode. Such a consideration seemingly involves a contradiction of requirements, but it is possible, although at the price of greater complexity and cost.

The two-transformer inverter

In the self-oscillating inverters covered thus far, the single transformer determines the switching rate and couples the resultant chopped wave to the load circuit. If these functions are accomplished separately, definite operating advantages result. This, however, is at the expense of added components and complexity, for two transformers are then required. One transformer, a relatively small one, operates in its saturating region and is connected in the base circuit. The other transformer is connected in the output circuit, but it operates in its linear region. Thus, the transformer that handles the high power levels does not develop the high hysteresis loss that is inevitable when core operation encompasses magnetic saturation. Other things being equal, you can expect higher efficiency from the two-transformer inverter, and, at lower frequencies at least, this is readily obtained. (At higher frequencies, the transistor switching losses might be appreciable. But this is a problem in all inverters where high efficiency is desired.)

A simple two-transistor, two-transformer inverter is shown in Fig. 2-5. In this arrangement, the output transformer, T1, does not saturate. Rather, saturable-core governed oscillation takes place because of transformer T2 in the base circuit. Because T2 is relatively small and carries a small fraction of the currents in T1, its core losses are less than in the single-transformer inverter. Even more important, spikes in the output circuit are at a much lower level than is readily achieved in the single-transformer inverter.

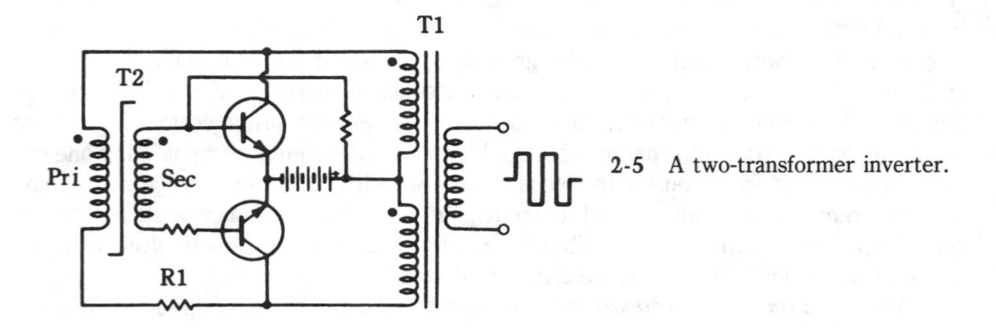

2-5 A two-transformer inverter.

In Fig. 2-5, frequency control can be had by varying resistance R1. This changes the voltage impressed across saturating transformer T2. An even better arrangement for achieving frequency control is the circuit of Fig. 2-6A. Moreover, if oppositely polarized zener diodes (Fig. 2-6B) are associated with the primary winding of saturable

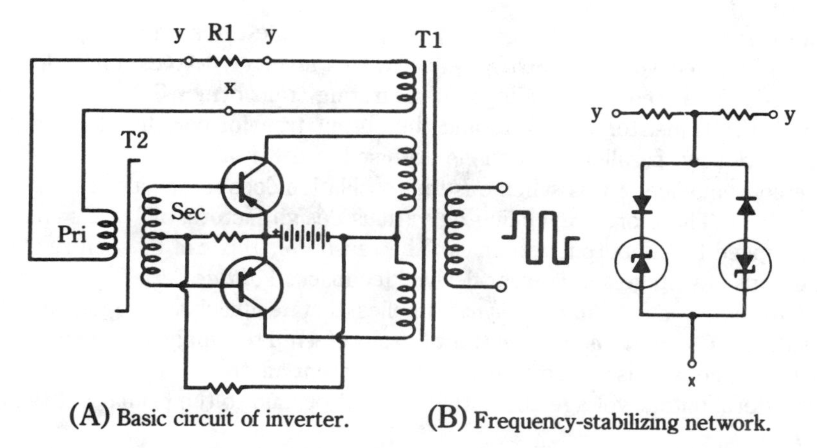

(A) Basic circuit of inverter. (B) Frequency-stabilizing network.

2-6 Another two-transformer inverter circuit.

transformer T2, the frequency can be greatly stabilized in the face of load and power-supply variations. Low-voltage avalanche (LVA) reference diodes have extremely sharp knees in the 4.5- to 9-V region and are excellent for this purpose. Best results are obtained when R1 is approximately centertapped and the reference diodes are accompanied by ordinary diodes in a network such as that in Fig. 2-6B.

Two-transformer inverter with current feedback

From initial inspection, the inverter illustrated in Fig. 2-7 might appear to be simply another circuit variation of the self-oscillating, two-transformer inverter. In a sense, it is, but this arrangement involves some unique operating principles. Whereas the other

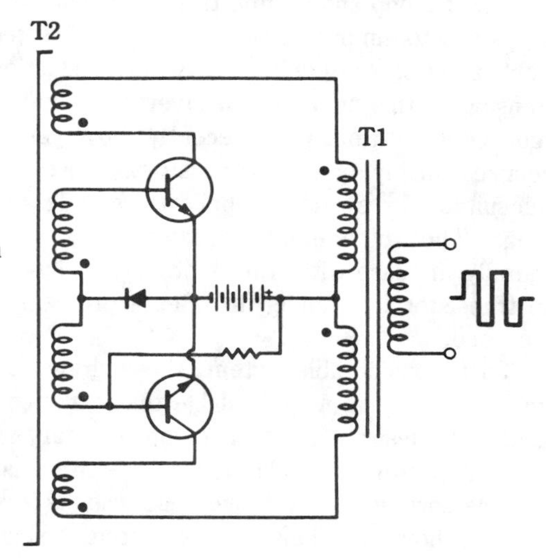

2-7 A two-transformer inverter with current feedback.

two-transformer inverters discussed (as well as the single-transformer types discussed previously) involved voltage-derived feedback, this inverter derives its feedback on a current basis. The feedback windings on saturating transformer T2 are connected in series with the transistor collectors and the output transformer. In all arrangements covered previously, parallel connections are used.

One consequence of this scheme is that oscillation does not occur readily, if at all, without a load. Therefore, depending on various design factors, at least some loading must be applied to the output terminals. The reason for this can be seen readily from the definitions of voltage and current-derived feedback. *Voltage-derived feedback* (often referred to simply as *voltage feedback*) implies that feedback is proportional to the output voltage. *Current-derived feedback* (often referred to simply as *current feedback*) implies that feedback is proportional to the current in the output circuit. In all of these inverters, output voltage and output current pertain to the primary of the output transformer.

Another unusual feature of the inverter with current feedback is that it will continue to oscillate with a short-circuited load. Unless special provisions are made, such short-circuit operation is likely to prove unsafe if allowed to continue. In contrast, voltage-feedback inverters shut down as a consequence of a shorted output. The circuit merely ceases to oscillate. Such a situation can usually continue indefinitely. And, upon removal of the fault, the circuit again commences operation. The lack of inherent short-circuit protection must be counted as a disadvantage of the current-feedback inverter.

This is not the whole story, however; it happens that the current-feedback inverter also displays a very compelling feature. Whereas the voltage-feedback types attain high operating efficiency only in the vicinity of full load, the current-feedback types provide high efficiency over a wide load range.

The complementary-symmetry inverter

The use of a pnp and an npn transistor instead of a pair of transistors of a single type often leads to simplified circuit design. This technique has been used for the output amplifiers in stereo equipment, where reduction of costs assumes high priority. The extension of this approach to inverters is natural because of the basic similarities in circuit configuration. Until recently, however, it has not been easy or economical to select reasonably matched transistors. Those intended for stereo use have not generally been suitable for switching applications, unless great care was exercised with regard to ratings. The vulnerability to damage from second breakdown was greater than is generally the case with transistors especially designed for switching circuits. Also, the pnp transistor has tended to offer a problem insofar as power-handling capability is concerned.

To a considerable extent, these drawbacks have been overcome, and npn-pnp transistor pairs are now available that make complementary-symmetry inverters both feasible and desirable. A basic complementary-symmetry inverter is shown in Fig. 2-8. This is a push-pull oscillator, and operation is quite similar to that of conventional saturable-core inverters. Spike generation is less than in the inverters with like transistor types, however. This is because the transistor that exerts clamping action on the

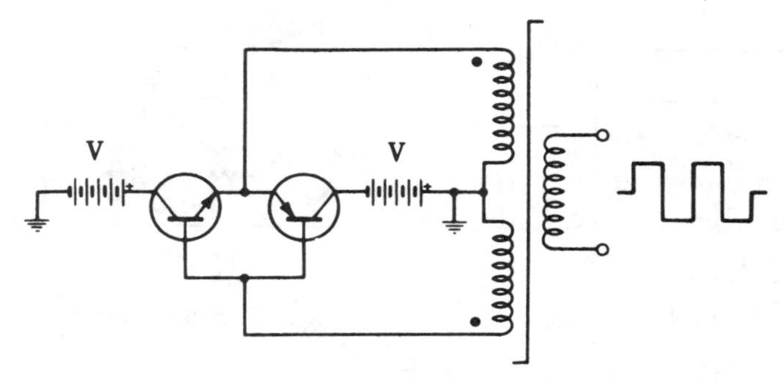

2-8 A complementary-symmetry inverter.

spike is connected directly to the transistor that has just turned off. Another operational feature of this arrangement is that common-mode conduction of the transistors is greatly reduced.

Transformer design and construction for the complementary-symmetry inverter is simplified not only because of the circuit configuration, but also because the spike problem is less severe than in ordinary inverters. This inverter is especially worthy of consideration when one does not wish to use a toroidal transformer or expensive core material. Applications for this circuit are still limited by power considerations, for it remains true that the power capability of npn transistors exceeds that of the pnp transistors at the higher power levels.

Single-transistor circuits

Each of the circuits shown in Fig. 2-9 makes use of a single transistor, rather than the push-pull arrangements previously covered. They are shown in the format of converters, rather than inverters because the loading of such inverters is either critical or is involved in the oscillatory mode. These circuits can bring about cost and component savings when matched to specialized applications. In general, however, they tend to be less efficient and less flexible than the two-transistor types. Although the four circuits have a definite configurational resemblance to each other, their operational modes differ. Accordingly, each circuit has its unique application area.

The circuit in Fig. 2-9A is a single-transistor version of the self-excited, two-transistor, saturable-core inverter. It is as if you simply removed one transistor from a two-transistor circuit. Notice the load circuit. At first glance, it appears that an error has been made in the location of capacitor C. This capacitor, however, functions neither as a filter nor as a resonator in the usual implication of these terms. To be sure, it cannot help forming an LC circuit, with respect to the load winding. However, no sustained interchange of electric and magnetic energy is involved. Instead, the energy stored in the capacitor when the transistor turns off is "dumped" back into the load winding. This resets the core so that the cycle can repeat. In this manner, capacitor C substitutes for the missing transistor. As might be expected, the relationships between the load and capacitor C are somewhat critical. Application areas include hobbyist, toys,

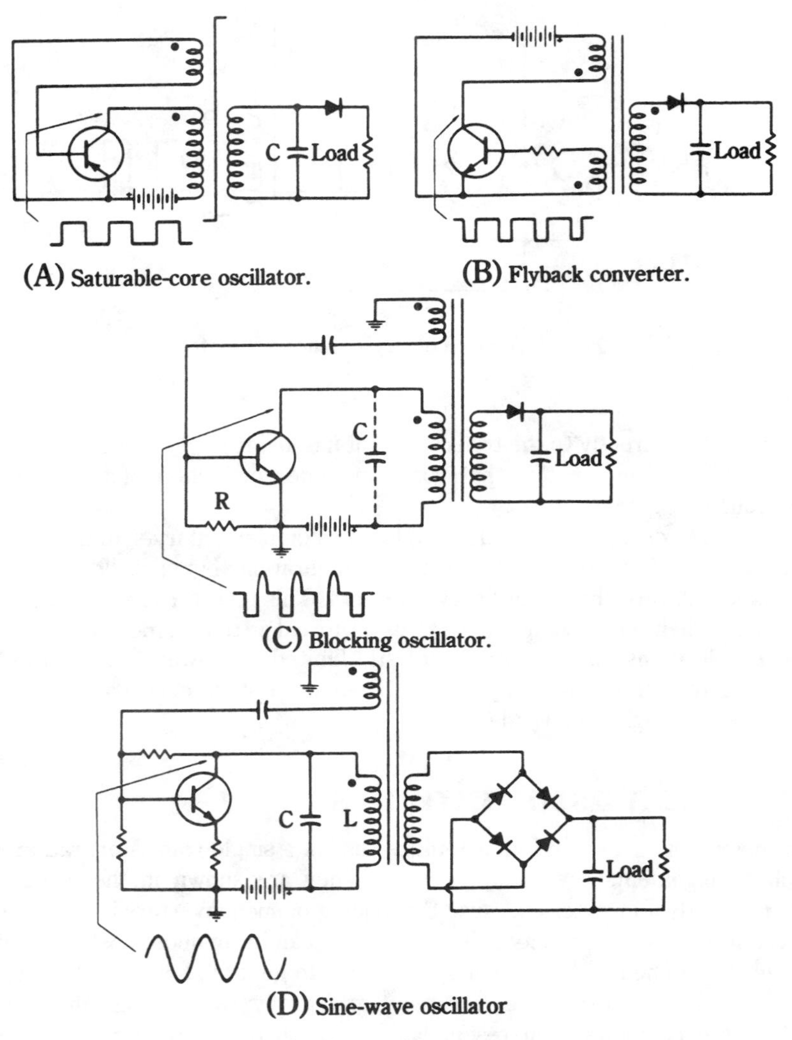

(A) Saturable-core oscillator.

(B) Flyback converter.

(C) Blocking oscillator.

(D) Sine-wave oscillator

2-9 Single-transistor converters.

and general purpose, where efficiency has low priority. The output is pulsating dc. One appropriate application is as a battery charger; another is fluorescent lighting.

The *flyback converter* shown in Fig. 2-9B is also known as *ringing choke* type. Actually, it is a blocking oscillator in which the recovery time is governed by an L/R time constant, rather than the more familiar RC time constant used in circuits formally designated as blocking oscillators, such as the circuit of Fig. 2-9C. The transformer core of the flyback converter shown in Fig. 2-9B does not saturate. However, in some designs core saturation does occur. In either case, the diode in the secondary circuit isolates the converter from its load as the current ramp is developing in the primary winding during the transistor on time. This current ramp eventually terminates regeneratively when the transistor can no longer supply the demanded current (or, in alternate designs, when core saturation occurs). The collapsing field then induces secondary current that, because of its polarity, is delivered to the load. A unique feature of this circuit is that the peak voltage exceeds that corresponding to the transformer turns ratio—ordinarily by a factor of three or four, but sometimes by as much as eight. This is a very good converter for capacitor-charging techniques, such as in photoflash units. It is also useful for fluorescent lighting and Geiger counters.

The circuit of Fig. 2-9C is a conventional blocking oscillator. Capacitance C usually is not caused by a physical capacitor, but is the stray capacitance of the winding and the associated circuitry. The converter starts to operate as a feedback oscillator at a relatively high frequency that is determined by the resonant "tank" in the collector circuit. It would generate a sustained train of sine waves, but for the fact that the positive feedback to the base is purposely made large. Also, the time constant of the resistance and capacitance in the base circuit is relatively great compared to the resonant frequency of the collector circuit. As oscillation commences, the base is driven rapidly into its forward-bias region. This reinforces the buildup of collector current, but collector-current saturation is quickly reached. Electromagnetically coupled feedback suddenly ceases, removing the forward bias in the base-emitter circuit. The base capacitor is then left in a charged condition, but polarized to apply cutoff bias to the base. Operation is halted abruptly, and the circuit remains quiescent until the charge in the base capacitor depletes itself through the "base leak," resistance R. Then, the action in the collector circuit repeats. Paradoxically, it can be seen that this type of oscillator never accomplishes its "intent"—to operate as a sine-wave oscillator—because of its highly vigorous starting action.

The width of the repetitive pulses generated is $\pi\sqrt{LC}$, where L is the inductance which the collector winding presents to the stray capacitance, C. The time interval between pulses is approximately equal to the time constant of the capacitor and the resistance, R, in the base circuit. The waveform is an aborted collector pulse accompanied by a half-cycle of collector tank-circuit "overshoot." The half-wave rectifier can be polarized to utilize either section of this waveform.

As depicted, this is not a useful converter if any appreciable power must be supplied. It is best suited for circuit applications in which a voltage is needed for some high-impedance biasing, hold off, or logic function. It can be constructed conveniently from a small pulse transformer and several components. As an example of the kind of service this low-power converter is suited for, one could be used to generate a few volts of cutoff bias for a FET stage in order to provide control of an electromagnetic relay in the output circuit. With the recent advent of power MOSFETs, this converter could conceivably be utilized for producing much higher power levels. The repetition rate could be made fast enough to produce output power with reasonable efficiency (if attempts are made to produce much power with bipolar transistors, the base capacitor becomes impractically large).

The converter shown in Fig. 2-9D has a configuration similar to that of Fig. 2-9C, but the circuit constants are tailored differently. The feedback is made sufficient to sustain sinusoidal oscillation; the time constant of the RC network in the base circuit is made sufficiently small to preclude the possibility of accumulating sufficient charge to disrupt oscillation. Class-A operation is the most straightforward way to obtain a good sine wave, if this is an objective. In order to attain better efficiency, feedback and reverse bias can be increased to move operation into the class-B or class-C region. More emphasis is then required on the *Q* and impedance level of the resonant tank if near-sinusoidal operation is to be maintained. Of course, the waveshape might not have high priority in such a converter. In any event, the RFI produced by this type of converter is likely to be very much less than from the relaxation circuits. Inasmuch as RFI and switching noise can be a severe problem with saturating-core and flyback converters, the sinusoidal feedback oscillator often merits consideration. Push-pull versions with modern ferrite-core tuned circuits and Schottky rectifiers can achieve overall efficiencies on the order of 70 percent to 75 percent. Here, too, the new power MOSFETs, because of their low gate-drive requirements, should prove useful.

Single-transformer, series-connected inverters

The circuit shown in Fig. 2-10 is representative of a novel technique that makes it possible for transistors with economical voltage ratings to be operated from a high-voltage dc source. The scheme can be extended to accommodate more than the two pairs of

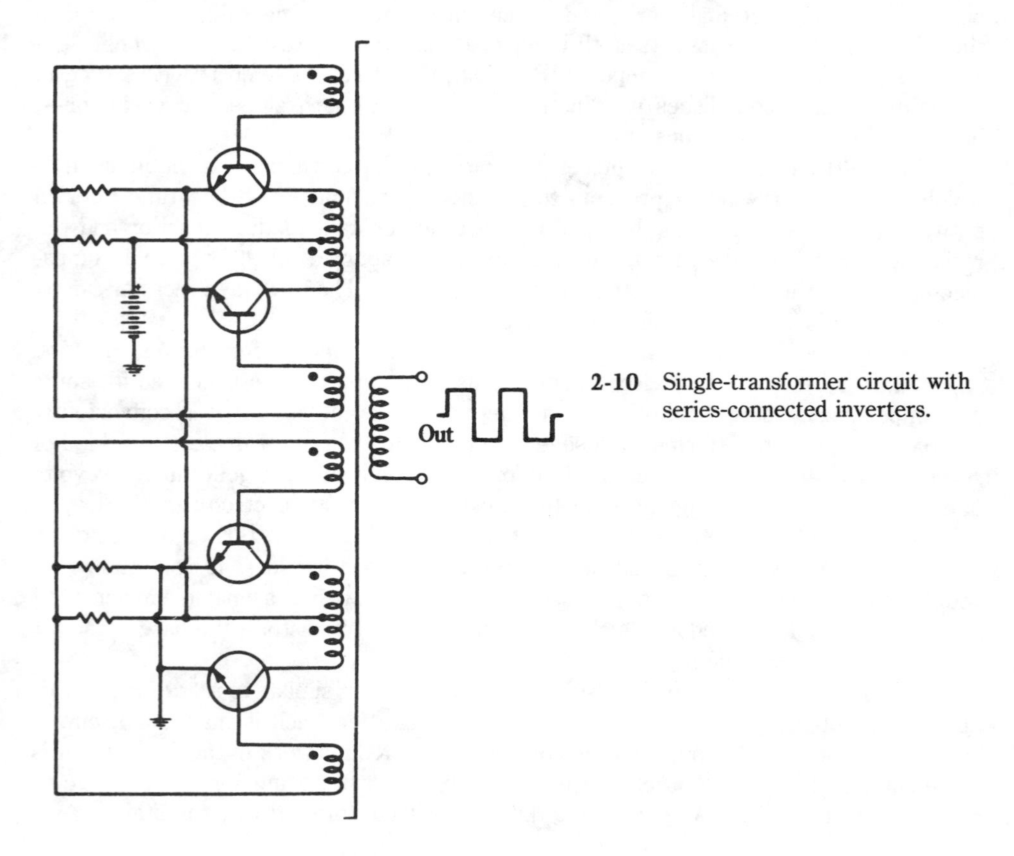

2-10 Single-transformer circuit with series-connected inverters.

transistors depicted. The basic idea is that pairs of transistors are connected in individually complete inverter circuits, but are *series-connected,* with respect to the dc supply. If individual inverters are properly functional in their own right, no trouble will generally be experienced in getting them to work harmoniously together. Synchronization and voltage sharing are automatic because of the common magnetic flux. Indeed, it is probably easier to double the output of a given transistor pair by this method than it is by parallel-connecting transistors. Overall core volume and weight will compare favorably with a transformer designed for a single transistor pair rated for the same power capability. The latter approach is schematically simpler, but it is often considerably more expensive when operation is to be from a high-voltage dc source. With bifilar windings, the construction of the transformer in Fig. 2-10 is simplified, and operational equality of the two inverters is attained.

Bridge inverters

Another way to utilize the power capability of four transistors is by means of bridge circuits, such as the one shown in Fig. 2-11. Notice that no centertap is required in the output transformer, which operates in the linear mode. The individual transistors are exposed to the dc supply voltage plus spikes, rather than approximately twice this amount as in conventional two-transistor inverters. In this respect, bridge inverters deserve consideration alongside series-connected inverters when a problem exists with the voltage ratings of the transistors. Bridge circuits have the potential for providing the most output power per dollar of all the inverter circuits. To a considerable extent, this stems from the way in which the transformer is used. In centertapped, two-transistor circuits, each half of the primary of the output transformer is used only half of the time. In a bridge circuit, the entire primary winding is in use all of the time.

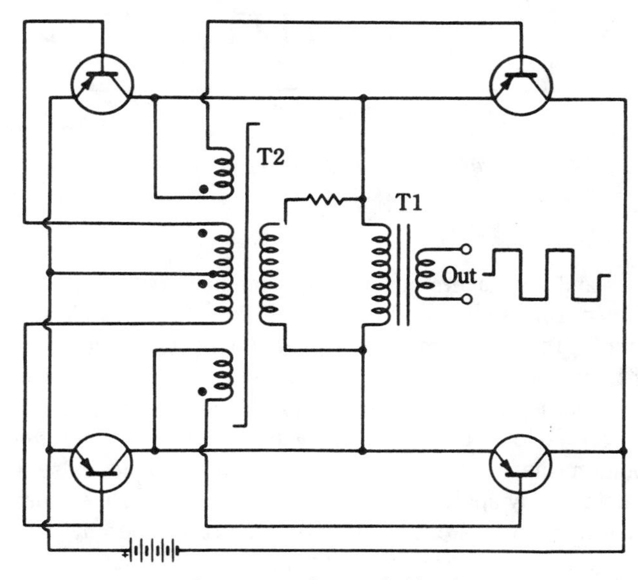

2-11 A two-transformer bridge inverter.

Complementary-symmetry bridge inverter

The use of complementary transistor pairs in the bridge configuration produces considerable simplification of circuitry and often brings about superior performance. The bridge inverter shown in Fig. 2-12 utilizes complementary transistor pairs. It is schematically simpler than complementary transistor inverters because the transformer does not have to be centertapped. The gain in simplicity over four-transistor inverters is even more obvious. This circuit, because of reduced common-mode conduction, performs well at high frequencies. The natural phase inversion that occurs in the opposite type transistors tends to make the effects of transformer leakage inductance less detrimental. This results in decreased spike generation compared to the conventional single-type transistor circuit. That is one reason why Fig. 2-12 is shown as a single-transformer inverter—the implication is that the single-transformer, complementary-symmetry inverter can provide about the same quality of performance as the conventional two-transformer bridge inverter. To the extent that this is true, the reduction in complexity and the savings in cost are even more dramatic.

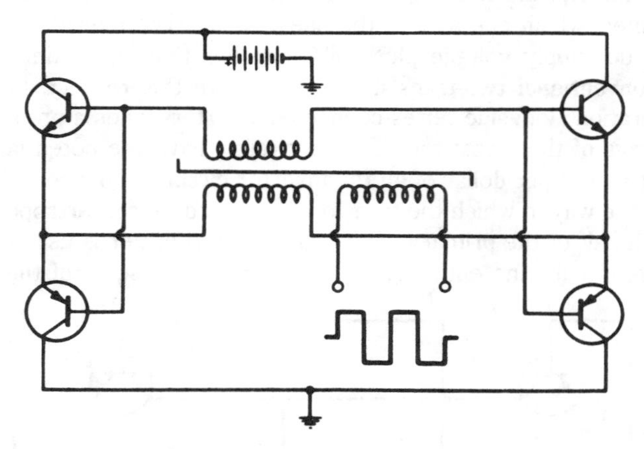

2-12 Complementary-symmetry bridge inverter.

Driven inverters

All the inverters and converters covered thus far have been self-oscillatory. Most of them can be driven from an external oscillator if their feedback circuits are disabled or omitted. In essence, the inverters then become amplifiers. Most often, they are designated as class-D amplifiers because they reproduce square waves and operate in a switching mode. That is, they are either driven to collector-current saturation, or they are completely turned off. Thus, the opportunity for high operational efficiency is retained. In addition, there are other advantages. For example, pulse-width modulation is readily accomplished by applying appropriate logic to the drive source. Also, greater flexibility of frequency control exists. A driven inverter generally uses a linear output transformer; this greatly diminishes the spike problem. Inasmuch as core losses are less in linear transformers than in saturating transformers, operating efficiency tends to be high. Finally, such an arrangement neatly circumvents problems associated with starting.

When a sine-wave output is desired, the driven inverter often takes the form of a push-pull class-B power stage. The theoretical maximum efficiency of 78.5 percent is, of course, less than can be attained from the switching mode, but minimal RFI is readily achieved.

Several examples of driven inverters are illustrated in Figs. 2-13 through 2-17. The circuit shown in Fig. 2-13 is essentially a push-pull power amplifier. The salient feature of such an arrangement is the use of linear transformers, particularly the output transformer. The driving source can be a multivibrator or any of a variety of oscillator circuits. Such driven inverters find considerable use in switching-type regulators because of the convenience with which a duty-cycle-modulated pulse source can be associated with the regulation feedback loop.

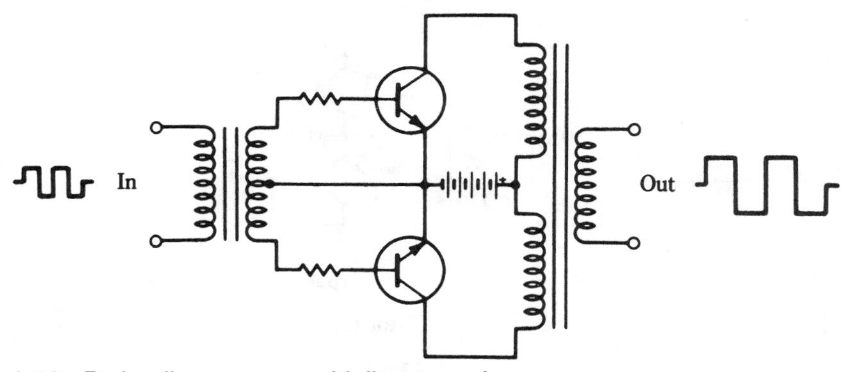

2-13 Push-pull power stage with linear transformers. General Electric Semiconductor Products Dept.

The circuit of Fig. 2-14 is functionally similar to the inverter output stage just described. However, simplified design results from the use of complementary-symmetry transistors. As can be seen, it is possible to dispense with the output transformer altogether. However, if isolation, rather than cost, space, or weight, is the prime requisite, a transformer can be connected between the output and the load. As shown, two dc power sources are needed.

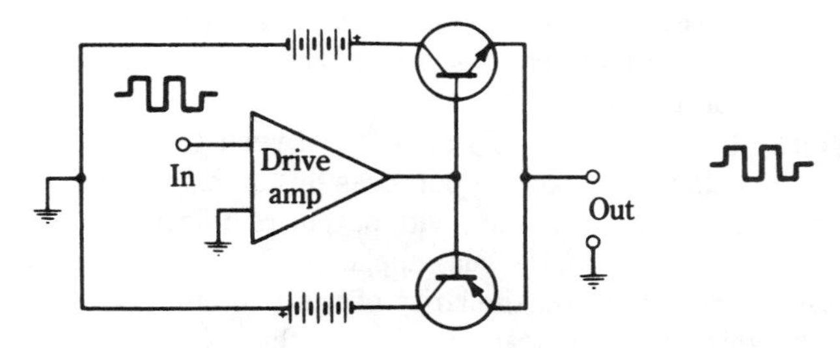

2-14 Push-pull power stage with complementary-symmetry transistors. General Electric Semiconductor Products Dept.

In Fig. 2-15, a saturable-core oscillator drives a power amplifier stage. This arrangement requires a higher parts count than the two-transformer inverter that it resembles. However, greater operational flexibility is achieved because the load is isolated from the oscillator. Thus, inductive or capacitive loads are less likely to interfere with starting or the maintenance of a symmetrical duty cycle. Another advantage of such an arrangement is that it lends itself conveniently to the implementation of regulation. For example, the relatively small power supply of the oscillator section can be a programmable voltage-regulated type that receives its control signal from a sensing circuit associated with the load. In such an instance, the square wave from the output stage is first rectified and filtered.

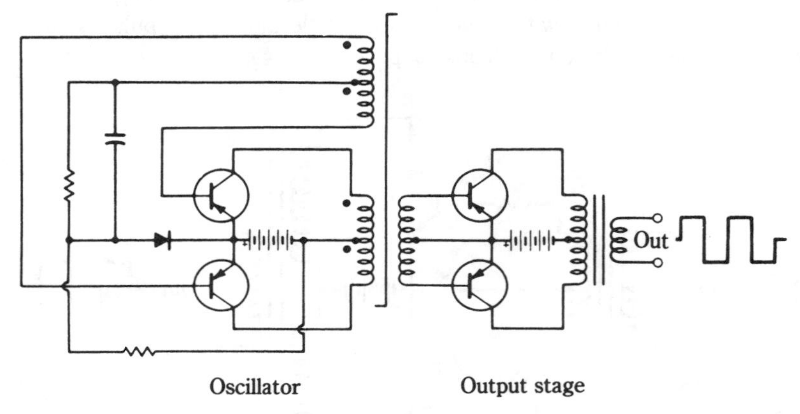

2-15 Inverter with saturable-core oscillator driving output stage. General Electric Semiconductor Products Dept.

A more sophisticated type of driven inverter is depicted in Fig. 2-16. Here, the input square wave is generated by a Type 556 IC dual timer. Much better precision and stability are obtainable from such an oscillator than from multivibrators designed around discrete devices. This alleviates common-mode conduction problems. The driven inverter of Fig. 2-17 is a flyback circuit in which the conduction of Q1 is governed by the waveform of a relaxation oscillator that uses a programmable unijunction transistor (PUT). The PUT simulates the operation of a conventional unijunction transistor in many respects, but it has better output and drive capabilities. It is actually an SCR with an anode gate. In the relaxation oscillator of Fig. 2-17, the duty cycle of the PUT is determined by R3. Appropriate versions of this type of driven inverter are used in the high-voltage section of TV sets.

At this point, it is apropos to distinguish between the flyback and the forward inverter. This is best done for practical purposes by comparing both types of switching circuits in their roles as converters, i.e., with output rectification so as to deliver direct current to the load. This is their usual application. The comparison is particularly relevant because of the schematic similarity of the two circuits.

Figure 2-18 depicts the important portions of the flyback and forward converters. Notice the difference in phasing of the transformer connections. Also, it will be seen that no filter choke is provided for the flyback converter. These differences stem from

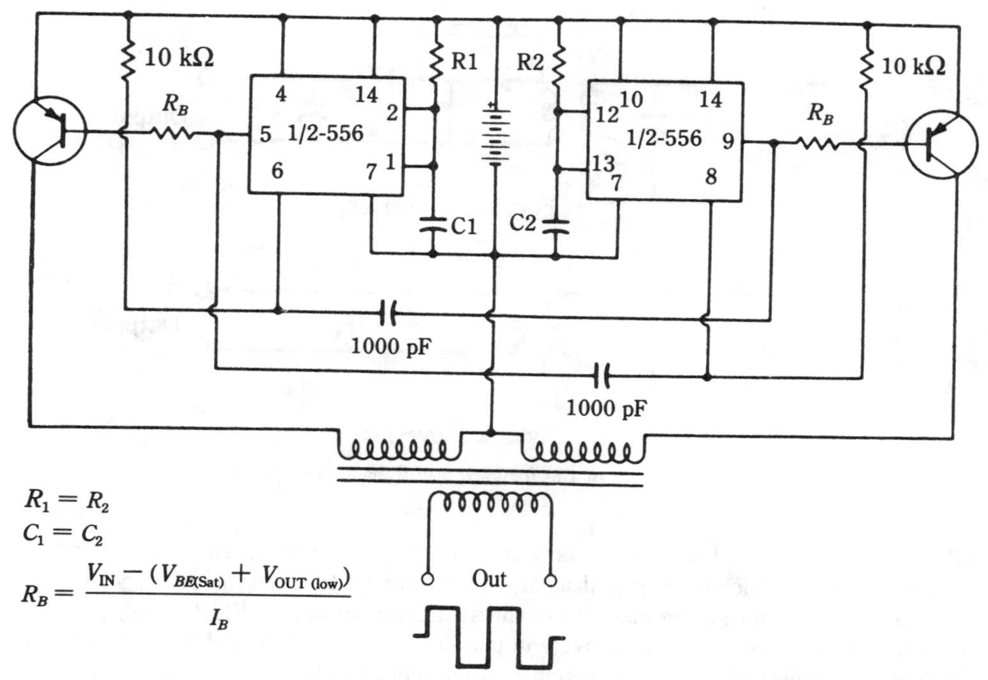

2-16 Inverter with integrated-circuit square-wave driver. General Electric Semiconductor Dept.

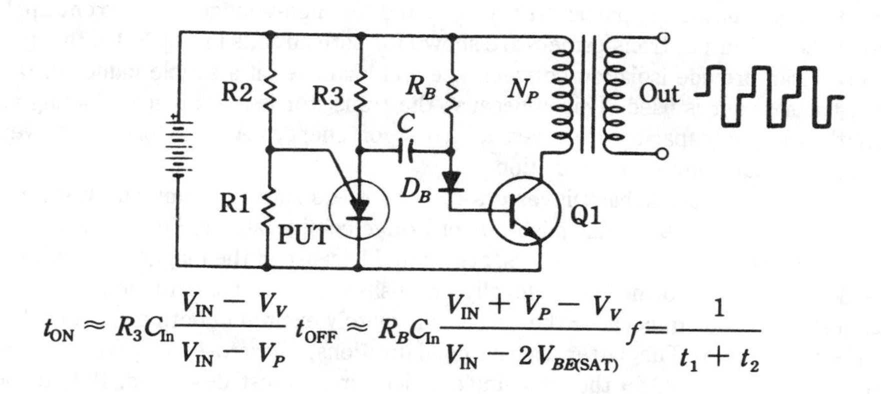

2-17 Flyback inverter with put relaxation oscillator as driver. General Electric Semiconductor Dept.

the different operating modes of the two circuits, not from optional considerations such as attenuation of output ripple.

In both circuits, energy is stored in the primary winding of the output transformer during conduction time. What distinguishes the operational mode of these two converters is the way in which this energy is discharged into the load. In the forward converter, the stored electromagnetic energy produces load current during *both* on and off-time of the switching transistor. Electrically, the load and the primary winding are

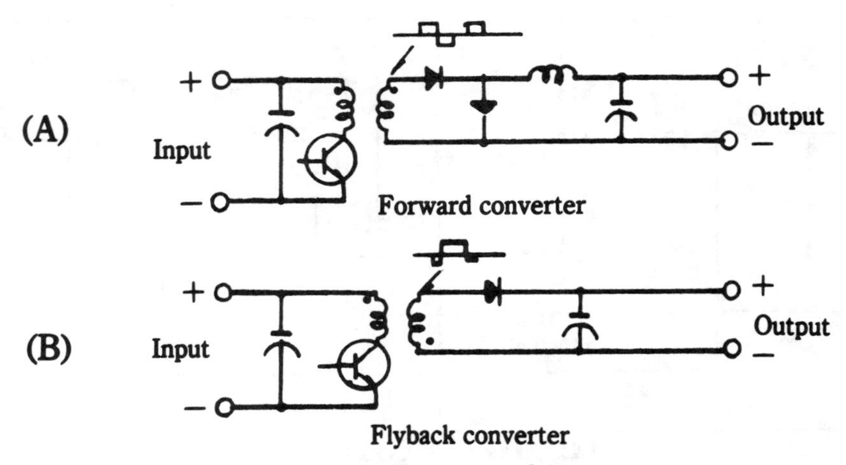

2-18 Comparison of the forward and flyback converter.

effectively in series. Thus, the forward converter can be viewed as a series circuit. Conversely, the flyback converter discharges the energy stored in the primary winding into the load *only* during the off-state of the switching transistor. Electrically, the load and the primary winding are effectively in parallel. Analysis of the flyback converter is accordingly predicated on the operation of a parallel circuit.

The forward converter provides better quality dc than the flyback circuit. Flyback converters are, however, particularly well-suited for high-voltage low-current applications. Although output transformers are shown for both circuits in Fig. 2-18, the flyback converter can provide isolation between line and load even if a simple inductor, rather than a transformer is used. This is because the transistor is in its nonconducting state when the output capacitor receives its burst of energy. A transformer, however, facilitates output voltage manipulation.

Both forward and flyback inverters or converters suffer a common disadvantage when contrasted to the various push-pull or bridge configurations. Power capability, as well as other performance parameters, are limited because of the magnetic saturation of the inductor or transformer core. Ideally, in push-pull or bridge arrangements, this is much less of a problem because the core is alternately excited in both directions during an operating cycle. The latter circuit configurations, therefore, require physically-smaller magnetics than do the two single-ended circuits just described. But, for low-power applications, flyback and forward inverters/converters merit consideration for their simplicity and economy.

3
CHAPTER
Inverter and converter design

THE INTENT OF THIS CHAPTER, AS WAS THE CASE WITH THE PRECEDING material is to provide useful insights into solid-state power electronics. This, and not step-by-step exposition of design procedure, is the objective of this book. With an understanding of the modern techniques used in inverters and converters, the applications presented in chapters 4 and 5 will become useful beyond their narrow performance ratings. If one is a designer, so much the better, but the hobbyist, experimenter, technician, or serviceman can also adapt these circuits and systems to his unique needs if there is awareness of some of the important design considerations. The inverters and converters that will be covered in the final two chapters have invested in them millions of dollars of corporate research and development funds, as well as the efforts of some of the nation's topmost engineers. It certainly should prove useful to probe into some of the aspects of design they have employed.

A decade ago, a chapter of this kind would not have had as much relevancy to the art as it does now. Then, you could make an acceptable inverter from general-purpose germanium power transistors and a radio power or filament transformer. Today, even a simple inverter or converter is expected to yield performance superior to that of such "thrown-together" arrangements. Also significant is the fact that some of those apparently simple configurations that are now state-of-the-art involve sophistications not discernible from the connection diagrams alone. It will be found both instructive and interesting to look into some of these matters.

Start-up of saturable-core inverters

Despite the extra features and the improved operating characteristics of other types of inverters, the self-oscillating saturable-core circuit retains basic importance in solid-state energy conversion. This is primarily because of the simplicity and favorable economics inherent in the circuit. Its shortcomings are progressively being made less formidable as better design techniques evolve and improved components become available. Even though the schematic diagrams have not undergone much change during the past decade, the performance of the saturable-core inverter and converter has been continually upgraded.

Not much has been said yet about the matter of *starting* such inverters. In dealing with the theory of oscillation, it is convenient to assume either that starting has already occurred or that it is assured. This is acceptable in that the reader is not distracted from the main topic under discussion. In the real world of hardware, it is often necessary to devote *specific attention* to the initial start-up, as well as to the recovery of oscillation from a nondestructive circuit fault.

An interesting situation is shown in Fig. 3-1. In this experimental setup, it is found that the saturable-core inverter simply shuts down when the load is short-circuited. This, of course, is desirable. Both transistors revert to their "off" states, or at worst to a low-current idling condition as determined by the bias networks. This inherent self-protection is reliable, and with appropriately designed circuitry, it can continue indefinitely. However, when the short is removed, the ***inverter will commence normal operation.*** Notice that the removal of the fault had not produced a current inrush in the primary winding, as happens when the dc supply is initially connected. Why should the inverter become active when the dc supply is already connected and you merely open the short circuit across the load winding?

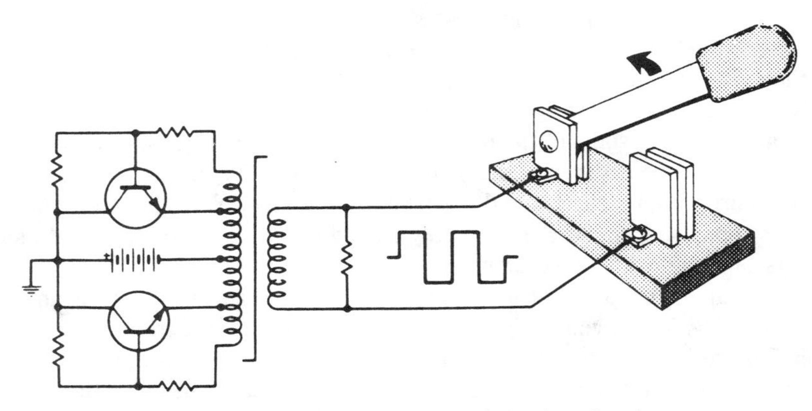

3-1 Experiment concerning start-up behavior of an inverter.

Insight into this phenomenon is provided by the circuit illustrated in Fig. 3-2A. This is a push-pull Hartley feedback oscillator. Notice the similarity between this circuit and the inverter circuit of Fig. 3-2B. The inverter does not have a physical resonating capacitor connected across its primary winding, but you can assume the existence of stray capacitance. The self-resonant frequency of the "tank" in Fig. 3-2B must be very high — certainly much higher than the square-wave frequency of the inverter. Although the Q of such a tank circuit is very low, this is offset by the extremely high transconductance of the power transistors. There is little else to prevent the inverter from behaving as a high-frequency feedback oscillator. The reason this mode of operation has

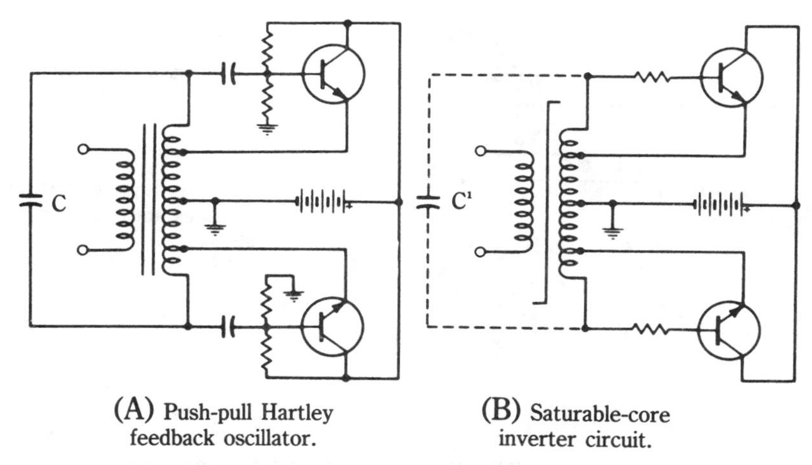

(A) Push-pull Hartley feedback oscillator.

(B) Saturable-core inverter circuit.

3-2 Intended and inadvertent Hartley-oscillator circuits.

been elusive is that it is so short-lived. A relatively few cycles of operation in the "linear" mode suffice to transfer the circuit activity to switching-mode oscillation. Note that one transistor alone constitutes a single-ended Hartley oscillator. Thus, when one transistor has more forward bias than the other, or when one transistor is on and the other is off, the opportunity for the previously described phenomenon to occur still prevails. This *transfer* of the oscillatory mode from the LC type of the switching type starts the inverter when a short is removed from the load winding.

In different inverter circuits, the equivalent LC feedback oscillator may not be the Hartley circuit. This does not change the start-up mechanism, however. And, there are other circumstances in which this aborted start as an LC oscillator assumes importance. For example, the application of dc supply voltage to the inverter may be gradual, rather than abrupt. Such a situation could arise from large filter or bypass capacitors or from filter chokes. Moreover, the turn on of some dc supplies is intentionally made slow to prevent destructive transients elsewhere in the powered equipment. The fact that the inverter need not necessarily be "shocked" into operation is not always appreciated. Probably, most starting provisions are intended to enhance the effects of a sudden application of dc voltage. However, the use of high-gain transistors helps promote LC oscillation. Fortunately, most of the techniques intended for starting from sudden energization by the dc supply also make conditions favorable for start-up from momentary LC oscillation. In any event, the worst case for starting generally occurs with full load and low temperatures.

Four commonly used start-up techniques for saturable-core inverters are shown in Figs. 3-3 through 3-6. Although these are depicted for a common-emitter circuit, the same ideas are applicable to all self-oscillating saturable-core inverters. Essentially, the objective is to provide a small amount of forward bias to at least one of the transistors. In Fig. 3-3, the forward bias is applied to the base of both transistors by resistance network R1/R2. An objection might be made here that this could lead to a latched-up condition with both transistors in a state of heavy condition. This possibility was mentioned in the previous chapter, where it was pointed out that such an occurrence is

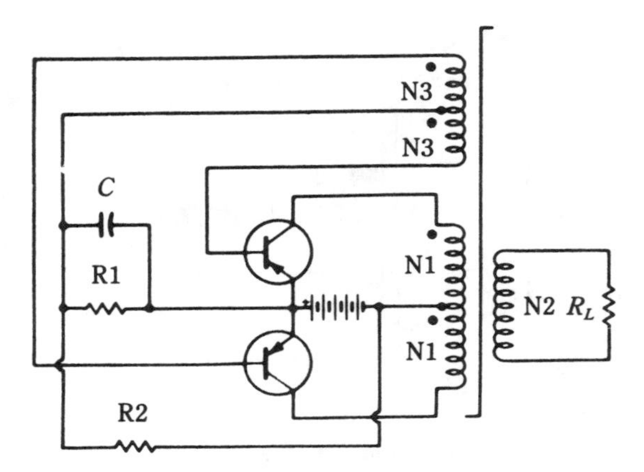

3-3 Single voltage divider for both bases as start-up provision. Motorola Semiconductor Products, Inc.

unlikely because of the improbability of perfect balance in the two halves of the push-pull circuit. The worst that could happen would be a momentary state of simultaneous conduction in the two transistors; a sustained latch-up could not happen because the requisite forward bias for transistor saturation is derived from a *dynamic*, rather than a static circuit condition. That is, the circuit would have to be in a state of oscillation, and therefore it would not be latched up.

The approximate values for R1 and R2 are easily calculated. Then, optimization of the inverter starting and operating characteristics can be achieved empirically by minor modifications of the calculated values. The base starting bias, V_B, for germanium transistors can be assumed to be 0.3 V; for silicon transistors, it is about 0.5 V. If we let V_{CC} represent the dc supply voltage, R1 and R2 can be obtained from the equation:

$$V_B = \frac{R_1 \times V_{CC}}{R_1 + R_2}$$

To proceed, first select a value for R1 that does not exceed the value of resistance obtained from the equation

$$R_1 = \frac{V_{FB} - V_{EB}}{I_B}$$

Where:

V_{FB} is the voltage from the feedback winding,
V_{EB} is the emitter-base voltage needed to saturate the transistor,
I_B is the resultant base current.

The need for this procedure stems from the fact that R1 limits the base current. If I_B is not known, a reasonable value may be obtained from the equation:

$$I_B = \frac{2I_p}{h_{FE}}$$

Where:

I_P is the collector current under full-load conditions,
h_{FE} is the minimum forward current gain over the load range of the inverter.

Notice that the essential philosophy of these "quickie" computations is to be certain to obtain sufficient base drive.

The best start-up network is one in which R1 is small enough to allow collector-current saturation under worst conditions, and R2 is no smaller in value than is needed to ensure worst-condition start-up. The resistance network will then dissipate the minimum power consistent with its circuit functions.

Capacitor C is usually referred to as a speed-up capacitor. As its name implies, it tends to promote faster switching transitions. It accomplishes this by shortening the turn-on time. (This is a common technique encountered in digital electronics where the rapid response of logic gates is important. Speed-up capacitors effectively neutralize the effect of charge in the input circuit of a switching device.) In an inverter, such speed-up can actually enhance the operating efficiency by reducing switching losses during rise time. It also happens that the speed-up capacitor aids the starting characteristics. This is particularly true when the initial application of dc voltage to the inverter is abrupt. Under such conditions, the speed-up capacitor provides the base with a momentary pulse of forward bias. To be really effective, the speed-up capacitor should be a little larger than is required to merely shorten the rise time when the transistor turns on.

The diode start circuit of Fig. 3-4 allows a high forward-bias voltage to be applied initially to the bases of the transistors. The initial base current is limited by both R1 and R2. As soon as the inverter becomes oscillatory, the bases derive their actuating current from the feedback winding, N3. This changeover is automatic and occurs as follows. Prior to oscillation, no feedback signal is available from N3. Base current for the transistors then comes through R1 and R2 from the dc power supply. The diode, being reverse-biased, is effectively out of the network. When oscillation sets in, the diode becomes forward biased and conducts base current supplied from the feedback winding. Because the feedback winding constitutes a much lower-impedance source than does the dc power supply in conjunction with R2, normal operation of the inverter ensues. That is, base current for the transistors is supplied mostly by the feedback winding.

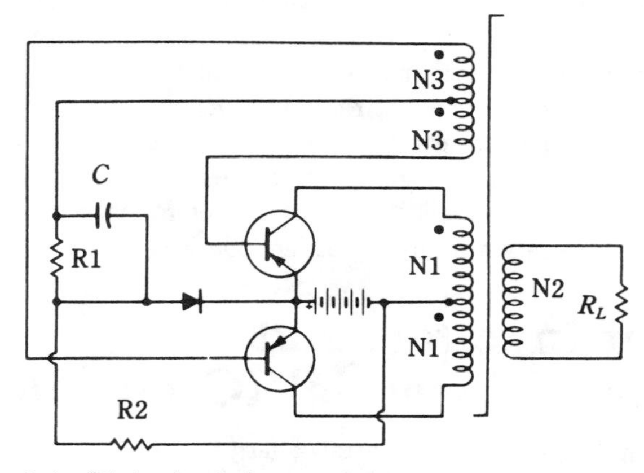

3-4 Diode circuit for start-up. Motorola Semiconductor Products, Inc.

Resistance R1 can be calculated in the same way as for the circuit in Fig. 3-3. Resistance R_2 will be much larger than R_1. The disparity between R_1 and R_2 is the feature responsible for the automatic selection of the base-current source. The actual value of R_2 can be determined empirically. Its value will depend on the power-supply voltage, but a value about ten times that of R1 is reasonable for initial evaluation.

The start-up technique of the inverter shown in Fig. 3-5 is similar to that of Fig. 3-3, except that individual bias networks are provided for the two transistors. By this means, one transistor can be biased more heavily into conduction than the other. This ensures a predictable starting mode. In itself, this is usually not of great consequence. However, the fact that much less reliance is placed on the imbalance of transistor characteristics makes this circuit a sure-fire starter. Its use merits consideration when the inverter must be started with loads, such as incandescent lamps or motors. The calculation of the resistances labeled R_1 is similar to the procedure outlined for Fig. 3-3. The same is true for one of the resistances designated as R_2. The other resistance shown as R_2 can then be made several times this value. The "starting" transistor will be the one associated with the lower R_2 value.

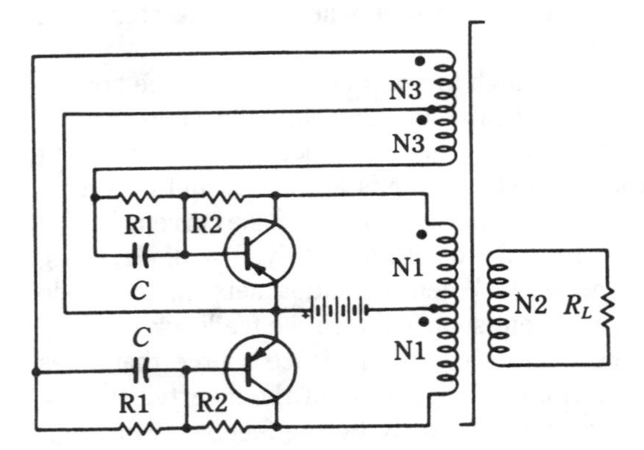

3-5 Individual base voltage dividers to provide start-up. Motorola Semiconductor Products, Inc.

The scheme depicted in Fig. 3-6 permits a strong forward bias to be applied to the transistors prior to oscillation. After oscillation begins, the voltage derived from the rectified output of an additional transformer winding cancels a substantial portion of this bias. The amount of cancellation is governed by the value of resistance R_1. Thus, when the circuit is in oscillation, transistor bias is principally provided by the base-feedback windings. By means of this technique, resistance R_2 can be chosen for best starting action without having to be concerned over degradation of operating efficiency. Notice that prior to starting, the full-wave rectifying diodes are effectively out-of the circuit. Operating base current can be adjusted by means of R_1 because of its effect on the amount of voltage cancellation occurring at junction X.

Control of inverters and converters with special IC modules

Inverters and converters are often parts of larger overall systems, such as power supplies, regulators, motor drives, etc. In such applications, their outputs are subject to

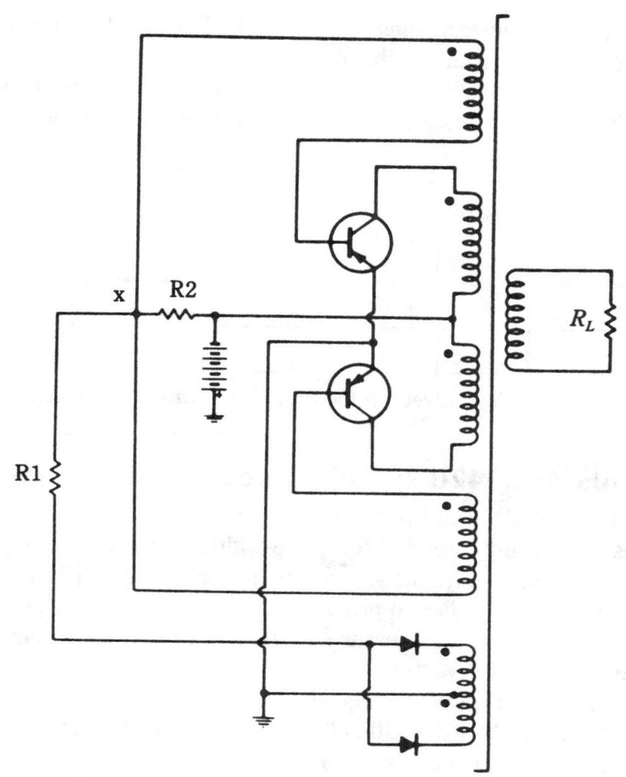

3-6 Bias-cancelling start circuit. Motorola Semiconductor Products, Inc.

control. The control function may be manual or automatic. One of the most difficult design tasks has been the implementation of low-level and logic circuitry for achieving this control. All manner of problems beset the designer when such control circuitry is made from discrete components. Moreover, the complexity and resultant cost of such control circuitry often tends to be considerable. This often comes as a surprise, because it is natural to think that most of the engineering effort rightfully belongs to the power circuit. In order to obtain reliability, reproducibility, reasonable packaging volume, and operational flexibility, it has often been necessary to be reconciled to less-than-desired overall performance. For example, the control circuitry should provide such features as soft starting, overload protection, pulse-width modulation, and variable dead time. Here we are dealing with driven, rather than self-oscillatory inverters.

The full potential of modern transistors, diodes, transformers, and capacitors cannot be realized in the face of such common control malfunctions as jitter, lack of dead time, unsymmetrical duty cycle, and limited or absent pulse-width modulation capability. These problems can be overcome through the use of special ICs for the control of inverters and converters. Two of these are covered.

The feature of dead time alone would make these IC modules valuable. This is because one of the difficulties encountered with the otherwise desirable driven inverter is the tendency toward ***common-mode conduction.*** This arises from the long turn-off time of transistors, from jitter in the drive oscillator, and from jitter in the drive oscillator, and from the effects of reactive loads. A clean solution to this problem is actuation from a stepped wave, such as that shown in Fig. 3-7, which is available from these ICs.

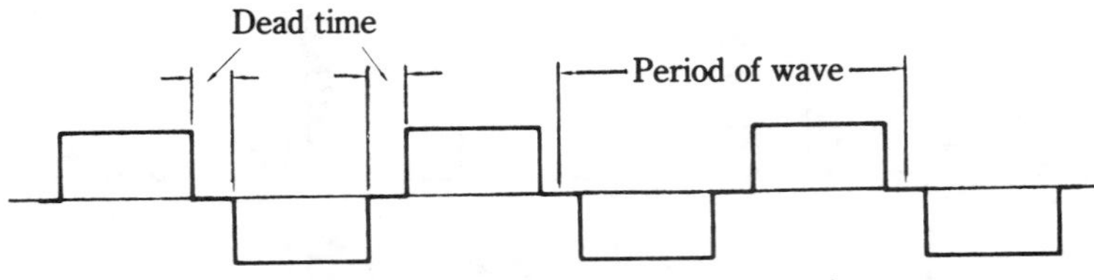

3-7 Ideal waveform for actuation of driven inverters.

The Motorola MC3420 switchmode regulator control circuit

The MC3420 is a 16-pin dual-in-line module available in either plastic or ceramic packaging. It is specifically intended for pulse-width-modulated drive of two external power transistors. Its output frequency ranges from about 2 to 100 kHz. Two or more MC3420s can be slaved together to provide additional synchronized control signals for multiple-transistor inverters. This feature should also make these modules useful in the gate-logic circuitry of SCR inverters.

The electrical characteristics for the MC3420 are listed in Table 3-1. The temperature range specified for the MC3420 is 0 to +70°C. The specified temperature range for the MC3520, otherwise similar, is −55 to +125°C. A block diagram of the MC3420 is illustrated in Fig. 3-8. In order to provide maximum circuit flexibility, most of the internal functions are brought out to the pins. The sequence of events within the module is indicated by the waveforms in Fig. 3-9.

Figure 3-10 indicates RC combinations that can be associated with terminals 1 and 2 in order to produce various ranges of output frequencies. Because a semilog graphical format is used to accommodate the frequency range, it is not immediately apparent that the frequency is inversely proportional to either R_{ext} or C_{ext}. For example, with C_{ext} selected as 0.01 μF, the output frequency is about 11 kHz when R_{ext} is 5 kΩ. Then, doubling R_{ext} to 10 kΩ causes the output frequency to become half its previous value, or 5.5 kHz. Also, as is evident from the waveform diagram of Fig. 3-9, the frequency of the ramp generator is actually double that of the output.

The Silicon General SG2524 regulating pulse-width modulator

The SG2524 is a monolithic subsystem that represents another manufacturer's solution to the control requirements of driven inverters and converters, as well as switching regulators, choppers, and similar circuits. The block diagram of this 16 pin dual-in-line IC module is shown in Fig. 3-11. The output frequency is determined by a resistance connected from pin 6 to ground and a capacitance connected from pin 7 to ground. The SG2524 is one of a family of similar ICs differing in temperature specifications. For example, the SG2524 is rated for operation over a temperature range of 0 to +70°C. The SG1524 is rated for a more extensive range, from −55 to +125°C.

Table 3-1 Electrical characteristics of MC3420 and MC3520.

Characteristic	Symbol	Min	Typ	Max	Unit
Supply Voltage	V_{CC}	10	—	30	V
Supply Current	I_{CC}	—	—	16	mA
Output Frequency Range	f_o	2.0	—	100	kHz
Frequency Stability ($T_A = T_{high}$ to T_{low}, $10 < V_{CC} < 30$ V)	—	—	4.0	—	%
Voltage Reference	V_{ref}	—	7.9	—	V
Temperature Coefficient of Voltage Reference ($I_{ref} = 400\ \mu A$)	TCV_{ref}	—	0.006	0.02	%/°C
Output Voltage	V_{OL}				
($I_{OL} = +40$ mA)		—	—	0.5	V
($I_{OL} = +25$ mA)		—	—	0.3	
Output Blocking Voltage	—	—	—	40	V
Oscillator Output Voltage ($I_{OL} = +5$ mA)	V_{osc}	—	—	0.5	V
Temperature Coefficient of Dead Time	TC_{DT}	—	0.15	—	%/°C
Inhibit I_{IL} ($V_{IL} = 0.7$ V)	—	—	—	−0.2	mA
Inhibit I_{IH} ($V_{IH} = 2.4$ V)	—	—	—	40	μA
Minimum Dead Time	—	0	—	—	μs

NOTES: $V_{CC} = 15$ V and $T_A = 25°C$ unless otherwise noted
$T_{low} = 0°C$ for MC3420, −55°C for MC3520
$T_{high} = +70°C$ for MC3420, +125°C for MC3520

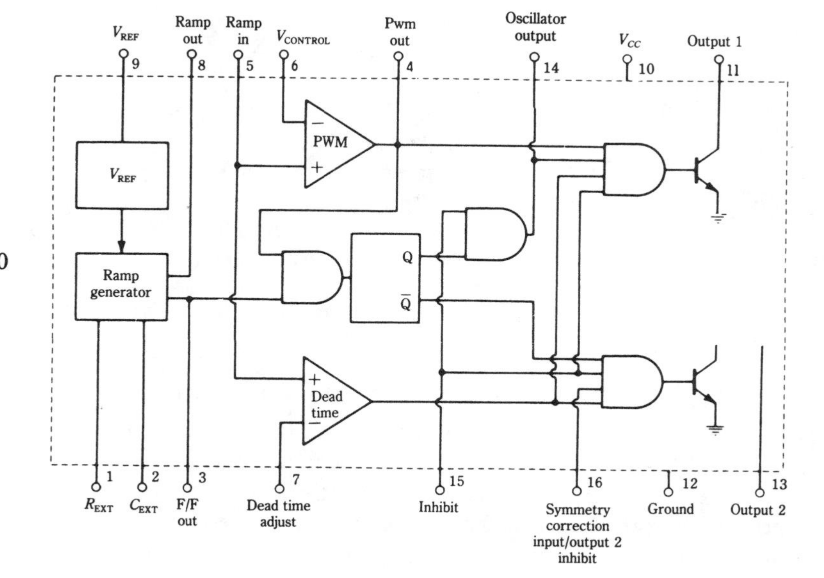

3-8 Block diagram of MC3420 inverter-control IC.

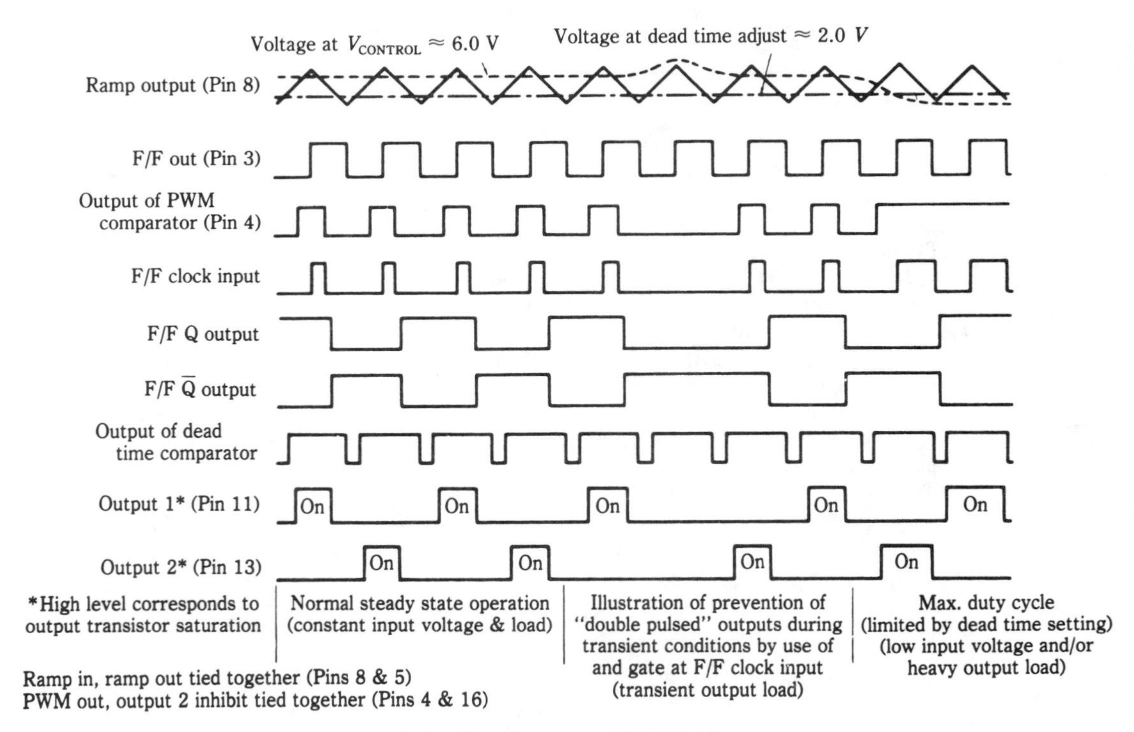

3-9 Waveforms in MC3420 IC.

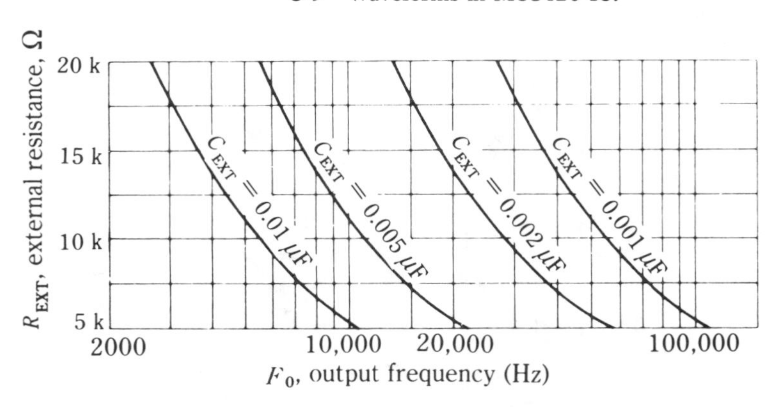

3-10 Frequency-range curves for MC3420 IC.

An example of the use of the SG2524 IC is illustrated in Fig. 3-12. This converter is voltage regulated by means of the pulse-width modulation provided by the SG2524. It will be observed that the simplicity of this arrangement compares favorably with that of linear regulators based on the Type 723 IC. Here, however, high operational efficiency prevails over the entire load range inasmuch as the control technique does not make use of dissipation. The output transformer operates without saturation, adding to the

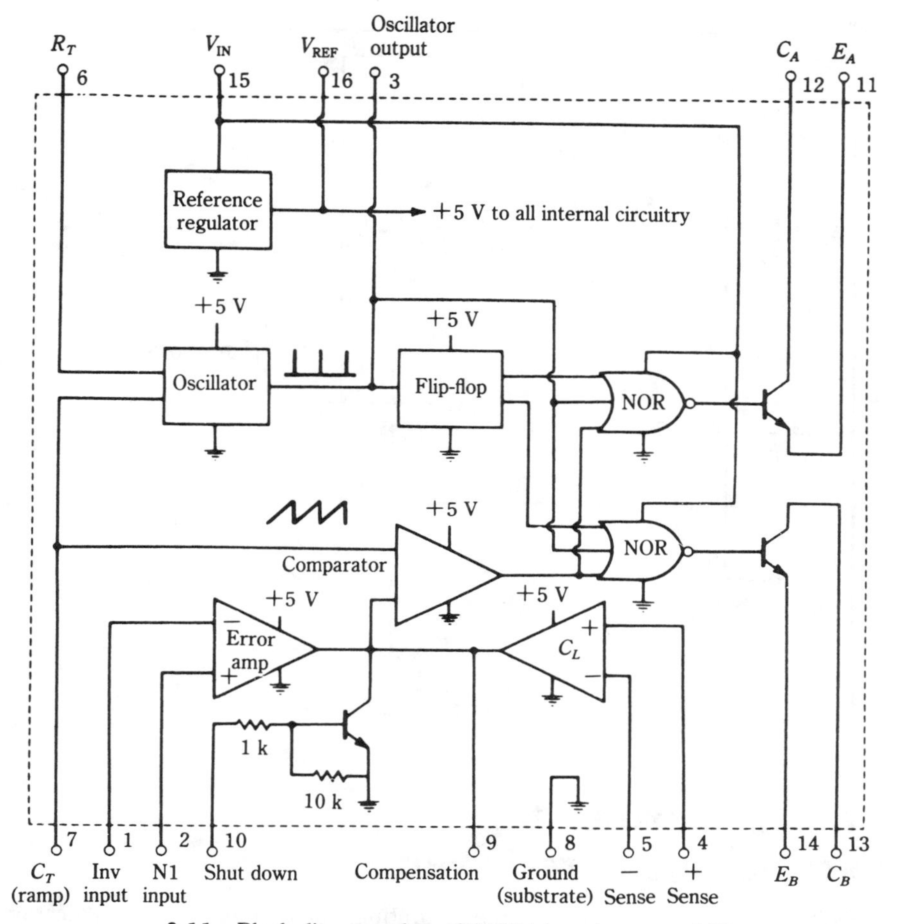

3-11 Block diagram of the SG2524 inverter-control IC.

overall efficiency of the converter. Such operation is readily achieved because of the symmetrical duty cycle applied to the inverter transistors by the IC. When the two transistors operate at unequal duty cycles—a common problem with control circuitry using discrete elements—a dc component appears in the transformer. This leads to earlier magnetic saturation than would otherwise be the case.

As with the Motorola unit, the control of frequency is accomplished by means of external timing resistance and capacitance. This is shown in Fig. 3-13A. The oscillator period is one half that of the output waveform. In the SG2524, a convenient way to obtain a predictable deadband is also by selection of the external timing capacitance. This is shown in Fig. 3-13B. Thus, for a given frequency, if you want a greater deadband, you choose an RC combination in which *C* has a relatively high value. Greater deadbands than those indicated can be obtained by connecting capacitance from pin 3 to ground (up to 5 μs, corresponding to 1000 pF, is possible by this means). Conversely, deadband control is achievable from external circuitry connected to pin 3.

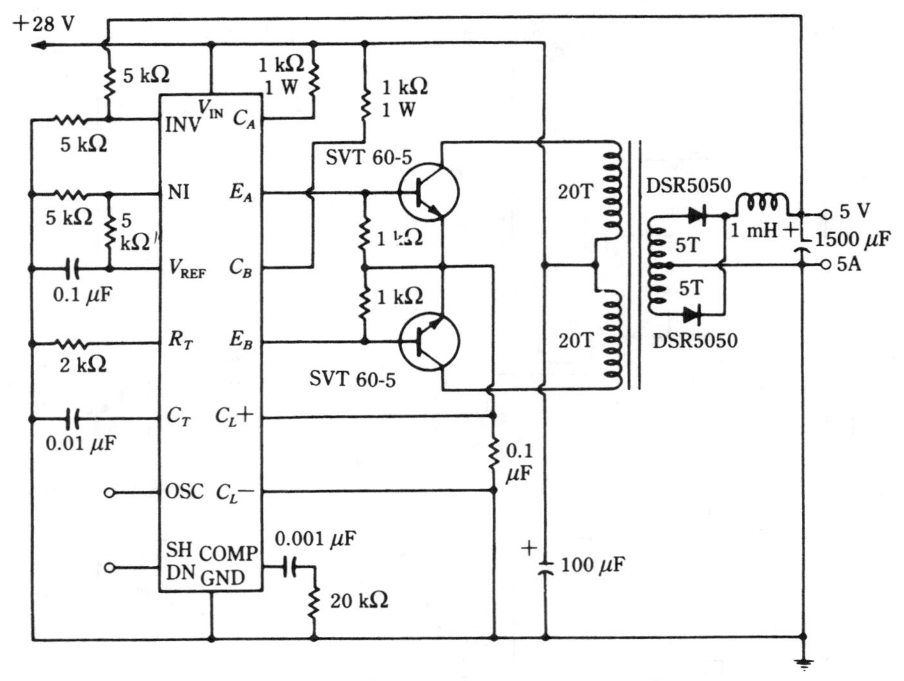

3-12 Driven converter using the SG2524 IC.

The transformer for the saturable-core inverter

The traditional design procedure for the saturating transformer was a modification of the processes one would go through to design a filament or bell-ringing transformer. Because most of the early inverters operated at 60 or 400 Hz, and performance was limited by the transistors and other components, this was more or less satisfactory. However, if it is true that the design of ordinary sine-wave transformers is a science and an art, the design of saturating transformers was more akin to witchcraft. For example, in the classical transformer equation (modified for square waves) the number of turns, N, on a winding is given by:

$$N = \frac{E \times 10^8}{f \times 4 \times B_{SAT} \times A}$$

We can specify the voltage, E, the frequency, f, and the core area, A, but for a core made up of silicon-steel laminations, how does one specify the saturating flux density, B_{SAT}? Because the onset of magnetic saturation is relatively gradual in silicon steel, B_{SAT} is nebulous. And there are other unknown parameters when this approach is used. For example, it is difficult to estimate the hysteresis and eddy-current losses in such cores.

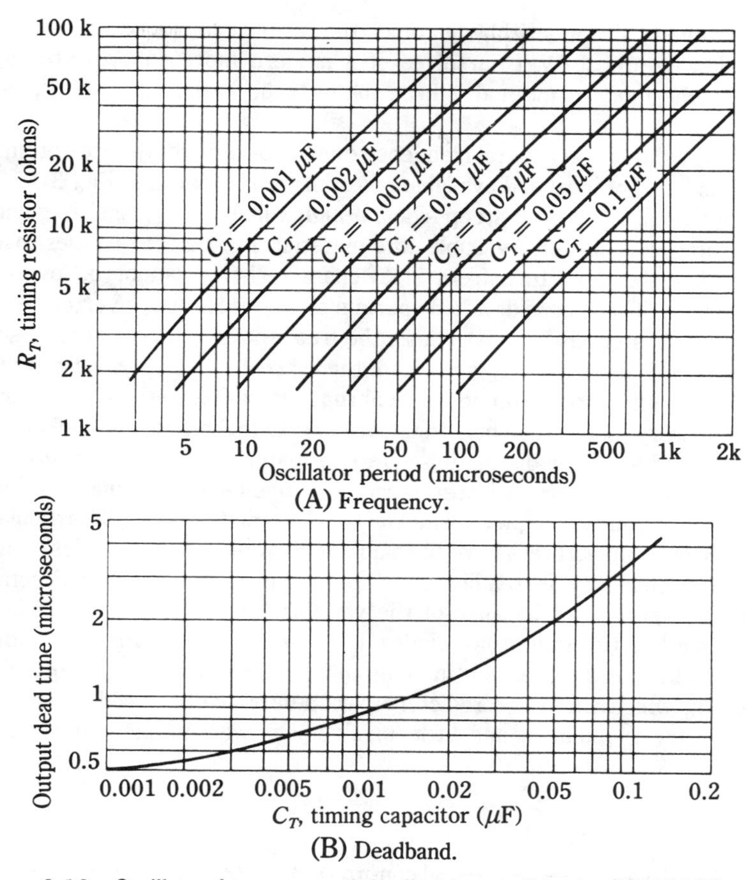

(A) Frequency.

(B) Deadband.

3-13 Oscillator frequency and deadband control in the SG2524 IC.

These losses, which have a significant effect on the operating efficiency at frequencies greater than 400 Hz, also have other adverse effects.

It has long been known that optimization of transformer efficiency occurs when the core and copper losses are equal. How to bring about this condition without extensive data acquisition and computer-aided design has been a problem beyond the capabilities of most workers in the field. Also, the incompatibilities of available window space for the windings and heat-removal requirements further complicate matters. Naturally, such unpredictability led to the attainment of design objectives through considerable empirical work. Seldom, however, was real optimization of performance achieved, to say nothing of reproducibility on the production line. And, extensions of performance often eluded simple proportioning or scaling techniques—one was again forced to modify by means of cut-and-try procedures.

This is not intended to cast aspersion on such design procedures. They were workable and well accepted. Indeed, for many hobbyist and cost-limited applications, it remains altogether acceptable to use them. And, if one uses a nickel-iron tape core rather than ordinary silicon-steel laminations, the results can be quite satisfactory. But, for best results we must avail ourselves of a modern more sophisticated approach. Paradoxically, the modern method will be found to be easier than the time-honored procedure.

Figure 3-14 shows the general frequency domain of commonly used core materials. The hysteresis loops of ideal and available material are compared in Fig. 3-15. An ideal core has a hysteresis loop with steep sides, a high value of B_{SAT}, and negligible area. Magnetic saturation is abrupt. Available core materials have these qualities to a reasonable degree, but the departure from ideal characteristics cannot be overlooked in practical inverters. These considerations pertain to self-oscillating inverters—in driven inverters, where hysteresis losses are less, the frequencies can be extended somewhat before unreasonable dissipation sets in. Also, there are ferrites specifically intended for driven inverters. Such ferrites do not have abrupt saturation regions, but their hysteresis loops are very narrow. When you consider both components of core loss—hysteresis and eddy-current dissipation—ferrite material can be considered superior for all frequencies. Also, ferrite material can be conveniently manufactured in shapes with optimum mechanical, magnetic, and electrical characteristics. Of particular significance, the ferrite manufacturers have been able to provide easy-to-use design data from which optimum efficiency is readily attainable. This is not easy to accomplish by previous design approaches because of the variables of core loss, copper loss in the windings, device loss (in transistors, diodes, etc.), window space for the windings, and operating temperature. By means of nomographs, you can now design ferrite inverter transformers in which these contradictory parameters are balanced for optimum results. Cut and try is minimal, and little involvement with mathematical "games" is required.

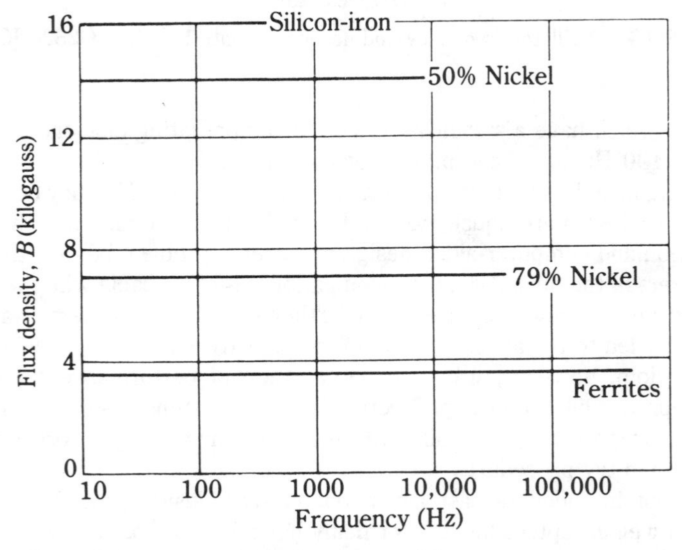

3-14 General frequency range of core materials. General Electric Semiconductor Products Dept.

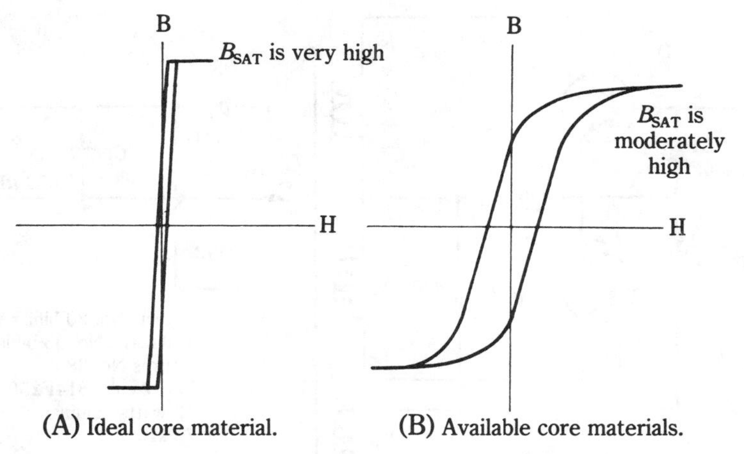

(A) Ideal core material. (B) Available core materials.

3-15 Hysteresis loops of core materials for oscillating inverters.

An objection could be raised that nomograph design techniques are "blind" because one no longer has the basic insight of the procedure. However, as long as one is aware of the implication of Faraday's law for inverters (the formula for N given earlier), mental grasp of the situation is not lost. It is simply a case in which the manufacturer has been able, by computer-aided design, to include the effects of variables not incorporated in this basic equation.

By the same token, it should not be surmised that the older method of applying the equation was "classical" in a rigorous sense. In most cases, you also had to make use of assumptions and rules of thumb. As an example, one widely used procedure starts on the premise that the number of turns on one-half of the primary winding is about three and one-half times the value of the dc supply voltage. This is merely a convenient assumption used to circumvent the problem of too many unknowns, but it usually provides a starting point for a successful design.

In order to illustrate these matters, three design procedures for inverter transformers will be covered briefly. The first is a "streamlined" version of the older approach. The second and third are nomograph methods.

Transformer design by calculation

The following calculations, based on an example furnished by the General Electric Semiconductor Products Department, pertain to the converter transformer of the circuit in Fig. 3-16. Circuit parameters relevant to the transformer are as follows:

Power-supply voltage V_{IN}: 6 V
Output power, P_{OUT}: 12 W
Transistor saturation voltage, $V_{CE(SAT)}$: 0.8 V
Saturation flux, B_{SAT}: 3500 gauss for final core
Frequency, f: 75 kHz for initial design procedure
Current density in windings: 750 circular mils per ampere

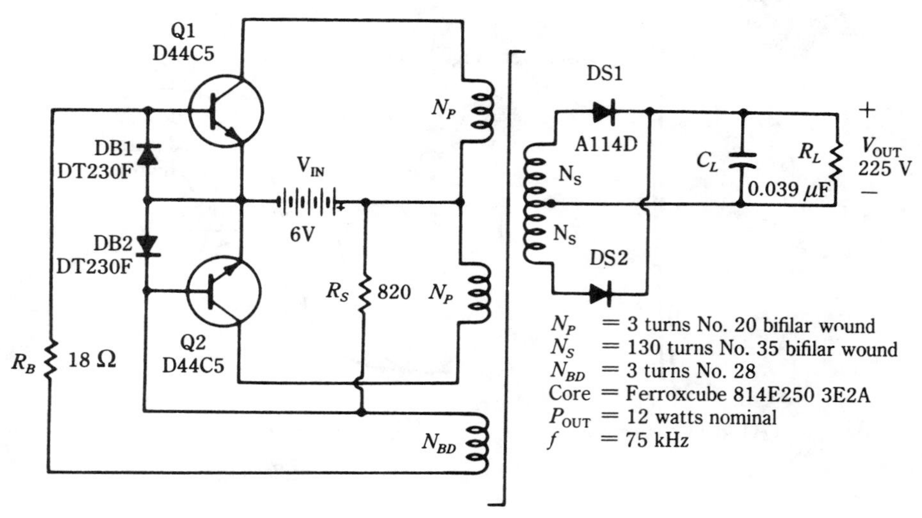

3-16 A converter circuit for transformer calculations.

The total power the transformer must handle (p') is the output power plus the losses in the output rectifiers plus the losses in the transformer.

The computations are described in the paragraphs that follow. For convenience, they are summarized in Chart 3-1.

Chart 3-1 Transformer design computations.

A. Material: Ferroxcube 3C8, chosen for considerations of efficiency, frequency response, and Curie temperature.
Configuration: E core for good heat dissipation and ease of building.

B. $B_{SAT} = 3800$ gauss

C. $B_{MAX} = B_{SAT} = 3800$ gauss

D. Wire Size (Based on Nominal Load Currents)

Primary:

$A_w = 750 \times 3.4 \times 0.5 = 1275$ cmil (Calculated)

Use No. 20 = 1022 cmil (Actual)

Secondary:

$A_w = 750 \times 0.053 \times 0.5 = 19.9$ cmil (Calculated)

Use No. 36 = 25 cmil (Actual)

Base-Drive:

$A_w = 750 \times 0.225 \times 1.0 = 169$ cmil (Calculated)

Use No. 28 = 159 cmil (Actual)

E.

$$K = 0.45\frac{1022}{1275} + 0.45\frac{25.0}{19.9} + 0.1\frac{159}{169} \approx 1$$

F.

$$p' = 12 + (0.004)(20) + (0.2)(20) = 16.1 \text{ watts}$$

$$W_aA_c = \frac{(2.4 \times 10^5)(16.1)(1.0)}{(75{,}000)(3800)(0.7)} = 0.019 \text{ cm}^4$$

Core No. 814E250 has $W_bA_c = 0.034$ cm^4, is available in 3E2A material only.

For 3E2A material:

$B_{MAX} = 3500$ gauss @ 70°C

$$B_{MAX} = 4100 \text{ gauss @ } 25°C$$

Then:

$$W_aA_c = 0.019 \frac{3800}{3500} = 0.021 \text{ cm}^4$$

Specifications of core No. 814E250:

$$W_b = 0.171 \text{ cm}^2$$

$$A_c = 0.2 \text{ cm}^2$$

$$l_w = 3.42 \text{ cm}$$

G. Copper Loss $\approx (0.13)(0.12)(3.42) = 0.053$ watt

H. Number of Turns:

Primary:

$$N_p = \frac{(5.2)(10^8)}{(4)(75{,}000)(3500)(0.2)}$$

$$= 2.5 \text{ turns (Calculated)}$$

$$N_p = 3 \text{ turns, No. 20, Bifilar (Actual)}$$

$$\text{Expected Frequency} = 75{,}000 \left(\frac{2.5}{30}\right) = 62{,}500 \text{ Hz @ } 70°C$$

$$= 62{,}500 \left(\frac{3500}{4100}\right) = 53{,}400 \text{ Hz @ } 25°C$$

Secondary:

$$N_s = 3 \left(\frac{225}{5.2}\right) = 130 \text{ turns (Calculated)}$$

$$N_s = 130 \text{ turns, No. 36, Bifilar}$$

$$= 260 \text{ turns Center Tapped (Actual)}$$

Base-Drive:

$$N_{BD} = 3 \left(\frac{4}{5.2}\right) = 2.3 \text{ turns (Calculated)}$$

$$N_{BD} = 3 \text{ turns, No. 28 (Actual)}$$

I.

$$V_{BD} = V_p \frac{N_{BD}}{N_p}$$

$$= (6.0 - 0.8)\frac{3}{3} = 5.2 \text{ volts}$$

J. Window Area:

$$W_a = 18{,}700 \,(5.06 \times 10^{-6})$$

$$= 0.09 \text{ cm}^2 \text{ (Needed)}$$

$$W_b = 0.171 \text{ cm}^2 \text{ (Available)}$$

K. Core Loss $= (0.55 \text{ cm}^3)(5 \text{ watts/cm}^3)$
$= 2.75$ watts

L. Transformer Loss $= \dfrac{2.75 + 0.05}{27} \approx 10\%$

Core selection

The first steps are choosing the frequency of oscillation and the selection of the core. Because the circuit of Fig. 3-16 is intended as a portable power supply for gas-discharge displays, its physical size should be small. For this reason, a high frequency, 75 kHz is chosen.

The choice of the high operating frequency limits selection of the core material to a ferrite. Initially, Ferroxcube 3C8 material appears to fill the bill, inasmuch as it is intended for high-frequency work. This choice is recorded at A in Chart 3-1. An "E" core configuration is selected because it is easy to wind and allows good heat removal. Later in the computational procedure, it will be found that this choice of material is not readily implemented because of demands made on the geometry of the core. A change is then made to another, but similar, core material, and the appropriate computational corrections are made. Such "cut-and-try" steps are often necessary because you might not know initially whether the required core dimensions will be found in the manufacturer's catalog. The idea is to achieve a reasonable fit, with the departure from required window size being slightly on the plus side.

Inasmuch as we have started with the choice of 3C8 core material, we record its saturation flux density from the manufacturer's catalog as 3800 gauss (B in Chart 3-1). For saturable-core inverters, $B_{MAX} = B_{SAT}$ when the B-H curve of the material is reasonable "square" (C in Chart 3-1).

Wire sizes

The next step is to calculate the required wire sizes for the three windings. These calculations are summarized in part D of Chart 3-1.

Primary winding The formula for calculating the wire size is:

$$A_W = current\ density \times I_{IN} \times duty\ cycle$$

Where:

A_W is the wire size in circular mils,
I_{IN} is the maximum current in the primary winding.

The current density is chosen as 750 circular mils per ampere. This is known to be a good compromise between temperature rise and compactness. Depending on such factors as cost, heat removal, and physical dimensions, the current density of copper windings for inverter transformers generally ranges from 600 to 1000 circular mils per ampere.

The magnitude of I_{IN} is determined by dividing the input power by the input voltage. In turn, the input power is the output power divided by the efficiency. You know the output power is 12 W. The immediate need is to find the efficiency. Then we can determine the input power, P_{IN}, and finally the wire size in the primary winding. In this example, graphical techniques are used to estimate the losses and the efficiency of the overall circuit.

To find the overall efficiency of the circuit, we need to assign reasonable values to the various losses in the circuit. For this purpose, the graphs in Figs. 3-17, 3-18, and 3-19 provide "guesstimates" of the transistor switching losses (p_{SW}), transistor "on" losses ($p_{Q\text{-}ON}$), rectifier losses (p_D), and transformer losses (p_{TRF}). The values (read from the graphs as a fraction of the input power) are totaled as follows:

$p_{SW} = 0.045$ (From Fig. 3-17)
$p_{Q\text{-}ON} = 0.16$ (From Fig. 3-18)
$p_D = 0.004$ (From Fig. 3-18)
$p_{TRF} = 0.2$ (From Fig. 3-19)
Total losses $= 0.41$, or 41% of total input

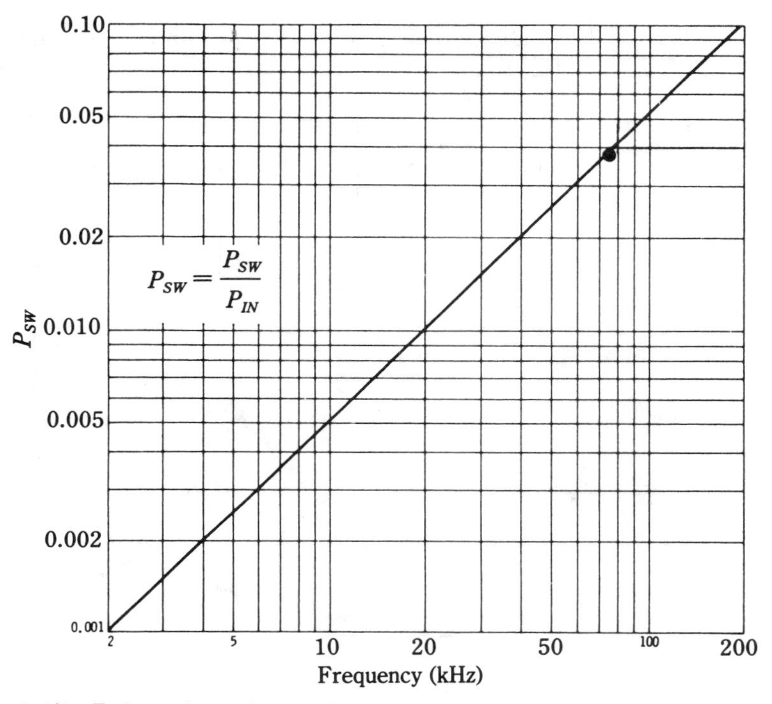

3-17 Estimated transistor switching losses. General Electric Semiconductor Products Dept.

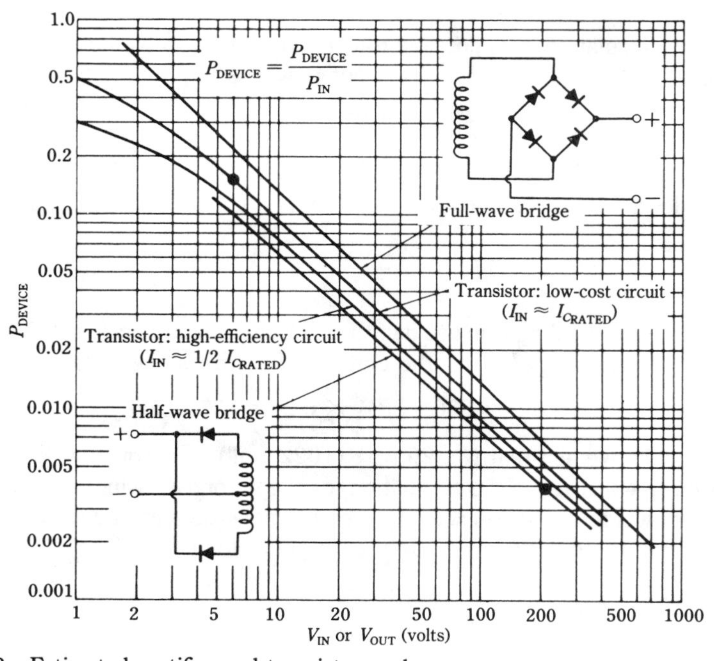

3-18 Estimated rectifier and transistor on losses. General Electric Semiconductor Products Dept.

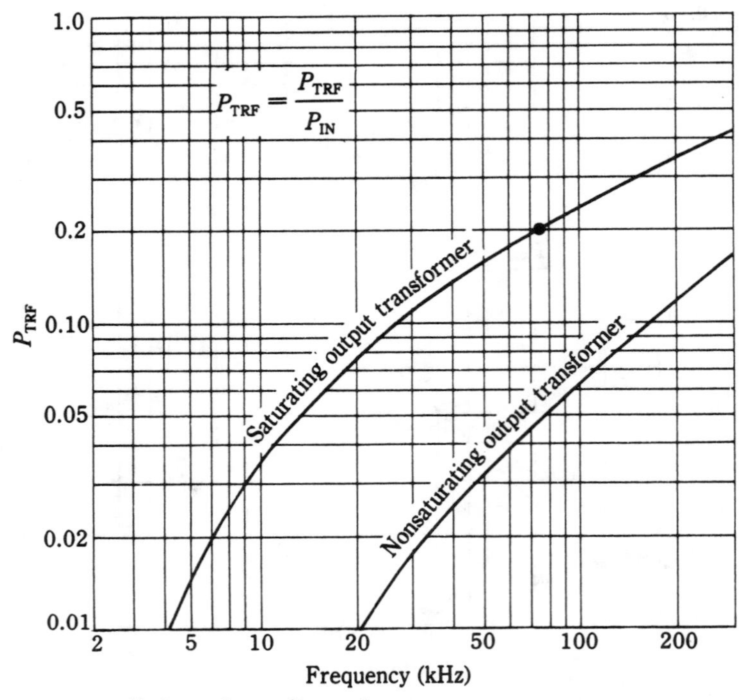

3-19 Estimated transformer losses. General Electric Semiconductor Products Dept.

With this information, it is possible to compute the overall efficiency of the circuit, which is 100 − 41 = 59 percent. The input power is:

$$P_{IN} = \frac{12 \text{ W}}{0.59} = 20.3 \text{ W}$$

The magnitude of I_{IN} can be determined to be:

$$I_{IN} = \frac{20.3 \text{ W}}{6 \text{ V}} = 3.4 \text{ A}$$

Each half of the winding is in use half the time, so the duty cycle is 0.5. Substituting this value and the value of I_{IN} into the equation for wire size gives:

$$A_W = 750 \times 3.4 \times 0.5 = 1275 \text{ cmil}$$

From a wire table, we find that No. 20 wire (1022 cmil) is reasonably close to this size.

Secondary winding The calculation for the secondary winding is similar to that for the primary winding. In this case you need to know the output current, I_{OUT}. This is found by dividing the output power by the output voltage:

$$I_{OUT} = \frac{12 \text{ W}}{225 \text{ V}} = 53 \text{ mA}$$

The wire size for the secondary is:

$$A_W = 750 \times I_{OUT} \times duty\ cycle$$

Because of the center-tapped secondary configuration, each half of the winding is active for half the time, and the duty cycle is 0.5. Therefore:

$$A_W = 750 \times 0.053 \times 0.5 = 19.9 \text{ cmil}$$

From a wire table, you would select #36 wire at 25 cmils. This selection compensates somewhat for the slightly skimpy wire in the primary.

Base-drive winding The calculation of wire size for the base-drive winding is similar to the previous two calculations, except that the duty cycle is 100%. This follows from the fact that the entire winding is driving one base or the other during the entire oscillation cycle.

The base current is estimated as follows: The transistors are assumed to have a current-gain factor (beta) of no less than 20 under worst operating conditions. Then, the base current is:

$$I_B = \frac{Collector\ current}{Beta}$$

$$= \frac{3.4 \text{ A}}{20} = 170 \text{ mA}$$

Because of the nature of the load the converter will be supplying, this value is increased by 33 percent. In initial design efforts, it is generally wise to try to make certain that *sufficient* base drive current is available. Thus:

$$I_B = 1.33 \times 170 \text{ mA} = 225 \text{ mA}$$

From this result and from the previous procedures, the wire size for the base-drive winding readily follows:

$$A_W = 750 \times 0.225 \times 1.0 = 169 \text{ cmil}$$

From a wire table, you would select #28 wire at 159 cmils.

Adjustment factor

Factor K (E in Chart 3-1) is an adjustment factor to help obtain an acceptable fit of the windings. The formula is:

$$K = 0.45\left(\frac{Selected\ pri\ wire\ area}{Calculated\ pri\ wire\ area}\right) + 0.45\left(\frac{Selected\ sec\ wire\ area}{Calculated\ sec\ wire\ area}\right) + 0.1\left(\frac{Selected\ base\text{-}drive\ wire\ area}{Calculated\ base\text{-}drive\ wire\ area}\right)$$

When the appropriate substitutions are made (see Chart 3-1), factor K is very nearly unity. Accordingly, 1.0 will be used for K in subsequent calculations.

Core size

Up to this point, the losses have been dealt with as a fraction of the total input power. But now, it's time to evaluate the total actual power in watts, p', that the transformer must handle. This is one of the main factors governing the required physical size of the core. Mathematically,

$$p' = \textit{power output}$$
$$= (p_D)\,(\textit{power input}) + (p_{TRF})\,(\textit{power input})$$

All of the quantities in this equation have already been evaluated. Therefore, substituting these values gives:

$$p' = 12 + (0.004)\,(20) + (0.2)\,(20) = 16.1\text{ W}$$

The calculation for W_aA_c is a significant one in the selection of core size. It is evident that there are two dimensions involved here. One of them, W_a, is the window area required for the windings. The other is the cross-sectional area of the core, A_c. Although W_a is determined by the space requirements of the windings and A_c is derived from Faraday's equation for inverters, these two geometric quantities when multiplied together yield W_aA_c, which can be considered a "size factor" for the core. The practical significance of W_aA_c is that it is often listed in the core maker's catalog. You must select a core in which this factor is large enough to be compatible with your calculations.

For convenience in distinguishing between ***required*** and ***available*** window-area dimensions, the value derived from our calculations is termed "required" and is designated W_a. The window area listed in the catalog is termed "available" and is designated W_b. Thus, we calculate W_aA_c and then seek an appropriate W_bA_c in the list of core specifications.

The equation used in step F of Chart 3-1 is:

$$W_aA_c = \frac{(2.4 \times 10^5\,(p')\,(K)}{(f)\,(B_{SAT})\,(k)}$$

Where:

2.4×10^5 is a factor for converting circular mils/ampere to square centimeters/ampere. It also takes into account the:

50 percent cycle of the major windings,
K is unity (step E in Chart 3-1),
p' is the total transformer power,
f is 75 kHz (chosen initially),
B_{SAT} is the saturation flux density for the chosen core material,
k is an assumed winding efficiency (70 percent for a hand-wound bobbin).

When these values are substituted in the equation (step F of Chart 3-1), the result is $W_aA_c = 0.019\text{ cm}^4$.

There is a no W_bA_c listing for type 3C8 "E" cores that is useful for your needs. Does this mean that all of our calculations have been in vain? Not necessarily. Fortunately, you can proceed by making a minor mathematical adjustment. This follows from the discovery that core #814E250 made of 3E2A material is similar to the initially selected 3C8 material and is available with an appropriate W_bA_c value. However, this material has a saturation flux density of 3500 gauss, as opposed to 3800 gauss for the 3C8 core. As can be seen in Chart 3-1, a simple proportion is set up to account for this difference, and the new W_aA_c value is calculated to be 0.021 cm^4.

Summing up, the second-try selection of core #814E250 appears to be a reasonable one. From the Ferroxcube catalog, $W_bA_c = 0.034$ cm^4, and from our latest calculation $W_aA_c = 0.021$ cm^4 (the quantity W_b is the actual window area available on the winding bobbin). If all goes well, the windings will occupy 0.021/0.034 of the available window area, or approximately 70 percent.

Copper losses

The calculation of the copper losses (G in Chart 3-1) is not essential to this problem because the copper losses are overwhelmed by the core losses. However, this is hindsight—you had no way of knowing this initially. The formula for copper losses is:

$$P_{Cu} = (J^2)\,(W_a')\,(r)$$

Where:

P_{Cu} is the copper losses in watts,
J is the current density,
W_a' is the available window area adjusted by the winding efficiency,
r is the resistivity of copper at 70° C,
l_w the mean winding length of the core window.

The current density is the reciprocal of 750 circular mils per ampere. For the equation, it is also necessary to apply a conversion factor to change the area from circular mils to square centimeters:

$$\frac{750\ \text{cmil}}{A} \times \frac{5.07 \times 10^{-6}\ \text{cm}^2}{cmil} = 3.8 \times 10^{-3}\ \text{cm}^2/\text{A}$$

Then:

$$J = \frac{1}{3.8 \times 10^{-3}} \approx 260\ \text{A/cm}^2$$

The value of r is 1.9×10^{-6} ohm-cm. For convenience, J^2 and r can be combined into a single factor:

$$J^2r\ \text{-}\ (260)^2\ (1.9 \text{ x } 10^{-6}) = 0.13$$

From the manufacturer's catalog, $W_a = 0.171$ cm^2 and $l_w = 3.42$ cm. This winding efficiency is 70 percent, so $W_a = 0.171 \times 0.70 = 0.12$ cm^2. Now you have all the values for substitution into the formula for P_{Cu}:

$$P_{Cu} = (0.13)\,(0.12)\,(3.42) = 0.053\ \text{W}$$

Number of turns

Before the number of primary turns can be calculated, the actual voltage, V_p, across each half of the primary winding must be determined. This is the dc supply voltage, less the collector-emitter saturation voltage of the transistors, V_{CE}. Inasmuch as V_{CE} has been assumed to be 0.8 V, it follows that $V_p = 6 - 0.8 = 5.2$ V. Then, use Faraday's equation to calculate N_p, one-half of the total primary winding. The equation and the value substitutions are as follows:

$$N_p = \frac{(V_p)\,(10^8)}{(4)\,(f)\,(B_{SAT})\,(A_c)}$$

$$N_p = \frac{(5.2)\,(10^8)}{(4)\,(75{,}000)\,(3500)\,(0.2)} = 2.5 \text{ turns}$$

Fractional turns are not feasible. Moreover, experience indicates that one- and two-turn windings often cause problems in leakage inductance. It appears that good electromagnetic coupling to other windings commences with at least three turns in inverter transformers. Accordingly, you "round off" the calculation to 3 turns. Also in the interest of low leakage inductance, the two 3-turn primary sections are bifilar wound. From part D of Chart 3-1, the wire size is #20.

As a consequence of the "fudging," the actual oscillation frequency will be about (2.5/3.0) (75,000) = 62,500 Hz at 70° C. However, this is acceptable because 75 kHz is only a "ball-park" value, and there is nothing critical about the frequency in a converter such as this one.

The oscillation will be slowed down even more at lower temperatures. This follows from the variation of B_{SAT} with temperature. Thus, at 25° C, B_{SAT} is 4100 gauss compared to 3500 gauss at 70° C. From Faraday's equation, the oscillation frequency at 25° C will be (3500/4100) (62,500) = 53,400 Hz. This value is still satisfactory, despite its departure from the initially assumed frequency.

The number of secondary turns, N_8, is calculated by using the voltage ratio between the secondary and primary windings. Thus:

$$N_8 = N_p \frac{V_s}{V_p}$$

$$= 3 \times \frac{225}{5.2} = 130 \text{ turns}$$

Because of the full-wave center-tap rectifier circuit, the *total* secondary winding consists of 260 turns. It, too, is bifilar wound. From part D of Chart 3-1, the wire gauge is #36.

The number of base-drive turns, N_{BD}, is based on the assumption that between 3.5 and 6.0 V should be developed. These limits have been derived from practical experience. If the induced voltage is much less than 3.5 V, difficulty can be experienced in obtaining sufficient base current under worst operating conditions. A voltage greater than 6.0 exceeds the safe operating range of the transistors. Also, efficiency is impaired by the use of higher base-drive voltages.

The calculation of the number of turns in the base-driven winding is first attempted with the objective of developing a nominal base-drive voltage of 4.0 V. Thus:

$$N_{BD} = 3\frac{4}{5.2} = 2.3 \text{ turns}$$

If you round off 2.3 to 2 turns, the induced voltage will be about 3.4 V. This is considered marginal from the standpoint of obtaining sufficient base drive. If we round off to 3 turns, 5.2 V will be developed, and this value appears reasonable. Perhaps a slight increase in the value of resistance R_B might be in order if excessive base drive is produced. In any event, 3 turns is the better of the two choices. As listed in part D of Chart 3-1, #28 wire will be used.

The turns calculations are summarized in part H of Chart 3-1. The calculation of the base-drive voltage is recorded in part I.

Window area

The needed window area, W_a, is obtained by first summing up the circular-mil area of copper in all of the windings. In a given winding, this is, ideally, the product of the circular-mil area of the wire and the number of turns. Inasmuch as the actual winding efficiency is assumed to be 70 percent, the grand total is divided by 0.7. Refer to Table 3-2, where the total (ideal) area is found to be 13,109 cmils. Applying the winding efficiency factor results in:

$$\frac{13{,}109}{0.7} = 18{,}700 \text{ cmil}$$

Converting to square centimeters:

$$18{,}700 \times 5.06 \times 10^{-6} = 0.09 \text{ cm}^2$$

Thus, W_a, the required window area is 0.9 cm². Earlier, you determined W_b, the available window area of the #814E250 core, to be 0.171 cm². This implies that 0.09/0.171, or 50 percent of the available window area will be filled. This is both reasonable and practical.

Next, calculate the core loss. From the maker's catalog, we find that the volume of this core is 0.55 cm³. Also, you can anticipate core losses of 5 W per cubic centimeter. Therefore, core losses = (0.55) (5) = 2.75 W.

Table 3-2 Summation of circular-mil areas in design example.

Winding	Number of turns	Wire size	Circular-mil area of wire	Total circular mils
Primary	2 × 3 = 6	No. 20	1022	6132
Secondary	2 × 130 = 260	No. 36	25	6500
Base-Drive	3	No. 28	159	477
				Total 13109

Losses

The overall transformer loss is the sum of the core loss and the copper loss. From parts G and K of Chart 3-1, 2.75 + 0.05 = 2.8 W. What percentage is this of the input power? To be realistic in view of the type of load used, the calculated input power of 20 W is increased by 33 percent:

$$P_{IN(Max)} = 1.33 \times P_{IN(full\ load)}$$
$$= 1.33 \times 20 = 27\ W$$

The transformer loss under worst conditions is 2.8/27, or approximately 10 percent. This is quite acceptable. Notice the great disparity between the copper and core losses. This is typical of such design procedures in which the prime goal is the reasonable, rather than the optimum, selection of core, wire size, etc. Indeed, you can breathe a sigh of relief when it is determined that the windings will fit the core window!

Notice the near equality of the amount of copper in the primary and secondary windings. Other things being equal, this condition is also one of the requisites of optimum transformer efficiency.

Transformer design by nomographs

The manufacturers of ferrite cores provide ***nomographs*** to facilitate the design of inverter transformers. By the use of these design aids, you can circumvent considerable amounts of mathematical labor and empirical techniques. Most important, however, you can achieve design optimization much more readily than with formal "crank-grinding" methods. Each manufacturer has developed his own unique "road map," and a little preliminary study is needed to develop skill with a particular method. The detailed instructions will not be given here, but sufficient information will be provided to convey the basic idea.

The nomograph illustrated in Fig. 3-20 pertains to the Indiana General IR-8100 family of "inverter-rated" ferrite cores. These cores develop maximum efficiency in the vicinity of 50° C. The arrows represent the design of a driven-inverter transformer. The inverter has the following parameters:

Operating frequency: 225 kHz
Power-supply voltage: 160 V
Output voltage: 40 V
Output current: 1 A
Ambient temp: 40° C
"Hot-spot" temp: 70°

A cross-type core is selected to provide easy mounting on a printed-circuit board. Also, the cross core has a reasonable low external field and is easier to wind than a toroid. From these specifications, you can deduce the temperature rise to be 30° C and the power output to be 40 W. The design path commences at points A and B on the 30° C line. The intersections of this line with the $M = 1$ lines comes about from the fact that $M = 1$ implies equal core and copper losses—a requisite for optimum efficiency. The vertical projection to C entails the intersection of the 160-V and the 25-kHz lines (because 25 kHz corresponds to a half-period, T1, of 20 μs). Projecting C horizontally to the number-of-turns axis locates 250 turns for the full primary winding.

Next, project horizontally from one-half the primary turns, that is, from 125 turns, and intersect with the vertical projection from B. This intersection D, yields power output as a function of input voltage. Read off the number 0.3 and multiply this by 160 to obtain 48 W. So far, so good, for the design objective is a nominal 40-W output capability (1 A at 40 V).

A second check is now made to ascertain that all is going well. You must determine the strength of the magnetic driving force, as given by $V_p(t)/N$. This is straightforward.

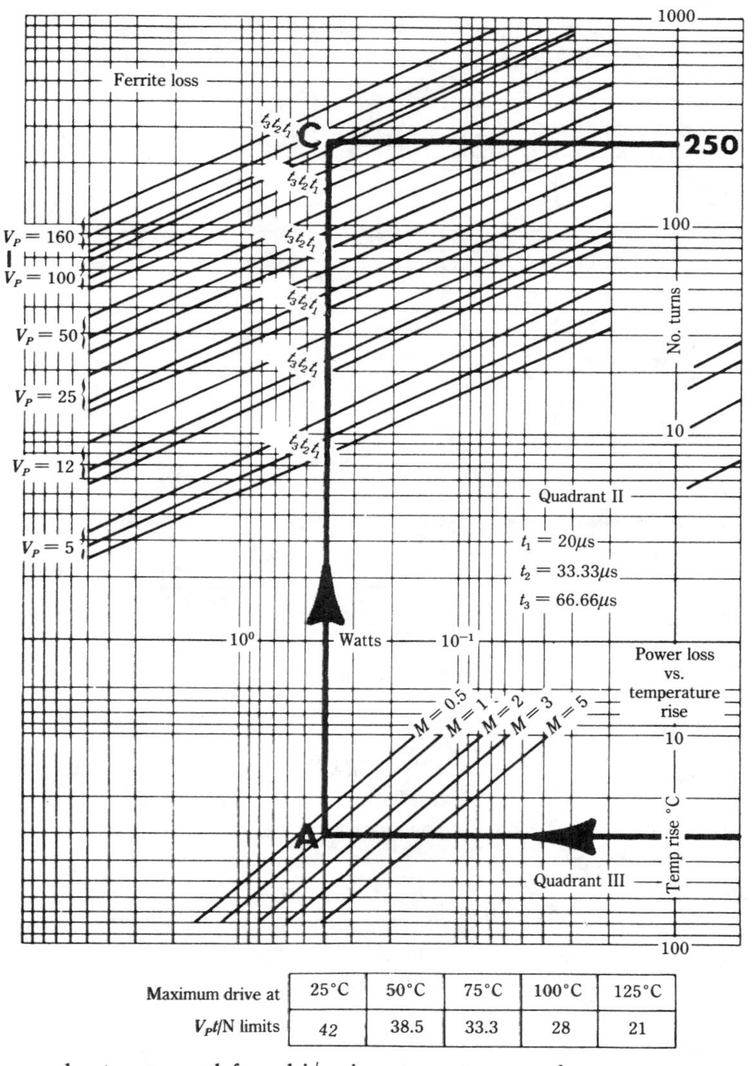

Maximum drive at	25°C	50°C	75°C	100°C	125°C
V_pt/N limits	42	38.5	33.3	28	21

3-20 Four-quadrant nomograph for a driven-inverter output transformer. Indiana General Electronic Products.

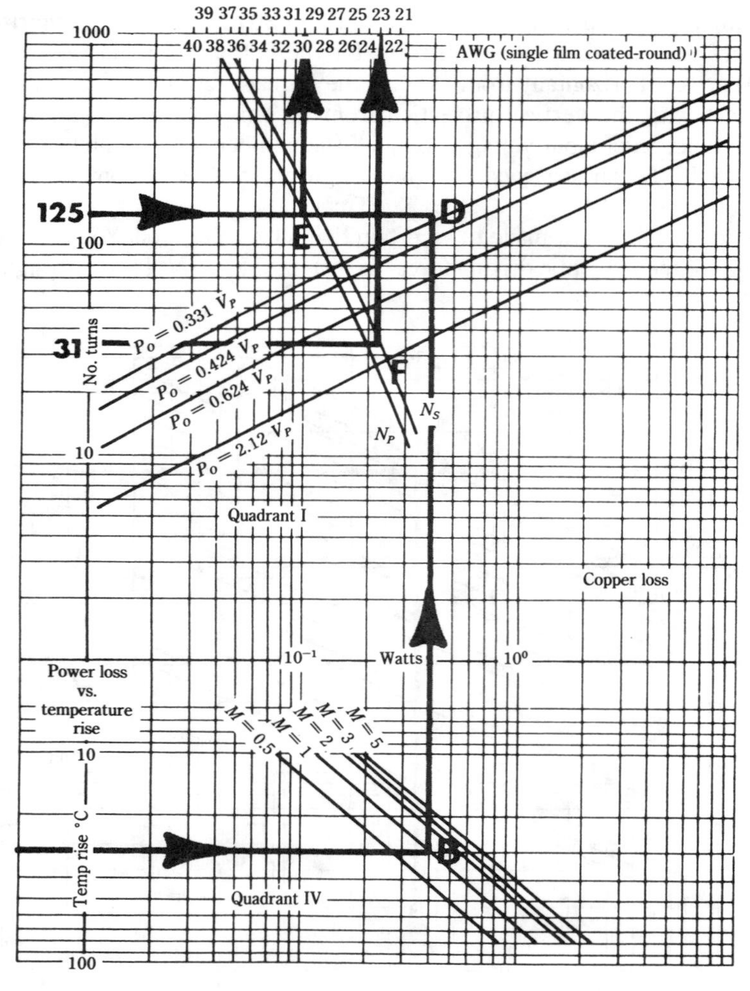

3-20 Continued

From the given and derived data, (160) (20)/125 = 25.6 volt-microseconds per turn. The quantity is now checked in the table below the nomograph. Because it is found that a maximum drive of 33.3 volt-microseconds per turn corresponds to 75° C, this transformer design is acceptable.

To determine the wire gauges for the primary and secondary windings, first establish intersections E and F. These are the appropriate crossovers of the N_p and N_s lines with the horizontal lines showing number of turns (125 for one-half of the primary and 31 for the secondary). The upward vertical projections of points E and F then locate #30 wire for the primary and #23 for the secondary.

State-of-the-art inverter design would entail the use of this sophisticated nomograph approach in conjunction with one of the inverter-drive ICs previously described.

Another nomograph is shown in Fig. 3-21. This one is furnished by Allen-Bradley for their R03 series of ferrite cores. The superimposed arrows represent the procedural steps leading to the design of a transformer for the saturable-core inverter shown. The inverter is specified as having an oscillation frequency of 10 kHz, a dc supply of 45 V, and an output capability of 26 VA (for all practical purposes, this means an output of 26 W). The nomenclature used in both the inverter and the nomograph is as follows:

f = frequency of the converter (Hz)
N = one-half the total number of primary turns
A_e = effective cross-sectional area (cm^2) of core

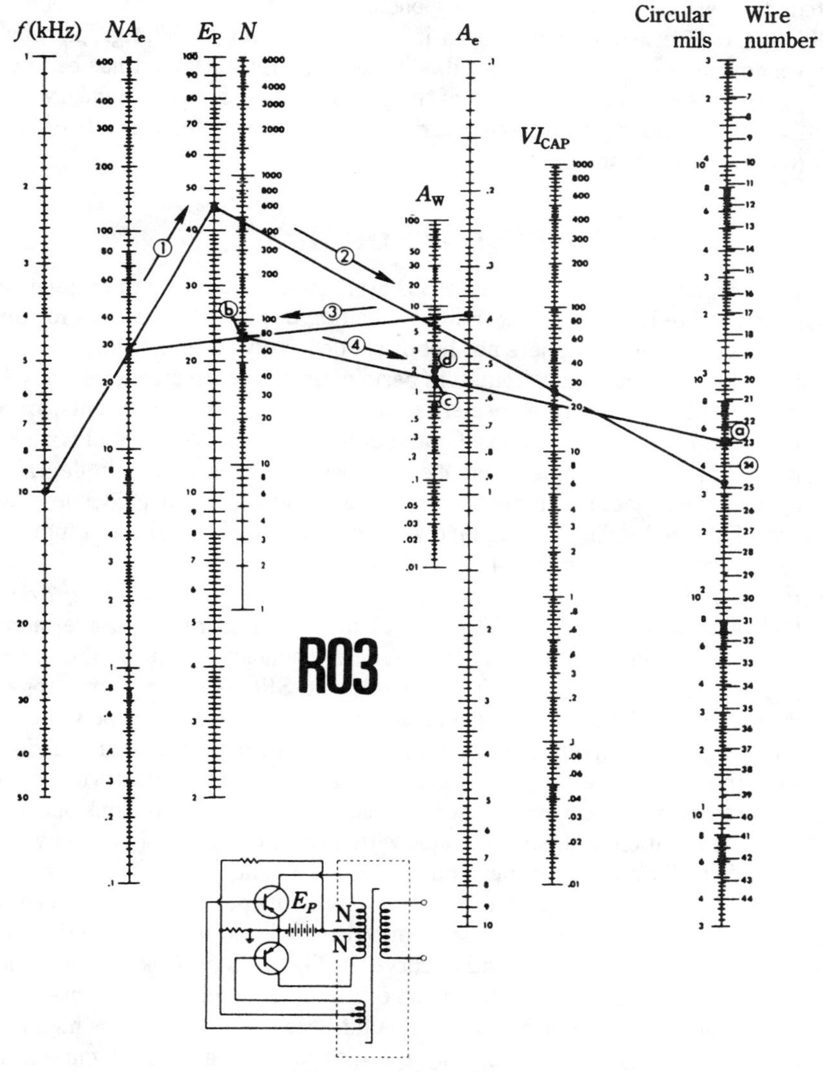

3-21 Nomograph for design of a self-oscillating inverter transformer. Allen Bradley Co.

A_W = window area (cm²) of core
NA_e = cross-sectional-area-turns product (turns: cm²)
E_p = peak dc or battery voltage
VI_{cap} = volt-ampere handling capacity of system

The nomograph is designed to allow for 70 percent of the window area to be filled. It also automatically provides that one-half of the filled window area is primary wire and the other half is secondary wire. In order to account for wire insulation, the wire is considered a gauge heavier than that actually selected. #24 is selected in this problem because it is assumed safer than "rounding off" to #25. But, for graphical purposes, the selected #24 is treated as #23, corresponding to point a. The core selected is the T0960H001A. The listed window area is 1.76 cm² (point d). The graphically determined window area is 1.55 cm² (point d). Therefore, a reasonable fit has been obtained with 75 turns of #24 wire on each half of the primary winding. The secondary winding is determined from ordinary transformer principles. The space for it has been automatically provided by the nomograph.

Selection of power transistors

If enough information is available, a rigorous procedure can be used to achieve optimum balance between reliability and cost in the selection of power transistors for inverters and converters. Neither designers nor hobbyists ordinarily are in a favorable position to learn the values of the needed parameters pertaining to thermal characteristics, leakage inductance of the inverter transformer, short-duration energy transients, or various anomalies of transistor specifications. On the other hand, the selection of transistors by "brute-force" approaches leads to poor economics and often results in degradation of performance. A satisfactory method consists of using empirical guidelines, but with awareness of the underlying factors involved. In this way, the "rules of thumb" will be seen as logical deductions, rather than blind stipulations.

First, you should understand the significance of the *safe-operating area* (SOA) concept. The basic rationale for SOA is that maximum current, voltage, and power ratings cannot be achieved simultaneously (some manufacturers use different nomenclature, such as SOC—safe operating curves—and SRO—safe region of operation). The curves of Fig. 3-22 define safe operating areas for the RCA-423 power transistor. If the load line of the inverter or converter lies within the graphical area bounded by the inner (i.e., the continuous, or dc, operation curve) the transistor will be in "safe territory." However, it is permissible for the load line to make excursions outside of this area if the time duration of the excursion is within the intervals designated by the outer curves. The SOA limits involve four boundaries—current, power, second breakdown, and voltage (first) breakdown. Actually, these statements pertain only to a case temperature of 25° C. Because practical heat-removal techniques are not likely to maintain case temperature at 25° C, the derating curves of Fig. 3-23 must be used to modify the SOA limits obtained from Fig. 3-22. In some instances, semiconductor manufacturers provide SOA curves that require no derating up to case temperatures as high as 75° C. Also, with some power transistors, only the dissipation-limited SOA curves need be derated for temperature.

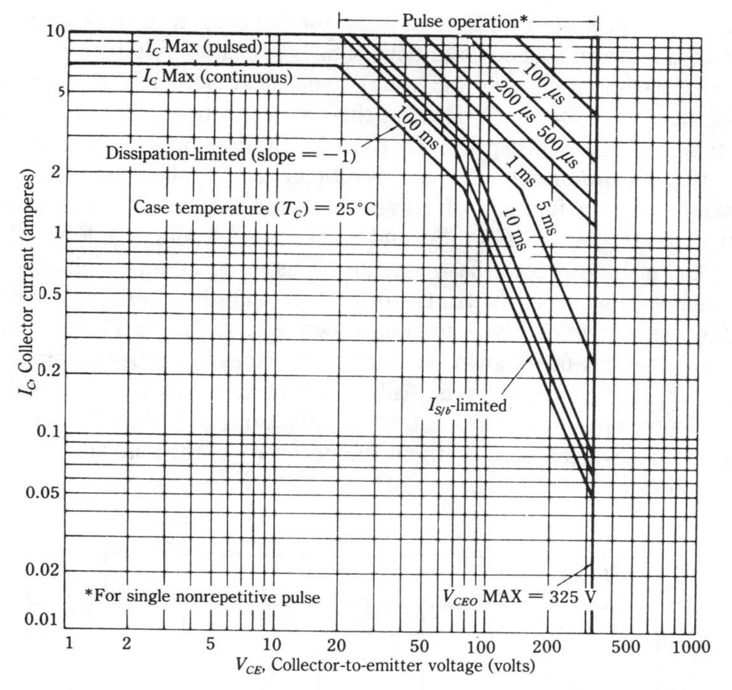

3-22 Typical set of safe-operating-area curves for a power transistor. RCA Solid State Div.

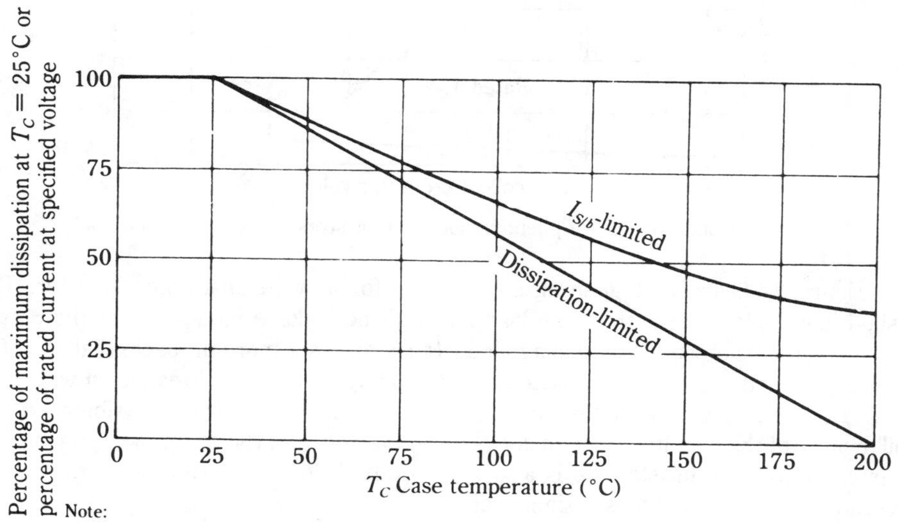

Note:
Current derating at constant voltage applies only to the dissipation-limited portion and the $I_{S/b}$ limited portion of maximum operating area curves. Do not derate the specified value for I_C MAX.

3-23 Temperature-derating curves. Courtesy RCA Solid State Div.

It should be appreciated that, were it not for the second-breakdown phenomenon, the SOA of a transistor would be essentially similar to that of a simple resistor. Practical resistors have maximum current and voltage specifications, and between these limits they are "dissipation limited." Notice that the 45° segments of the curves in Fig. 3-22 represent constant power dissipation (because of the log-log graphical plot).

Notice that ***second breakdown*** is an additional factor influencing the SOA curves. Second breakdown is a function of current, voltage, and time—that is, of energy. Temperature exerts a lesser effect. Second breakdown (at point x in Fig. 3-24) involves restricted areas of the collector-base junction. This leads to very high current density and to thermal destruction of the transistor even though the ***average*** junction temperature might be only nominal. Second breakdown is much more likely to destroy the transistor than the first breakdown resulting from excessive collector-emitter voltage. Generally, when sufficient energy is made available to cause second breakdown, the transistor will develop a collector-emitter short. Not obvious from the curve is the fact that second breakdown can occur without the necessity of the load line passing through the first-breakdown region.

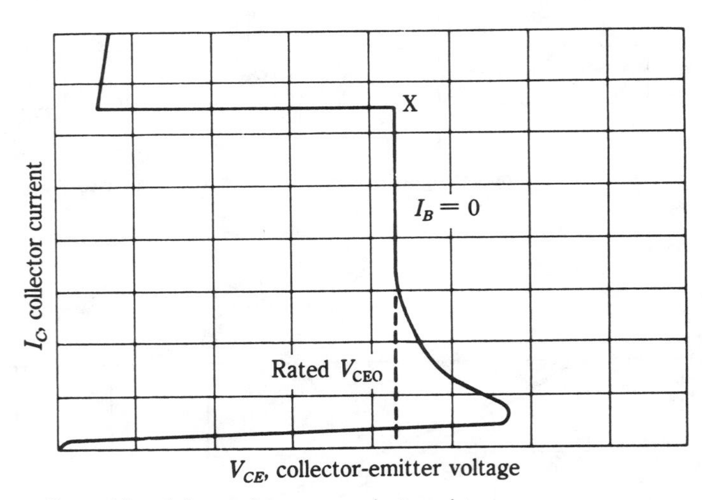

3-24 Second-breakdown phenomenon in transistors. Motorola Semiconductor Products, Inc.

Figure 3-25 shows a generalized load line for a two-transformer inverter. The dashed line depicts the path of the load line with no leakage inductance in the output transformer. It is obvious that leakage inductance makes it more probable that the SOA will be exceeded. Specifically, it becomes more likely that second breakdown will occur in the transistor. Other factors besides leakage inductance can contribute to the "billowing" of the load-line path. In converters, for example, the effect of the uncharged filter capacitors during start-up is also deleterious. In both inverters and converters, excessive feedback produces a similar effect.

It is a good idea to take a photograph of an oscilloscope display of the load line. This is desirable even for those who attempt a quantitive design approach with SOA data, although it cannot help with the initial design procedure.

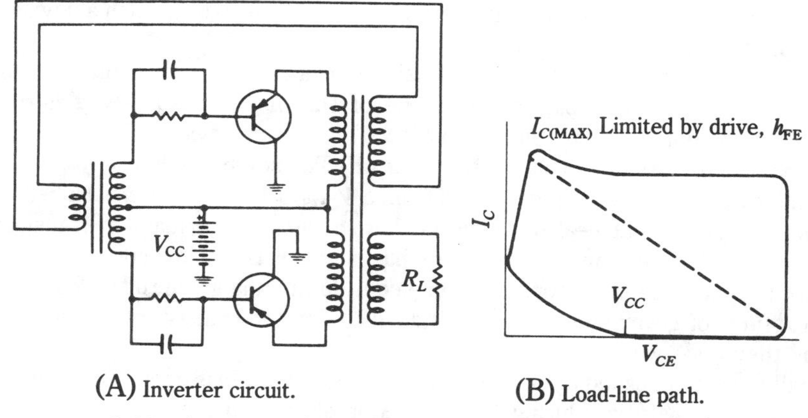

(A) Inverter circuit. (B) Load-line path.

3-25 A two-transformer inverter and its load-line path.

Fortunately, you can usually do quite well with an empirically guided selection approach. The first thing to decide is the voltage rating of the transistors. If there were no leakage inductance, this could be determined from the fact that each transistor is exposed to twice the dc supply voltage. With a small safety factor included, it would then suffice to specify transistors with V_{CEO} ratings of about 2.3 times the dc supply voltage. In fact, this is quite reasonable for driven inverters that use toroidal output transformers. For other inverters and converters, it is generally best to specify V_{CEO} at 2.5 to 3 times the dc supply voltage.

The collector-current capability of the transistor should be at least P/kE, where P represents the desired output power, k is the expected efficiency, and E is the dc supply voltage. Although higher efficiencies are often forthcoming, it is reasonable to assume a value of 75 percent for k. Current gain, β, should be high at the specified current capability. In the interest of operating efficiency, betas in excess of 20 are desirable. Modern transistors generally provide greater current gains than this, and with Darlingtons β values of several hundred are commonplace.

Although high-beta power transistors are likely to prove beneficial with regard to efficiency and possibly starting ability, you should not give undue consideration to this or any other transistor parameter. This is because a general trade-off exists among nearly all parameters. What is gained in one, is lost in others. Also, too much of a good thing often proves undesirable. As an example, it is best to find a transistor with moderate beta, but declining current gain past the specified current capability. Such a transistor will tend to reduce the intensity of voltage spikes.

Another instance of too much of a good thing is excessive gain-bandwidth product, f_t. This parameter is closely related to the switching speed of the transistor. It would generally be mistaken logic to choose a device with a rated 50-kHz switching capability over one specified at 20 kHz for a 25-kHz inverter. High gain-bandwidth product or short switching time is attained at the expense of the SOA. Other things being equal, it is usually unwise to seek frequency capability beyond that with which one can "squeak by." Of course, in this hypothetical situation, if it was found that another manufacturer offered a power transistor intended for service in 30-kHz inverters, such a device would certainly merit consideration.

A very desirable feature in power transistors is a low collector-saturation voltage, $V_{CE(SAT)}$. Emphasis on this parameter is one of the major reasons why germanium transistors have not been completely displaced by silicon devices. The lower the dc supply voltage, the greater is the impairment of efficiency by high $V_{CE(SAT)}$. One must ascertain, however, that efficiency gained from a $V_{CE(SAT)}$ is not counterbalanced by high collector leakage current—also a general characteristic of germanium transistors. A positive attribute of germanium transistors has been relatively low cost. This might or might not be a compelling feature when due consideration is given to the less favorable thermal ratings of germanium devices. And, silicon power transistors are no longer as costly as they once were.

Finally, most published SOA curves pertain to forward-bias operation. In inverters, the reverse-bias operation during the transistor off time makes the transistor more vulnerable to second breakdown. That is why many inverter circuits have diode clamps across the emitter and base connections. By the same token, feedback windings should not be designed to apply excessive-base-drive voltage. A wise decision is to use transistors specifically intended for service in inverters. These will have received the

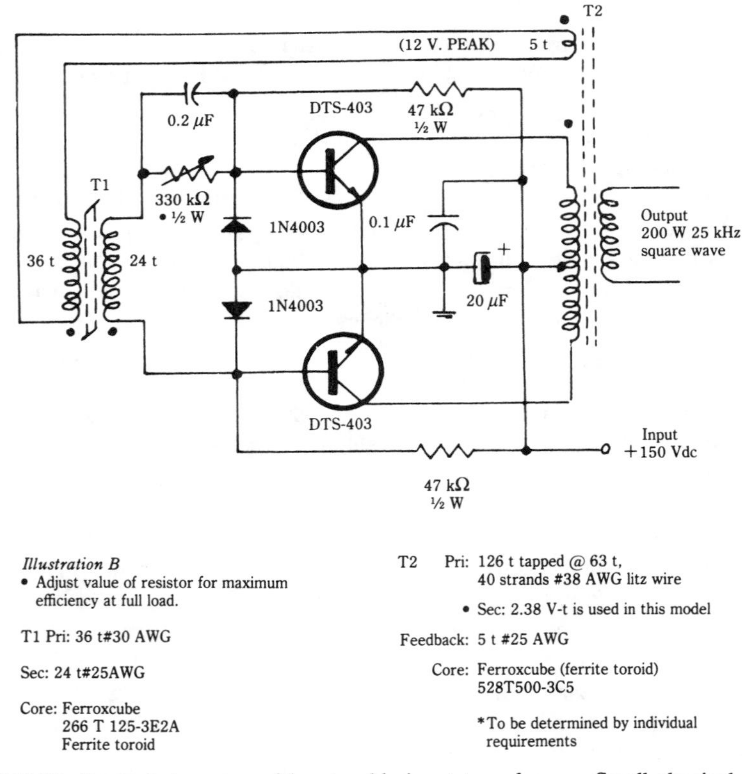

3-26 A 250-W ultrasonic inverter with saturable input transformer. Small physical dimensions are achieved with high-frequency operation. Delco Electronics.

manufacturer's special attention with regard to electrical ruggedness for operation in the switching mode.

A practical example of an inverter using some of the aforementioned design concepts is shown in Fig. 3-26. With 250-W capability at a nominal frequency of 25 kHz, this inverter is of the two-transformer varieties and uses transistors specifically intended for high-frequency operation. Both transformers are toroids, but only the input transformer is permitted to saturate. Notice the diode clamps in the base-emitter circuits. A single empirically adjusted base-current limiting resistance suffices inasmuch as it limits base current in both transistors. Pay heed to the phasing of the transformers, and an initial test should be carried out at a reduced supply voltage (with erroneous connections, high-frequency transistors can oscillate in an unintentional mode and self-destruct).

4
CHAPTER
Transistor inverter and converter applications

THIS CHAPTER WILL COVER A NUMBER OF USEFUL INVERTERS AND CONVERTERS that use power transistors as active elements. The technical expertise and monetary outlay invested by the large companies in the development of these circuits has been considerable. You can avoid much costly trial and error by paying heed to their solutions. Even though specific applications are likely to require experimentation and modification of the designs, it is often better to start with proven designs than to begin from scratch. Those who wish to innovate are also likely to fare better if the state of the art is first evaluated.

Insofar as possible, the values of the components for these circuits and systems are given. It is especially true with inverters and converters that a certain amount of empirical modification is necessary to optimize efficiency, reduce transients, enhance starting, etc. This experimentation inevitably involves the inadvertent destruction of costly components. By following the basic techniques to be found in the ensuing pages, you can take advantage of the empirical design "debugging" already accomplished by the manufacturers.

Simple 60-Hz inverter

The 60-Hz inverter shown in Fig. 4-1 is about as easy to construct and as inexpensive as you could desire. Yet, it is capable of providing some very useful services. Operating from an automobile battery, it can supply 50 W for the operation of such devices as an ac/dc radio, electric shaver, fluorescent lamp, small soldering iron, 40-W incandescent lamp, recorder, or portable phonograph. Its essential ingredients are a filament transformer and two general-purpose germanium power transistors. Although this is a saturable-core oscillator, no separate feedback windings are used. Rather, feedback is produced by cross-coupled connections in the manner of a multivibrator.

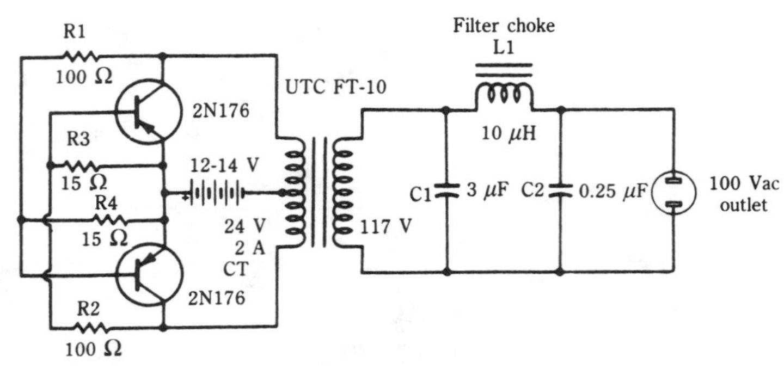

4-1 A simple cross-coupled 60-Hz inverter. Motorola Semiconductor Products, Inc.

At a full load, the efficiency is in the vicinity of 75 percent, and the output voltage is about 106 V. The "mild" pi-section filter despikes the output waveform and causes a trapezoid wave, rather than the usual square wave, to be available at the output. This makes the device more suitable for the operation of radios, recorders, and other electronic equipment.

In this type of circuit, the efficiency, frequency, output voltage, and starting ability are interdependent to a marked degree. Accordingly, some experimentation with the biasing resistances might be profitable. It is likely, however, that only one of them, such as R1, might have to be modified. Insofar as possible, the biasing networks for the two transistors should be approximately balanced. Otherwise, an unsymmetrical waveform, unequal transistor dissipation, and other malfunctions can result.

A "workhorse" 60-Hz inverter

When extreme economy of both money and time is not the predominant factor in producing a 60-Hz inverter, the circuit of Fig. 4-2 merits consideration. This 110-W ac supply is representative of the tens of thousands of inverters that have been installed in automobiles, boats, and even aircraft. Although the design might be described as being just short of becoming antiquated, this approach is time-proven and has the reputation of being capable of reliable and satisfactory service. It can be either scaled up or "beefed up" for higher power levels. Units with appropriate transistors and heatsinks can supply a half-kilowatt or so of 60 Hz ac. It is a better engineering approach to inversion than the admittedly simpler inverter of Fig. 4-1. Nonetheless, this one is also basically simple, and it occupies a close second place in the matter of cost.

The circuit is a common-collector configuration with individual feedback windings on the transformer. In most installations, this implies that the transistors can be mounted directly on either individual heatsink or a common heatsink. This method of avoiding the use of mica spacers or other electrical insulation is a thermal plus—especially for germanium power transistors. The low collector-emitter voltage drop inherent in germanium transistors enhances electrical efficiency and is, additionally, another favorable thermal feature. The transformer is fairly massive, as you would expect for 60-Hz operation. It is constructed with "old-fashioned" silicon-steel lamina-

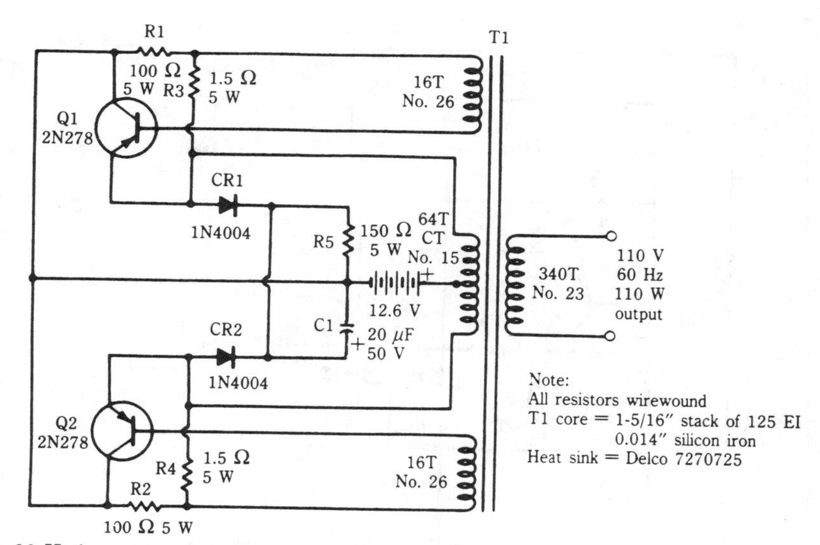

4-2 A 60-Hz inverter using silicon-steel transformer and electromagnetic feedback. Delco Electronics.

tions, but such a core is quite adequate for the purpose. Despiking is accomplished by diodes CR1 and CR2 in conjunction with a "snubber" network, R5 and C1.

The performance of this inverter is shown by the graph of Fig. 4-3 and is typical of single-transformer saturable-core inverters. The square-wave output will be suitable for most radios, TVs, and appliances. Some ac motors will operate in a satisfactory manner. In others, however, trouble might be encountered with excessive temperature rise. Also, the inverter cannot compete with power-line operation when high starting currents are required.

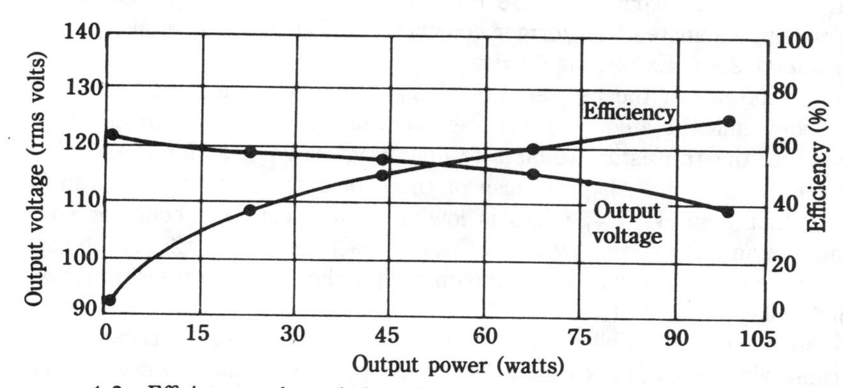

4-3 Efficiency and regulation of inverter in Fig. 4-2. Delco Electronics.

A 120-watt general-purpose inverter

The inverter illustrated in Fig. 4-4 is intended as a circuit or system building block, inasmuch as it does not produce the 60 Hz required by the preponderance of ac

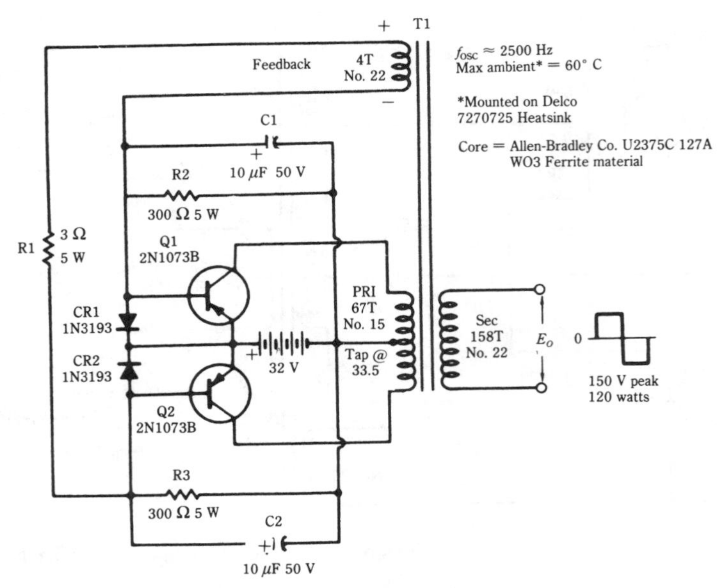

4-4 A 120-watt general-purpose inverter. Delco Electronics.

equipment. It could, of course, be used for soldering irons, incandescent lamps, fluorescent lighting, etc. However, its operation makes it a likely candidate for the beginnings of a converter, a regulated dc power supply, or an electronic ignition system. The oscillation frequency is in the vicinity of 2500 Hz—high enough to obtain good efficiency with economical components. The physical size of the transformer is drastically reduced from the size of 60-Hz designs. The ferrite core also provides an effective way of avoiding the acoustical noise problem that tends to plague audio-frequency inverters using laminated transformer construction. This problem is likely to be more bothersome at 2500 Hz than at 60 Hz.

Notice the use of the base-emitter clamp diodes, CR1 and CR2, in this circuit. These diodes limit the amount of reverse bias that can be applied to the transistors. This protects the transistors (their safe operating area is less for reverse-bias than forward-bias operation). Also, because of these diodes, the transistors can be hard driven so that their $V_{CE(SAT)}$> loss is low. In this circuit, the collector spikes are attenuated in an indirect way by the action of capacitors C1 and C2. These capacitors slow down the rise of base current at the time when the saturated transformer demands very high collector current.

In the interest of flexibility as a functional block in a more extensive system, there is sufficient window space left over in the transformer to permit the use of larger wire or the addition of more secondary windings. One such possibility is to add a high-voltage winding for a photoflash or strobe device, or for the cathode-ray tube of an oscilloscope.

Inasmuch as the feedback winding is not centertapped, you might wonder how a complete circuit is achieved for the base-emitter junctions of the transistors. This path is provided by diodes CR1 and CR2. Polarity conditions are correct for these diodes to pass current during the time in the cycle when the bases are forward biased. Thus, these diodes perform a double function. If it were decided that the protection offered by the diodes was not needed, it would not do to omit them from the circuit, for then the inverter would not operate.

A 100-watt two-transformer inverter

The inverter shown in Fig. 4-5 is self-oscillatory, but it utilizes two transformers. The output transformer operates in its linear region, and the much smaller input transformer is driven into magnetic saturation twice each cycle. A relatively small hysteresis loss results with this scheme, and the operating efficiency of the circuits is high. As can be seen from the performance curves of Fig. 4-6, the full-load efficiency is approximately 93 percent. The efficiency at half-load is comparable to that ordinarily achieved with single-transformer inverters at full load. As with the previous circuit, this inverter is intended as a building block for converters, switching regulators, dc power supplies, etc. Because of its 18-kHz nominal operating frequency, it is also useful for driving ultrasonic transducers.

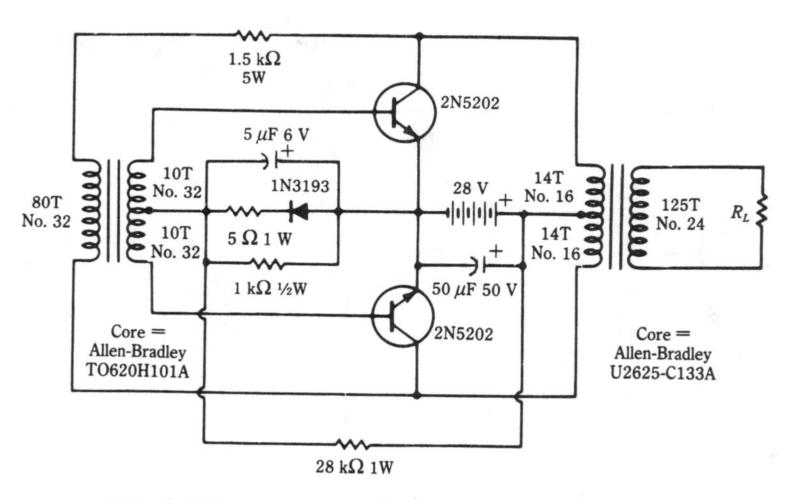

4-5 A 100-watt two-transformer inverter. RCA Solid-State Div.

The 2N5202 silicon epitaxial power transistor is especially suitable for high-switching-rate applications such as this one. The rise and fall times of this transistor are each less than 0.5 μs, and the storage time is on the order of 0.75 μs.

The 50-μF capacitor across the dc source should be physically situated at the inverter if the connecting leads are more than 6 inches long. Otherwise, the lead inductance can allow larger switching transients in the output waveform than need be. This expedient has not been mentioned before because of its lesser importance at lower operating frequencies. Also, the relatively large spikes produced in single-transformer inverters are not highly responsive to this remedy. But the performance of even single-transformer, low-frequency inverters can be affected if the impedance of the dc

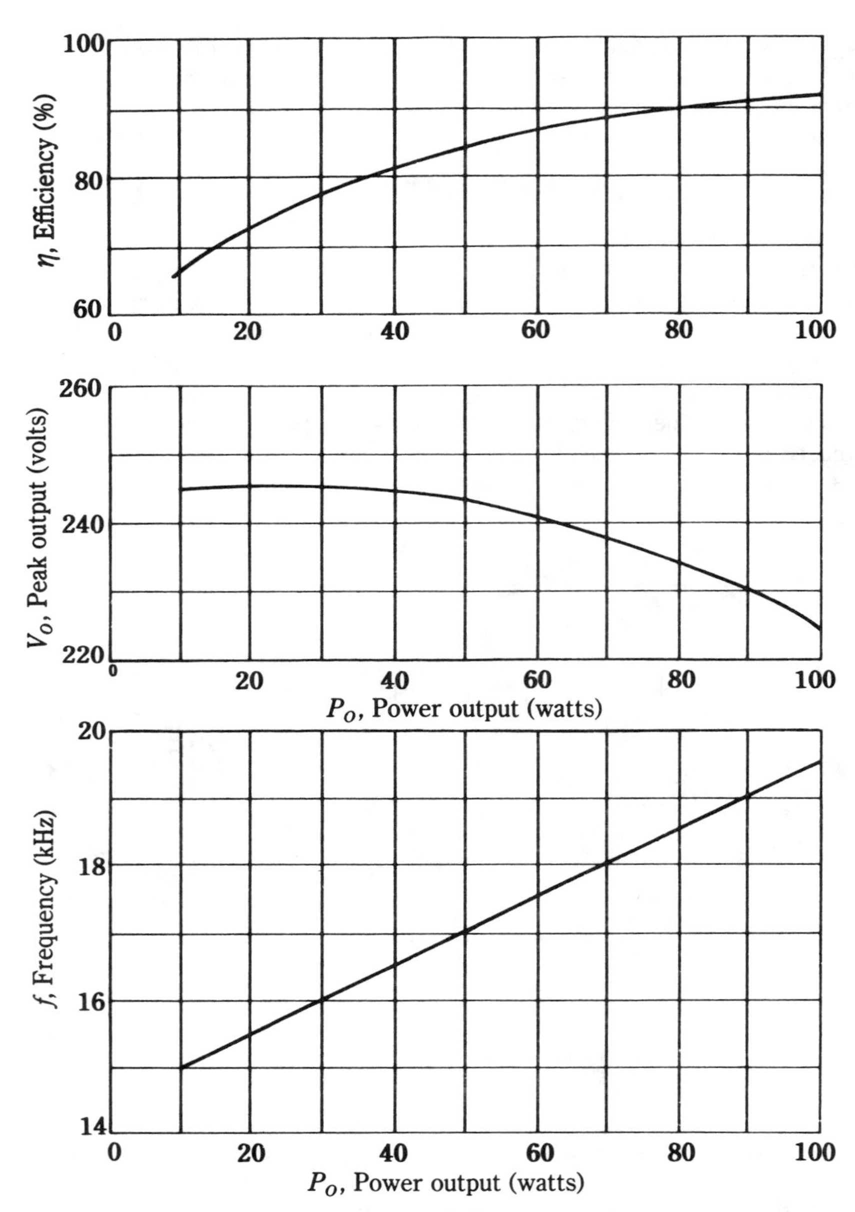

4-6 Performance characteristics of inverter of Fig. 4-5. RCA Solid-State Div.

supply is not sufficiently low. A large capacitor at the inverter dc-input terminals is a good idea for most installations. When such a capacitor is additionally paralleled with a smaller mylar or ceramic capacitor, the higher-frequency components of RFI generated by the inverter can often be attenuated considerably.

Low-input-voltage inverter

Because a number of two-transistor inverters have been investigated thus far, it might appear initially that the circuit of Fig. 4-7 is just another version of a common theme.

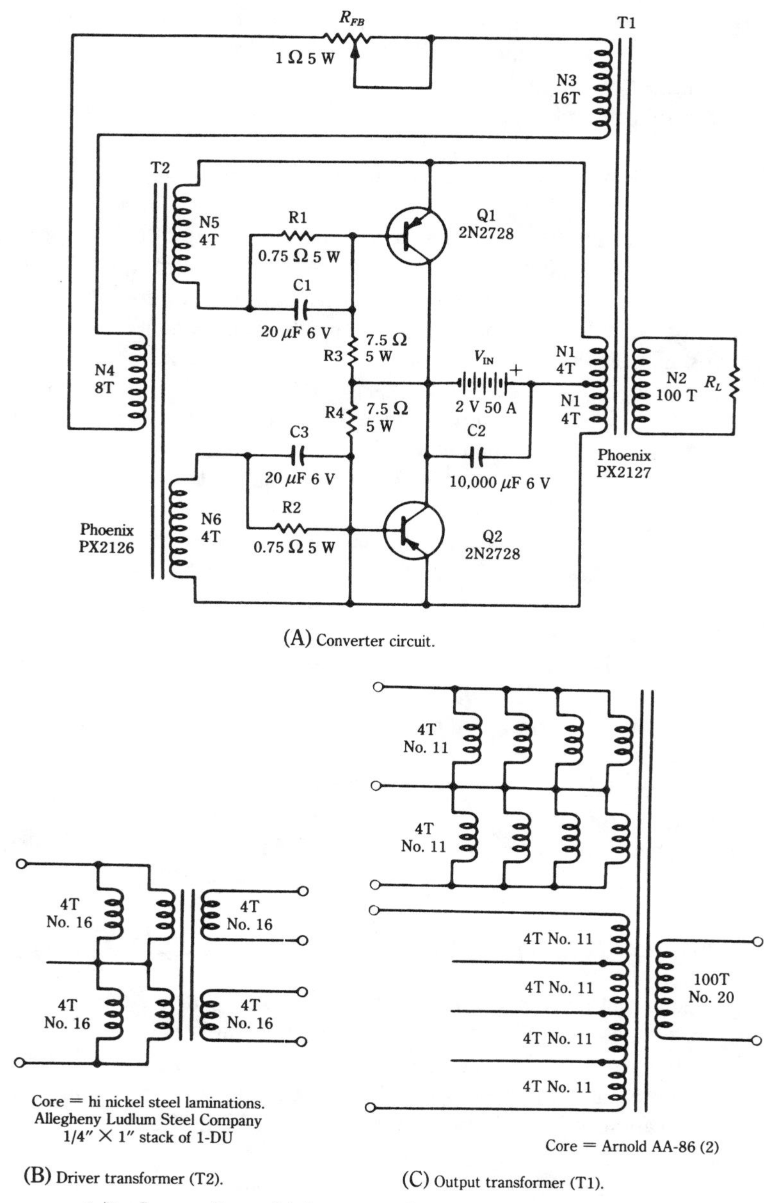

(A) Converter circuit.

(B) Driver transformer (T2).

(C) Output transformer (T1).

4-7 Low-voltage, high-current inverter. Motorola Semiconductor Products, Inc.

This inverter, however, displays a unique feature—it will deliver about 80 W into appropriate load when the dc supply voltage is only 2 V. With the development of solar cells, thermoelectric power sources, and fuel cells, such operational capability assumes practical importance. Also, small sealed lead-acid cells with attractively high ratios of energy output to weight have recently become available. One application for these cells could be in the uninterruptable power supplies needed for computer systems with volatile memories. The cells are inexpensive, can be used in any position, and can be discharged and charged up to 1000 times. Their nominal 2-V output renders them tailor-made for the inverter of Fig. 4-7. One of these cells, the size of an ordinary D flashlight cell, can deliver 50 A to this inverter for several minutes with ease. This compares to the tens of milliseconds of operation available from the capacitors conventionally used for brownout protection.

In order to bring about efficient 2-V inverter operation, type 2N2728 germanium power transistors are used. These have a collector-emitter saturation voltage on the order of 0.1 V. The two-transformer configuration is a natural here. By its use, the hysteresis loss in the output transformer is kept low—only the small input transformer saturates. The oscillation frequency is about 1 kHz and is adjustable by means of potentiometer R_{FB}.

Another expedient used in this low-voltage high-current inverter for the sake of efficiency is the paralleling of transformer windings. This will be noted in the primaries of both the input and output transformers. The centertap on the primary of the input transformer is not used; neither are the inner taps on winding N3 of the output transformer. At the voltage and current levels in this inverter, the copper losses constitute the source of greatest power dissipation. Obviously, too much emphasis cannot be placed on the connecting leads between the dc source and the inverter. An example of the magnitudes involved in this circuit is the 0.1 V collector-emitter drop that occurs when a transistor is driven into saturation; this represents a resistance of less than 0.002 Ω, which is roughly equivalent to a one-foot length of #12 copper wire. Both transistors are mounted on a 3-inch × 1-inch × 1/8-inch copper plate without insulating spacers.

The performance characteristics of this inverter are shown in Fig. 4-8. It is important that the load resistance (Fig. 4-8B) not be lower than 20 Ω. At lower values, there is heating of the transistors because they do not operate under saturated conditions.

Inverter technique for brushless dc motor

An interesting experimental project is illustrated in Fig. 4-9, which shows a shaded-pole motor drive from a dc source. Shaded-pole motors are widely utilized in applications requiring low starting torque, such as clocks, fans, and various instrumental uses. A salient feature of the arrangement shown in Fig. 4-9 is the reduced likelihood of RFI compared to that often accompanying the arcing and sparking of brushes in commutator-type dc motors. Although the saturable-core oscillator is certainly not without interference potential, the harmonic spectrum of the relatively low switching rate, for example 30 to several hundred hertz, is likely to be less troublesome in the radio-frequency bands than is the hash from brushes. Moreover, various snubbing and

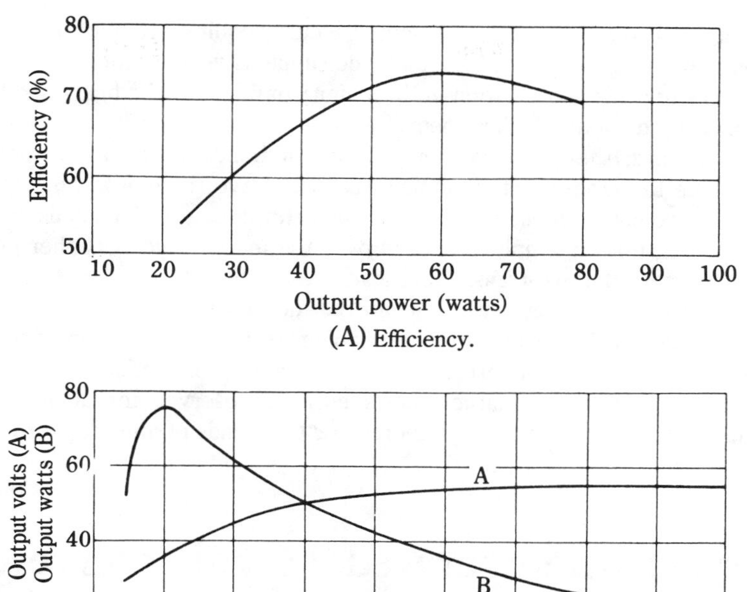

(A) Efficiency.

(B) Output.

4-8 Performance characteristics of inverter of Fig. 4-7. Motorola Semiconductor Products, Inc.

despiking techniques can be applied to the "bare-bones" circuit in order to slow down the switching transitions because efficiency would not generally have a high priority in such an application.

The tricky part of producing such a "brushless dc motor" is to satisfy simultaneously the requirements of both the inverter and the motor. The most difficult project

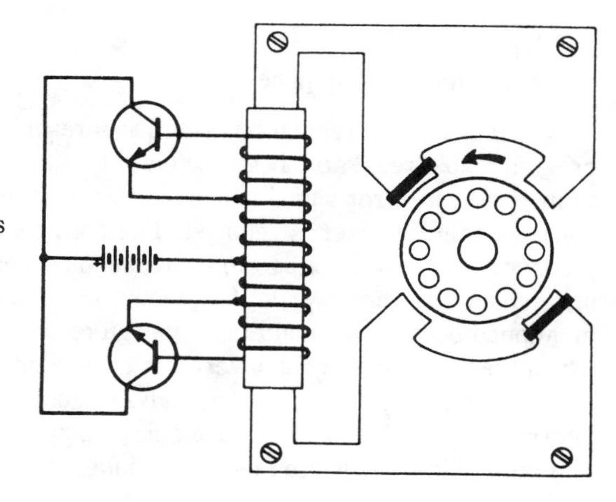

4-9 Inverter application for a brushless dc motor.

would be a clock drive because the motor would require a highly accurate 60-Hz frequency. A well-regulated, but adjustable, dc supply is very useful here, for then it is only necessary to achieve oscillation in the vicinity of 60 Hz — the final operating speed is determined by dc voltage adjustment.

For small fans, the exact speed of the motor is not so important, and there is relaxed latitude for experimentation. You can even leave the motor winding as is and associate it with a complementary-symmetry inverter circuit, which needs no taps. The feedback windings(s) can simply be placed over the motor winding. Other possibilities are bridge inverters and two-transformer inverters. In the latter instance, the burden of frequency control is removed entirely from the motor winding.

This only skims the surface of a developing field, the application of inverters to large motors. By means of inverters, the industrial-type induction motor can be endowed with variable speed characteristics hitherto reserved for dc motors. Here, eliminating the necessity of passing hundreds or thousands of amperes through sliding contacts pays worthwhile dividends.

225-watt line-operated 15-volt converter

The arrangement shown in Fig. 4-10 is actually a regulated dc power supply. By means of this circuit technique, 60-Hz magnetic components are completely eliminated, and dramatic reductions in size and weight are brought about. For example, this converter weighs about 2 pounds, as contrasted to 15 pounds for a conventional supply using a 60-Hz input transformer. The savings are accomplished essentially because 20-kHz magnetic components have been utilized in place of 60-Hz components. A 20-kHz transformer is minuscule compared to a 60-Hz transformer of the same power-handling capacity. The five composite sections of this regulated converter are clearly identifiable from Fig. 4-10. They are:

- Power-line bridge rectifier
- Programmable series-pass voltage regulator
- Two-transformer inverter
- Output bridge rectifier
- Error-feedback amplifier

In this system, regulation occurs as a result of the effect of the error signal on the programmable regulator. If the dc output voltage developed across the load, R_L, tends to increase, an error signal is applied to the base of transistor Q4, and the dc voltage applied to the inverter is reduced. The converse action takes place in the event of a tendency of the load voltage to decrease. This regulatory action occurs with relatively little change in the inverter frequency, and none in its duty cycle. (These facts are mentioned because of an alternate design technique, called ***pulse-width modulation,*** in which the duty cycle of the inverter is controlled. The nearly constant frequency of the circuit in Fig. 4-10 provides two advantages: relative ease of obtaining good output filtering, and predictability of harmonic energy. On the other hand, the overall efficiency of a pulse-width-modulated system is inherently better.)

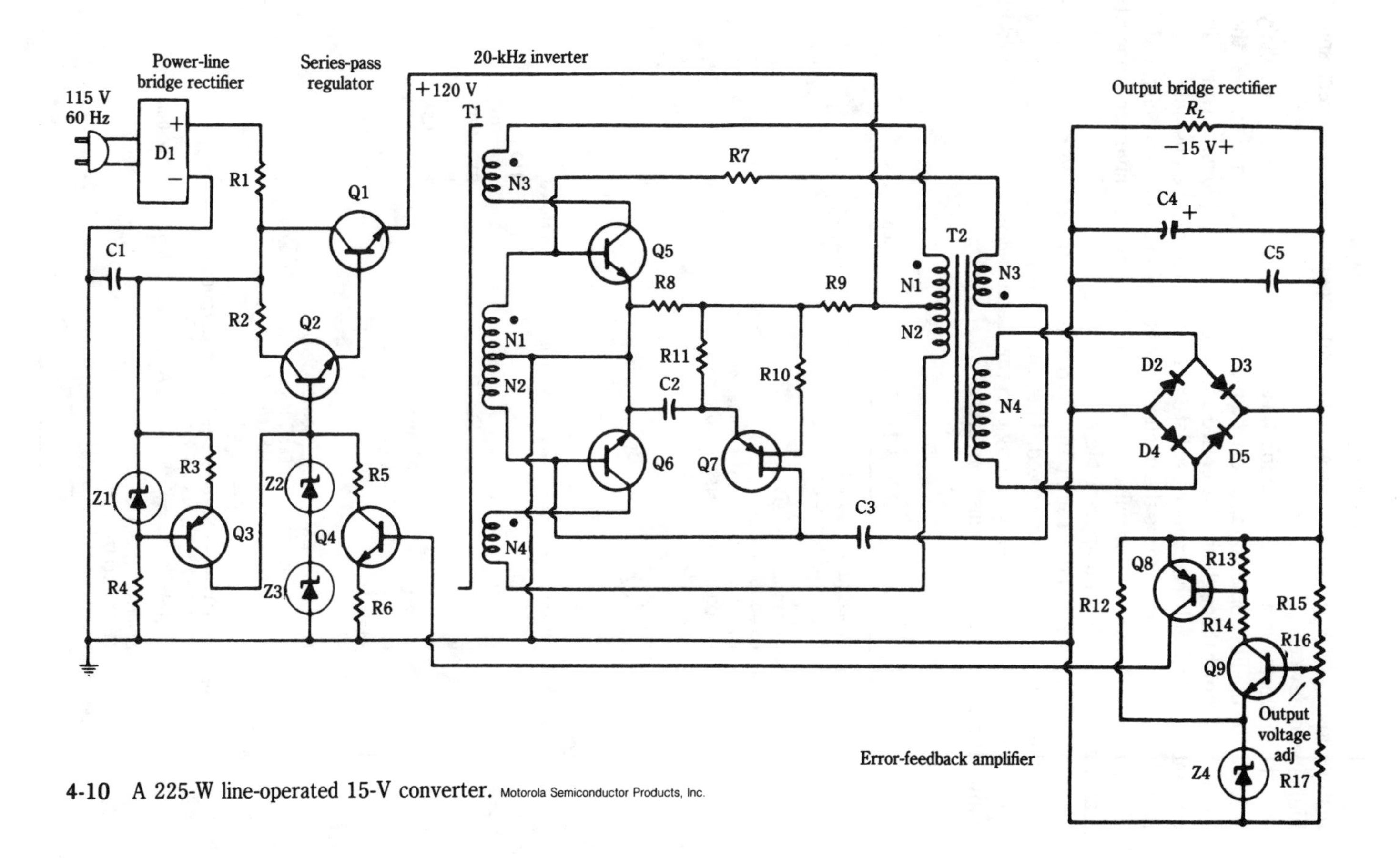

4-10 A 225-W line-operated 15-V converter. Motorola Semiconductor Products, Inc.

The output bridge rectifier consists of four Schottky "hot-carrier" diodes. These feature lower forward voltage drop than conventional junction diodes, and they are high-speed devices.

The inverter is a current-feedback type, but it also incorporates auxiliary winding N3, which supplies some voltage-derived feedback. This technique enables the inverter to be oscillatory throughout the entire range of load conditions. (Inverters with only current-derived feedback cannot oscillate at no load and can be unstable at light loads.) Associated with the inverter is the unijunction relaxation oscillator configured about Q7. This circuit provides starting pulses for the inverter. Once the inverter is operative, this starting circuit exerts no appreciable effect.

Zener diodes Z2 and Z3 limit the dc output voltage of the series-pass regulator when the converter is first energized. Thereafter, the error signal assumes control, and these zeners are deprived of operating current. Table 4-1 is the parts list, and Fig. 4-11 shows the performance for this converter.

Table 4-1 Converter parts list.

C1—2500 μF, 350 V Electrolytic
C2—0.1 μF Disc Ceramic
C3—0.1 μF Paper
C4—10 μF Electrolytic
C5—0.25 μF Paper
D1—MDA-980-4 Bridge Rectifier Assembly
D2, D3, D4, D5—1N5826, 20 V, 15 A
Q1, Q5, Q6—2N6307
Q2, Q4—2N5052
Q3—2N5345
Q7—2N4870
Q8—2N3905
Q9—2N3903

All Resistors in Ohms and ½ W Unless Otherwise Noted

R1—1, 10 W
R2—100
R3—82
R4—22K
R5—1.5K, 15 W
R6—200
R7—15
R8—4.7K
R9—51
R10—1K
R11—10K
R12—270
R13—1K
R14—7.5K
R15—2.5K
R16—5K
R17—3.5K

T1—Core—Magnetics Inc. 80623—½ D—080
N1, N2—20 Turns Each, No. 30 AWG (Bifilar)
N3, N4—3 Turns Each, No. 20 AWG

T2—Core—Arnold 6T 5800 D1
N1, N2—100 Turns Each, No. 20 AWG (Bifilar)
N3—7 Turns No. 26 AWG
N4—12 Turns Each, No. 12 AWG (No. 16 AWG, 3 in Parallel)

Z1—1N4733, 5.1 V
Z2, Z3—1N4760, 68 V
Z4—1N4736

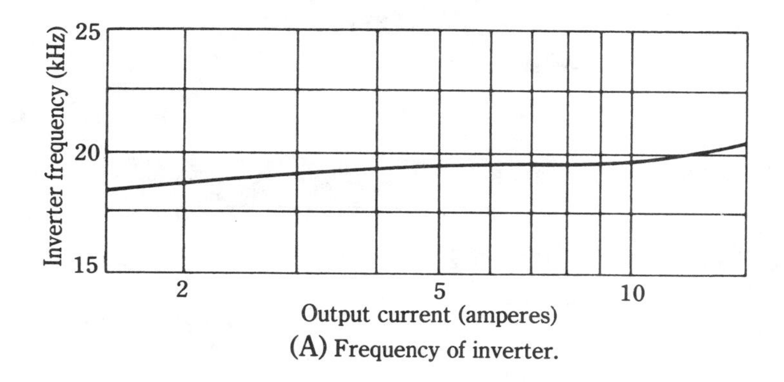

(A) Frequency of inverter.

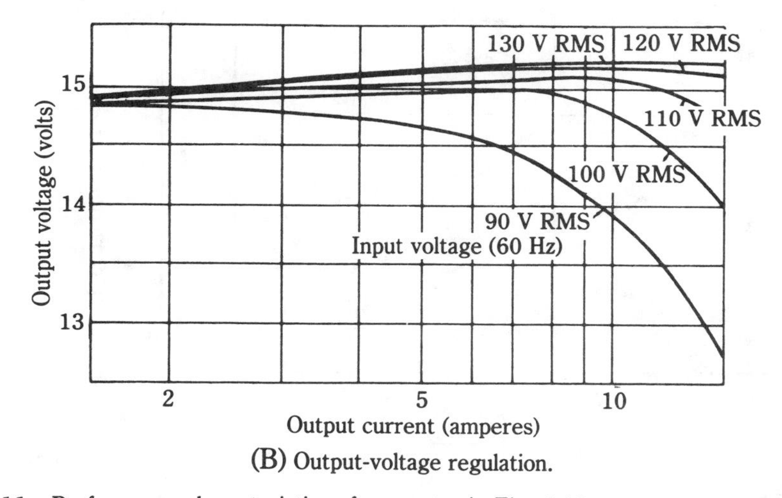

(B) Output-voltage regulation.

4-11 Performance characteristics of converter in Fig. 4-10. Motorola Semiconductor Products, Inc.

A 500-watt regulating converter

The converter shown in Fig. 4-12 accomplishes output voltage regulation by means of pulse-width modulation. Other things being equal, this approach to control of the output has the greatest potential for high efficiency of any of the schemes thus far considered. For example, whereas control is achieved by dissipation in a series-loss transistor in the converter of Fig. 4-10, this converter does not rely on any deliberately induced power dissipation. Rather, the output is controlled by varying the duty cycle of the push-pull Darlington stage that contains the four DTS-712 power transistors. The power stage is, in turn, driven by cascaded monolithic Darlingtons arranged in the appropriate configuration to preserve the push-pull nature of the circuit. It will be seen that each of the two cascades makes use of a DTS-2000 input stage and a DTS-1020 driver stage. The eight active devices mentioned make up the *driven inverter.* The remainder of the system involves the logic for supplying appropriate drive pulses to this driven inverter, as well

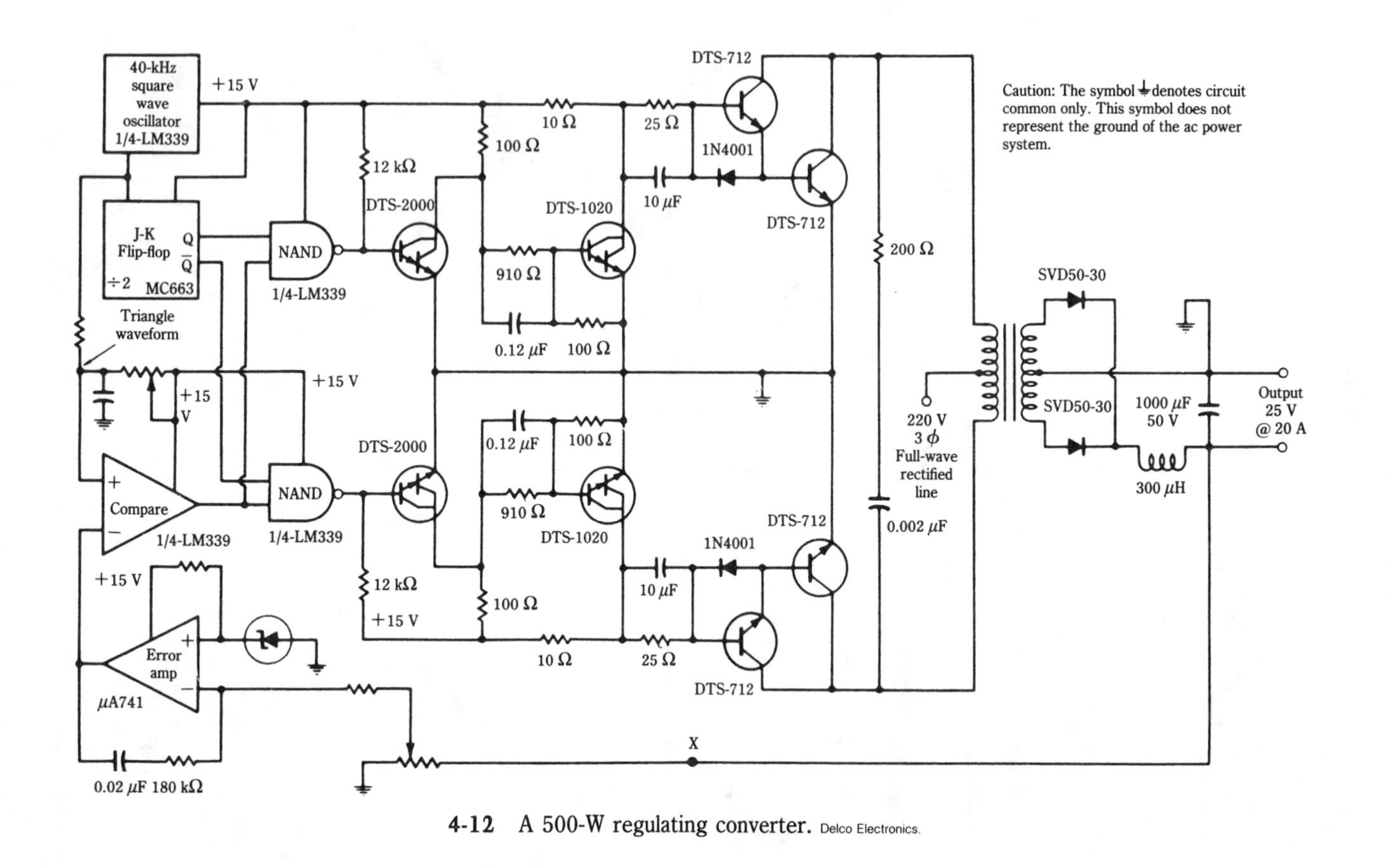

4-12 A 500-W regulating converter. Delco Electronics.

as other functions, such as rectification and filtering, output-voltage sensing, feedback and comparison, and generation of the basic operating frequency.

The basic operating frequency is generated by one of four similar sections of the LM339 IC module. Square waves at 40 kHz are produced in this circuitry, which is essentially a free-running multivibrator. This "clock" frequency actuates an MC663 JK flip-flop, which divides the input frequency by two and provides the 20-kHz drive rate for the inverter. The important function performed by this flip-flop is the generation of two outputs, Q and $\overline{Q}$, that bear a complementary relationship to each other. This satisfies the drive requirements of the push-pull inverter.

It will be observed that the two complementary 20 kHz waves from the flip-flop drive the inverter through NAND gates, each of which is a section of the LM339 IC. It is through the control of the NAND gates that pulse-width modulation is achieved. To see how this is done, focus your attention on the "compare" circuit (also a section of the LM339). Specifically, determine the nature of the two signals applied to its inputs and its resultant output signal that controls the states of the NAND gates.

Because of an RC integrating network associated with the 40-kHz square-wave oscillator, one of the inputs to the comparator (the "compare" function block) is a triangular wave. The other comparator input signal is a dc voltage determined by the output and fed back from the error amplifier. The consequence of this arrangement is that the NAND gates are actuated sooner or later in the 20-kHz driving wave, depending on the magnitude and polarity of the error signal. In the overall system, the tendency is always for the 20-kHz duty cycle to assume an on-to-off ratio such that substantially constant voltage is maintained across the load. In the event that the output voltage attempts to increase, the on-to-off ratio of the inverter is caused to decrease by the amount needed to restore the output voltage to its set value. The opposite sequence of events occurs in the event of a tendency for the output voltage to decrease. Significantly, during such regulatory action, the inverter power stage is always either on or off—operation never occurs with the power stage in linear mode. Ideally, such control is 100 percent efficient; practically, the efficiency is reasonably high.

This converter is suitable for certain industrial applications, where knowledgeable people are aware of the dangers lurking in the lack of input-output isolation. Or, it can be packaged and incorporated inside equipment in such a manner that ordinary usage cannot expose an operator to the hazard of electrical shock. Such a hazard exists because of variable circumstances in the way three-phase industrial power can be provided. Unless you know what prevails, with respect to "grounds," "neutrals," or "commons" in a particular three-phase supply, this converter should not be operated. It can be used safely when attention has been given to these factors. Better still, the experimentally minded builder can render the use of the converter safe under any conditions of power-line grounding by one of the following modifications:

1. Insert an optoisolator in the feedback line at the point marked X in Fig. 4-12. This generally requires additional amplification, for example from an op amp. This is a not uncommon method for obtaining very effective isolation of an output circuit from the power line.
2. The present setup can be retained, except that the feedback can be derived from an extra winding (and small rectifier) on the output transformer. Regula-

tion can be somewhat degraded, but usually the operation remains satisfactory. The full isolation capability of the output transformer is then attained.

3. In industrial applications, there might not be objection to a three-phase isolation transformer. (Of course, for many applications, an attractive feature of this converter is that it possesses the potentiality of dispensing with the use of such a transformer.)

A 3.5-A +15-V power supply is required for the driver stages and the logic. A simple type will suffice. A suitable supply for the power stage of the converter is illustrated in Fig. 4-13. The three-phase bridge rectifier produces a ripple frequency of 6*f*, or 360 Hz for a 60-Hz line. The RMS value of this ripple is only 4.2 percent of the average dc output voltage. Therefore, little if any filtering is required. (By contrast, the RMS ripple in the output of a conventional single-phase, full-wave rectifier is 48 percent of the average dc output voltage.) Capacitor C is useful for attenuating incoming line transients.

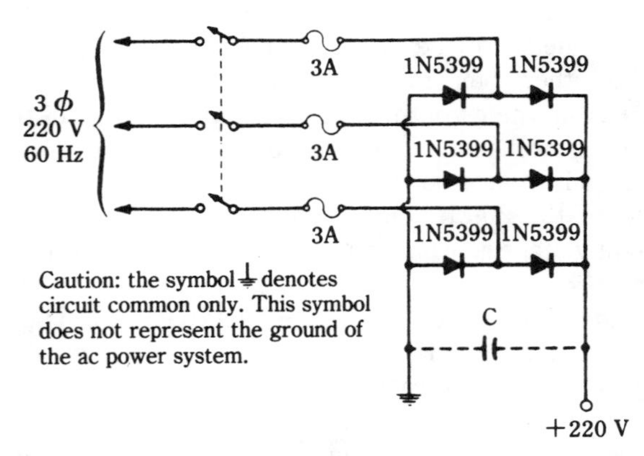

4-13 Three-phase rectifier suitable for converter in Fig. 2-12.

Driven flyback converter

The driven flyback converter of Fig. 4-14 performs the useful function of supplying ±15 V from a positive 5-V source. The need for such dc level translation is often encountered in digital logic systems where the logic elements are powered from a master 5-V supply, but various peripheral and auxiliary function blocks require 15-V supplies. For example, linear op amps and comparators generally operate from positive and negative 15-V sources and have current requirements of from several milliamperes to several tens of milliamperes.

The IC block represents one of the Silicon General family of regulating pulse-width modulators, such as the SG1524, the SG2524, or the SG3524. The driven flyback inverter is the SVT 60-5 npn transistor. Instead of being self-oscillatory, this transistor is turned on and off by the regulating pulse-width modulator. Note the feedback connection from the +15-V output, through the 25-kΩ resistance, to the Invert terminal of the pulse-width modulator. The output voltage is thereby sensed and

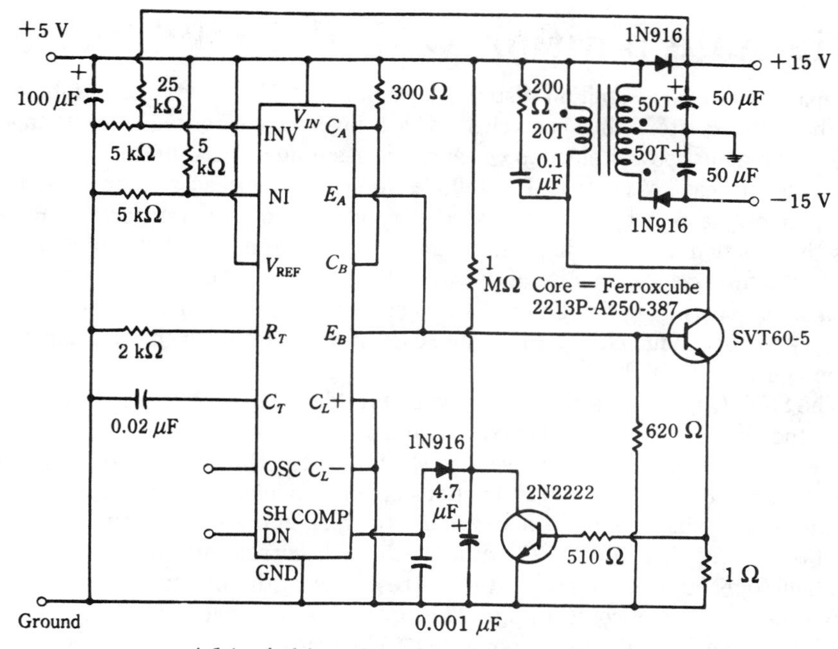

4-14 A driven flyback converter. Silicon General Inc.

internally compared with a stable reference voltage. The resultant error signal controls the duty cycle of the internal oscillator in such a way as to keep the flyback inverter on longer if the output voltage requires boosting. Conversely, if the output voltage requires lowering, the duty cycle of the internal oscillator is decreased, thereby reducing the turn-on time of the flyback inverter.

Regulation occurs because the energy stored in the primary (collector) winding of the transformer is proportional to the length of time current passes through it. And, the more energy, the higher is the voltage induced in the secondary winding when the inverter transistor is turned off.

The 2N2222 transistor and its associated components make up a current-overload protective circuit. If there is excessive current through the 1-Ω sampling resistance, this transistor turns on. In so doing, it inhibits the drive signal from the pulse-width modulator. The nature of the protection is such that repetitive attempts to resume normal operation are made at a rate governed by the 4.7-μF capacitor. As long as the overload persists, such attempts merely result in shut-off of the drive. But, if the fault is cleared, normal operation automatically resumes. The 1N916 diode protects the "Comp" terminal so that the 4.7-μF capacitor does not affect internal frequency compensation.

It should be noted that the output rectifier consists of two oppositely polarized half-wave circuits. This configuration stems from the nature of the flyback pulse and should not be confused with the more usual full-wave, center-tapped circuit, which it resembles schematically.

Solid-state ignition with flyback inverter

The capacitor-discharge ignition system shown in Fig. 4-15 derives many of its features from the self-oscillatory, but controlled flyback inverter configured around stages Q1 and Q2. The actual power stage is Q1, but it is associated with emitter follower Q2 in order to secure ease of control. The telltale sign of the flyback inverter is the hold-off diode, generally associated with the secondary winding of the transformer. Here, D2 fulfills the function of "unloading" the transformer while energy is being stored in its core. The natural oscillatory rate of the Q1/Q2/T1 inverter circuit is much greater than the maximum spark-plug firing rate. This is so because a number of flyback pulses are necessary to apply a full charge to storage capacitor C2, from which the primary of the ignition coil is energized.

The SCR, Q5, acts as a controlled switch for "dumping" the charge from capacitor C2 into the primary of the ignition coil when a plug must be fired. The SCR, in turn, is triggered by emitter-follower transistor Q4. Transistor Q3 is a control stage for the flyback inverter circuit. One of the functions of Q3 is to inhibit inverter operation while the SCR is on. Thus, the inverter never has to work into a short-circuited load. Such control is initiated through the R7 path, which feeds current into the base of Q3 from the initially opened breaker points; this corresponds to the on state of the SCR.

Another function of control stage Q3 is to "meter" the inverter pulses so that C2 is maintained at the desired voltage, nominally 350 V, prior to spark-plug firing. A sensing circuit consisting of R8, R6, and zener diode D4 is connected from storage capacitor C2 to the base of transistor Q3. As a consequence of this control path, Q3 is caused to conduct when sufficient voltage builds up across capacitor C2. When this occurs, part of the base drive to Q2, and therefore to Q1, is shunted to ground. This causes the inverter circuit to oscillate less vigorously than would otherwise be the case, but at a higher frequency. The charge in C2 is then maintained at its intended level prior to firing of a spark plug. This regulatory action is clearly seen in the top waveform of Fig. 4-16A. Output ignition pulses are displayed in Figs. 4-16B and 4-16C. The waveforms in Fig. 4-16 are for a V-8 engine operating at 2000 rpm.

Commutation of the SCR occurs because the SCR is deprived of current when the storage capacitor has given up its charge to the primary of the ignition coil. Diode D3 prevents any ringing current from keeping the SCR on or retriggering it. Also, the shut-down of the inverter by control transistor Q3, described previously, is also an important factor in commutating the SCR. After the SCR is turned off, the stage is then set for a subsequent inverter operational cycle when the ignition breaker points close. Notice that closure of the breaker points also holds the base of Q4 at ground potential so that the commutated SCR cannot inadvertently be triggered. During the time the breaker points are closed, the action shown in Fig. 4-16A recurs.

Parts, other than T1, for the circuit in Fig. 4-15 are listed in Table 4-2. Transformer T1 is wound as follows: A 1/2-inch bobbin and E-I stack of grain-oriented silicon steel are used. First, 150 turns of #28 wire are wound and labeled S1 and F1 on the winding. Second, 50 turns of #24 and #30 wire are wound bifilar and labeled S2 and F2. Third, 150 turns of #28 wire are wound and labeled S3 and F3. All windings are wound in the same direction. A total air gap of 70 mils (35-mil spacer) is used. Connections are made as shown in Fig. 4-17.

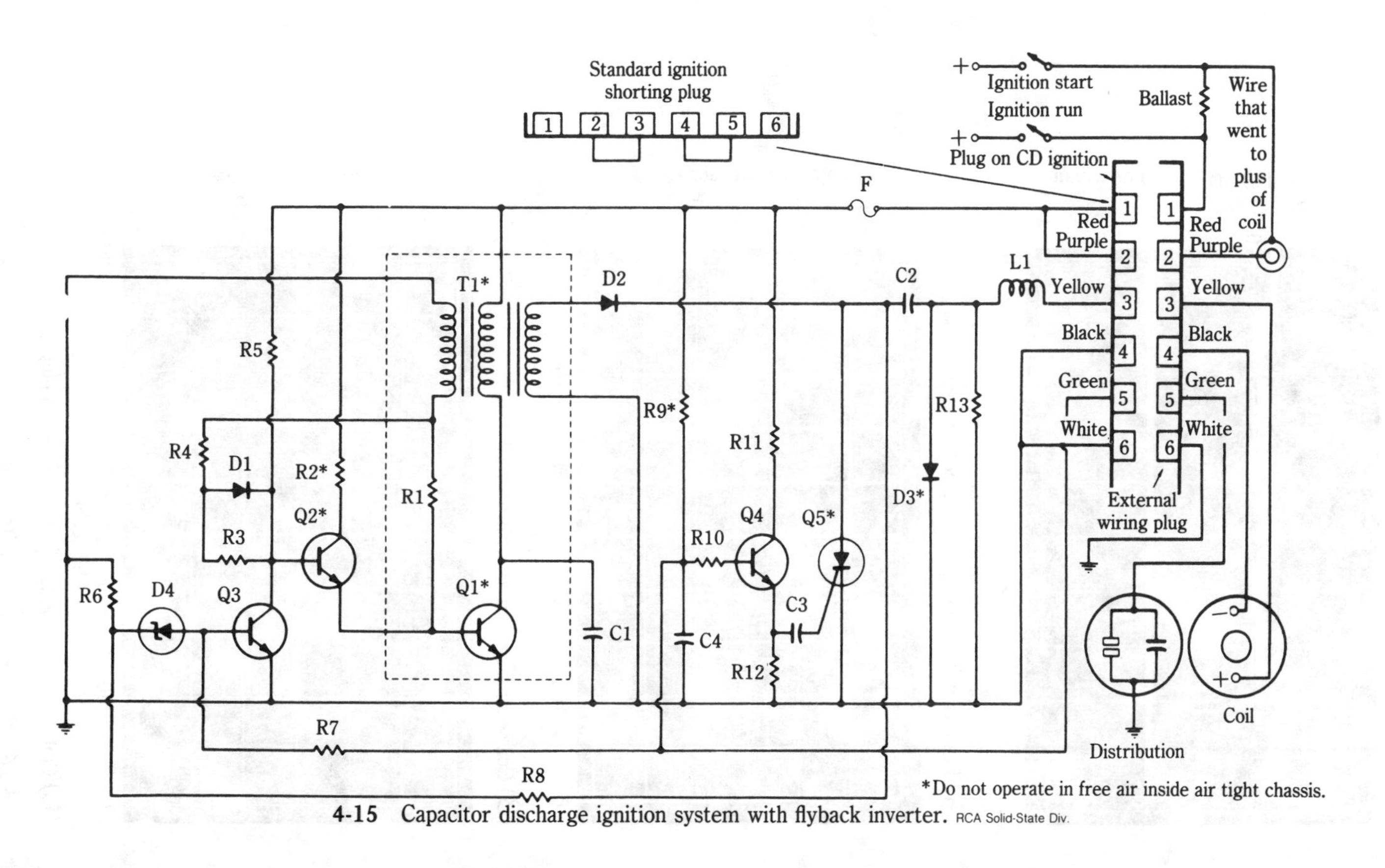

*Do not operate in free air inside air tight chassis.

4-15 Capacitor discharge ignition system with flyback inverter. RCA Solid-State Div.

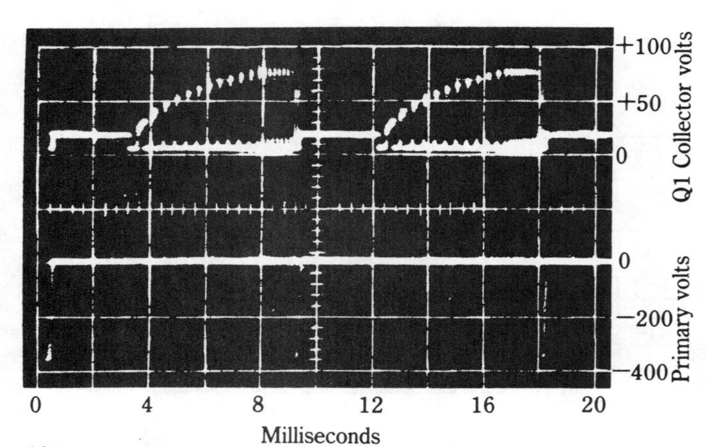

(A) Primary voltage and Q1 collector voltage.

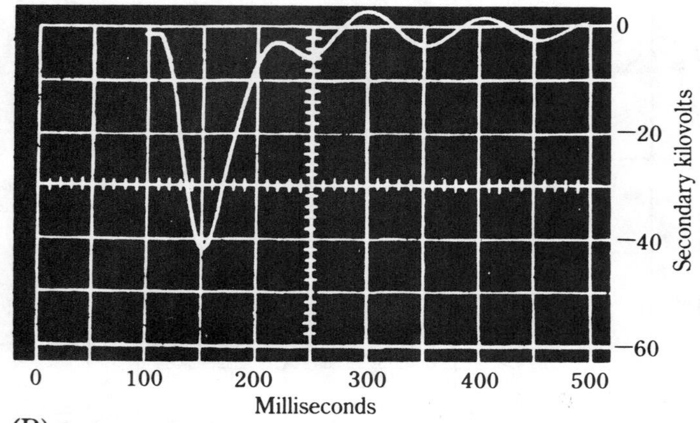

(B) Coil secondary voltage with open circuit.

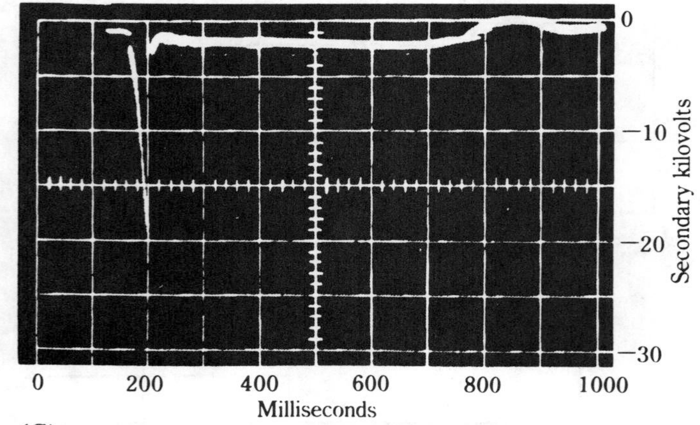

(C) Coil secondary voltage, properly gapped plug.

4-16 Voltage waveforms in circuit of Fig. 4-15. RCA Solid-State Div.

Table 4-2 A parts list for Fig. 4-15.

Part
Q1—2N3055
Q2—2N3053
Q3—2N3241
Q4—2N3241
Q5—RCA 40657
D1—1N3193
D2—1N3195
D3—1N1763A
D4—12 V, ¼ W
C1—0.25 μF, 200 V
C2—1 μF, 400 V
C3—1 μF, 25 V
C4—0.25 μF, 25 V
F—5A
L1—10 μH, 100 Turns of No. 28 Wire Wound on a 2-W Resistor (100 Ohms or More)
R1—1000 ohms, ½ W
R2—35 ohms, 5 W
R3—22,000 ohms, ½ W
R4—1000 ohms, ½ W
R5—18,000 ohms, ½ W
R6—15,000 ohms, ½ W
R7—8200 ohms, ½ W
R8—0.39 megohm, ½ W
R9—220 ohms, 1 W
R10—1000 ohms, ½ W
R11—68 ohms, ½ W
R12—4700 ohms, ½ W
R13—27,000 ohms, ½ W

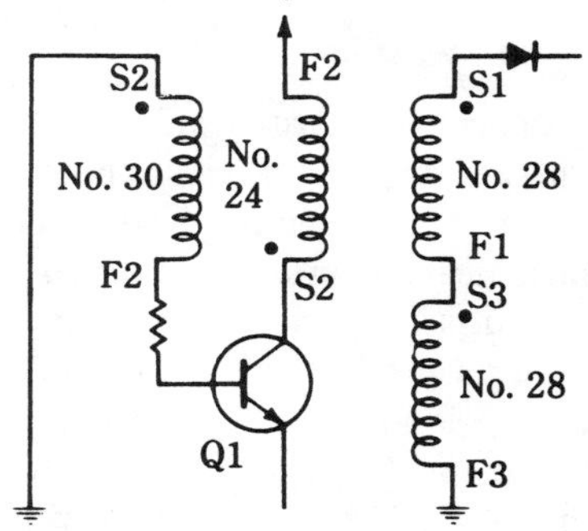

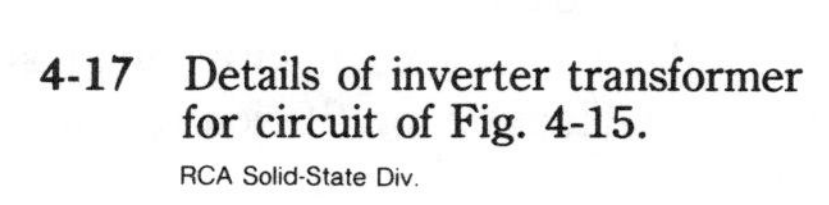
4-17 Details of inverter transformer for circuit of Fig. 4-15.
RCA Solid-State Div.

Photoflash-capacitor charger utilizing a driven converter

The circuit shown in Fig. 4-18 makes it possible to charge a 480-μF capacitor to 500 V in 15 seconds for the operation of a photoflash lamp. The energy available for the lamp is:

$$\tfrac{1}{2}\, CV^2 = \frac{(480 \times 10^{-6})\,(500)^2}{2} = 60 \text{ joules}$$

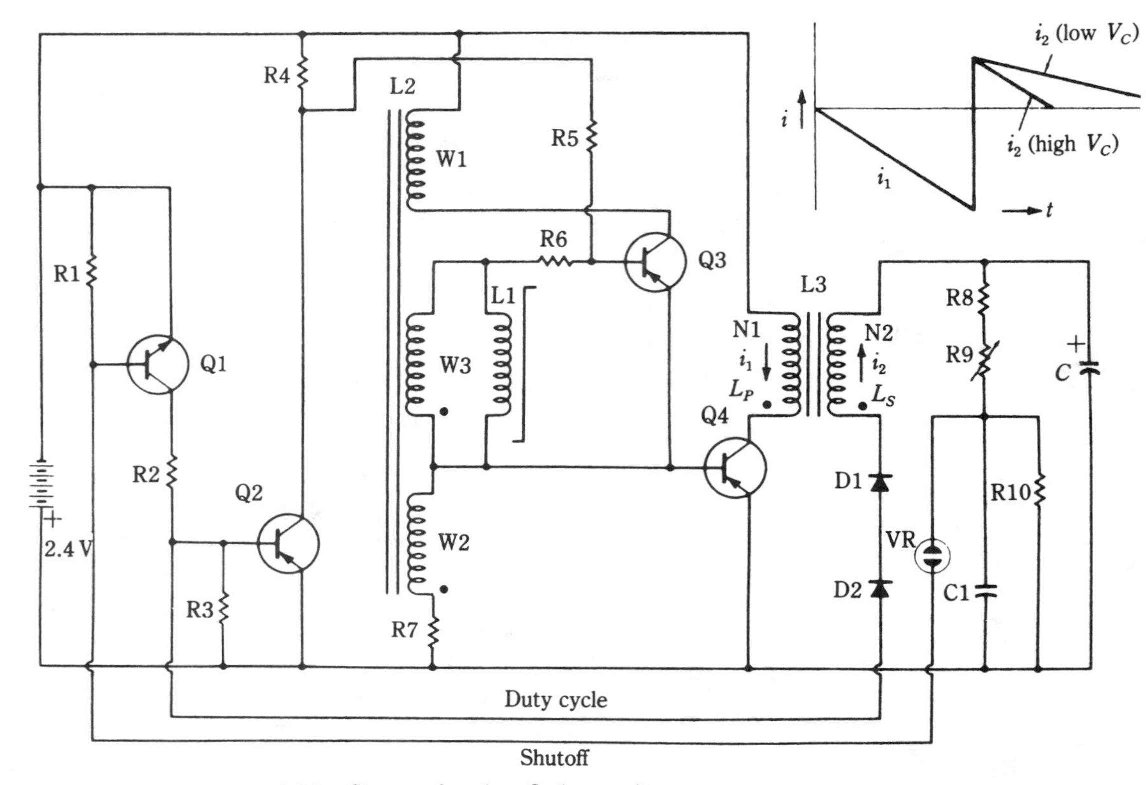

4-18 Charger for photoflash capacitor. Motorola Semiconductor Products, Inc.

This is a respectable energy level, one easily capable of producing lethal consequences with haywire lash-ups and careless operators. You could easily be misled by the fact that this unit operates from two tiny size AA nickel-cadmium cells, each with a modest rating of 500 milliampere-hours, but the particular combination of 60 joules and 500 V qualifies the output capacitor as a danger to life. Even so, this system is representative of a consumer's product, and it is possible to construct and operate it safely. The important point is to practice caution in experimentation with and operation of equipment of this nature.

The heart of this circuit is the driven power stage, Q4, and the blocking oscillator, Q3. Stages Q1 and Q2 are involved in the control of the blocking oscillator. The control is such that the nominally 20-kHz blocking oscillator generates a long off time when the output capacitor has little charge, but reduces its off time as the capacitor accumulates charge. By means of such a variable duty-cycle charging source, the capacitor can be brought to its 60 joule energy content in minimal time. This is an important feature from the photographer's viewpoint.

The blocking oscillator, Q3 and its associated components, is unconventional because of the presence of the saturable-core inductor, L1. Ordinarily, the characteristics of such a circuit would be predominantly governed by the blocking-oscillator transformer, L2 in this case. In this circuit, the abrupt saturation of L1 terminates the on time of the oscillator. This makes the circuit behave in a manner somewhat analogous to that of a free-running monostable multivibrator; that is, constant-width pulses are generated. The pulse width is determined from Faraday's transformer equation as follows:

$$T = \frac{2NAB_{SAT} \times 10^{-8}}{E_{W3}}$$

Where:

T is the on time in seconds,
N is the number of turns on L1,
A is the cross-sectional core area of L1 in square centimeters,
B_{SAT} is the saturation flux density of L1 in lines per square centimeter,
E_{W3} is the peak voltage applied to L1 from winding W3 on L2.

Because of the way in which L1 is used in this circuit, the factor 2, rather than the usually encountered 4 appears in the above equation.

Although the on interval of the blocking oscillator is constant, the off interval is controlled by stage Q2. This control is achieved by diverting the base drive of Q3. Specifically, what happens is as follows: The current that the "flyback" voltage induces in the secondary of L3 circulates through D1, D2, and R3; it depends on the amount of charge accumulated in the storage capacitor, C. If C is virtually depleted of charge, it requires a relatively long time to absorb the energy from L3. Accordingly, the charging current through R3 causes stage Q2 to divert base drive from Q3. During this time, Q3 remains quiescent. Only when the storage capacitor has substantially absorbed the energy from the flyback pulse is Q3 allowed to generate a subsequent pulse. Thus, the duty cycle of the blocking oscillator is modulated in compliance with the required time for energy transfer from L3 to the storage capacitor. As a consequence of this action, the oscillator frequency gradually increases from a low to a higher value as the storage capacitor becomes charged. The range of frequencies is from about one to several kilohertz.

As thus far described, the system would be operational and essentially satisfactory if operated from the ac power line. However, it is desired to maximize the number of flashlamp operations obtainable from the battery power supply. To serve this objective, additional circuitry is added to shut down the blocking oscillator when the storage capacitor is fully charged. A voltage-sensing network consisting of R8, R9, R10, VR, and C1 is connected across the storage capacitor, C. The resistances are proportioned so that the network can function in two modes. When the storage capacitor is just short of its fully charged state, the neon lamp, VR, will be intermittently ionized (notice that the voltage-sensing network has the essential configuration of a simple relaxation oscillator). When the storage capacitor is fully charged, the neon lamp will conduct continuous current because conditions will no longer be favorable for alternate ionization and deionization. (The "latch-up" condition that plagues the usual uses of these relaxation circuits is deliberately used in this application.)

The current that passes through the neon lamp when it ionizes affects the off time in a way similar to the duty-cycle modulation previously described, except that the effect is greater because of the additional amplification provided by stage Q1. If VR is blinking, the inverter will be rendered inoperative for longer times than is possible from the duty-cycle modulation alone. When VR becomes latched in its ionized state, the blocking oscillator is completely deprived of base drive, and the inverter shuts down. Action automatically resumes when the storage capacitor loses charge from leakage or following an operational cycle of the photoflash lamp. Notice that in addition to performing an important circuit function the neon lamp also provides visual indication that the photoflash unit is ready to use.

Parts for the circuit are listed in Table 4-3. Details of the inductors are given in Chart 4-1.

Table 4-3 Parts list for Fig. 4-18.

C1—0.2 μF ±20%, 100 V
C (Load Capacitor)—480 μF, 500 V
D1, D2—MR814 (Fast-Recovery Rectifier)
Q1—MPS6520 (Selected)
Q2—MPS6563 (Selected)
Q3—MPS6562 (Selected)
Q4—MP3613 (Selected)
VR—Neon Lamp (Selected 5 AG)
R1—39K
R2—100Ω
R3—1.0K
R4—120Ω
R5—150Ω
R6—270Ω ±5%
R7—7.5Ω ±5%
R8—1.0 MΩ
R9—2.0 MΩ Pot
R10—390K ±5%
Note: All resistors ±10%, ¼ W, Unless Otherwise Specified

Chart 4-1 Inductors for Fig. 4-18.

L1: Timing Inductor

Core:	Ferroxcube 266T125-3E2A
Winding:	145 Turns, No. 36 Wire

L2: Drive-Oscillator Transformer

Core:	Ferroxcube No. 18/11PL00-3B7
Bobbin:	1811F2D
Air Gap:	0.005 in
Windings:	W1: 40 Turns, No. 28 Wire
	W2: 20 Turns, No. 30 Wire
	W3: 140 Turns, No. 36 Wire

L3: Output Transformer

Core:	Ferroxcube No. 26/16P-L00-3B7
Bobbin:	Ferroxcube No. 26/16F2D
Windings:	N1: 11 Turns, No. 18 Wire
	N2: 1100 Turns, No. 38 Wire
Air Gap:	0.030 in

A 750-watt inverter

The inverter, or frequency converter, shown in Fig. 4-19 is representative of what can be accomplished with modern power transistors and circuit techniques. This system is essentially a driven inverter; it derives dc power by rectification of the ac from the power line and delivers three-phase power at a frequency determined by a logic circuit. It will operate from either a single-phase or three-phase line, and it can be made to accommodate a line voltage of either 120 V or 208 V. Moreover, the line frequency can be from 47 to 1250 Hz. The three-phase output power can be varied over the range of 380 to 1250 Hz, regardless of the nature of the input power. Output voltage can be either 120 or 208 V. The output waveshape is stepped and approximates a sine wave, rather than a square wave.

This inverter can be used to provide 400-Hz power for aircraft or for laboratory simulation of the electrical environment in aircraft. It can be used to produce variable speed from three-phase induction motors, but in this application a provision should be made to adjust the motor voltage in direct proportion to the frequency. Thus, a motor nominally rated for, for example, 110 V at 400 Hz should have about 220 V applied to it when operated at double speed with 800 Hz. Also, when the inverter is used for powering motors, consideration should be given to starting conditions so that it is not subjected to undue stress from inordinately high inrush current.

For convenience, the schematic diagram of the inverter has been broken down into three parts. Figure 4-19 shows the output stages. These are driven from the variable oscillator and six-stage ring counter (Fig. 4-20) by way of the matrix and driver section (Fig. 4-21). Figures 4-20 and 4-21 make up the *logic circuitry* of the inverter. Parts values are shown on the diagrams; transformer design information is given in Chart 4-2.

Figure 4-19 includes the power-line rectifier, the switching transistors, the input transformers (T1, T2, and T3), and the output transformer (T5). Inasmuch as the output circuit has three-phase connections, the basic problem in making the inverter operative is to provide the proper signal format to the input transformers. The logic for accomplishing this will be described in subsequent paragraphs. Notice that a strapping provision is made to permit operation from either a three-phase or a single-phase power line. The 11,000-μF filter capacitor is adequate for either mode of operation.

The logic circuitry commences with a tunable unijunction-transistor oscillator, Q2 in Fig. 4-20. The 75-kΩ variable resistor in its emitter circuit permits adjustment of the ultimate inverter output frequency through the range 380 to 1250 Hz. The following 2N3053 transistor (Q3) is a driver stage. The subsequent array of transistors constitutes a six-stage ring counter. The output pulses from the ring counter are available at terminals 1 through 6.

Both this portion of the logic circuitry and that depicted in Fig. 4-21 derive their dc operating power from a small voltage-regulated power supply involving transformer T4, a 2N2102 series-pass transistor (Q1), and associated components. The power-line frequency range of the inverter (47 to 1250 Hz) is essentially established by the small power transformer, T4.

The portion of the logic circuitry shown in Fig. 4-21 contains a diode matrix, a pair of 2N3053 transistors as driver stages for each of the three phases, and the primary windings of input transformers T1, T2, and T3 (each of these transformers has two

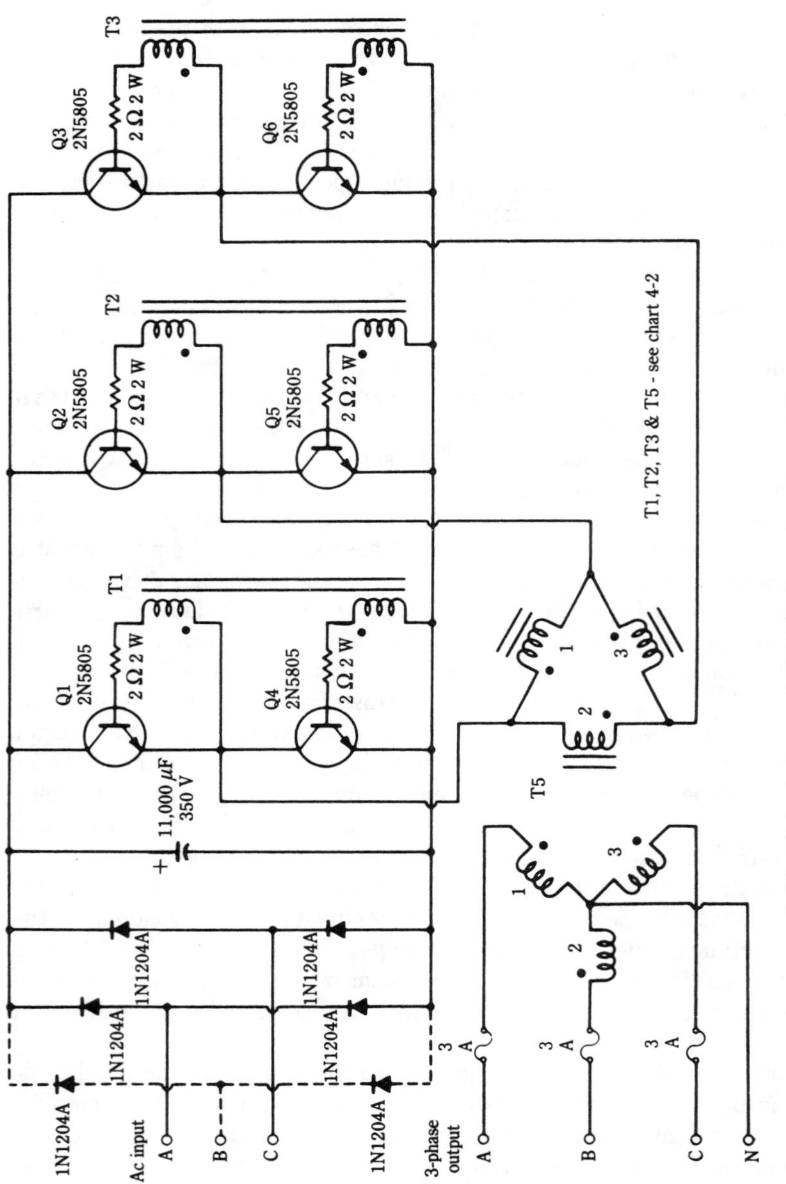

4-19 Output stages of 750-W, three-phase inverter. RCA Solid-State Div.

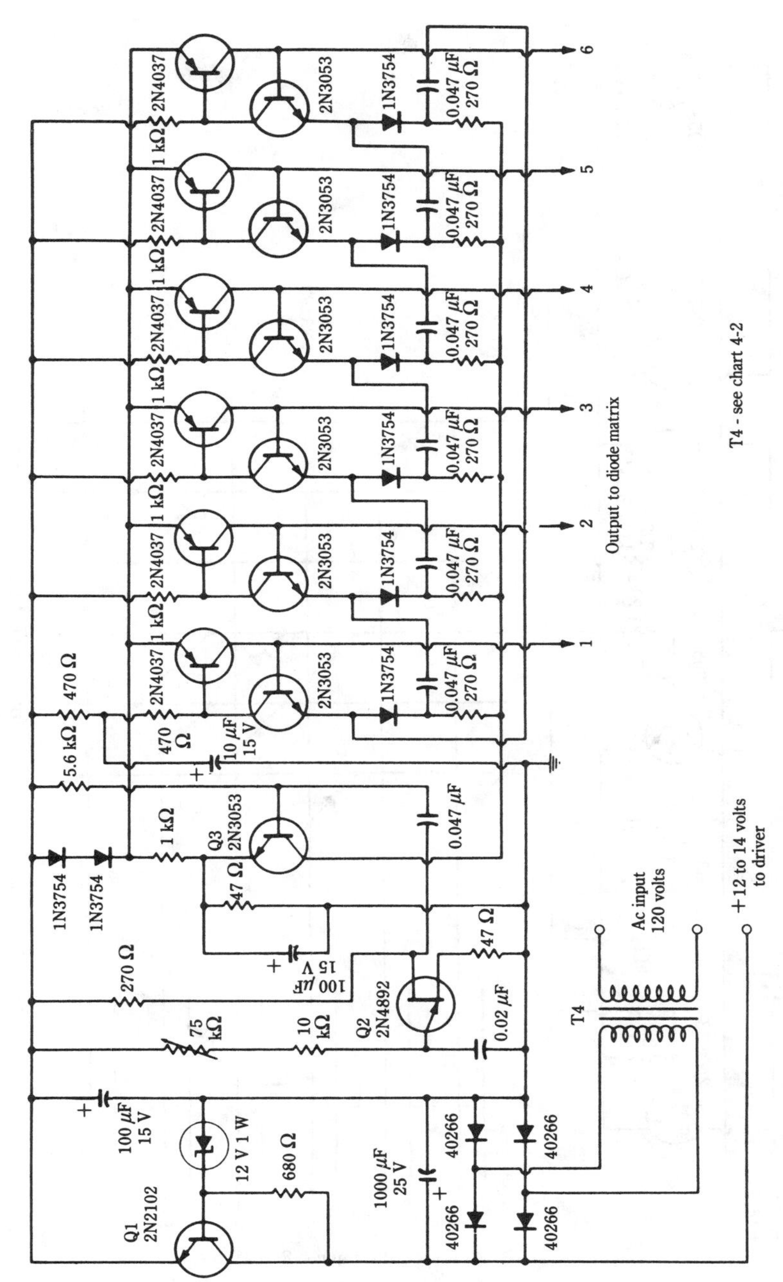

4-20 Variable oscillator and six-stage ring counter.

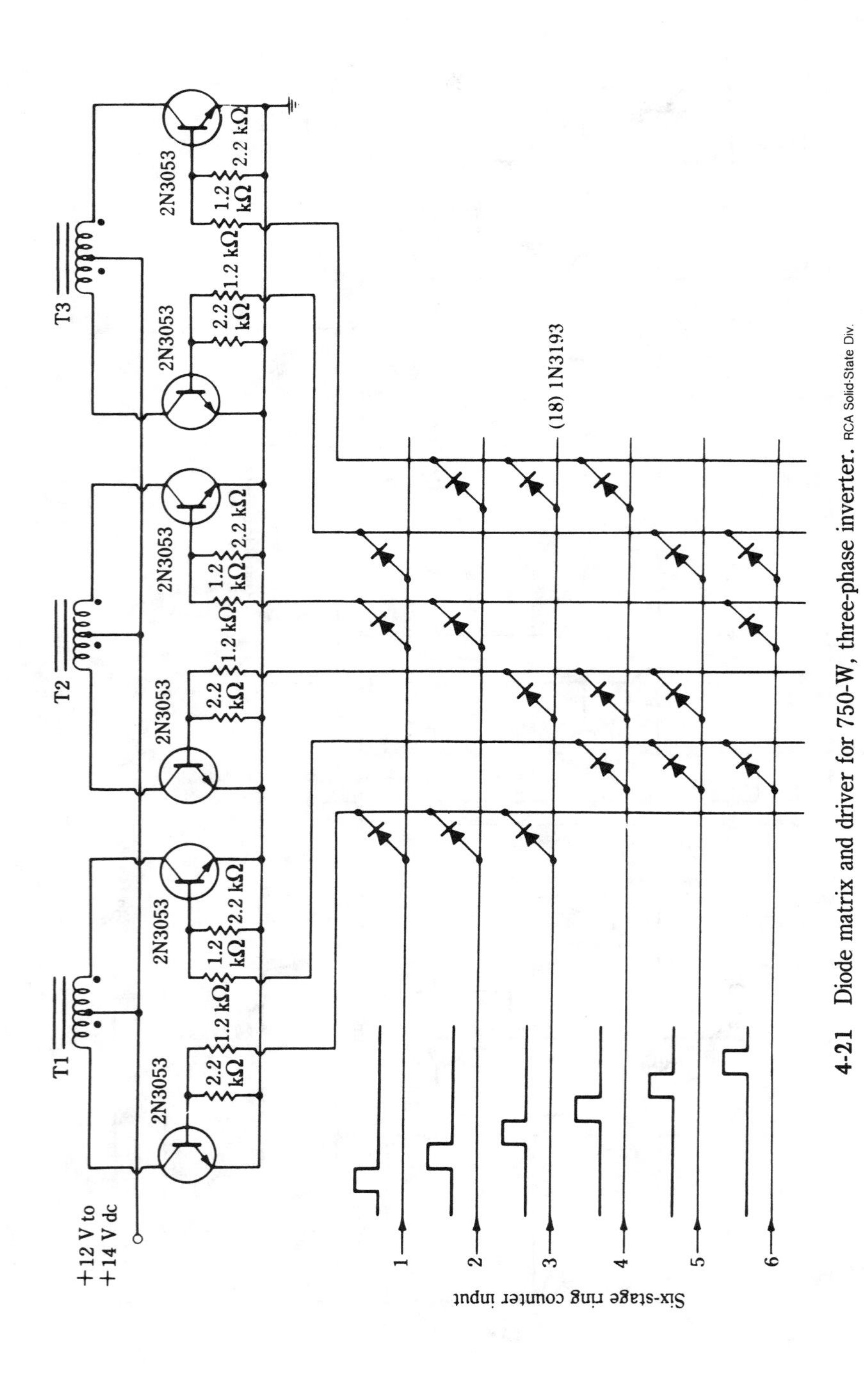

4-21 Diode matrix and driver for 750-W, three-phase inverter. RCA Solid-State Div.

secondary windings, as can be seen in Fig. 4-19). The function of the diode matrix is to guide the incoming pulse sequence to the driver stages in such a manner that the combined pulses form an array of three-phase currents in the input transformers (T1, T2, and T3).

Transformer design information is given in Chart 4-2. The polarity, or phasing, of the input and output transformer connections is extremely important and must comply with the information conveyed by the phasing dots in Figs. 4-19 and 4-21. The identities of the leads can be easily confused during winding; it is strongly recommended

Chart 4-2 Transformer design data for 750-W, three-phase inverter.

T1, T2, T3—Driver Transformer	
CORE	—Square Stack 21EI Microsil (0.006) Magnetic Metals Co. 21EI3306
PRIMARY	—14 Volts 140 Turns Bifilar No. 29 Wire (in Series) 20 Turns Per Layer 7 Layers
SECONDARY	—4 Volts 52 Turns Bifilar No. 29 Wire 13 Turns Per Layer 4 Layers
T4—Stepdown Isolation Transformer for Logic Circuit Supply	
CORE	—Square Stack 75EI Microsil (0.006) Magnetic Metals Co. 75EI3306
PRIMARY	—120 Volts 1200 Turns No. 32 Wire 100 Turns Per Layer 12 Layers
SECONDARY	—12 Volts 128 Turns No. 22 Wire 32 Turns Per Layer 4 Layers
T5—Output Transformer	
CORE	—Square Stack 1.2EI3φ Microsil (0.006) Magnetic Metals Co. 1.2EI3φ3306
PRIMARY (DELTA)	—120 Volts 188 Turns No. 17 Wire 47 Turns Per Layer 4 Layers OR —208 Volts 325 Turns No. 19 Wire 55 Turns Per Layer 6 Layers
SECONDARY (WYE)	—120/208 Volts 200 Turns No. 17 Wire 50 Turns Per Layer 4 Layers

that proper phasing be confirmed by low-level electrical tests prior to any attempt at operation from the power line.

The voltage waveform across one of the output phases is shown in Fig. 4-22. The scale of the oscilloscope display is 200 volts per division. This stepped wave has total harmonic distortion of 24 percent and, for most purposes, constitutes a satisfactory approximation to a sine wave. In the event that a smoother waveform is desired, the requisite filtering will be much less than would be needed for the usual square-wave inverter output.

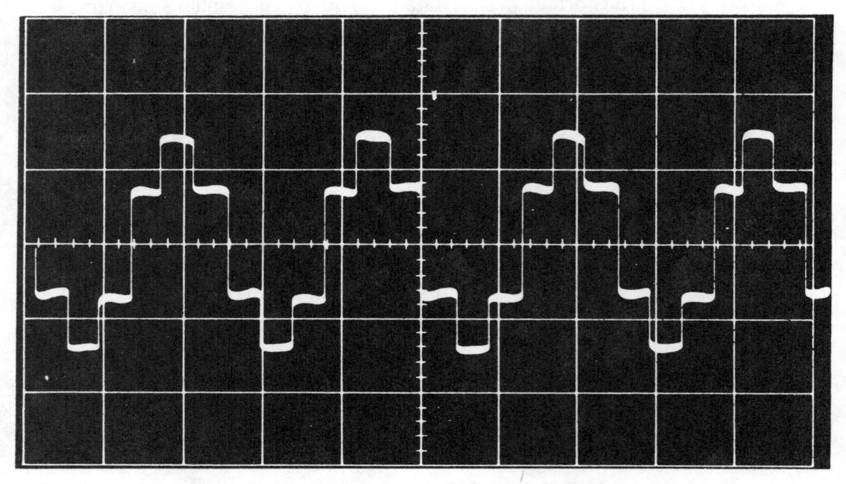

4-22 Output voltage waveform of one phase of 750-W, three-phase inverter. RCA Solid-State Div.

The output characteristics of the inverter are shown in Fig. 4-23. This performance is for a three-phase, 208-V power line and the inverter delivering 400 Hz to a resistive load. At rated output, the output voltage is 208 V, and the load current is 750/(1.732 × 208), or 2.08 A.

Digital-logic PWM inverter

The inverter to be described is quite different from the oscillator and driven types hitherto discussed. The sine-wave output is not generated directly; rather, discrete pulses with constant amplitudes, but with varying widths and positions are produced. The energy content in a sequence of such pulses is such that the effect of a sine wave is produced in the load. Indeed, if a small amount of inductance is present in the load, the pulses will be integrated, and a reasonably good sine wave of current will be produced. Inasmuch as this inverter is intended for the variable-speed operation of ac motors, such a "synthesized" current sine wave can be expected because of the inductive nature of motors. The electromagnetic torque of motors derives from current. Therefore the applied voltage pulses can be considered as a means to an end. A block diagram of this

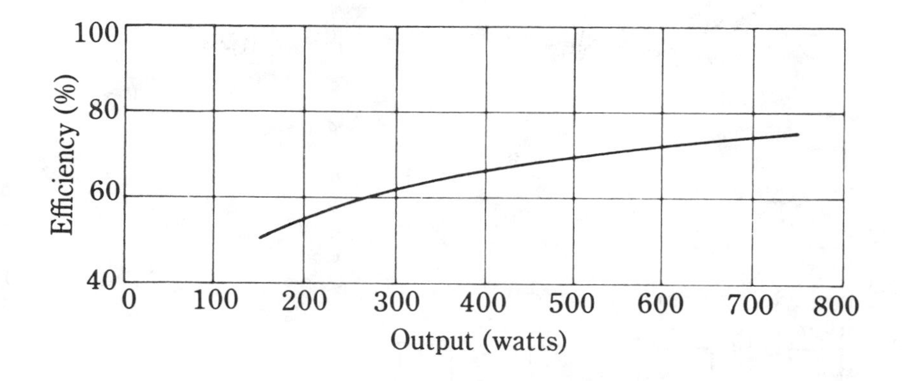

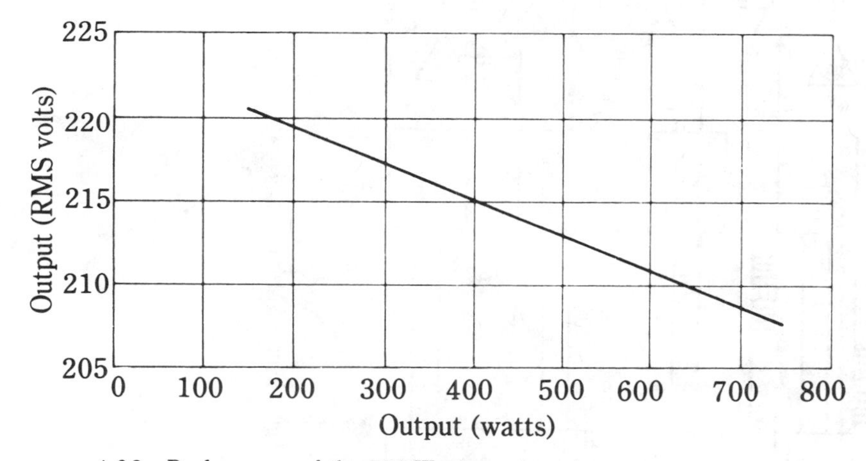

4-23 Performance of the 750-W, three-phase inverter. RCA Solid-State Div.

inverter is shown in Fig. 4-24. Notice that it has outputs for the operation of single-phase, two-phase, and three-phase motors.

The heart of the scheme involves the programming and addressing of the 1024 bit read-only memory (ROM). This CMOS memory outputs four serial words, each 256 bits long. An up-down count pattern in the addressing logic enables full sine wave information to be extracted from memory—even though sine-weighted pulse trains are stored only for a 90° portion of a wave. The ROM coding format is shown in Fig. 4-25. The formation of half sine-wave segments by memory address is illustrated in Fig. 4-26; at 180° and 360°, a polarity switch inverts the waveform. The format in the mask-programmable ROM appears in Fig. 4-27. Notice the 256 $\times$ 4 bit organization.

A logic diagram of the inverter is shown in Fig. 4-28 (a list of integrated circuits is given in Table 4-4). Output amplifiers are not shown in Fig. 4-28. It can be seen that considerable simplification would result if the single-phase output were sufficient for the purpose at hand. However, the added versatility of the system in its entirety is attained relatively inexpensively with the incorporation of a few more ICs. In Fig. 4-28, the frequency adjustment on the clock oscillator is appropriately designated "RPM CONTROL" (the speed of induction motors is proportional to frequency).

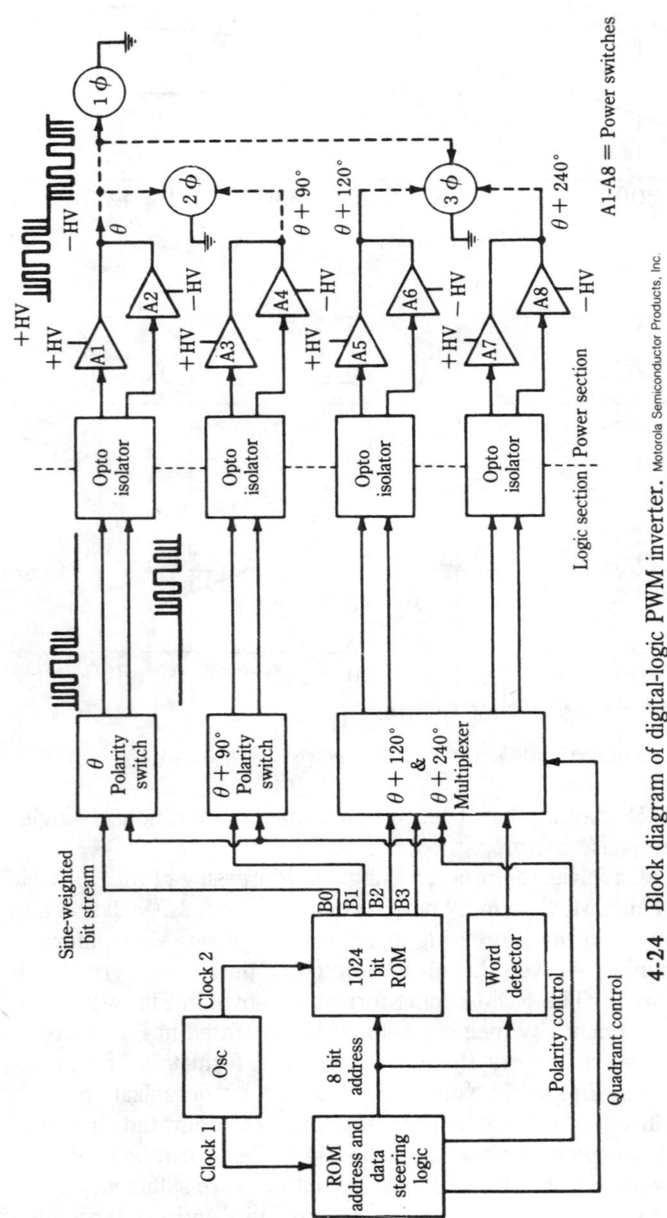

4-24 Block diagram of digital-logic PWM inverter. Motorola Semiconductor Products, Inc.

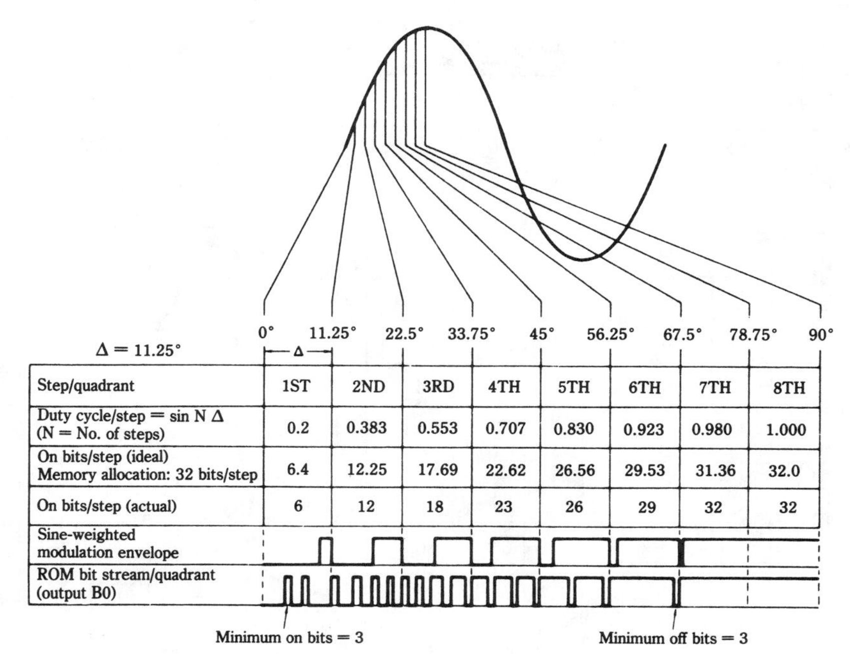

Step/quadrant	1ST	2ND	3RD	4TH	5TH	6TH	7TH	8TH
Duty cycle/step = sin N Δ (N = No. of steps)	0.2	0.383	0.553	0.707	0.830	0.923	0.980	1.000
On bits/step (ideal) Memory allocation: 32 bits/step	6.4	12.25	17.69	22.62	26.56	29.53	31.36	32.0
On bits/step (actual)	6	12	18	23	26	29	32	32
Sine-weighted modulation envelope								
ROM bit stream/quadrant (output B0)								

4-25 Coding pattern for ROM. Motorola Semiconductor Products, Inc.

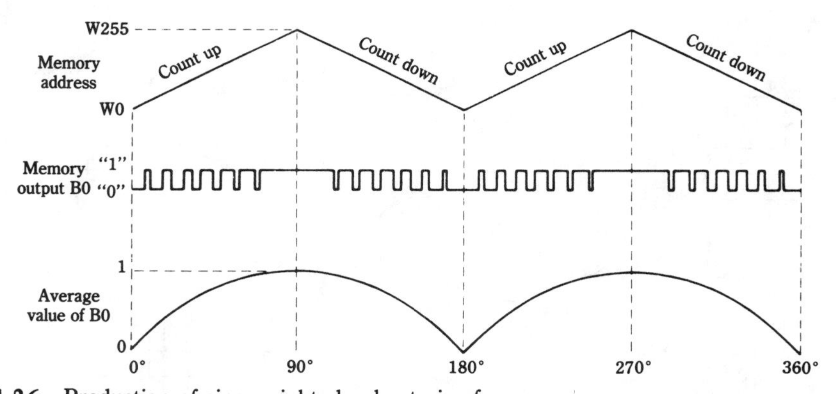

4-26 Production of sine-weighted pulse trains from memory. Motorola Semiconductor Products, Inc.

A single-phase power stage is shown in Fig. 4-29. The 4N26 optocoupler at the input in this diagram also appears at the output in Fig. 4-28. The single-phase motor is a 1/50 hp, 1725 rpm (60 Hz) shaded-pole type. Smooth speed control extends down to about 90 rpm.

TV vertical-deflection circuit

The TV vertical-deflection circuit of Fig. 4-30, despite its close resemblance to an audio amplifier, bears relevance to the subject of inverters, for the power from the dc supplies is "inverted"—transformed into ac power.

8-bit ROM Address	
B3	Multiplexed to generate
B2	$\theta + 120°$ and $\theta + 240°$ outputs
B1	Generates $\theta + 90°$ output
B0	Generates θ output

	Bit					Bit					Bit					Bit					Bit			
Word	3	2	1	0	Word	3	2	1	0	Word	3	2	1	0	Word	3	2	`1	0	Word	3	2	1	0
0	1	1	1	0	51	1	1	1	0	102	1	1	1	1	153	1	0	1	1	204	1	1	0	1
1	1	1	1	0	52	1	1	1	0	103	1	1	1	0	154	1	0	1	1	205	1	0	1	1
2	1	1	1	0	53	1	0	1	0	104	1	1	1	0	155	1	0	1	1	206	1	0	1	1
3	1	1	1	0	54	1	0	1	0	105	1	1	1	0	156	1	0	1	1	207	1	0	1	1
4	1	1	1	0	55	1	0	1	0	106	1	0	1	1	157	1	0	1	0	208	1	0	0	1
5	1	1	1	0	56	1	1	1	1	107	1	0	1	1	158	1	1	1	0	209	1	0	0	1
6	1	1	1	0	57	1	1	1	1	108	1	0	1	1	158	1	1	1	0	210	1	1	0	1
7	1	1	1	0	58	1	1	1	1	109	1	0	1	1	160	1	1	0	1	211	1	1	0	1
8	1	1	1	0	59	1	1	1	0	110	1	0	1	1	161	1	0	0	1	212	1	1	0	1
9	1	1	1	1	60	1	1	1	0	111	1	1	1	1	162	1	0	0	1	213	0	0	1	1
10	1	0	1	1	61	1	1	1	0	112	1	1	0	1	163	1	0	1	1	214	0	0	1	1
11	1	0	1	1	62	1	1	1	0	113	1	1	0	1	164	1	0	1	1	215	0	0	1	1
12	1	0	1	0	63	1	1	1	0	114	1	0	0	0	165	1	0	1	1	216	1	0	0	1
13	1	1	1	0	64	1	0	0	1	115	1	0	1	0	166	1	0	1	1	217	1	0	0	1
14	1	1	1	0	65	1	0	0	1	116	1	0	1	0	167	1	0	1	1	218	1	1	0	1
15	1	1	1	0	66	1	0	0	1	117	1	0	1	1	168	1	0	0	1	219	1	1	0	1
16	1	1	1	0	67	1	1	1	1	118	1	0	1	1	169	1	0	0	1	220	1	1	0	1
17	1	1	1	0	68	1	1	1	0	119	1	1	1	1	170	1	0	0	1	221	1	0	1	1
18	0	1	1	0	69	1	1	1	0	120	1	1	1	1	171	1	0	1	1	222	1	0	1	1
19	0	1	1	0	70	1	1	1	0	121	1	1	1	1	172	1	0	1	1	223	1	0	1	1
20	0	1	1	0	71	1	1	1	0	122	1	0	1	1	173	1	0	1	1	224	0	0	0	1
21	1	1	1	1	72	1	1	1	1	123	1	0	1	1	174	1	0	1	1	225	0	0	0	1
22	1	1	1	1	73	1	1	1	1	124	1	0	1	1	175	1	0	1	1	226	0	1	0	1
23	1	1	1	1	74	1	0	1	1	125	1	0	1	0	176	1	0	0	1	227	1	1	0	1
24	1	1	1	0	75	1	0	1	1	126	1	0	1	0	177	1	0	0	1	228	1	1	0	1
25	1	1	1	0	76	1	0	1	0	127	1	1	1	0	178	1	0	0	1	229	1	0	0	1
26	1	0	1	0	77	1	1	1	0	128	1	1	0	1	179	1	1	0	1	230	1	0	0	1
27	1	0	1	0	78	1	1	1	0	129	1	1	0	1	180	1	1	1	1	231	1	0	0	1
28	1	0	1	0	79	1	1	1	0	130	1	0	0	1	181	0	1	1	1	232	1	0	1	1
29	1	1	1	0	80	1	1	1	1	131	1	0	1	1	182	0	0	1	1	233	1	0	1	1
30	1	1	1	0	81	1	1	1	1	132	1	0	1	1	183	0	0	1	1	234	1	1	1	1
31	1	1	1	0	82	1	0	1	1	133	1	0	1	1	184	1	0	0	1	235	0	1	0	1
32	1	1	1	1	83	1	0	1	1	134	1	0	1	1	185	1	0	0	1	236	0	1	0	1
33	1	1	1	1	84	1	0	1	1	135	1	1	1	1	186	1	0	0	1	237	0	1	0	1
34	1	1	1	1	85	1	1	1	0	136	1	1	1	1	187	1	0	0	1	238	1	1	0	1
35	1	1	1	0	86	1	1	1	0	137	1	1	1	1	188	1	0	1	1	239	1	0	0	1
36	1	1	1	0	87	1	1	1	0	138	1	0	1	1	189	1	0	1	0	240	1	0	0	1
37	1	1	1	0	88	1	1	1	1	139	1	0	0	1	190	1	0	1	0	241	1	0	0	1
38	1	1	1	0	89	1	1	1	1	140	1	0	0	1	191	1	1	1	0	242	1	1	0	1
39	1	1	1	0	90	1	0	1	1	141	1	0	0	0	192	1	1	0	1	243	1	1	0	1
40	1	1	1	1	91	1	0	1	1	142	1	0	1	0	193	1	1	0	1	244	1	1	1	1
41	1	1	1	1	92	1	0	1	1	143	1	0	1	0	194	1	0	0	1	245	0	1	1	1
42	1	0	1	1	93	1	0	1	0	144	1	0	1	1	195	1	0	0	1	246	0	1	1	1
43	1	0	1	0	94	1	1	1	0	145	1	0	1	1	196	1	0	0	1	247	0	0	0	1
44	1	0	1	0	95	1	1	1	0	146	1	1	1	1	197	0	0	1	1	248	1	0	0	1
45	1	1	1	0	96	1	1	0	1	147	1	1	1	1	198	0	0	1	1	249	1	0	0	1
46	1	1	1	0	97	1	1	0	1	148	0	1	1	1	199	0	0	1	1	250	1	1	0	1
47	1	1	1	0	98	1	0	0	1	149	0	0	1	1	200	1	0	0	1	251	1	1	0	1
48	1	1	1	1	99	1	0	1	1	150	0	0	0	1	201	1	0	0	1	252	1	1	0	1
49	1	1	1	1	100	1	0	1	1	151	0	0	0	1	202	1	1	0	1	253	0	1	0	1
50	1	1	1	1	101	1	0	1	1	152	1	0	0	1	203	1	1	0	1	254	0	1	0	1
																				255	0	0	0	1

4-27 Memory format in mask-programmable ROM. Motorola Semiconductor Products, Inc.

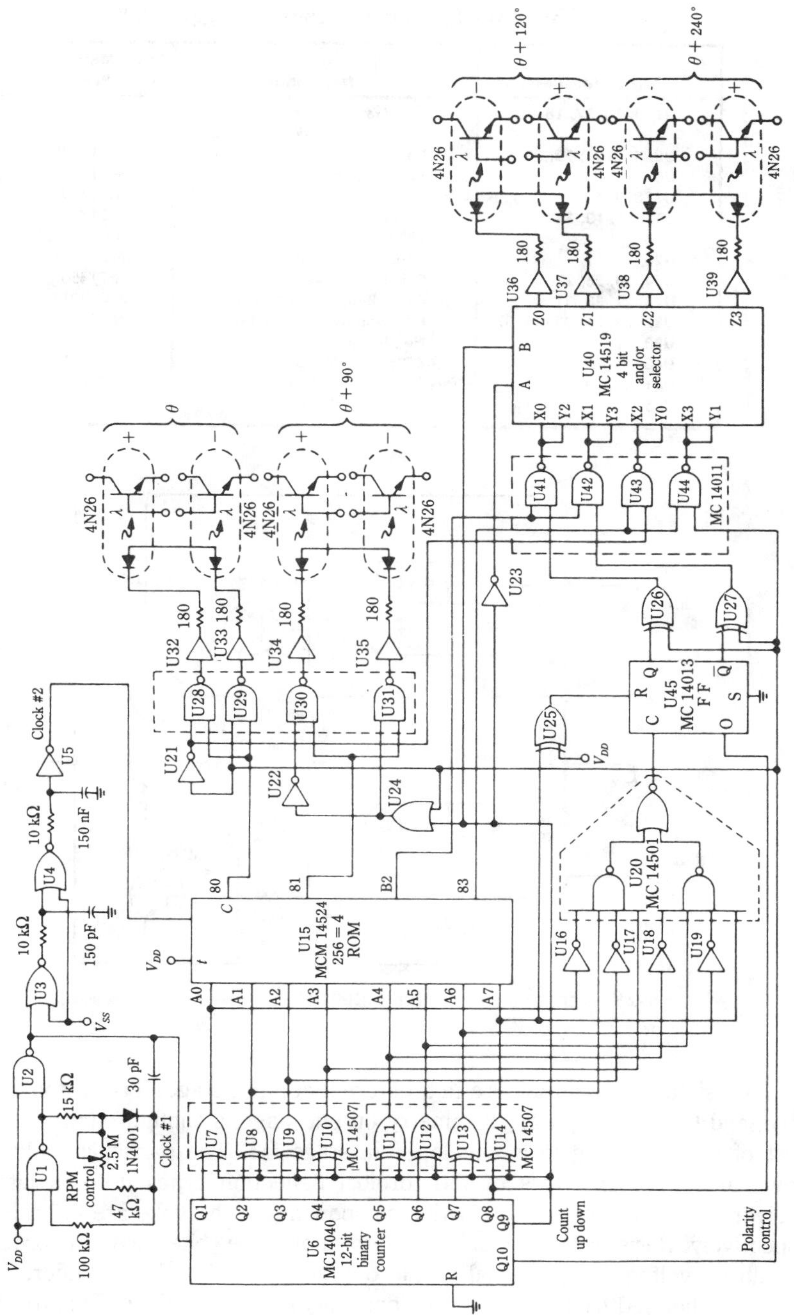

4-28 Logic diagram of digital-logic PWM inverter. Motorola Semiconductor Products, Inc.

Table 4-4 Integrated circuits for Fig. 4-28.

Circuit Identifiers	Description	Motorola Part No.
U1, 3, 16, 17, 18, 19	Hex Gate (1 2-Input NAND, 1 2-Input NOR, 4 Inverters)	MC14572
U2, 4, 5, 21, 22, 23	Hex Gate	MC14572
U6	12-Bit Binary Counter	MC14040
U7, 8, 9, 10	Quad Exclusive OR Gate	MC14507
U11, 12, 13, 14	Quad Exclusive OR Gate	MC14507
U15	1024-Bit ROM	MCM14524
U20	8-Input AND Gate	MC14501
U24, 25, 26, 27	Quad Exclusive OR Gate	MC14507
U28, 29, 30, 31	Quad 2-Input NAND Gate	MC14011
U32, 33, 34, 35, 36, 37	Hex Noninverting Buffer	MC14050
U38, 39	Hex Noninverting Buffer	MC14050
U40	4-Bit AND/OR Selector	MC14519
U41, 42, 43, 44	Quad 2-Input NAND Gate	MC14011
U45	Dual Type-D Flip-Flop	MC14013

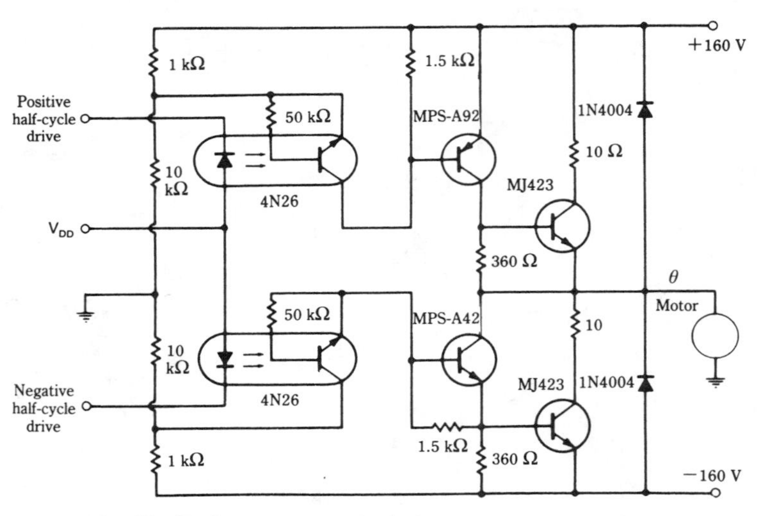

4-29 Single-phase power output stage. Motorola Semiconductor Products, Inc.

Transistors Q1 and Q2 are in a differential input stage. Transistor Q3 is a driver stage, and the contemporary-symmetry output stage is configured around Q4 and Q5. Most of the voltage gain is developed in stage Q3. The particular yoke parameters shown are representatives of the toroidal deflection types used in late-model TV receivers. The inductive and resistive components of the yoke are 2.5 mH and 3.0 Ω, respectively. Resistor R8 (0.5 Ω) is used to sample the yoke current in order to develop a feedback voltage proportional to the output current of the amplifier. The sampled voltage is then fed back to Q2 of the input differential amplifier. This results in a very linear current ramp through the yoke.

Notice that this circuit dispenses with the awkwardly large electrolytic capacitor often used in the output circuit. The overall voltage gain is about 40. The npn power

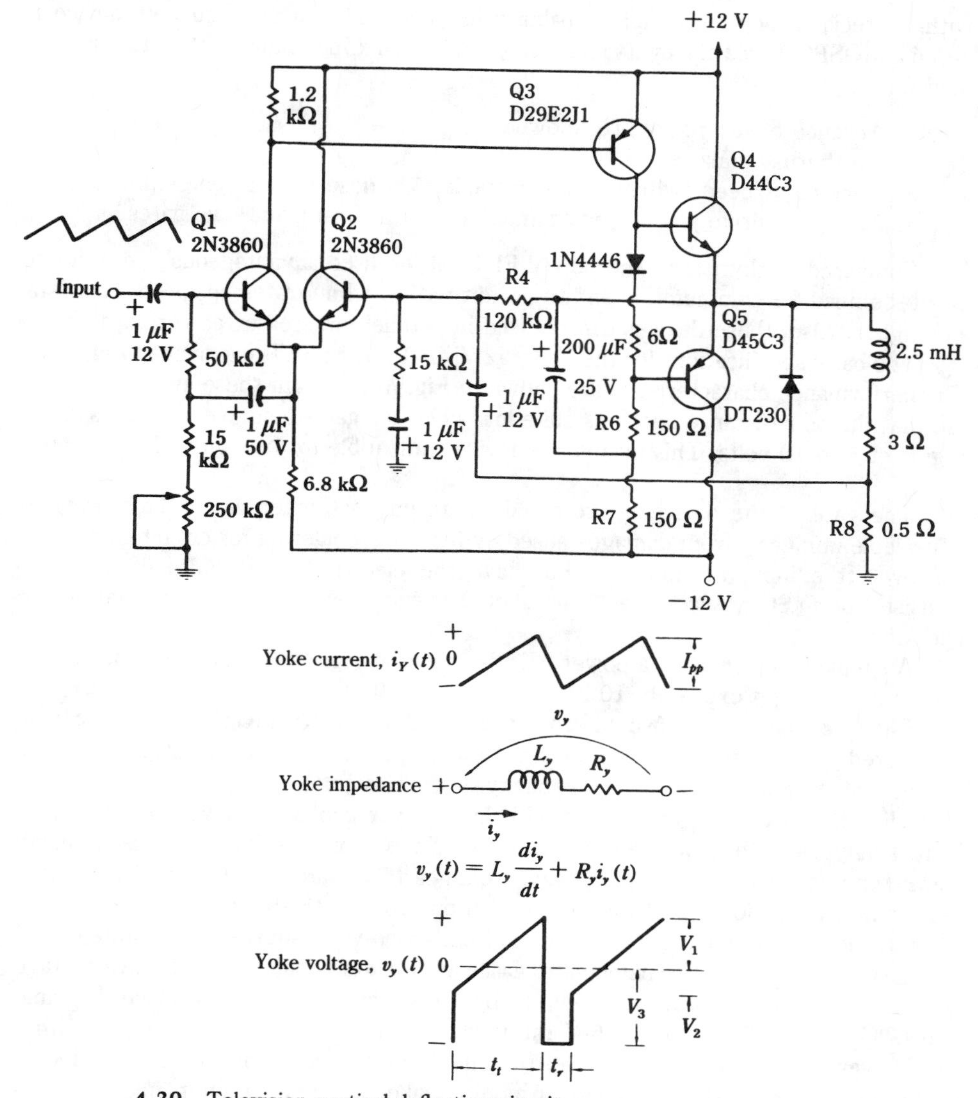

4-30 Television vertical-deflection circuit. General Electric Semiconductor Products Dept.

transistor supplies a peak-to-peak current output of 3.4 A. The pnp power transistor supplies about 4 A peak-to-peak. These transistors require small aluminum heatsinks with an area of approximately 4.5 in.[2]

The power MOSFET

Until recently, field-effect transistors were, because of their nature, considered to be devices for use only in applications involving signal-level power. Now, a type of FET with appreciable power-handling capability has been introduced. The new device is called a MOSPOWER FET by its manufacturer, Siliconix Inc. Significant features of the device are:

1. Absence of secondary breakdown
2. No thermal runaway
3. Very high speed switching, for example, 5 nanoseconds at one ampere
4. Negligible drive power, for example, can be driven from logic gates

Compared to thyristors, the power FET can be used advantageously in order to avoid the need for commutation and to perform at switching rates up to several MHz, and higher. Also, these devices can be used in parallel for greater power capability.

The basic specifications for the Siliconix VMP-1 power FET are given in Fig. 4-31. The performance characteristics are shown in Fig. 4-32. From these curves, you can see that the on resistance becomes quite low when the gate-to-source voltage is in the region of 5 to 10 volts. This makes the device compatible for use with CMOS control logic.

The gate of the device is protected by an internal zener diode. This provides reasonable immunity from damage caused by handling, soldering, or circuit transients. However, the input section does not have the electrical ruggedness of a bipolar transistor or an SCR. Accordingly, care should be exercised in both implementation and operation.

A simple inverter using a power FET is shown in Fig. 4-33. The frequency of the square wave is approximately 10 kHz when C is 100 pF.

Figure 4-34 shows a push-pull configuration. This class-B MOSFET circuit can be considered when either sine-wave or square-wave power is needed. A transformerless output circuit enhances the basic simplicity of the design.

With the advent of transistors with high-power capability, inverters are being increasingly used to power and control electric motors. As has been seen, many transistor inverters deliver square-wave voltages to their loads. The question arises regarding the relationship of the inverter output voltage to the rated motor voltage. Most ac motors are intended for operation from a sine-wave source and the rated RMS voltage is indicated on the nameplate. Calculation shows that a square-wave voltage should exceed an RMS sine-wave voltage by 10 percent in order to produce the same effect. Thus, a motor with a 115-V rating should, theoretically, be impressed with a square wave of 126.5 V. In practice, the effects of eddy currents, hysteresis, inductance, and torque interference from harmonics make experimentation mandatory. With some motors, the 110-percent square causes excessive heating and must be reduced in value.

Adapting an audio amplifier to perform as a driven inverter

Inverters often prove very useful in vehicles and boats. Usually, the basic objective is to convert battery voltage to 60-Hz, 120-V so that a variety of loads can be conveniently accommodated. These include electric shavers, TVs and radios, fluorescent lights, and a variety of appliances. Such inverters tend to be most useful if they can deliver power levels in the 20 to 100-W range. This suggests the use of an audio amplifier equipped with an appropriate output-transformer and fed from a 60-Hz oscillator. The idea is

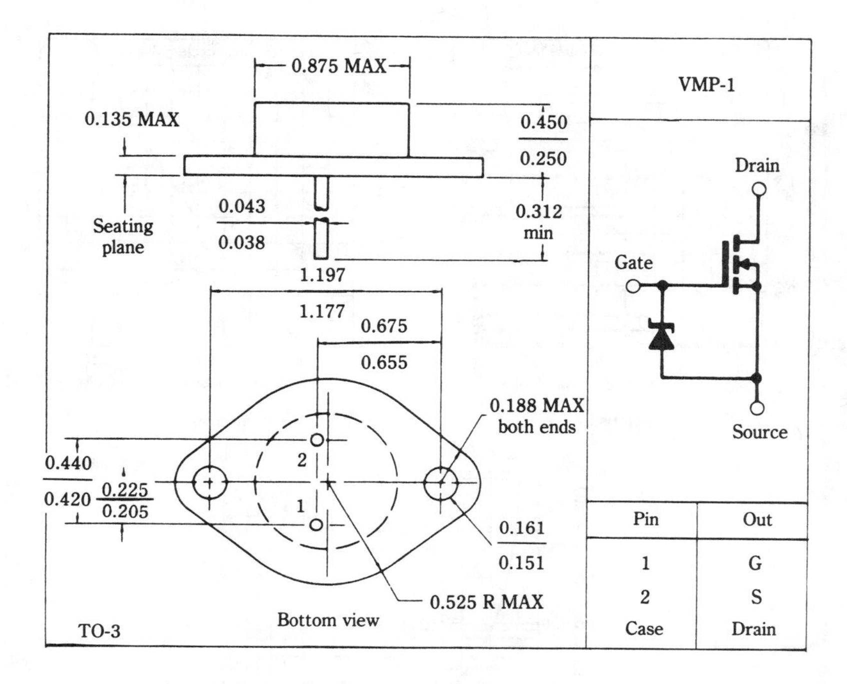

ABSOLUTE MAXIMUM RATINGS

Maximum drain-source voltage	60 V
Maximum drain-gate voltage	60 V
Maximum drain current	2.0 A
Maximum gate (Zener) current	10 mA
Maximum gate (Zener) voltage	15 V
Maximum dissipation @ 25°C case temperature	35 W
Thermal derating factor	3.5°C/W
Temperature (operating and storage)	−55 to +150°C

ELECTRICAL CHARACTERISTICS (25°C unless otherwise noted)

		Characteristics		Min	Typ	Max	Unit	Test Conditions
1	Static	BV_{DSS}	Drain-source breakdown	60			V	$V_{GS} = 0$ V; I_A
2		$V_{GS(th)}$	Gate threshold voltage	0.8		20	V	$V_{GS} = V_{DS}$; $I_D = 1$ mA
3		I_{GSS}	Gate-body leakage			0.5	μA	$V_{GS} = 15$ V; $V_{DS} = 0$ V
4		$I_{D(off)}$	Drain cutoff current			0.5	μA	$V_{GS} = 0$ V; $V_{DS} = 24$ V
5		$I_{D(on)}$	Drain ON current*	1			A	$V_{GS} = 10$ V; $V_{DS} = 24$ V
6		$I_{D(on)}$	Drain ON current*	0.3		1.2	A	$V_{GS} = 5$ V; $V_{DS} = 24$ V
7	SW	$r_{DS(on)}$	Drain-source ON resistance*		1.9	2.5	Ω	$V_{GS} = 10$ V; $I_D = 1$A
8	Dynamic	g_m	Forward transconductance*	200			m℧	$V_{DS} = 24$ V; $I_D = 1$A
9		C_{iss}	Input capacitance		38		pF	$V_{GS} = 0$ V; $V_{DS} = 24$ V
10		C_{rss}	Reverse transfer capacitance		7		pF	$V_{GS} = 0$ V; $V_{DS} = 24$ V
11		C_{oss}	Common source output capacitance		33		pF	$V_{GS} = 0$ V; $V_{DS} = 24$ V

*Pulse test

4-31 Specifications of Siliconix VMP-1 power MOSFET. Later power MOSFETs tend to dispense with Zener gate protection. Siliconix Inc.

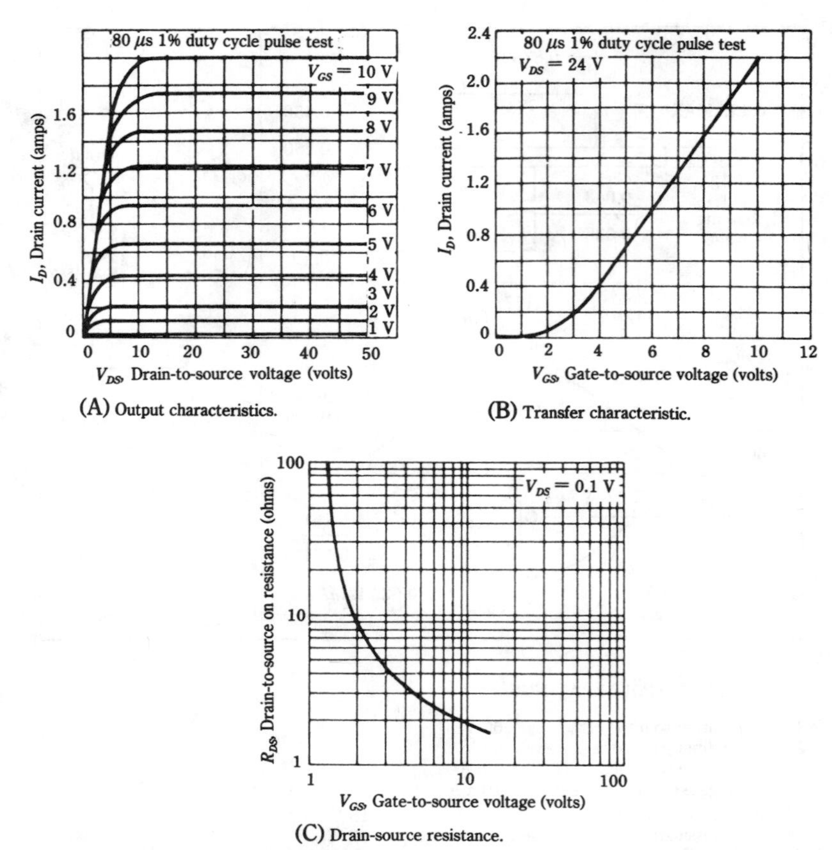

(A) Output characteristics.

(B) Transfer characteristic.

(C) Drain-source resistance.

4-32 Performance curves for Siliconix VMP-1 power MOSFET. Although the VMP-1 has been superceded by more powerful MOSFETs, these characteristic curves are representative of all of such devices. Siliconix Inc.

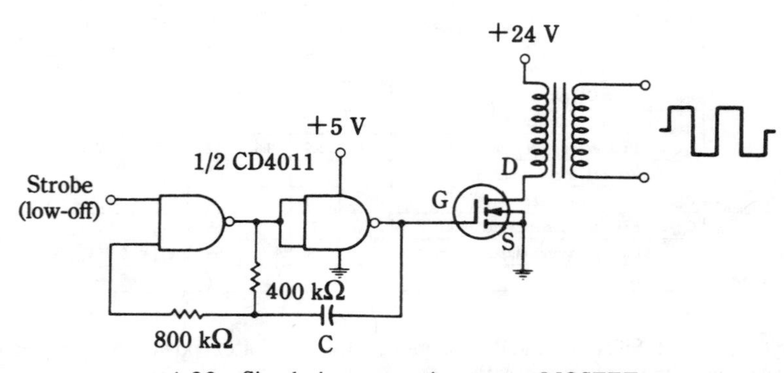

4-33 Simple inverter using power MOSFET.

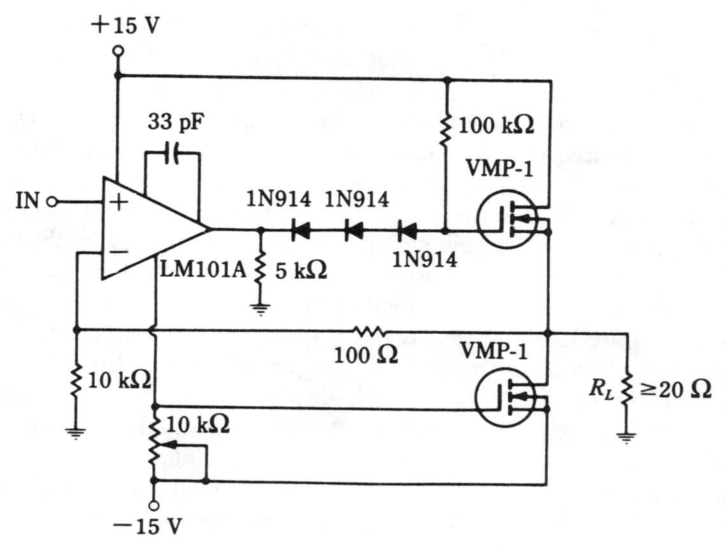

4-34 Push-pull amplifier suitable for use as a driven inverter. Siliconix Inc.

particularly attractive because audio and stereo amplifiers are readily available at low cost from ham auctions and flea markets.

When an audio amplifier is pressed into the type of service described, it becomes a driven inverter. One of the important things to realize is that such a project is best reserved for the experimentally inclined. For, although the use of an audio amplifier in this mode of operation appears to be simple and straightforward, the basic fact is that we really know little about the power capability of a commercial audio amplifier. This is because the final phase of its "design" was done in the sales department, not the engineering laboratory. Thus, a so-called 100-watt amplifier might only be capable of providing 100-watts of output power on infrequent and short-duration peaks of the music waveform. An attempt to coax 100-watts of sustained power for any length of time from the amplifier might well be the prelude to a costly replacement of power transistors and other over-loaded components. On the other hand, some better-quality amplifiers incorporate conservative design which might enable a 100-watt rated unit to actually deliver continuous lower at that level; or if not, such an amplifier might be brought up to its claimed performance by a few minor changes, such as beefing up the heat removal and/or upgrading a few passive components.

The next thing to consider is the output transformer, the primary of which substitutes for the speaker voice coil (most modern amplifiers have either direct feed to the speaker, or interpose a large output-coupling capacitor). An assumption can be made here that works out quite well in practice. This assumption is that a "filament transformer" can be readily obtained that will provide a sufficiently close impedance and voltage ratio to do the job satisfactorily. A wide variety of these transformers are generally available. Secondary voltages of 5, 6.3, 7.5, 10, 12.6, 24, 36, and 48 are common. Inasmuch as these windings are often center-tapped, and because the primary winding can have several taps, the experimenter enjoys a good deal of latitude in determining optimum windings-ratios. Notice that the secondary of the filament transformer becomes the primary when used as an output transformer. It doesn't hurt to select a huskier transformer than is dictated by its power ratings. I^2R heating is not the problem, but magnetic saturation can sometimes be a problem if square waves are used.

Figure 4-35 is representative of commonly-encountered audio amplifiers intended for operation from an automotive battery-system. Instead of being constructed from discrete transistors, many of these amplifiers have been much simplified through the use of dedicated power ICs. The basic circuit of Fig. 4-35 is that of a bridge amplifier. A bridge amplifier is comprised of two push-pull amplifiers driven so that their outputs are phased to be series-aiding—while one end of the load is being driven in the positive direction, the opposite end is being driven in the negative direction. (For stereo operation, two such channels are needed, i.e., two pairs of push-pull amplifiers.) In order to obtain inverter-mode operation, we will deal with just one such channel. The experimenter could double power by getting two channels into the act, but not without considerable attention given to proper phasing and balancing.

The single-channel amplifier of Fig. 4-35 can put about 100 watts into a 4-ohm load. Notice carefully the input and feedback connections of the two Sprague ICs. From

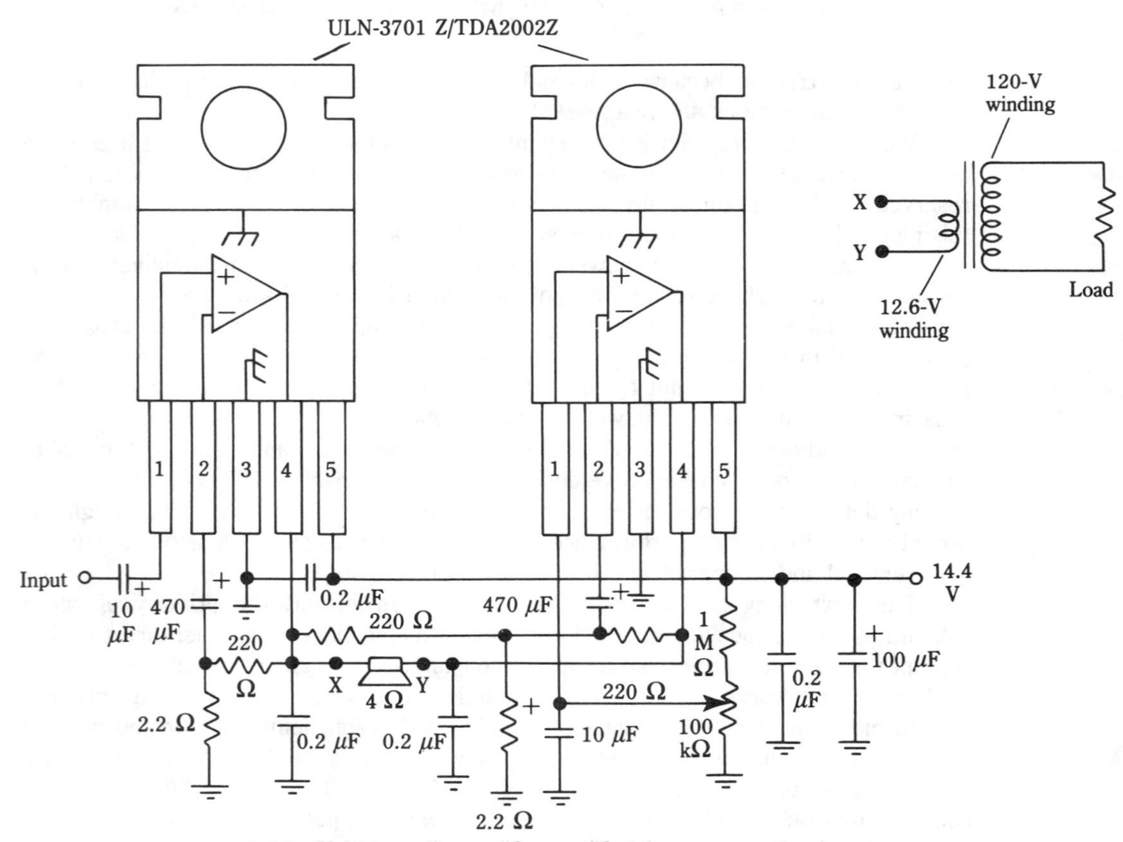

4-35 Vehicle audio amplifier modified for use as an inverter.

op-amp theory, you can see that each IC is set up to develop a voltage gain of 100. The audio input is fed to the left-hand amplifier which does not invert polarity. The output of this amplifier undergoes potentiometric division by a factor of 100 and is then fed to the right-hand amplifier, which functions as a polarity inverter. Because of these connections, the load is driven simultaneously in both positive and negative directions. Load power is twice what it would be with only a single IC. The adjustable resistance is for achieving zero quiescent current with no audio signal.

Our task is to substitute a suitable filament, or output transformer, for the speaker voice coil, and to provide an oscillator or audio signal generator for driving the amplifier at 60 Hz. This will enable the amplifier to become a 120-volt, 60-Hz driven inverter. With adequate heatsinking, about 100 watts of load power can be provided. When driven by a sine-wave signal, an inverter of this type has certain advantages over saturable-core inverters. With the sine-wave operation, RFI and EMI will be negligible. There is no start-up problem. Also, it is much easier to produce a precise and stable frequency. Where it is desired, voltage regulation can be accomplished by fairly straight-forward circuit techniques. Electric motors are usually "happier" when applied with a sine-wave voltage; ordinary square waves contain enough harmonic energy to appreciably increase eddy-current and hysteresis losses in motors, and sometimes to cause torque problems. Finally, the commercial availability of the ICs, and suitable filament transformers greatly simplifies the implementation task.

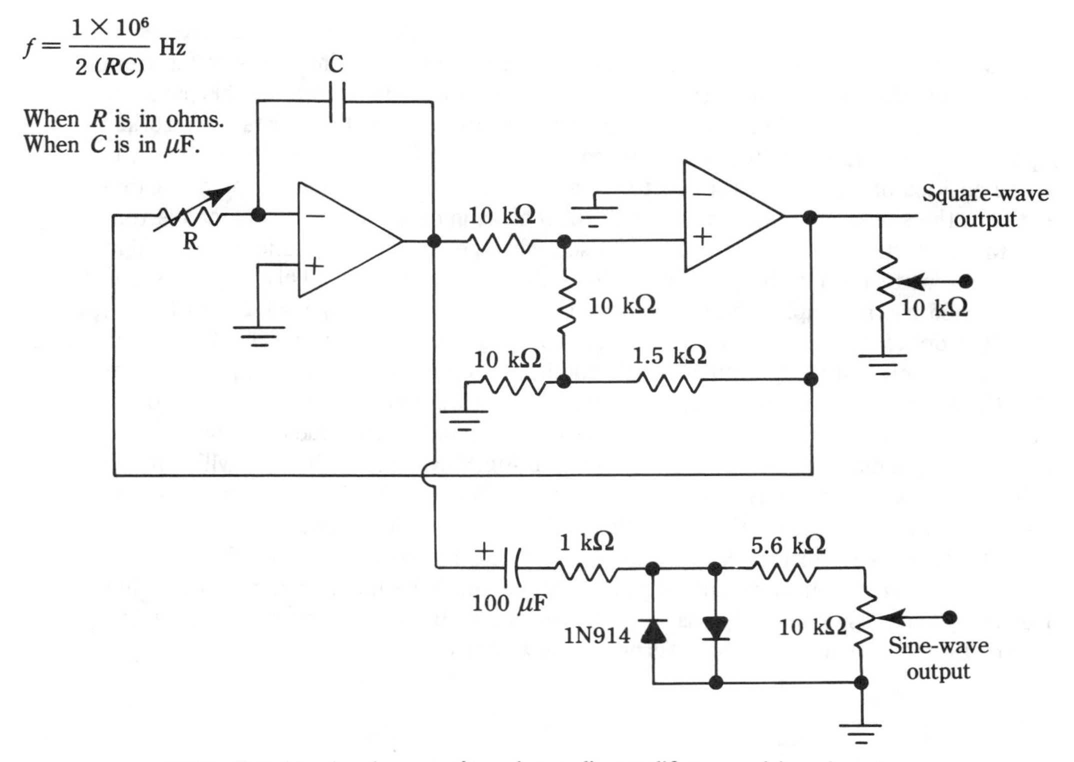

4-36 Suitable signal-source for using audio amplifier as a driven inverter.

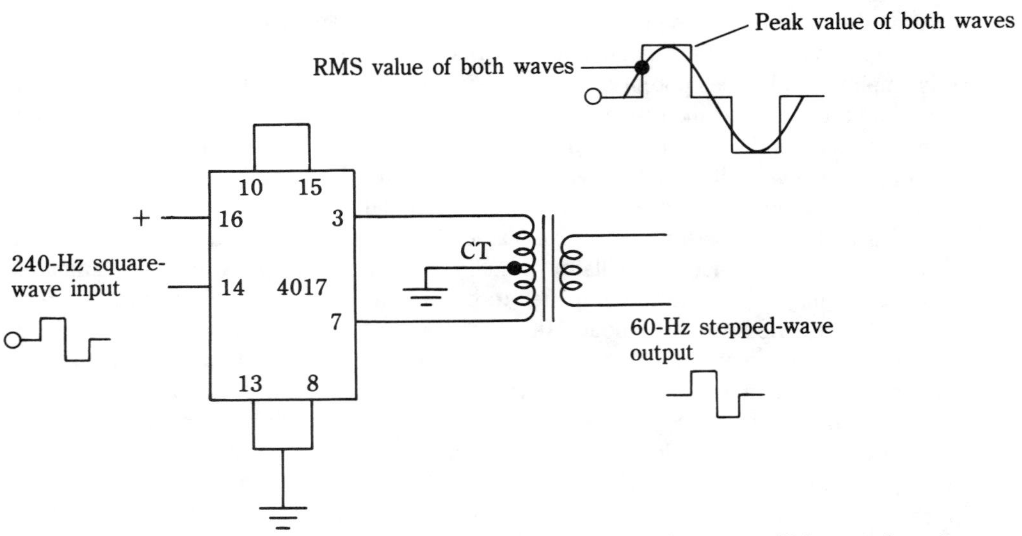

4-37 Unique stepped-waveform, which behaves simlarly to a sine wave. This waveform is likened to a sine wave having the same RMS and peak values. It can be generated by a 4017 decade-counter in conjunction with a small, center-tapped audio transformer.

The audio signal generator shown in Fig. 4-36 is suitable for driving the inverter. It provides choice of either a square- or sine-wave output. The sine wave is far from "pure" from the audio enthusiast's viewpoint, but is quite satisfactory for this purpose. A suitable RC combination for the frequency-selective network consists of a 10-kΩ "pot" and a 1.0-μF capacitor. With this combination, 60 Hz should occur at about the 8-kΩ position of the pot. It is suggested that all preliminary experimentation be carried out with the sine-wave output and at 100 Hz. This is in order to help avoid inadvertent saturation of the transformer. Also, it is wise to first get the inverter functioning well at about 50-watts output, before approaching the 100-watt level. This, of course, is determined by the amplitude of the drive. Use a resistive dummy load during tests.

If a decade counter is inserted between a square-wave source and the driven inverter, an interesting operational mode obtains with the square-wave frequency set at 240 Hz, and when the division factor is four. The 60-Hz stepped waveform at the output of the counter has the unique property that the peak and RMS values are the same as for a corresponding sine wave. This is shown in Fig. 4-37. Thus, the load will (almost) "think" it is being driven by a sine wave. Such a waveform will usually exert less stress on the inverter transistors than an ordinary 50% duty-cycle square cycle.

Whatever waveform is decided upon, it is often a good idea to use 90 or 100 Hz instead of 60 Hz. A bit more output power will generally be forthcoming at the higher frequency and most loads will behave as well as with 0 Hz. Some electric shavers will be speeded up; this will usually be interpreted as a plus feature.

5 CHAPTER

Thyristor inverter and converter applications

THE POWER-HANDLING CAPABILITY OF MODERN THYRISTORS OFTEN COMES AS A surprise to those who have not kept pace with the art. Units with simultaneous ratings of several thousand volts and several thousand amperes have become available. To see a few such door knob-sized devices assume tasks that previously required physically massive apparatus is to get a genuine feeling for technological progress.

Unfortunately, thyristors have some shortcomings along with their impressive voltage and current ratings. The most serious of these is the need for commutation, that is, the need for some means to turn the device off. This is not at all analogous to the turn on and turn off of transistors. Unlike transistors, the thyristor can no longer be controlled by its control electrode (gate) once the device has been triggered into conduction. The only way it can then be turned off is by interruption of the main conduction current. (Some progress has been made in the development of thyristors that *can* be turned off by appropriate gate signals. These have great potential, but thus far they have been successfully developed only for low and moderate power levels. To this extent, they merit consideration as competitors with power transistors.)

Failure to commutate properly can be more than merely a cause of improper performance. It often leads to destruction not only of the noncommutating thyristor, but of associated thyristors and other components as well. And closely allied with turn off failure is inadvertent turn on. It is not always easy to protect against these two modes of faulty performance; you must be able to anticipate the general nature of line transients and load behavior under both usual and unusual circumstances.

Despite the extra care needed in design, and often in application, the use of thyristors in inverters and converters increases steadily. Not only do designers find the ratings and operational efficiency of thyristors attractive, but much confidence has been gained with respect to circuit techniques for securing reliable performance. In addition, the manufacturers have enhanced the predictability of their performance.

SCR substitution for a thyratron tube

For educational and instructional purposes, the SCR is often compared to the thyratron tube. Analogous, if not similar, mechanisms apparently do occur in both the gaseous and solid-state domains. However, usually the devices cannot be freely interchanged in actual practice because the characteristics of SCRs generally are not immediately compatible with thyratron circuitry.

Although direct substitution of an SCR for a thyratron is not practical, combinations of devices can be used to replace thyratrons. The circuit shown in Fig. 5-1 is an example of a solid-state "synthesized" thyratron. It is specifically intended as a direct substitution for the type C3J thyratron, which was often used in inverter and converter circuits.

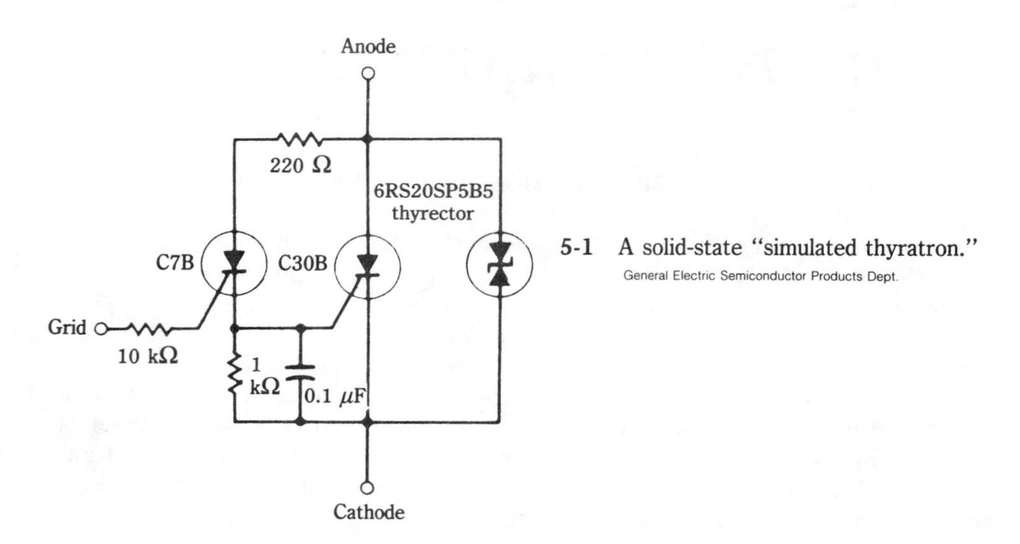

5-1 A solid-state "simulated thyratron."
General Electric Semiconductor Products Dept.

Because thyratrons characteristically have high input impedances and high power amplification, the simulated circuit utilizes a cascaded arrangement with a small SCR driving the medium-current C30B SCR. The 0.1-μF capacitor and the thyrector are incorporated in the circuit to reduce the possibility of inadvertent triggering from transients or from high rates of change of voltage. The circuit can be connected to the base of an old C3J tube and simply plugged into the socket that previously received the thyratron.

The power amplification displayed by this circuit is much higher than ordinarily prevails with SCRs that have the load-current control capability of the Type C30B. For example, the simulated thyratron triggers with less than 20 μA of "grid" current. The negative input voltage that the old thyratron inverter or converter might supply is perfectly acceptable to the simulated circuit. Indeed, such negative voltage will provide added protection against false triggering.

Transformerless dc power supply

The circuit shown in Fig. 5-2 might well occupy a gray area, with respect to its qualifying as a true converter. However, it takes ac from the power line, rectifies it, chops it, and finally converts it to steady dc for the operation of low-power equipment that requires 10 to 15 V at 100 mA or so. Such a sequence of functions corresponds to that performed by conventional converter-type power supplies. This circuit is unique in that no core components at all are needed. Although the regulation is not outstanding, being dependent mainly on the size of filter capacitor C1, the efficiency can approach 80 percent. For applications in which cost, size, and weight are more important than regulation, this interesting circuit merits consideration.

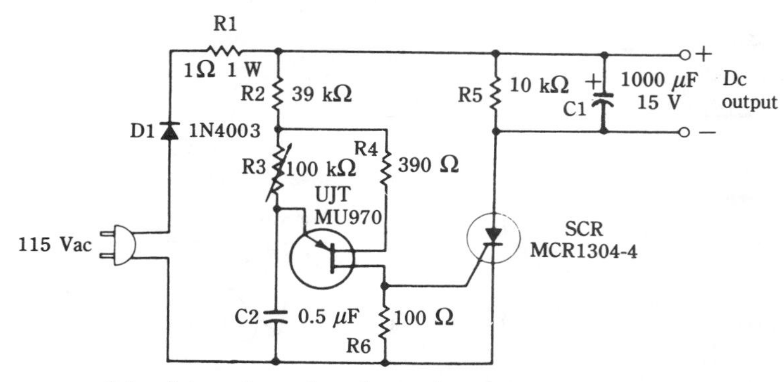

5-2 A transformerless dc supply. Motorola Semiconductor Products, Inc.

The UJT oscillator is synchronized to the 60-Hz line because of the discrete half sine waves provided by rectifier D1. Therefore, the UJT generates 60-Hz pulses, despite the time constant of its emitter RC network. The RC network does, however, govern the time of occurrence of the UJT pulses relative to the zero crossings of the 60-Hz input voltage. Thus, adjustment of R3 determines at what point in the input cycle the SCR will be triggered into conduction. With early triggering, the integrating action of the filter capacitor, C1, will develop a high average dc output voltage. The converse is true when R3 is adjusted for late triggering of the SCR. Ideally, this method of voltage manipulation involves no dissipation because the SCR is either fully on or completely off.

Commutation of the SCR is inherent in the nature of the circuit—the half-wave rectification provides voltage reversal and 180° of recovery time. At 60 Hz, no commutation or spurious triggering problems should be experienced. This scheme has interesting possibilities at higher frequencies, where the requisites of filtering and regulation are more easily met with practical-sized output capacitors. A disadvantage of this circuit for some applications is that there is no isolation from the power line.

A driven parallel-type SCR inverter

The SCR inverter shown in Fig. 5-3 is "driven" in the sense that it is not self-excited. Unlike the case in driven transistor inverters, however, the drive consists of narrow

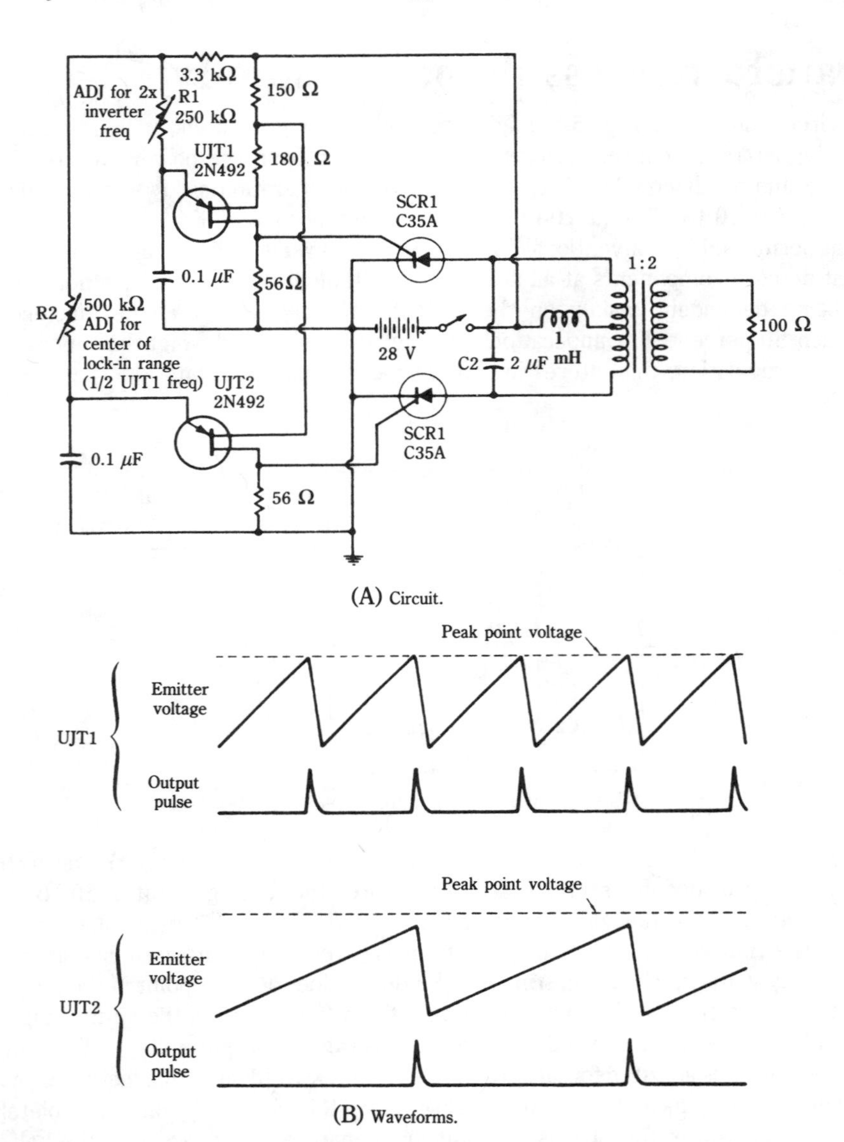

5-3 A driven push-pull SCR inverter. General Electric Semiconductor Products Dept.

trigger pulses. Once an SCR is triggered at its gate, it regeneratively makes the transition from its off to its on state. Subsequent triggers (if available) can produce no further effect. The SCR remains in its conductive state until it is turned off by a pulse received through the commutation capacitor, C2, when the alternate SCR turns on. This seems straightforward enough, yet there is more to the operation of this 60-Hz 112-W inverter than initially meets the eye.

It is obvious that two synchronized unijunction transistor oscillators act as the trigger source for the SCRs. Notice, however, that UJT1 has *twice* the oscillation frequency of UJT2. Specifically, UJT1 has a pulse-repetition rate of 120 Hz; UJT2 then divides this by two and delivers a pulse-repetition of 60 Hz. A quick appraisal of this situation could very well be that the SCRs would tend toward erratic or unstable operation in attempting to respond to two different triggering rates. Fortunately, this is not the case; stable performance at 60-Hz output is one of the salient features of this unique approach to inverter design.

The effect of the divide-by-two trigger source is the production of triggers appropriately displaced in time to produce proper push-pull inverter action at 60 Hz. It is true that every other 120-Hz trigger pulse serves no useful purpose. On the other hand, it fortunately happens that the useless trigger causes no undesired response. It appears at the gate of SCR1 about the same time that this SCR is receiving a commutation pulse from SCR2, which has just turned on from a 60 Hz trigger. The commutation pulse takes precedence because it momentarily deprives SCR1 of its forward anode-cathode voltage. So, there is no conflict with regard to the state of SCR1 — it turns off. The next 120-Hz trigger pulse appearing at the gate of SCR1 will turn it on (and simultaneously turn off SCR2 via commutating capacitor C2). Summarizing, SCR1 is turned on by every other 120-Hz trigger pulse from UJT1, and SCR2 is turned on by every 60-Hz trigger pulse from UJT2. Each SCR is commutated to its "off" state when the alternate SCR is turned on. It should now be evident that this frequency-division technique provides precisely timed trigger pulses to cause the inverter to produce a 60-Hz square-wave output.

Actually, UJT1 is adjusted for 120 Hz, and the "natural" frequency of UJT2 is adjusted by means of R2 to be about 10 percent lower than its intended 60-Hz rate. Under such circumstances, UJT2 will lock onto the 120-Hz synchronizing frequency and thereby deliver exactly half of the sync rate, or 60 trigger pulses per second. In this circuit, 90 percent overall efficiency is attained with no saturation in the transformer.

Capacitor-discharge ignition system with SCR energy switch

The circuit shown in Fig. 5-4 is the second solid-state ignition system to be covered. In the ignition circuit of Fig. 4-15, the focus of interest was on the transistor inverter. In this circuit, it will be more relevant to direct major attention to the SCR energy switch. Note that both circuits use transistor inverters, and both also use SCRs to deliver the energy stored in a capacitor to the primary winding of the ignition coil. Otherwise, the circuits are quite different.

The circuit of Fig. 5-4 uses a straightforward two-transistor saturable-core inverter (considering the association of the inverter with the bridge rectifier, you can just as well designate the circuitry associated with transistors Q1 and Q2 as a converter). About 175 Vdc is developed across capacitor C1. This, however, is only indirectly the source of the electrostatic energy that is "dumped" into ignition coil. When the system is in operation, choke L1 becomes part of a resonant circuit involving capacitor C2. However, because of diode CR1, neither a sustained nor a damped train of oscillations can occur. Instead, only an initial quarter-cycle of the attempted oscillation can get under way. This suffices to double the voltage developed across capacitor C2 to about

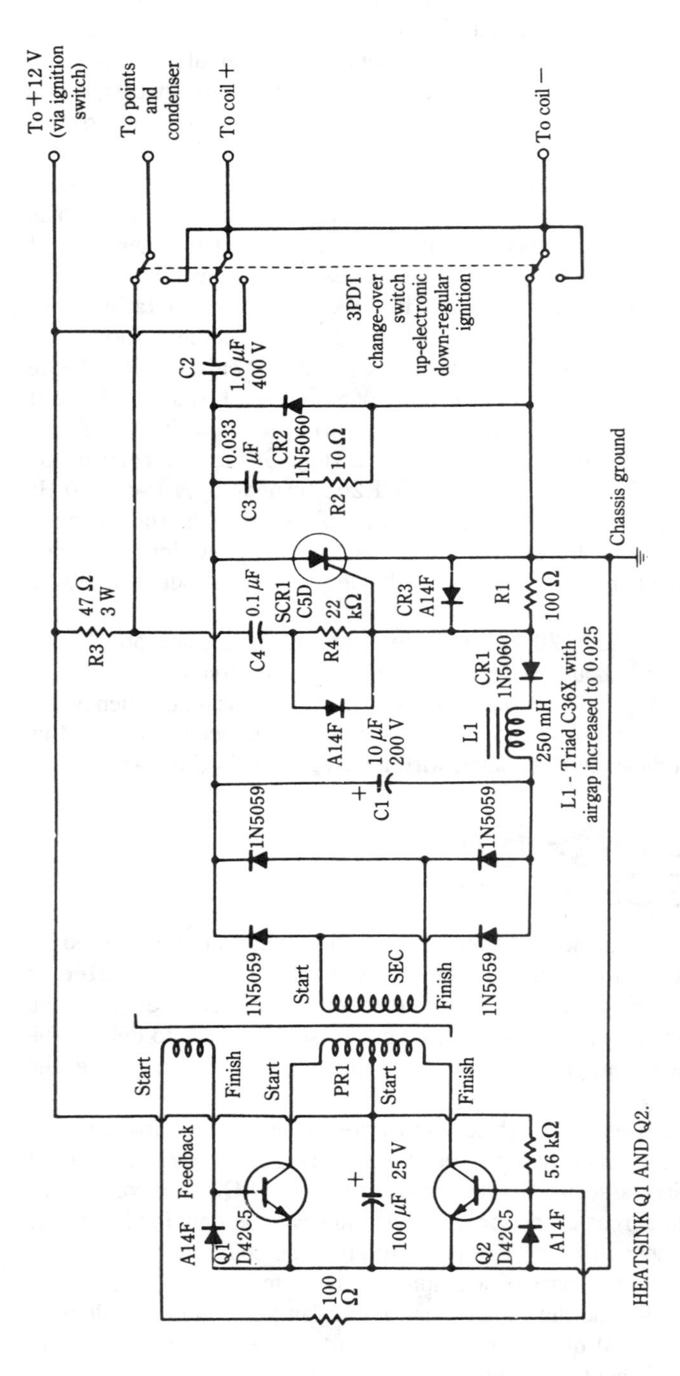

5-4 Capacitor-discharge ignition system with SCR energy switch. General Electric Semiconductor Products Dept.

350 V. This charge is then trapped in C2 by the action of the stand off diode, CR1. When the breaker points open, capacitor C4 is free to charge from the battery; it is this charging current that triggers the gate of the SCR. The SCR then dumps the energy from storage capacitor C2 into the primary of the ignition coil. The proper spark plug is fired through the same distributor action that prevails in conventional (Kettering) ignition systems.

This sequence of events might appear a bit unusual when first encountered. However, the technique is straightforward and is not new. Those with radar backgrounds should recognize it as an adaptation of the dc resonant-charging circuits used in radar modulators. The salient feature of this method of voltage doubling is that it can be operated over a very wide range of repetition rates. This is desirable in both the radar and the auto-ignition applications.

When the breaker points open, the SCR remains on long enough to fire a spark plug. However, the SCR is part of a resonant circuit formed by capacitor C2 and the primary inductance of the ignition coil. At the first reverse-voltage overswing of this resonant "tank," the SCR turns off. Thus, commutation is a consequence of the circuit arrangement itself.

SCR series-type inverter in an induction cooking unit

The basic phenomenon of induction heating in metals is not new; it has been used for decades in industrial processes, such as case-hardening of gears and other machine parts. Hitherto, however, the concept of adapting it to the heating of cooking vessels has been greatly hampered by obstacles involving cost, size, and reliability. Such inverter devices as rotary machinery, thyratrons, ignitrons, or vacuum tubes could be operated and maintained in industrial applications, but they were hardly appropriate in the home. And, thus far, power transistors have proven too expensive for this use.

Shown in Fig. 5-5 is an experimental, but practical, circuit for an induction cooking unit. From a functional approach, three basic sections can be identified. First is the SCR series inverter involving SCR2. In this circuit, shock-excited oscillation is produced in the series "tank," L1/C4, when the SCR is caused to switch on and off periodically. The oscillation is not sustained, but is produced in single-cycle "bursts" (as illustrated in Fig. 5-6). The triggering rate of the SCR is in the vicinity of 18 kHz, just above the threshold of audible sound. The resonant frequency of L1C4 is higher, in the neighborhood of 35 kHz. The difference between the drive and LC-resonating frequencies is necessary to allow reliable commutation of the SCR.

Actually, L1 is the transducer. It consists of a spiral-wound copper-tubing coil with an inductance of 6 μH. The metallic cooking pot is placed above this coil and becomes heated by eddy current. If the pot is iron or steel, hysteresis also contributes to the heating. The copper coil remains relatively cool. The capacitance of C_4 is 3.45 μF in order to resonate with L1 at 35 kHz.

In Fig. 5-6, the first, or positive, half of each individual cycle is caused by current through the LC tank circuit and SCR2. As the current reverses, however, the SCR is turned off. The continuance of the cycle through its negative excursion is then caused

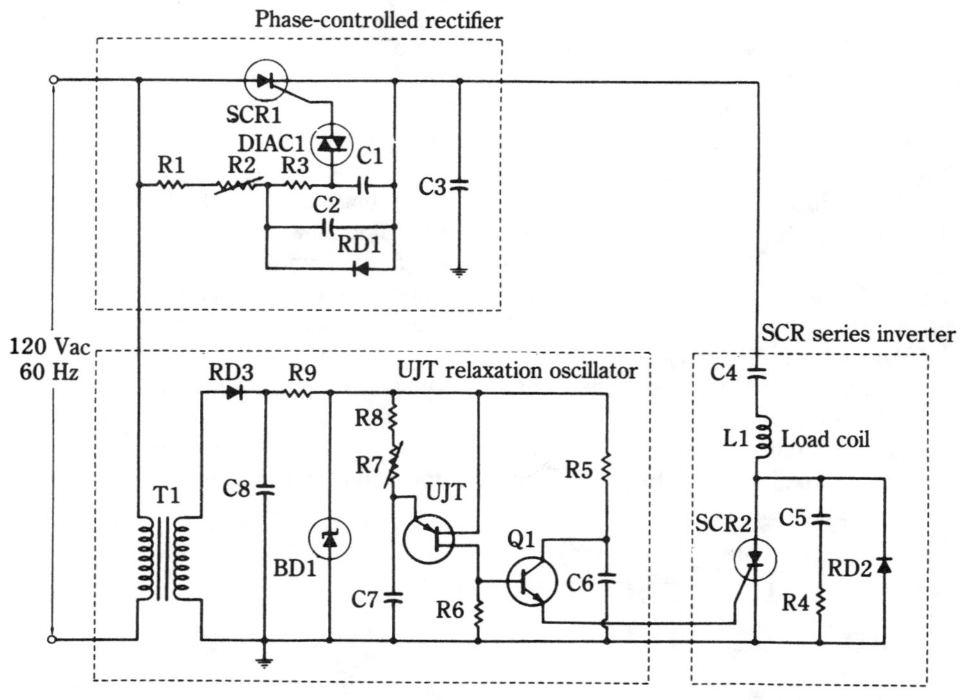

5-5 Induction cooling circuit with SCR series-type inverter. International Rectifier Corp.

by current through C4/L1 and diode RD2. Thus, SCR2 and RD2 cooperate to produce a single complete cycle. Prior to a subsequent gate trigger, the SCR is provided a suitable recovery interval to ensure reliable commutation. That is why there is a time gap between the two cycles in Fig. 5-6.

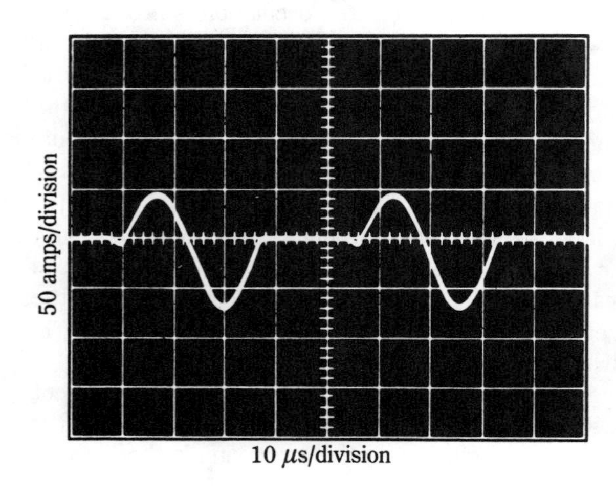

5-6 Current waveforms in the tank circuit of Fig. 5-5. International Rectifier Corp.

Resistor R4 and capacitor C5 form a snubber circuit to decrease the rate of voltage change across the SCR. This, too, is to help bring about reliable commutation and prevent false triggering.

The second section of the induction cooker is the UJT oscillator. Its function is to deliver a train of trigger pulses to the gate of SCR2. The pulse-repetition rate is about 18 kHz. This pulse rate can be attained by using a 0.01-μF timing capacitor for C7. Suitable values for R7 and R8 would then be 10 Ω and 1 Ω, respectively. Emitter follower Q1 provides current amplification for the UJT pulses. General-purpose signal-level transistors such as the 2N697 or the 2N3565 are suitable. The power supply for the UJT oscillator consists of T1, a 25-V filament transformer; half-wave rectifier RD3; C8, a 1000-μF filter capacitor; and a zener regulator network. The zener diode is a 25-V 3-W unit. Resistor R9 is a 150-Ω 5-W wire-wound unit.

The third section of the induction cooker is the main power supply, a phase-controlled rectifier that allows continuous control of its output from substantially zero to about 130 Vdc. Thus, a small variable resistance, R_2, can be used to adjust the amount of heat induced in the cooking pot. Half-wave rectification and output controllability are provided by SCR1, and the unidirectional output pulses are smoothed by filter capacitor C3. The dc thereby obtained is entirely suitable for the purpose of shock-exciting the resonant tank. This phase-controlled rectifier uses a double-time-constant trigger circuit in order to secure operating features relevant to the practical requirements of cooking equipment. For example, there should be freedom from the annoying "hysteresis" effect that plagues single-time-constant trigger circuits. (The earlier inexpensive light dimmers often displayed this characteristic—a given adjustment setting did not always produce the same light intensity.) Additionally, double-time-constant control provides wider adjustment range—from near zero to maximum output.

The experimenter might wish to use a full-wave rectifier in conjunction with a variable autotransformer. Such an arrangement would be justifiable from a performance standpoint, but it would not be competitive on a cost basis with the phase-controlled scheme shown.

Thyristor horizontal-deflection system for TV receivers

The horizontal-deflection system of a TV set is at the same time an inverter and a converter. As an inverter, it transforms the energy from a low-voltage dc source to the precisely "tailored" current sawtooth needed to produce the lateral scanning motion of the electron beam in the picture tube. As a converter, it develops high-voltage dc for the ultor and focus electrodes of the picture tube. The horizontal-deflection system is also more or less intimately associated with other important functional blocks of the TV receiver, such as the convergence circuitry, the color killer, the burst amplifier, and the AGC section. The reproduction of a good-fidelity video image on the picture-tube screen is strongly dependent on the performance of the horizontal-deflection system. It is also true that, because of the relatively high power levels together with high peak values of currents and voltages, the reliability of a TV set is considerably dependent on the reliability of the horizontal-deflection system

Until recently, the trend of solid-state designs had been essentially to substitute transistors for tubes. This was a natural approach and has resulted in the development of power transistors that are both technologically advanced and low in cost. However, it is now realized that the basic tube-oriented system possessed an inherent flaw—the current wave delivered to the deflection yoke was highly dependent on the operation of active devices. That is why even an "economy-model" black-and-white set had so many rear-chassis adjustments.

The horizontal-deflection system shown in Fig. 5-7 utilizes thyristors rather than power transistors, and it produces a deflection current wave that is dependent for its shape and timing integrity primarily on passive components. The active devices are ITRs, integrated thyristor (SCR) and rectifier units. These were developed after evaluation of SCR inverters revealed the frequent inclusion of a "free-wheeling" diode in the circuits. The development of a single device not only reduced manufacturing costs of TV receivers, but also eliminated the detrimental effects of connecting-lead inductance. (The circuit of Fig. 5-7 can be implemented with separate SCRs and fast-recovery rectifiers if their connecting leads are kept short. This was done in many TV sets prior to the advent of the ITR.

Many of the subtle details of the circuit operation will not be covered here; you should consult a specialized treatise on television techniques if you desire more details. The salient operational features of this deflection system will be outlined, however. One interesting aspect of Fig. 5-7 is its simplicity compared with horizontal-deflection systems of the past.

In order to see more clearly what goes on in the circuit, refer to the simplified circuit shown in Fig. 5-8. This minimal layout can represent either discrete SCRs and rectifier diodes, or ITR devices. One SCR-diode combination is designated the *commutating switch,* whereas the other similar combination is termed the *trace switch.* The trace switch is active for the production of the linear current ramp that scans the electron beam laterally across the screen of the picture tube. As you might infer from its name, the commutating switch is involved with the "off," or retrace, interval.

The sequence of events in the horizontal-deflection yoke is shown in Fig. 5-9. Those circuit elements involved in the formation of each part of the waveform are indicated. The current ramp that impels the spot across the screen is designated i_y. This ramp must be linear, and it must have a specific rate of rise aid peak amplitude. The repetition rate of these ramps is that of the pulses delivered by the horizontal oscillator (15,750 Hz for monochrome, 15,734 Hz for color).

In analyzing the circuit, remember that diodes become conductive upon the application of a forward-polarized voltage, and that the same is true of SCRs with the proviso that a trigger pulse must also be applied to the gate. Notice that the current-path arrows of Fig. 5-9 indicate the conventional direction of current, from positive to negative.

The trace portion of the sawtooth is the result of oscillatory action in the yoke inductance, L_y, and capacitance, C_y. The retrace portion of the sawtooth is primarily the result of oscillatory action in inductor L_R and capacitor C_R. Both oscillatory actions are limited to the initial part of a cycle—opportunity is never provided for a buildup of sustained oscillations.

As is often the case when cause-and-effect relationships produce a net overall

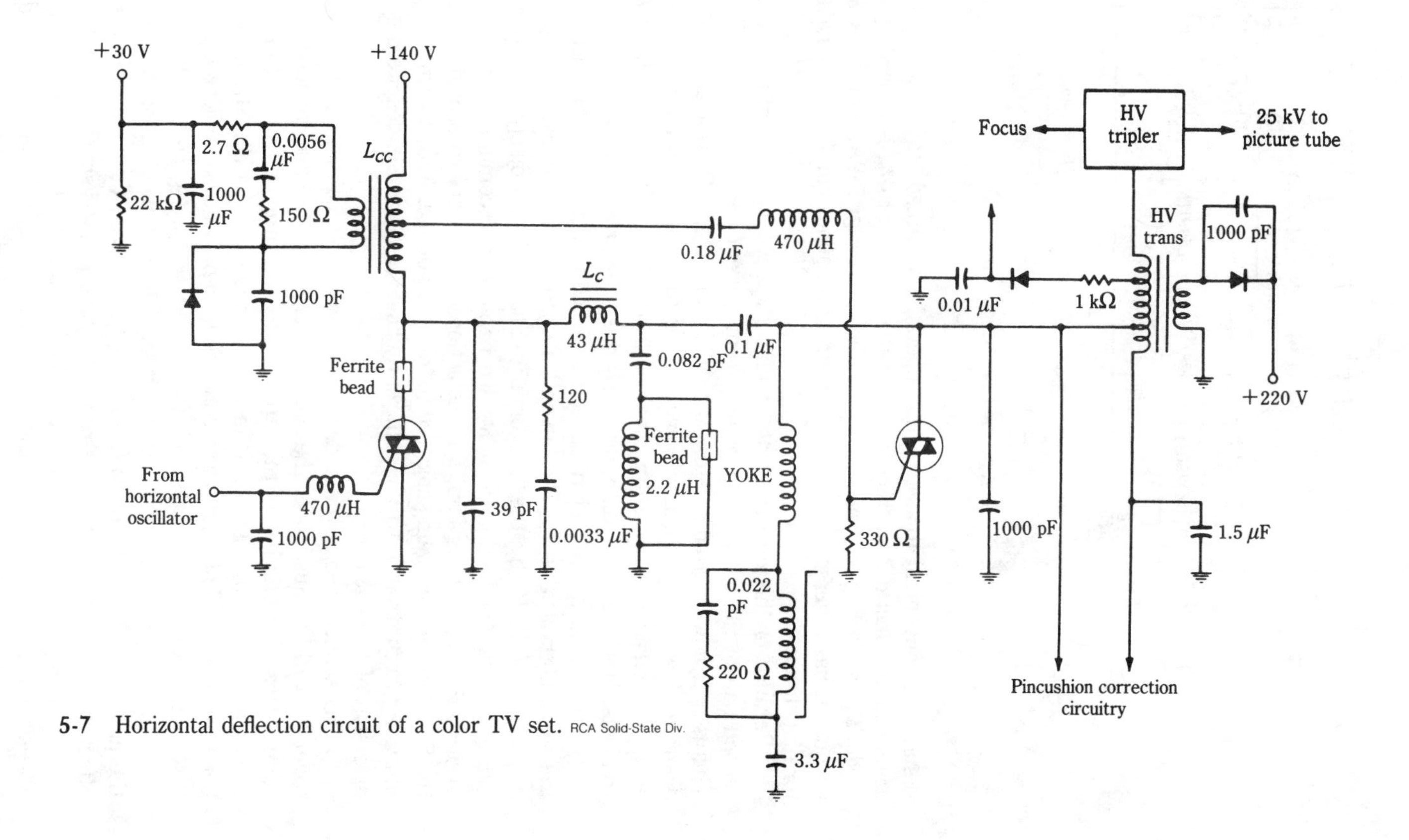

5-7 Horizontal deflection circuit of a color TV set. RCA Solid-State Div.

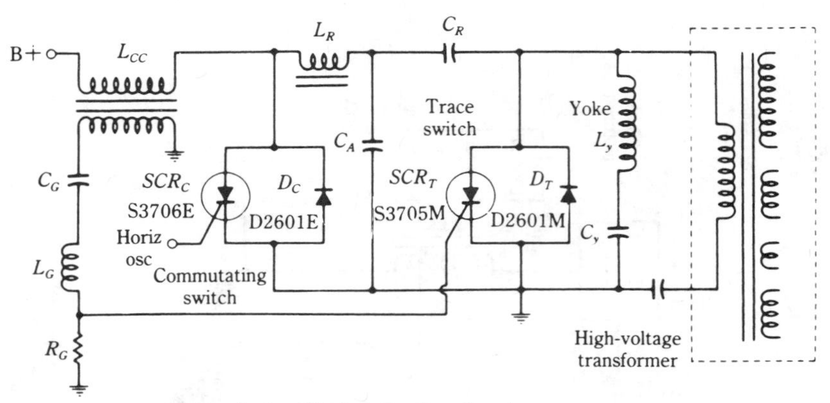

5-8 A simplified deflection circuit. RCA Solid-State Div.

result, it is convenient to assume that operation is already taking place. That being the case, the first half of the trace ramp, that is, from t_0 to t_2, is caused by conduction of diode D_T in the trace switch. This is shown in Fig. 5-9A, where SCR_T, is deleted because it has reverse voltage applied across its anode-cathode terminals and is therefore inactive.

Because of the zero crossing at time t_2, it would be natural to assume that the second half of the trace ramp results from current conduction of SCR_T in Fig. 5-9B. This is not, however, completely true. If it was, diode D_T would be deleted from this diagram. There is an additional event prior to attainment of time t_5. Because of this event, the final ramp current is diverted to the SCR_C circuit, and diode D_T once again conducts. However, this time, diode D_T does not provide a path for the yoke current.

This complication is clarified in Fig. 5-9C, which shows the current paths during the time interval t_3 to t_5. The "event" referred to in the preceding paragraph is the triggering of SCR_C by the horizontal oscillator. The objective is to commutate, or extinguish the conduction in SCR_T. To be sure, the commutation pulse occurs at t_3, prior to the end of the ramp, t_5. This action is necessary to provide recovery time for the SCR before it is again subjected to forward voltage. And, it turns out that the ramp of trace current from t_0 to t_5 is sufficiently linear and precise, despite this "last-minute" diversion of current paths.

The sequence of events described thus far results in commutation of SCR_T, together with exhaustion of the energy stored in the yoke inductance. Retrace now commences. The first half of the retrace time (the interval from t_5 to t_6) now occurs, as depicted in Fig. 5-9D. The SCR in the commutating switch, SCR_C, having previously been triggered at time t_3, remains conductive to provide the indicated current path. The resonant frequency of the "tank" (consisting of L_R, C_R, and L_y) now establishes the ramp time. Accordingly, the interval from t_5 to t_6 is much shorter than, for example, the portion of the trace ramp from t_0 to t_2. (Because C_y is much larger than C_R, the effect of C_y on the overall series-resonance frequency is small enough to be neglected.)

The second half of the retrace cycle (Fig. 5-9E) involves the time interval t_6 to t_0. The situation shown can be considered an extension of the action started in Fig. 5-9D. When the retrace ramp crosses the zero axis at time t_0, the current path shifts from

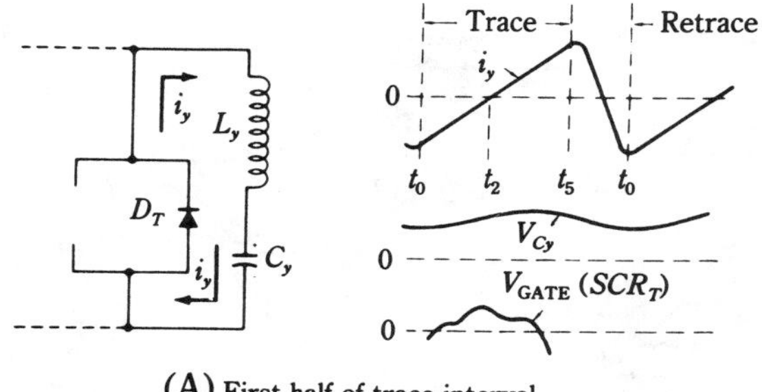

(A) First half of trace interval.

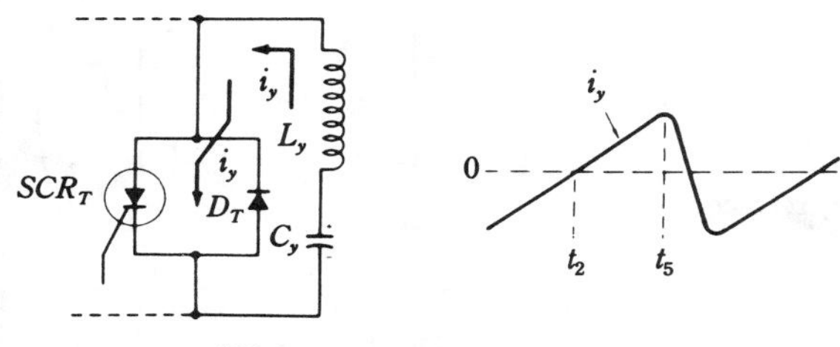

(B) Last half of trace interval.

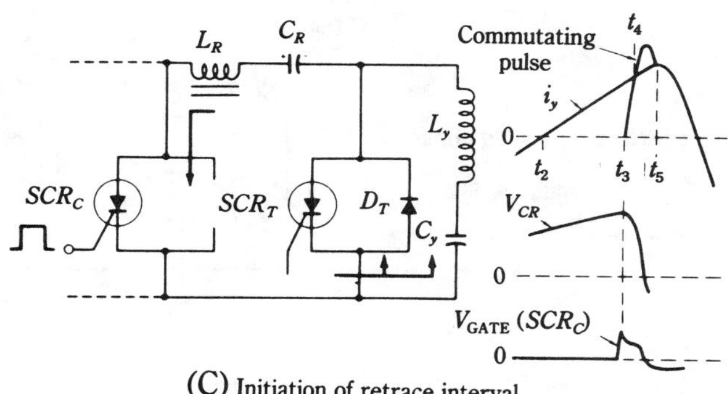

(C) Initiation of retrace interval.

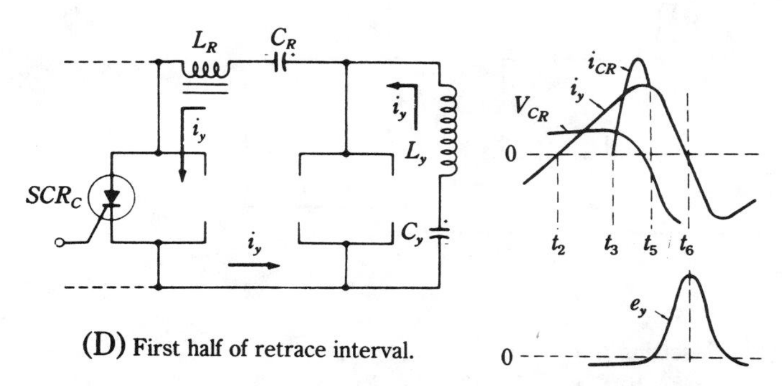

(D) First half of retrace interval.

5-9 Waveforms and equivalent circuits for a horizontal deflection circuit. RCA Solid-State Div.

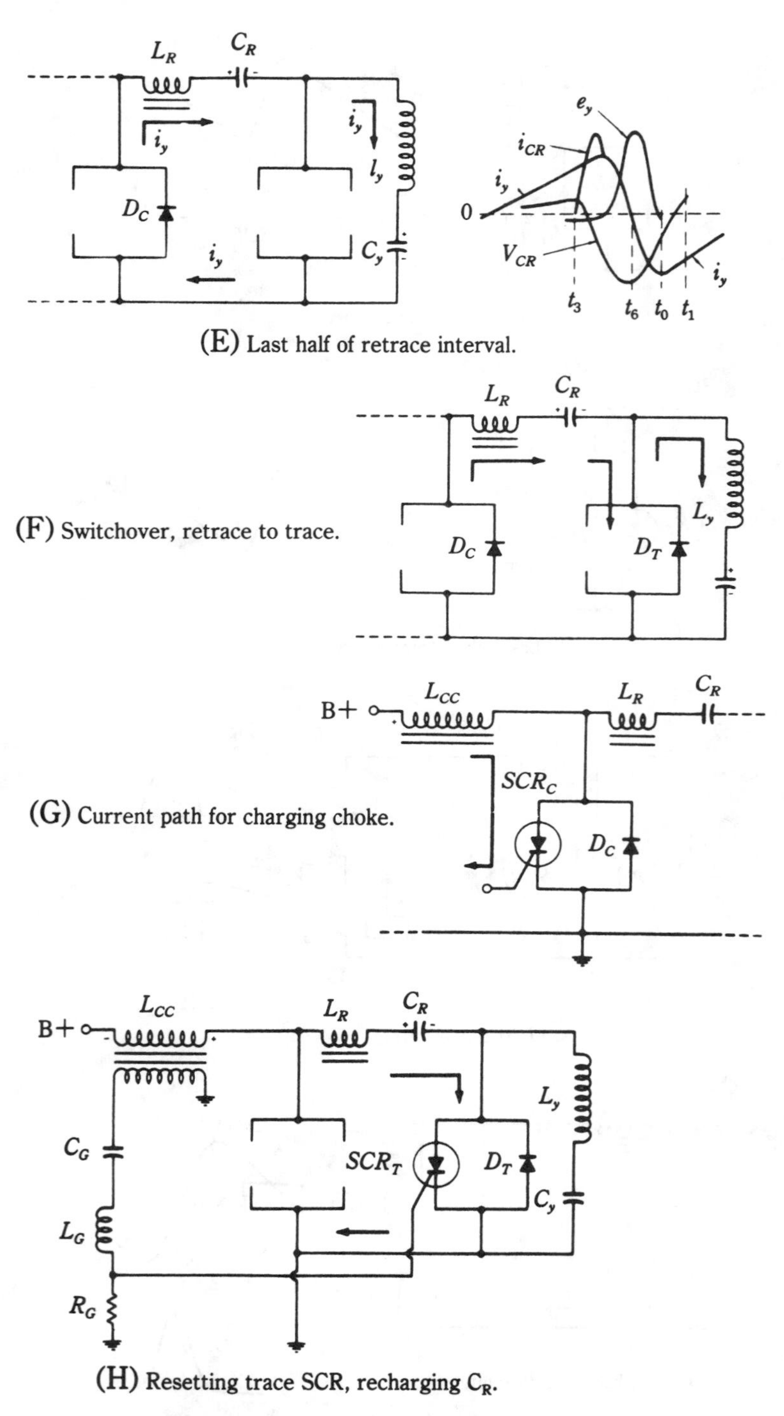

(E) Last half of retrace interval.

(F) Switchover, retrace to trace.

(G) Current path for charging choke.

(H) Resetting trace SCR, recharging C_R.

5-9 Continued

SCR_C to D_C. This is because of the polarity reversal that occurs place then, imposing reverse bias on SCR_C and forward bias on D_C.

As the portion of the retrace ramp from t_6 to t_0 terminates its excursion, the situation shown in Fig. 5-9F exists. This is the retrace-trace switchover, which occurs at t_0. For a relatively brief time, diodes D_C and D_T both conduct. At the next instant—that is, at the commencement of the trace ramp, t_0 to t_2—the situation again reverts to the circuit condition originally depicted in Fig. 5-9A.

The events indicated in Fig. 5-9 account for the formation of the current sawtooth in terms of four distinct segments. Each segment, it was seen, is characterized by a unique combination of conductive and nonconductive switching elements. However, the system is not self-oscillatory. In order to maintain its cyclic operation, circuit actions in addition to those described are also necessary.

From Figs. 5-9D, 5-9E, and 5-9F, it can be seen that there is current through inductor L_R during the entire retrace interval (t_5 to t_0). For half of this time interval, the L_R current path is completed through SCR_C; for the remaining half, the path is through D_C. The net result is storage of energy in the magnetic field of inductor LR. When the retrace ramp terminates at time t_0, the commutating switch (SCR_C and D_C) turns off. The magnetic field built up in inductor L_R then collapses, inducing an EMF that charges capacitor C_R. This charging process continues through the trace period until time t_4 is reached. Thus, the energy that will be needed to produce a new retrace ramp, t_5 to t_0, is stored in capacitor C_R well ahead of time. If this energy was not available, circuit operation would simply stop at the end of the trace ramp, t_5.

One more circuit action must be delineated to account for the repetitive production of the current sawtooth waveform. In Fig. 5-9H, the gate circuit of SCR_T is shown connected to a secondary winding on charging inductance L_{CC}. Because of this connection, SCR_T is made ready for conduction by the time it also receives forward anode-cathode voltage at time t_2. This situation will be made clear by reviewing Figs. 5-9A and 5-9B. In particular, the presence of the gate signal, V_{GATE}, is shown in the wave diagrams of Fig. 5-9A. Without this resetting of SCR_T, circuit operation would stop at t_2.

The generation of high-voltage dc stems from the very rapid action ("flyback") of the retrace ramp in its excursion from t_5 to t_0. The pulses of induced EMF that appear across the secondary winding of the high-voltage transformer are at a much higher voltage level than would be the case if the primary winding was energized with a sinusoidal wave. As shown in Fig. 5-7, still greater "step-up" of the voltage level is imparted by a dc voltage multiplier, a voltage tripler in this case. Insofar as the high-voltage dc circuit is concerned, the horizontal-deflection system described thus far is a straightforward flyback converter. Often, the high-voltage dc circuit is associated with a voltage regulator in order to stabilize the size and brightness of the picture in the face of such variables as temperature and ac line voltage.

A flasher with adjustable on and off times

Flashers are unique implementations of inverter techniques and are widely used in vehicular, aircraft, and maritime applications, and for danger warnings. The circuit shown in Fig. 5-10 is quite versatile because the on and off times can be adjusted

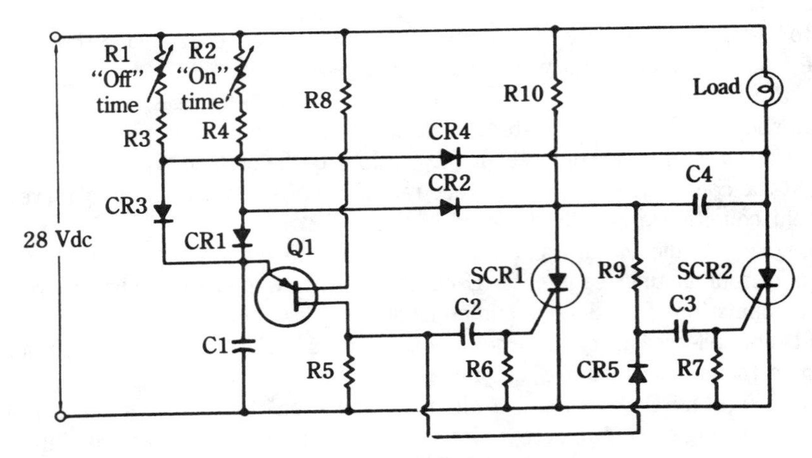

5-10 A flasher with adjustable on and off times. General Electric Semiconductor Products Dept.

independently. The capacitor-commuted flip-flop built around SCR1 and SCR2 is, in essence, a driven inverter. Drive comes from the trigger pulses generated across resistance R_5 by Q1, a UJT oscillator. Although such an SCR flip-flop triggered from an RC relaxation oscillator might appear to be a simple, even primitive, switching circuit, this flasher has some interesting subtleties and sophistications.

For example, it is necessary that only one SCR be triggered when the flasher is initially set in operation. This requisite is provided by the network composed of R9, C3, and CR5. It might appear that either or both SCRs could initially turn on. However, when the dc is first applied, the anode of SCR1 becomes positive, as does the cathode of diode CR5 (through R9). Thus, CR5 is reverse-biased and will not provide passage for a gate-trigger pulse for SCR2. On the other hand, the first pulse generated by UJT Q1 finds both the anode and gate conditions favorable for the triggering of SCR1. Therefore, SCR1 turns on; in so doing, it charges capacitor C4 through the lamp load. Notice that the SCR2 anode side of capacitor C4 becomes positive, while the other side of the capacitor is brought very nearly to ground potential. This is a prelude to the commutation of SCR1, which occurs with the turn on of SCR2.

The turn on of SCR1 changes two other circuit conditions: the reverse bias on diode CR5 is removed, and the R2/R4 charging path for timing capacitor C1 is bypassed because of the forward biasing of diode CR2. Therefore, capacitor C1 commences its charge cycle through path R1/R3/CR3. Delay, or "off" time (while sufficient voltage is being developed across C1 to actuate the UJT) is adjustable by means of R1. When the UJT generates another pulse, the state of the flip-flop circuit is favorable for the turn on the SCR2. That event then dumps the stored charge in commutating capacitor C4 across the SCR1, thereby turning that stage off. In turn, a new charging path for timing capacitor C1 is produced: CR4, now being forward biased, bypasses the R1/R3 charging path, and C1 commences its charge cycle through R2, R4, and CR1. The "on" time is set by R2.

The next trigger pulse then reverses the conductive state of the SCR flip-flop, and the action described becomes repetitive. A list of parts values for this flasher is given in Table 5-1.

Table 5-1 Parts list for Fig. 5-10.

R1, R2 – 500K Linear Potentiometer
R3, R4 – 750K, ½ W
R5 – 100Ω, ½ W
R6, R7 – 1K, ½ W
R8 – 270Ω, ½ W
R9 – 4.7K, ½ W
R10 – 250Ω, 5W
C1 – 0.47μF, 50 V
C2, C3 – 0.22 μF, 50 V
C4 – 4 μF, 50 V Nonpolarized
Q1 – GE 2N2646 UJT
SCR1, SCR2 – GE C106F
CR1-CR5 – GE A14F
Load – GE 50 C 1.4 A Lamp

Flasher using a gate turn-off SCR

Silicon-controlled rectifiers with gate turn-off characteristics have been available for a number of years. However, their applications have been somewhat limited because of low power ratings and because of lack of confidence in their circuit reliability. The gate turn-off feature is desirable because it would impart to inverters and kindred switching circuits such attributes as simplicity, low cost, operational flexibility, and freedom from the complications associated with the commutation process.

Semiconductor technology is making steady progress with such devices. Whereas they were more or less limited to signal-level power a few years ago, there are now gate-turn-off (GTO) SCRs that can switch more than a kilowatt with better than 95 percent efficiency at 20 kHz. These devices are switched on in the same manner as conventional SCRs. However, the application of a gate pulse that is negative 70 V, with respect to the cathode interrupts their conduction. As with turn on, turn off is regenerative.

A good way to get the feel of this device is to experiment with the flasher shown in Fig. 5-11. This flasher can handle 1 kilowatt of incandescent-lamp load when operated from a 220-Vdc source. When you consider the current surge accompanying the initial application of power to such a load, the intrinsic ruggedness of these devices can be appreciated.

This flasher is well suited to such purposes as advertising and display. Because intense illumination is obtainable from lamps operating at this power level, it is probable that this flasher can be put to good use as a warning light for hazardous locations or situations.

When the circuit is first connected to its dc source, timing capacitor C1 begins to charge. When it develops sufficient voltage across its terminals to break down diac D1, the GTO SCR is gated on, and the lamp load is energized. Now, capacitor C2 begins to accumulate charge until its voltage breaks down diacs 2 and 3. The resultant voltage pulse thereby applied to the gate is about -70 V, with respect to the cathode of the GTO SCR. This turns the device off, depriving the lamp load of current. Then the cycle

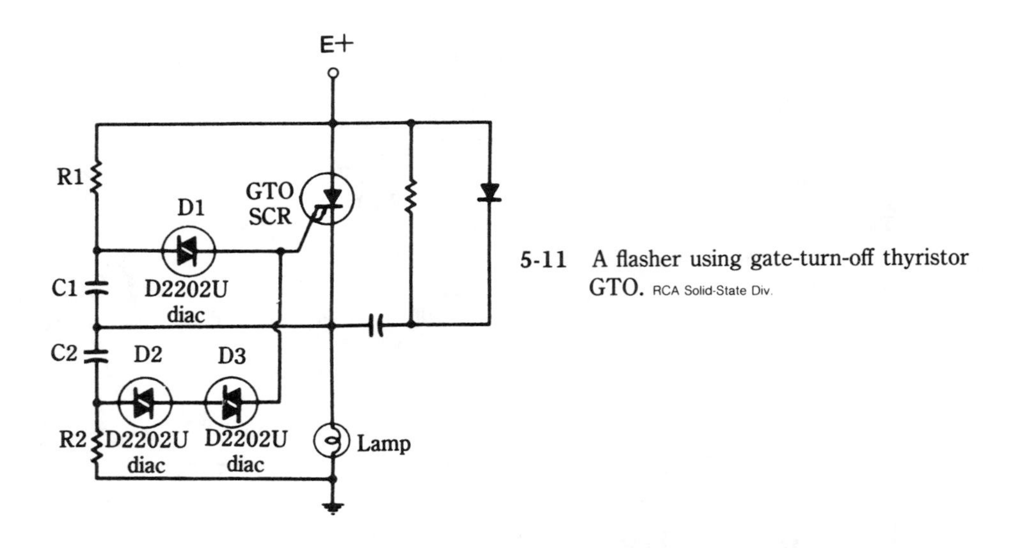

5-11 A flasher using gate-turn-off thyristor GTO. RCA Solid-State Div.

repeats. As a starter, C_1 and C_2 can each be about 0.47 μF; R_1 and R_2 can initially be in the vicinity of 6.8 MΩ. Use 100 Ω and 0.1 μF in the "snubbing" network.

A chaser with independent stage delays

A chaser differs from a flasher in that a turned-on lamp remains on until the last lamp has undergone its illumination cycle. Then, all lamps are extinguished and the entire series is free to repeat the turn-on sequence. In advertising applications, the effect is the illusion of a sign or display in motion. Similarly, automobile turn indicators utilizing chaser-operated rear lamps tend to be more attention-getting than simple flashing displays. Previously, the "chasing" was implemented by mechanical switching techniques. Such applications were straightforward enough, but they suffered from reliability problems. The switching of "cold" incandescent lamps at frequent intervals is not always a simple problem. These difficulties can be overcome by using solid-state electronic circuitry.

Figure 5-12 depicts a simple, but effective, iterative technique for cascading any number of stages to form a self-resetting flasher. Each stage consists of a programmable unijunction transistor (PUT) and an SCR. The PUT simulates the operation of a unijunction transistor (UJT). However, the PUT has a major advantage over the UJT in that its *intrinsic stand-off ratio,* rather than being fixed as with a UJT, is selectable by means of a pair of resistances. Such resistances are the 100-kΩ and 200-kΩ units shown connected to the anode gate of the first D13T1. This is tantamount to saying that these resistances determine V_p the peak-point voltage of the PUT. It will be recalled that V_p is the voltage at which the "emitter" of a UJT triggers.

Despite the analogy between the two devices, the PUT is structurally quite different from the UJT. The PUT is actually an SCR in which the anode gate, rather than the cathode gate, is brought out as the control electrode. It is surprising that the PUT can usually be substituted for the UJT, considering their internally dissimilar

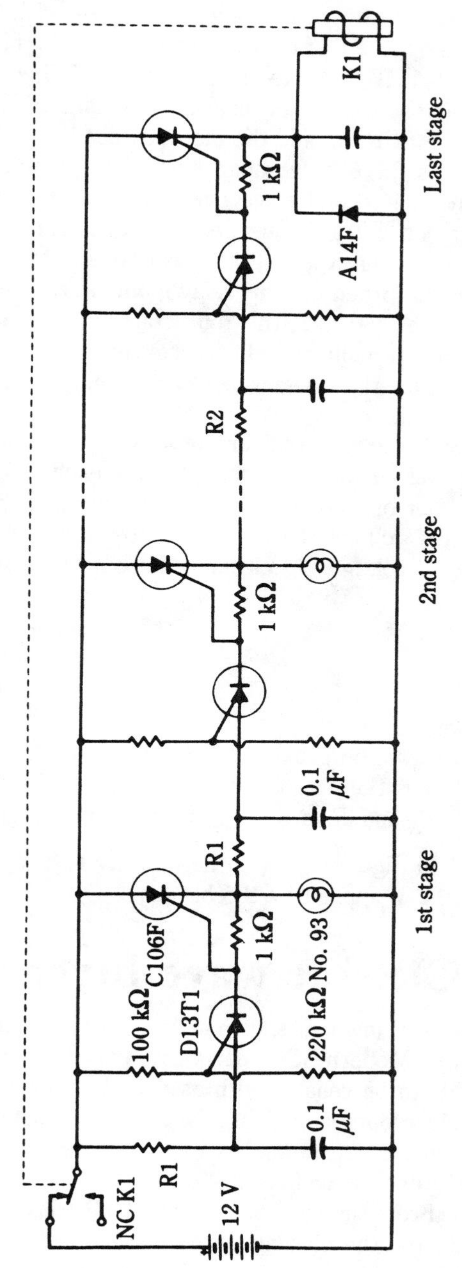

5-12 A chaser with independent stage delays. General Electric Semiconductor Products Dept.

structures. For example, the PUTs in Fig. 5-12 are connected as relaxation oscillators. In each stage, the RC network analogous to the charging circuit of conventional UJT oscillators consist of R1 in conjunction with the 0.1-μF capacitor.

Suppose that operating power has just been made available from the 12-V battery. The 0.1-μF capacitor in the first stage begins to charge; when it develops sufficient voltage across its terminals, the D13T1 PUT fires. In so doing, the PUT triggers the C106F SCR on, and the incandescent lamp of the first stage is illuminated. This event enables the 0.1-μF capacitor of the second stage to begin its charging cycle. Accordingly, the second stage eventually becomes active in the same manner described for stage one. The sequence progresses through the entire cascade of stages. Notice, however, that the last stage operates relay K1, rather than a lamp load. When the last stage is turned on, this relay momentarily interrupts the dc power source, whereupon all stages return to the "off" state. Because of this resetting, relay K1 is de-energized, and its normally closed contacts reconnect the dc power. Then, the 0.1-μF capacitor of the first stage begins a new charging cycle, and the entire sequence of events is repeated.

Notice that the turn on of one stage does not affect the circuit condition of the preceding stage. The delay time of each stage can be chosen independently if desired. Another operating feature of this circuit is that timing is substantially independent of battery voltage. This is because the firing voltage of the PUTs is a fixed fraction of the applied voltage. Specifically, these devices fire at a voltage, V_p, given by the formula:

$$V_p = \frac{E \times R_A}{R_A + R_B}$$

Where:

V_p is the peak-point (firing) voltage,
E is the dc voltage applied across the gate network,
R_A and R_B are the resistances in the gate network.

In Fig. 5-12, R_A is the 220-kΩ resistance, R_B is the 100-kΩ resistance, and E is the 12 V applied by the battery.

SCR sine-wave inverter

The SCR inverter shown in Fig. 5-13 is capable of generating sine waves at high power levels. Modern SCRs permit frequencies as high as 30 kHz and beyond. Commutation tends to be reliable with essentially resistive loads. The UJT oscillator is adjusted to produce optimally sinusoidal output. This will occur at a frequency determined by the circuit LC constants, but not at the resonant frequency calculated from $1/(2\pi\sqrt{LC})$ as with tuned amplifiers.

From the fact that a single UJT relaxation oscillator is used to trigger both SCRs, and from the phasing of the pulse-transformer windings, it is evident that both SCRs receive trigger pulses simultaneously. Inasmuch as the probability is remote that both SCRs will simultaneously commence current ramps at the same rate, you can assume that one of them initially displays higher activity than the other. Assume that SCR1

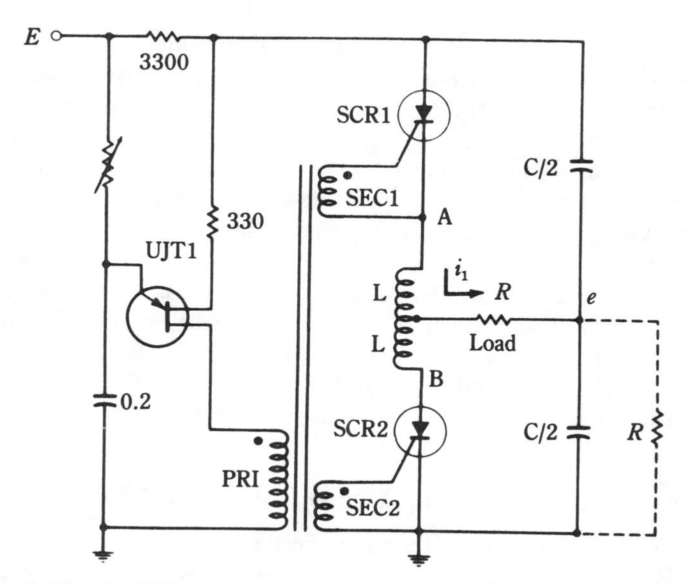

5-13 An SCR sine-wave inverter. General Electric Semiconductor Products Dept.

responds more strongly to a trigger pulse than does SCR2. That being the case, current i_1 will be present in the SCR1 half of the center-tapped inductor (the conventional direction of i_1 is indicated on the diagram). By autotransformer action, this will induce a voltage across the opposite half of the inductor, and this voltage will be so polarized as to back-bias SCR2. Thus, the SCR which initially turns on most vigorously—SCR1 in our assumed situation—immediately establishes circuit conditions such that only it can respond to the first trigger pulse.

The firing of SCR1 causes the lower capacitor C/2 to charge from the dc supply. But this back-biases SCR1, thereby turning it off. The next trigger pulse then finds SCR2 in a favorable circuit condition for triggering. This is because point B becomes positive through autotransformer action when SCR1 is turned off. Alternate turn-on and turn-off of the two SCRs occurs in this way, with the turned-on SCR always reinforcing the turned-off condition of the other SCR.

An operating frequency is given by the approximation:

$$f \approx \frac{1}{2\pi}\sqrt{\frac{1}{LC} - \frac{R^2}{L^2}}$$

This is the frequency of the shock-excited oscillation that would occur if the circuit action allowed ringing between trigger pulses. This equation for resonance differs from the simpler form, $1/(2\pi\sqrt{LC})$, encountered in communications circuits where the Q factor is much higher. In parallel-resonant circuits that have low Q, the frequency is quite dependent on the resistance in the circuit. The designation of the capacitors as $C/2$ is a strategem used to simplify mathematical treatment of the resonant circuit.

Figure 5-13 shows that the load, R, can optionally be connected across one of the resonating capacitors. When this is done, point e should be connected to the center tap of the inductor.

With the first method of connecting the load (R shown with dashed lines), the load resistance can be anything from infinity down to a certain minimum value. Insufficient load resistance will over-damp the resonant circuit and thereby prevent proper commutation of the SCRs.

With the second method of connecting the load (R shown with dashed lines), the load resistance can be anything from infinity down to a certain minimum value. Insufficient load resistance will over-damp the resonant circuit and thereby prevent proper commutation of the SCRs.

A convenient procedure for obtaining sinusoidal output is to monitor the voltage developed across the load and initially adjust the UJT oscillator for a very low frequency. Spaced output pulses will then be produced, but these will come together as the triggering frequency is increased. A triggering rate can be found that will result in the optimal approach to an ideal sinusoid of voltage across the load.

A 1-kW 10-kHz sine-wave inverter

Although the circuit in Fig. 5-14 bears a superficial resemblance to the one in Fig. 5-13, its operation is quite different. The unique features of this circuit are the "center tap" provided for the dc supply, and the fast-recovery diodes connected antiparallel to the SCRs. Two power supplies are generally used with this type of inverter. However, in Fig. 5-14, this need is circumvented through the use of the two 50-μF capacitors. Dynamically (at 10 kHz), these capacitors form the equivalent of a center tapped, or dual, source of operating power. For the following description, it will be helpful to refer to the waveforms in Fig. 5-15.

In this circuit, the SCRs are triggered alternately, rather than simultaneously, as in Fig. 5-13. Suppose that SCR1 is triggered. This causes the 4.7-μF resonating capacitor

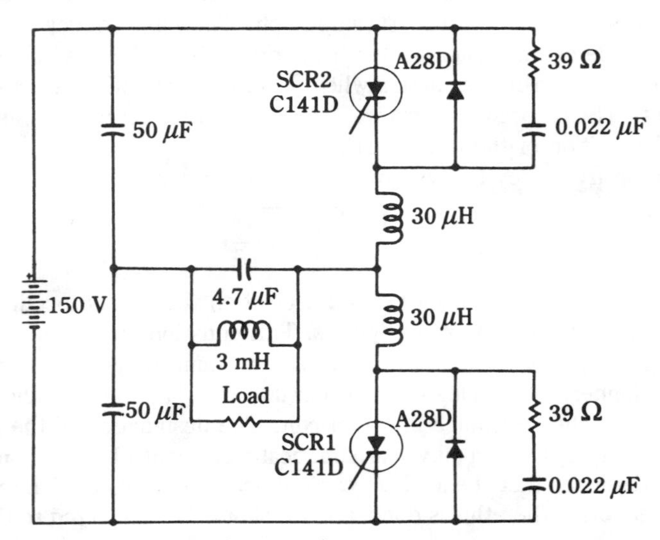

5-14 A 1-kW 10-kHz sine-wave inverter. General Electric Semiconductor Products Dept.

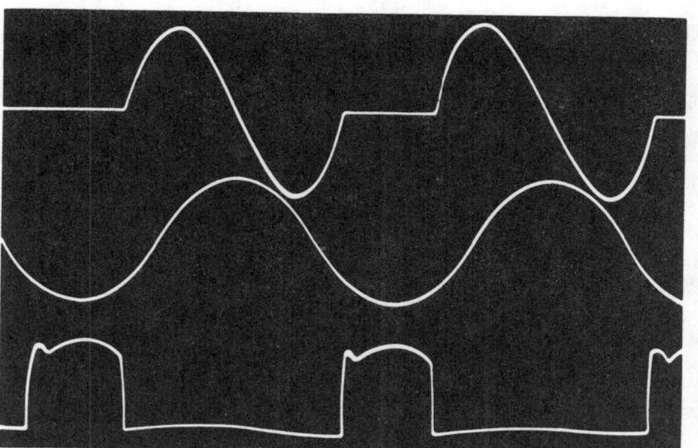

5-15 Waveforms in the inverter of Fig. 5-14.

Top trace: SCR and diode current
Middle trace: load voltage
Bottom trace: voltage across SCR1

to charge to nearly twice the dc supply voltage. Because the capacitor charging current tends to be oscillatory (because of the 30-μH inductance associated with SCR1), this current then reverses and finds a return path to the power supply through the diode shunted across SCR1. As a result, turn-off voltage is developed across SCR1. Notice that a single triggering pulse has produced half of an ac cycle. The initial quarter of this cycle is conducted by SCR1; the second quarter of the cycle is conducted by the diode.

Next, SCR2 is triggered, and the lower half of the circuit produces an action similar to that caused by the triggering of SCR1. The polarities of the currents and voltages are now reversed, however. Accordingly, smoothly continuous cycles are generated if the trigger pulses are appropriately spaced. It turns out that a fairly good sine wave is produced when the ratio of resonant frequency to triggering frequency is about 1.35:1.

The 3-mH inductor is included in Fig. 5-14 to simulate the primary of a linear output transformer. Such a transformer would be used to provide isolation and/or impedance matching for the load. This inductance is not critical, but values in the vicinity of a hundred times that of the resonating inductances reflect a good balance between cost and certain operational problems that exist when this inductance is too small. The RC networks shunted across the SCRs are snubber circuits to slow the voltage switching transitions so that false triggering will not occur. It is mandatory in this circuit that the diodes be fast-recovery types. This inverter has very good voltage regulation, and it is exceptionally tolerant of reactive loads.

A logic-controlled cycloconverter

Strictly speaking, the cycloconverter is neither an inverter nor a converter. Whereas an inverter has a dc input and an ac output, and a converter transforms dc to ac, but back to dc again, the cycloconverter operates from ac and delivers ac at a different (lower) frequency. Thus, the cycloconverter is actually a frequency converter. In practice, there is often an overlap among the functions of these three types of circuits, and the cycloconverter therefore falls within the domain of our interests.

A simple single-phase cycloconverter circuit is shown in Fig. 5-16. Various modes of performance can be obtained, depending on the switching logic applied to the gates of

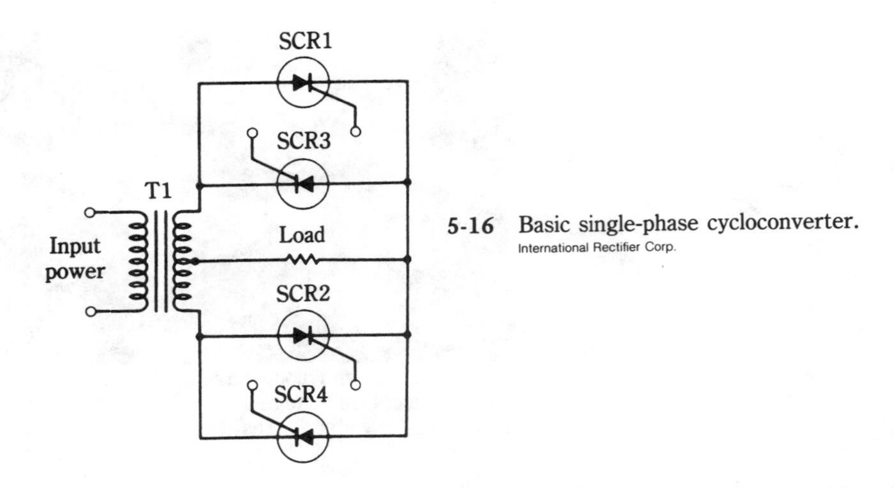

5-16 Basic single-phase cycloconverter.
International Rectifier Corp.

the SCRs. Two additional factors besides the "floating" gates contribute to the operational flexibility of this circuit. One of these is the self-commutation of the SCRs (because they are powered from an ac source). The other comes about through the use of a *zero-voltage switch* associated with the logic system. Basically, the zero-voltage switch makes it possible for incoming half-cycles from the ac power line to be extracted and subsequently reorganized into a different wave pattern—one of lower frequency. The zero-voltage switch and the gating logic make possible the use of these half-cycles as "building blocks."

The block diagram of the logic-controlled cycloconverter is shown in Fig. 5-17. The zero-voltage switch ensures that gate-trigger trains are synchronized to the power-line frequency. The logic network is a programmable counter. Ordinarily, it is profitable to feed the cycloconverter a high input frequency—from several hundred to at least several thousand hertz. However, operation from the 60-Hz power line is feasible if the application can use low output frequencies, such as 15, 20, and 30 Hz.

The basic cycloconverter circuit modified for use with a gate logic system is shown in Fig. 5-18. This arrangement provides isolation between logic and power. It also neatly circumvents awkward gating-circuit requirements of the basic cycloconverter of Fig. 5-16. In Fig. 5-18, V1, V2, and V3 can be simple (but separate) 9-V unregulated dc supplies; a current capability of 150 A is sufficient.

Cycloconverter waveforms are shown for three different output frequencies in Figs. 5-19, 5-20, and 5-21. In each case, the numbers shown in the output pulses indicate which SCR (Fig. 5-18) is turned on for each pulse. The four lower, rectangular waveforms represent gate signals delivered by the logic system of Fig. 5-22.

The output frequencies in Figs. 5-19, 5-20, and 5-21 are 1/2, 1/3, and 1/4, respectively, of the input (power-line) frequency. Although the outputs have the designated periodicity, the waveforms are obviously quite rough. In some cases, rudimentary filtering might be necessary. An output filter can be either of lowpass or bandpass configuration. The windings of electric motors generally provide much, if not all, of the necessary smoothing.

A three-frequency programmable logic system is shown in Fig. 5-22A. It consists

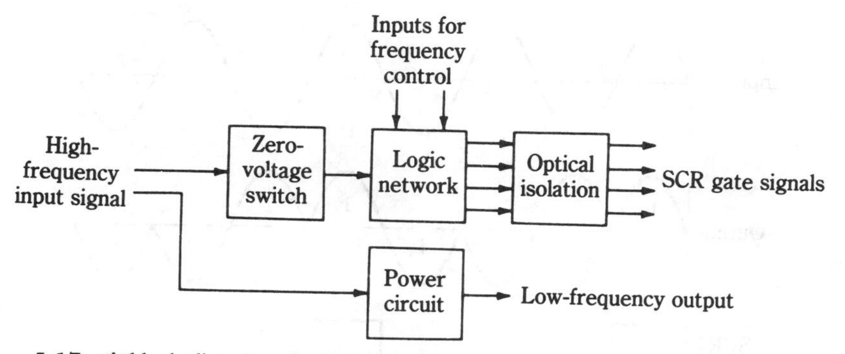

5-17 A block diagram of a logic-controlled cycloconverter. International Rectifier Corp.

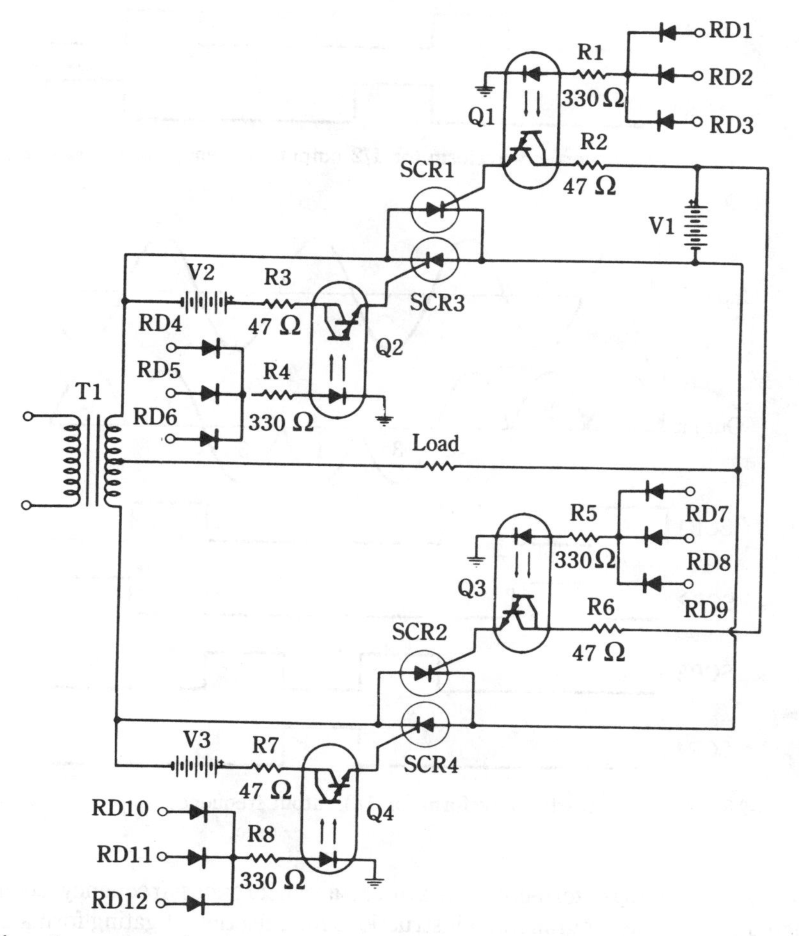

5-18 A cycloconverter that is modified for use with a gate logic-system. International Rectifier Corp.

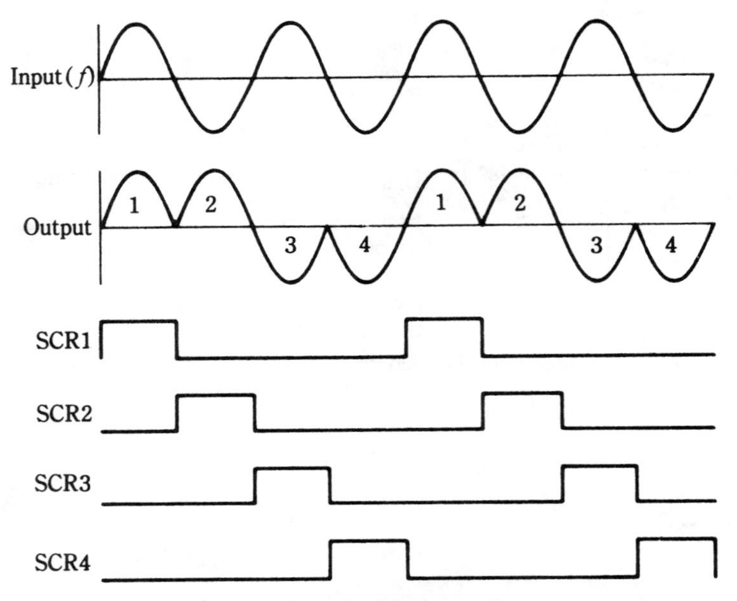

5-19 A cycloconverter waveform for 1/2 output frequency. International Rectifier Corp.

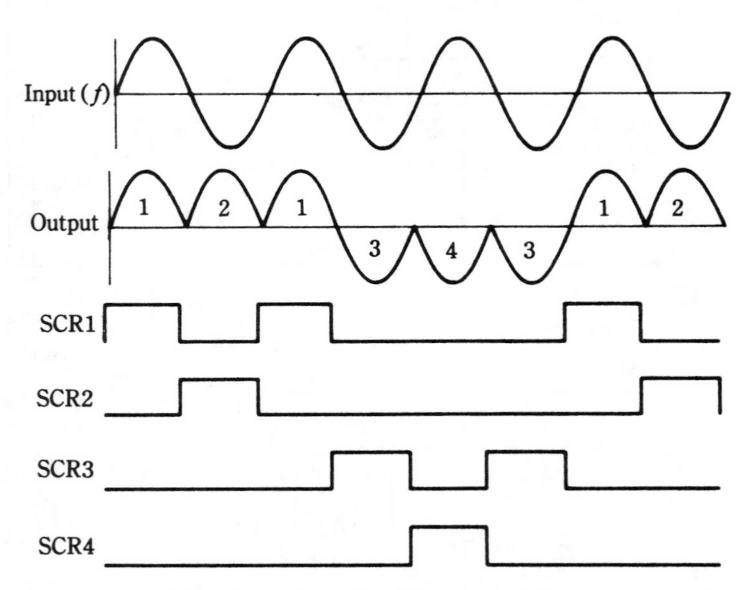

5-20 A cycloconverter waveform for 1/3 output frequency. International Rectifier Corp.

essentially of a voltage zero-crossing switch, a synchronous frequency counter, and associated gates. The programming instructions for selection of gating formats pertaining to the three cycloconverter output frequencies are listed in Fig. 5-22B.

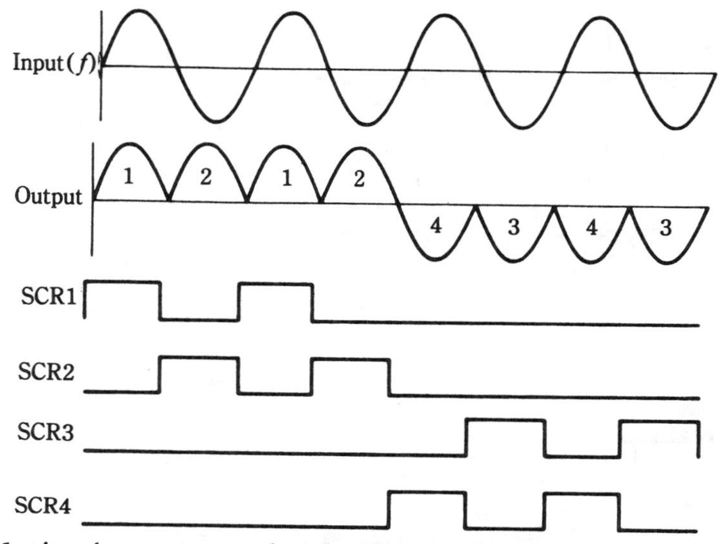

5-21 A cycloconverter waveform for 1/4 output frequency. International Rectifier Corp.

Gate-turn-off (GTO) thyristors in inverters

The inverter circuit shown in Fig. 5-23 is presented mainly to stimulate the interest of the experimentally inclined, and design details are, therefore, not included. Using recently developed GTO SCRs, this inverter has the very desirable feature that no commutation techniques are needed. Turn-on is initiated with a positive triggering pulse, as with conventional SCRs. Turn-off is achieved with an approximately 70-V *negative* pulse. Not only does this operational characteristic avoid commutation problems, but the device is suitable for use with logic control formats, such as pulse-width modulation. Simulated sine-wave outputs at frequencies as high as 30 kHz can be produced with such an inverter. The overall operating efficiency can be anticipated to be a little over 90 percent. For many applications, and particularly this one, the GTO SCR can be considered to combine the desirable features of SCRs and power transistors.

For frequencies up to 30 kHz, use the RCA G5001 series of GTO SCRs in this circuit. For frequencies up to 5 kHz, use the RCA G5002 series of GTO SCRs. For 60 and 400 Hz, use the RCA G5003 series of GTO SCRs. Where a GTO SCR with lower frequency capability will suffice, worthwhile savings in initial cost can be realized.

Some insight into the nature of presently available GTO SCRs for inverter applications can be obtained from Fig. 5-24. This simple arrangement can be considered a driven inverter. Its resemblance to one of the switching elements of Fig. 5-23 is particularly relevant. Figure 5-24B sums up the power dissipations in the device when turned on and off at a 20 kHz rate. Because 1200 W of 20-kHz power is switched in the load resistance, R_L, the GTO SCR used in this circuit represents a tremendous advance over earlier gate-turn-off thyristors. The overall efficiency is about 95 percent, and a good square wave is produced.

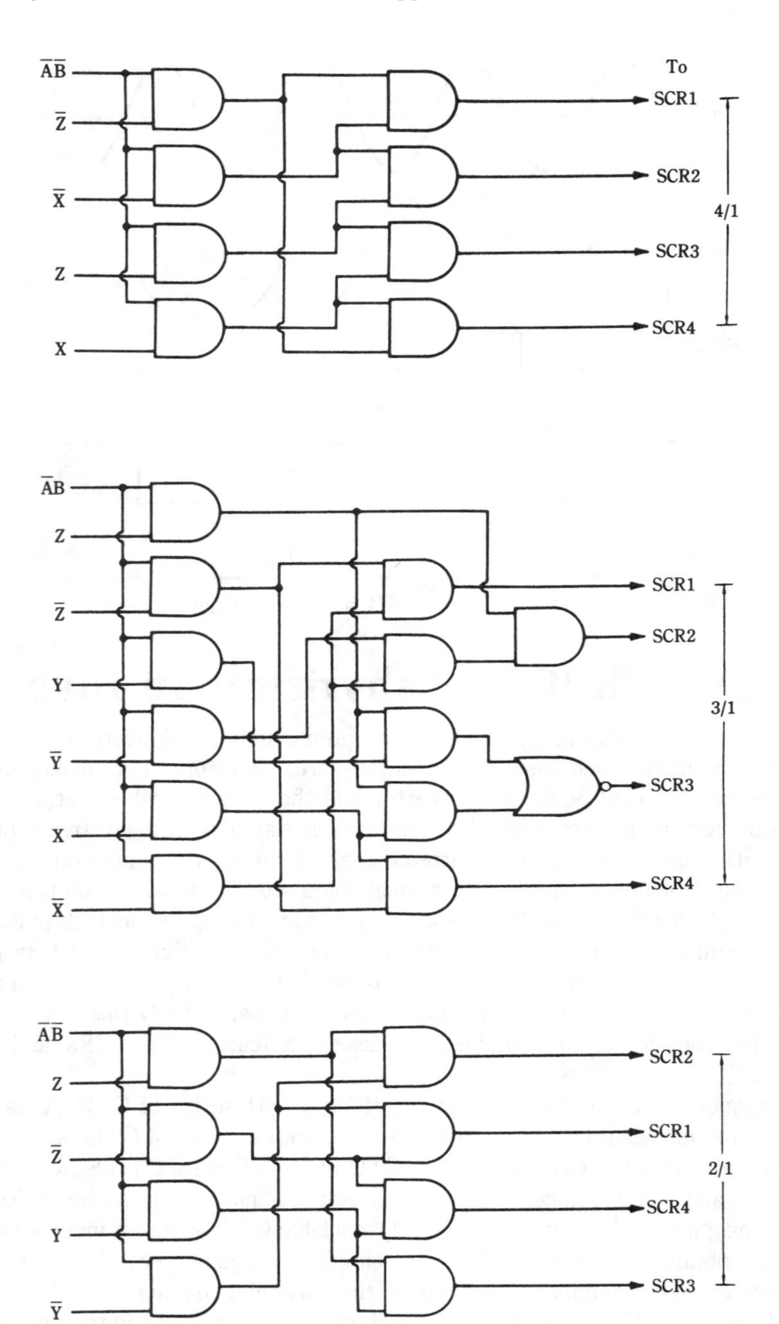

(A) System diagram.

5-22 A logic system for three selectable cycloconverter output frequencies. International Rectifier Corp.

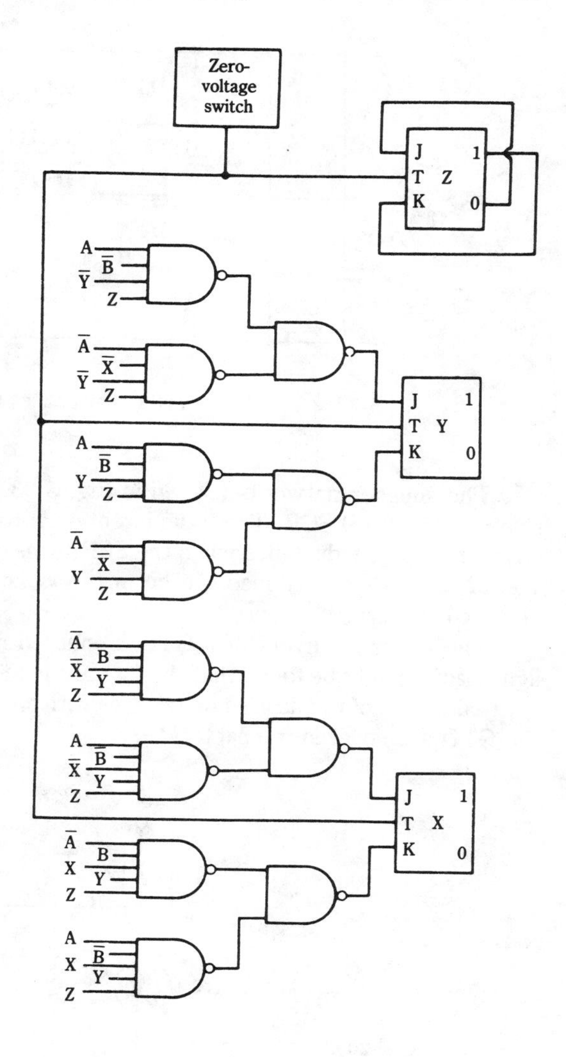

A B	Modulus	Frequency reduction
0 1	4	2/1
0 1	6	3/1
1 0	8	4/1

(B) Programming instructions.

5-22 Continued

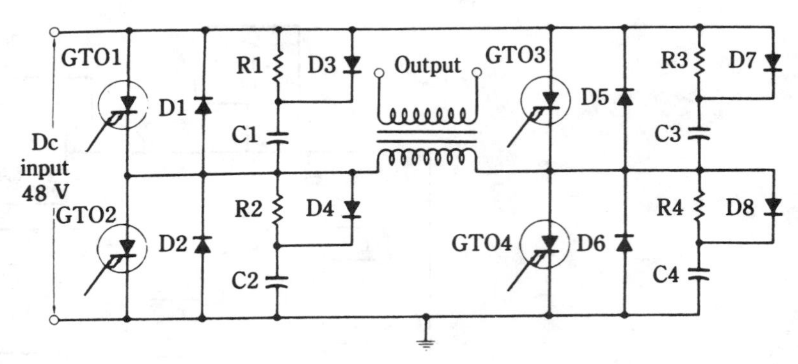

5-23 A bridge inverter that utilizes GTO SCRs. RCA Solid-State Div.

The snubber networks (shown in Fig. 5-23 and Fig. 5-24) are important for the safety of the GTO SCR. The basic intention is to absorb transient energy released by stray or leakage inductance when the SCR turns off. As a starter, the resistance can be 100 Ω, and the capacitance can be in the vicinity of 0.1 μF. The diode should be a fast-recovery type.

The latching current of a GTO is inherently high. Therefore, when switching a light load, it might be found that the forward gate-drive has to be maintained during the entire duration of the desired on time, as with a transistor. Specialized drive-circuits for the GTO are covered in chapter 19.

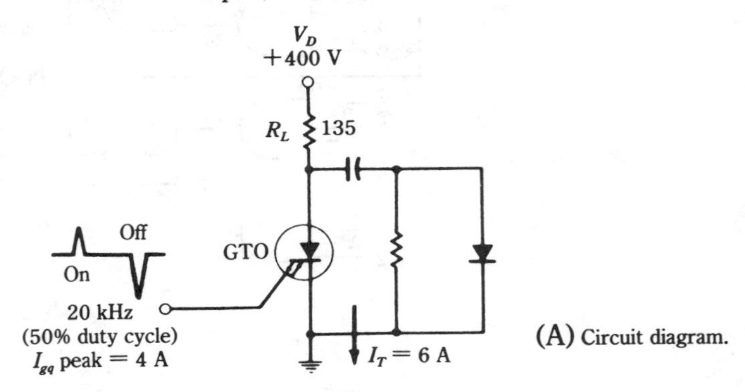

(A) Circuit diagram.

Power dissipation and temperature for 20 kHz resistive load switching with $I_T = 6$ A and $V_D = 400$ V

1200 watts

f (kHz)	T_J (°C)	P_D (on) (W)	P_D (off) (W)	P_{DC} (W)	P_{TOTAL} (W)	$T_J - T_C$ (°C)	T_C (°C)
20	100	10	17.4	7.3	34.7	52	48
20	125	10	23.0	7.0	40.0	60	65

(B) Power dissipation and temperature.

5-24 A circuit for demonstrating the inverter capabilities of the GTO SCR. RCA Solid-State Div.

Another way the GTO differs from ordinary SCRs is that it has negligible or no reverse-blocking capability. In practice, this generally is of no great consequence because of the almost universal use of the flywheel diode, as is seen in Fig. 5-23.

GTO thyristors are available from other manufacturers with different mixes of operating parameters. For example, Unitrode features devices that can be turned off with negative gate pulses on the order of 5 V and with several milliamperes. It should be kept firmly in mind that load-handling power, voltage rating, turn-off drive, switching speed, and temperature characteristics tend to be closely interrelated. Although overall performance improvements are constantly being made, the ongoing game of trade-offs remains the main impetus to advertised "breakthroughs." That, of course, is why so many different semiconductor devices are available—each boasts of ranges and modes of operation featuring superior performance over other devices. For the purposes of inverters and switching regulators, it is presently a fantasy to conjure up a GTO device with the power-handling ratings of inverter-type SCRs and the frequency capability of the power MOSFET.

With regard to thyristors in general, it is interesting to note the way that producers of consumers products have delved into the basics of inverters and regulators to develop circuits with somewhat different uses of components than their counterparts in instrumentation, transportation, space applications, etc. The unique aspect of the consumer's market is, of course, *cost.*

SCR regulator for color TV receiver

The color TV supply shown in Fig. 5-25 makes use of the fact that SCRs can simultaneously provide both rectification and control. Instead of a conventional UJT oscillator, this design uses the so-called ***programmable unijunction transistor (PUT),*** in this instance a MPU131. This device performs much the same function as the ordinary UJT in Fig. 1-10C. The MPU131 is connected as a relaxation oscillator which actually never runs freely, but is reset every time the ac line voltage crosses zero; in other words, the oscillator is synchronized to the line frequency. No special circuitry is needed to accomplish this, for it is brought about by the 120-Hz output ripple from the lightly filtered full-wave rectifier formed by the 1N4003 diodes.

Whether the PUT fires early or late in the ac cycle depends upon the dc voltage at its gate, which is obtained from the output of the supply (notice that the TV set is represented by R_L). If the dc output voltage starts to increase, the MPU131 starts to fire later, delaying the firing of the SCRs to lower the average dc output level. If the output voltage starts to decrease, the SCRs fire sooner. A given SCR can be triggered into conduction only during that portion of the ac cycle when its anode is positive relative to its cathode. Thus, the full-wave-rectifying property of the diode/SCR bridge is retained.

The usual output LC filter in a switching-type regulator is absent here. The filtering is provided instead by an RC network, despite the obvious I^2R power loss. This technique is cheaper to implement than the usual inductor/capacitor combination; however, small LC line filters are provided in order to minimize noise on the power line because of SCR turn-on pulses and other noise generated by the TV set itself, such as

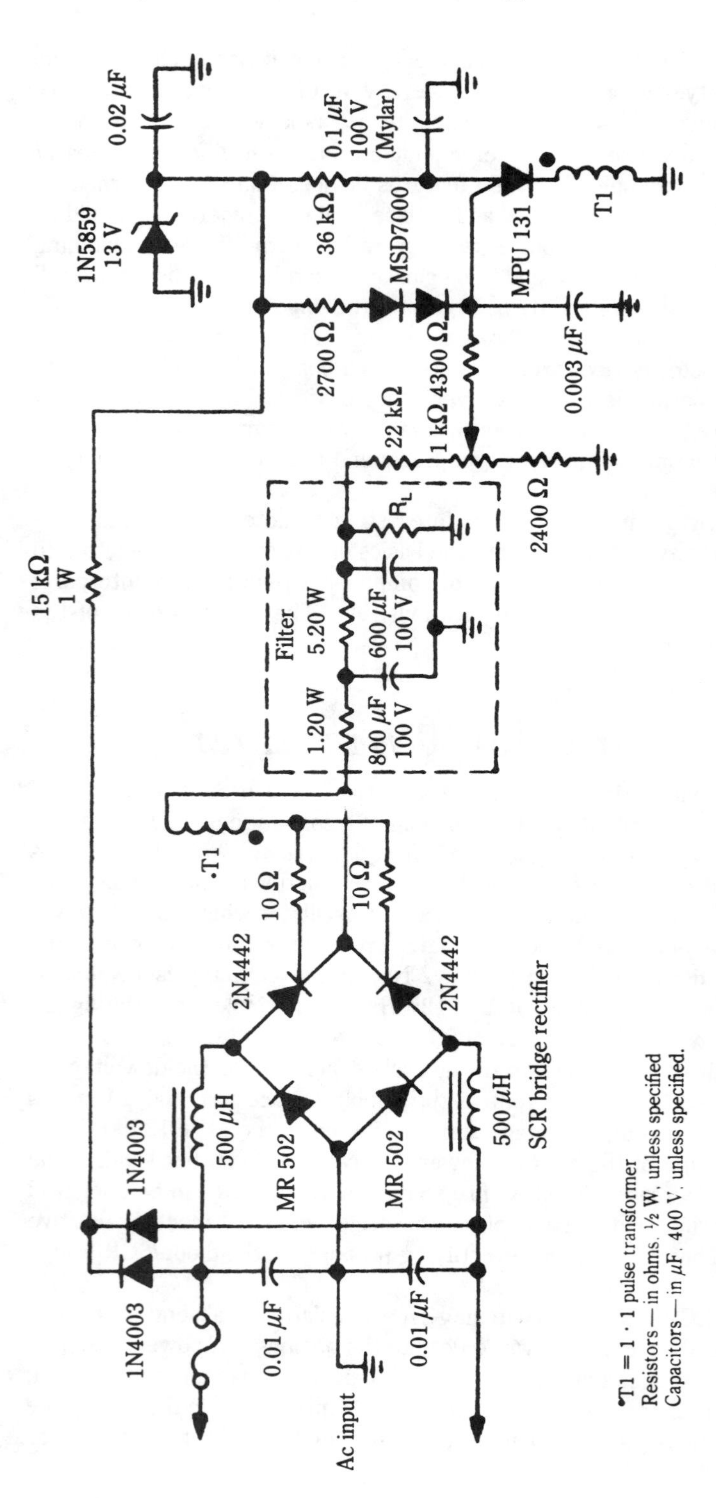

5-25 An SCR-switching-type supply for color TV receivers. This interesting supply provides a nominal 80-V output at currents up to 1.5 A. A minimum load of about 200 mA is required because of the SCR holding current. Motorola Semiconductor Products, Inc.

the vertical and horizontal frequencies. This supply can provide 80 V at 2 percent regulation over a line-voltage range of 105 to 140 V. Current capability is 1.5 A.

Temperature regulator using 0-V switching technique

The circuit shown in Fig. 5-26 is a switching-type power supply in which the load is an electric heater. The feedback path consists of air space, and the sensor is a thermistor. The novel feature of this arrangement is that the electrical power to the heater always comprises an integral number of cycles. When more power is needed, the pulse trains of voltage are applied to the heater and made to contain an additional cycle. The beauty of this switching technique is that virtually no RFI and EMI are generated. Accordingly, you will not see the customary high-frequency filter networks so prominent in conventional thyristor control and regulating circuits—those using phase control of individual cycles. The heart of the scheme is the General Electric PA424 module, which generates trigger pulses for triacs only when the line voltage crosses zero, and then only when commanded to do so. Therefore, the triac always delivers an integral number of cycles to the load. The operation of the temperature controller is easier to grasp if you think of the PA424 as simply another gate-trigger device, to be used in place of a *diac* (a 4-layer diode), a UJT, or a pulse transformer. In the functional diagram of Fig. 5-27, the zero-crossing detector can deliver a forward-bias pulse to the base of the output Darlington stage. In turn, a gate firing pulse is available from terminal 4. Operation is governed by the signal appearing at terminal 13. If transistor Q1 is held in its conductive state, no pulses appear at the output. The power supply consists of diode D6 in conjunction with an external capacitor, such as the 200-μF unit (Fig. 5-26). Whenever transistor Q1 is taken out of conduction, a gate-firing pulse appears at terminal 4 of the very next zero-crossing of the line voltage.

The modulator consists of a PUT (a 2N6027) in a relaxation oscillator having a period of about 30 seconds. This slow waveform is superimposed upon the signal sensed by the thermistor at terminal 13 of PA424. This simulates a temperature exclusion of about 2° F and causes the system to anticipate the temperature established adjustment of R_S. This results in less overshoot and undershoot (Fig. 5-27B) of the equilibrium temperature than would be the case without such modulation. Thus, this electronic thermostat displays much less thermal lag than a typical bimetallic switch.

Integral-cycle switching is one of the newer regulation concepts. Thus far, its implementation has been largely limited to 60-Hz applications. This scheme has interesting potential for high-frequency supplies, where it could emerge as a possible means to reduce electrical interference.

Motor speed regulation by phase-controlled switching

The simple circuit shown in Fig. 5-28 provides adjustment and regulation for the speed of dc and universal-type motors. Only a few parts are involved, but there is feedback,

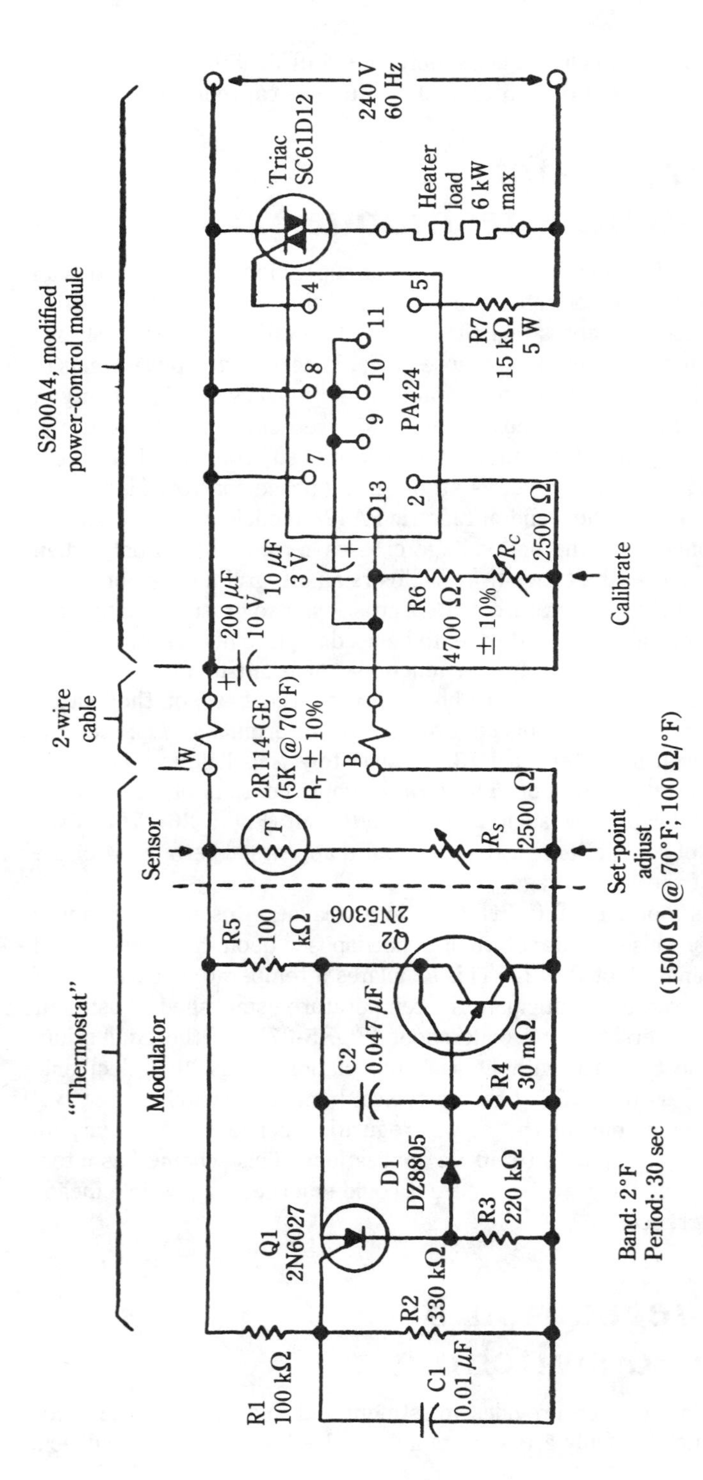

5-26 A temperature regulator that uses the zero-voltage switching technique. General Electric Semiconductor Products Dept.

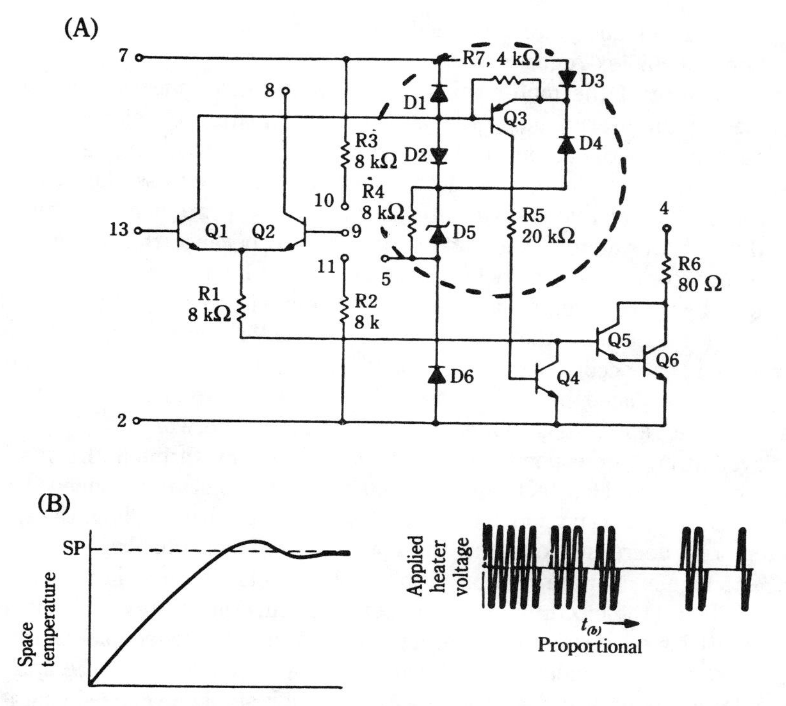

5-27 Zero-voltage switching that is accomplished with the use of a dedicated IC. The PA424 IC produces a series of complete cycles, thereby generating minimal switching noise. General Electric Semiconductor Products Dept.

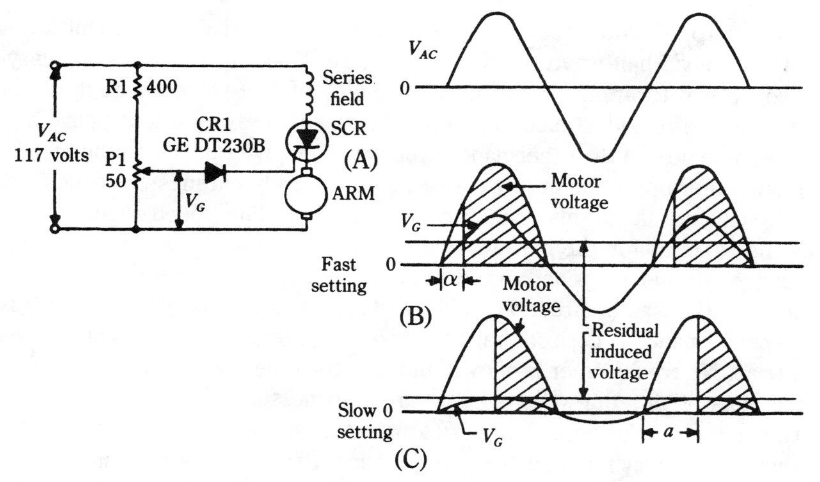

5-28 The application of phase-controlled SCR to regulate motor speed. General Electric Semiconductor Products Dept.

sensing, comparison, and pulse-time modulation occurring in this circuit, much like the servo action in complex regulators.

The armature of the motor spins in a magnetic field, and such motor action is accompanied by an internal generator action. The armature develops a counter EMF that opposes the flow of current into the motor. Under no-load conditions, the counter EMF approaches the value of the actual voltage applied across the armature. This means that only a relatively small internal voltage drop provides the force to push current through the motor. The current is not zero because the motor needs some current to overcome its own frictional losses. Under heavy loads, the counter EMF decreases and larger currents are passed to develop electromagnetic torque sufficient to carry the load. Because the armature is connected in the gate cathode circuit, the faster rotation that occurs under light loads inhibits firing of the gate because of the bucking effect produced by the counter EMF on the cathode. The motor speed is governed by the gate voltage obtained from the R1/P1 network.

Delayed firing corresponds to less average current through the motor, so its torque is decreased. This technique is used to reduce the motor speed if it tries to increase because of a relaxation in loading. Under greater shaft loading, the speed of the motor tends to decrease, lowering the counter EMF and enabling the SCR to fire earlier. The higher average current results counteracts the drop in speed.

The SCR and the motor must be compatible in current ratings, which is to say that the SCR must have sufficient current capability. Although this simple regulation technique has been applied widely for the control of small motors, it can be applied to the control of larger motors. It does not, however, exhibit stable low-speed characteristics, and additionally a large installation would also require adequate over-current protection.

Motor-speed regulator with full-wave control

Half-wave control of motor speed has enjoyed wide popularity. The circuit in Fig. 5-28 feeds the motor via half-wave rectification. There is no transformer secondary in this circuit, so core saturation by the dc component of half-wave rectification is not a problem. The motor inductance also tends to make the torque less pulsating than it might otherwise be. On the other hand, a full-wave supply of power to the motor results in much smoother operation at low speeds, and generally extends the control range. A circuit for accomplishing this is shown in Fig. 5-29. The speed-regulating principle remains the same as for the half-wave circuits.

A bridge rectifier is needed to route the current so that it is always passed through the motor in the same direction. This method is particularly advantageous when full-wave performance is desired, but the power requirements of the motor rule out the use of a triac. No transformer is shown, but one could be used for the sake of isolation.

Variations of this full-wave scheme are also possible. One, in particular, simply places the speed-regulating unit in series with the motor and the ac power line. This means that the primary function of the regulator is to maintain a constant average value of ac voltage across the motor. Although this technique does exert some control over the speed of the motor, it does not respond well to variations in motor loading.

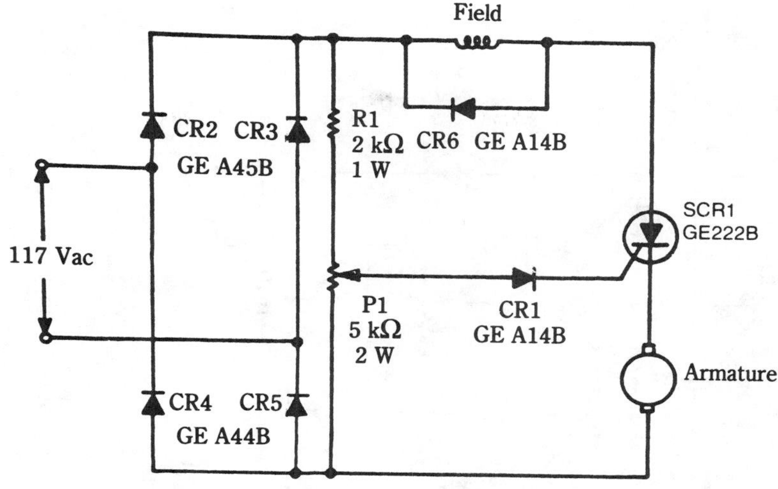

5-29 A full-wave speed regulator. General Electric Semiconductor Products Dept.

However, by modifying the control circuit, it would be possible to incorporate a tachometer and optoisolator to feed back speed information.

Using a switching-type power supply for speed control

Although the use of solid-state electronics for motor control is a well-known technique, it often happens that only those directly involved are aware of the many things that can be accomplished. By means of sensing, feedback, and pulse manipulation—the very techniques used in switching-type power supplies—the starting, speed, and torque characteristics of motors can be altered at will. An example is the synchronously controlled motor circuit shown in Fig. 5-30 in which the dc motor is caused to maintain its speed with the same constancy as an ac synchronous motor. At the same time, the starting and pull-in torque remain high.

In this circuit, the main SCR is Q2, which receives turn-on pulses from the UJT oscillator configured about Q1. Commutation or turn-off action is imparted to this SCR by Q3, which is controlled by the light-activated SCR (Q4). Light coming through the aperture of a motor-driven disk tends to turn Q2 off, and the oscillator tends to turn it on. If the motor tends to run slower than its "synchronous" speed—the rate of turn on pulses delivered to Q2—the duration of the pulses will be lengthened so that the average current passing through the armature and field is increased. This increases the electromagnetic torque and the motor accelerates. If it should accelerate to a speed in excess of its "synchronous" speed, the current pulses from the main SCR become too short to maintain such speed, and the motor falls back to its regulated rotational rate. If the load on the motor varies, the width of the pulses delivered by the main SCR vary in order to provide just enough average torque to the motor so that its pulse turn off rate equals the pulse turn-on rate.

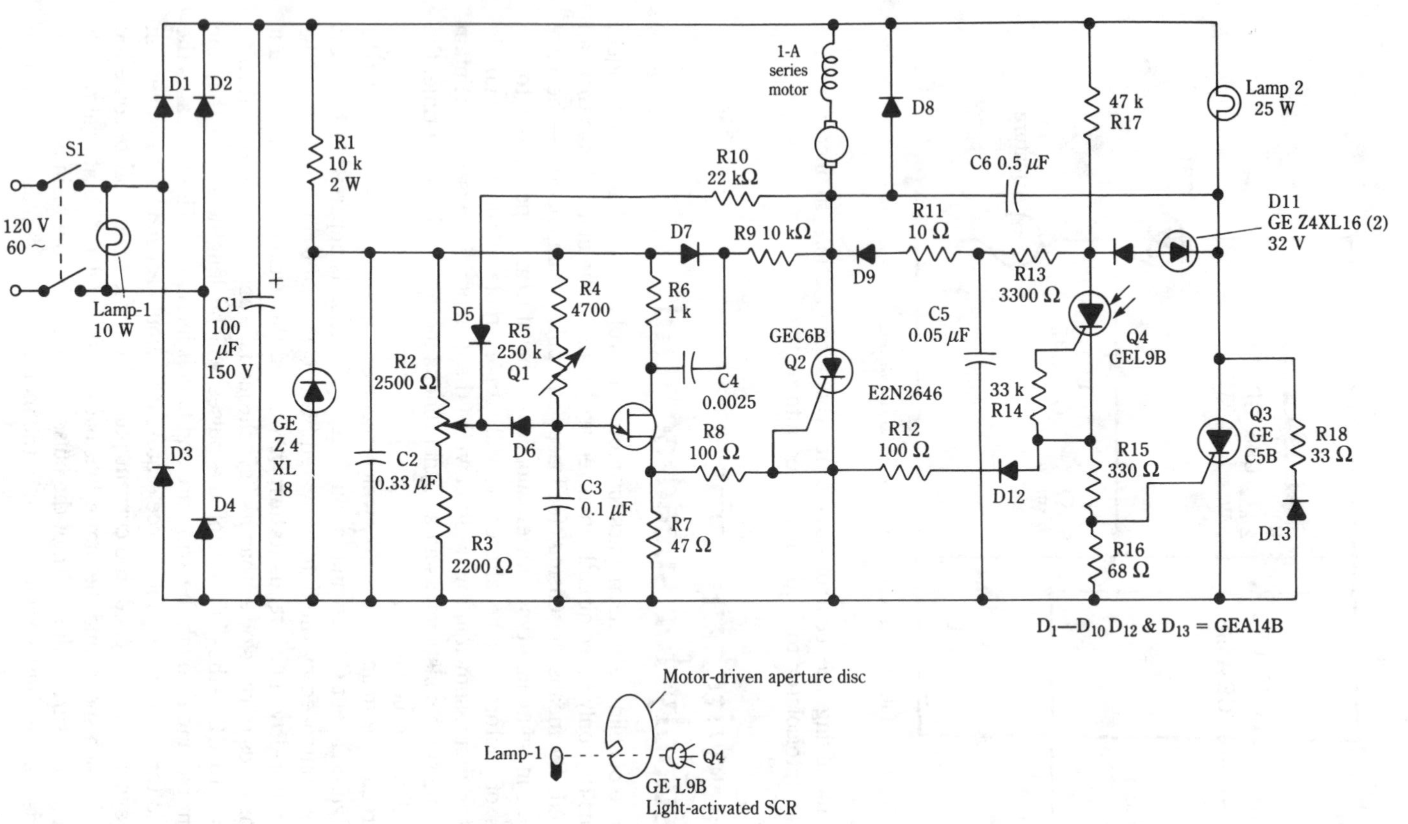

5-30 Precise regulation of dc motor speed with a switching-type power supply. This control technique gives the series-type dc motor the same speed characteristics as an ac synchronous motor. General Electric Semiconductor Products Dept.

In order that the motor action will be critically damped, delaying circuitry is inserted in the feedback path. Components R2, R3, R10, and D5 and D6 represent one such damping network. Diode D8 is equivalent to the free-wheeling diode in ordinary switching supplies, and it provides a current path from stored energy in the motor inductance during SCR off time. Notice that in this circuit, frequencies (rather than voltage levels are compared); therefore, no zener or other voltage-reference source is needed. This scheme has application to recorders and other instrumentation, particularly when operation must be obtained from battery power (the input bridge rectifier can operate from dc, as well as ac).

Phase-controlled ac voltage regulator

A tried and proven system is shown in Fig. 5-31. It is a flexible arrangement and readily lends itself to the use of other solid-state elements. For example, General Electric markets an extensive line of SCRs suitable for kilowatt operating powers. The basic circuit can also be scaled down to lower power levels. Unlike most of the regulators in this chapter, this one regulates the RMS level of its ac output, rather than providing voltage-stabilized dc. Its applications have included constant heat for controlling chemical and industrial processes, photocopying machines, ac motors, and special military uses. It is readily converted to an RMS ac current regulator (see Fig. 5-32). A lighter-weight 400-Hz version is possible with an appropriate transformer and a few other modifications.

The power-handling portion of the circuit consists of output auto transformer T3 and power thyristors SCR3 and SCR4. The gates of these devices are controlled by the smaller "pilot" SCRs. An optical-feedback path is provided by the lamp/photocell unit, designated respectively as L1 and P.C. The cadmium photocell is connected in the arm of a resistance bridge. The resistance of the cell governs the base-emitter bias of transistor Q1, which is the input stage of the differential amplifier. The conductive state of Q2 determines how quickly capacitor C2 can attain sufficient voltage to fire unijunction transistor Q3, which provides the triggering pulses for the pilot SCRs. The feedback loop is thus able to regulate the output voltage by phase control of each half-cycle. Although Q3 is connected as a relaxation oscillator, it is not free-running—rather, it is synchronized to the power line via the ripple in the unfiltered dc impressed across its bases. Every 1/120 second this oscillator is reset and C2 commences a fresh charging ramp. Interestingly, even though the output waveform in nonsinusoidal, it is the RMS value of the output voltage that is regulated, because the luminous intensity of the lamp is proportional to the RMS current through it, and therefore to the RMS voltage applied to it.

SCR inverters and converters

The use of SCRs in inverter and converter applications is attractive because of the high power-handling capability of these thyristors. A basic problem is commutation, for unlike the transistor, the thyristor must be forcibly turned off. Unfortunately, this cannot be done by means of the gate; at least not with conventional SCRs, and not at the

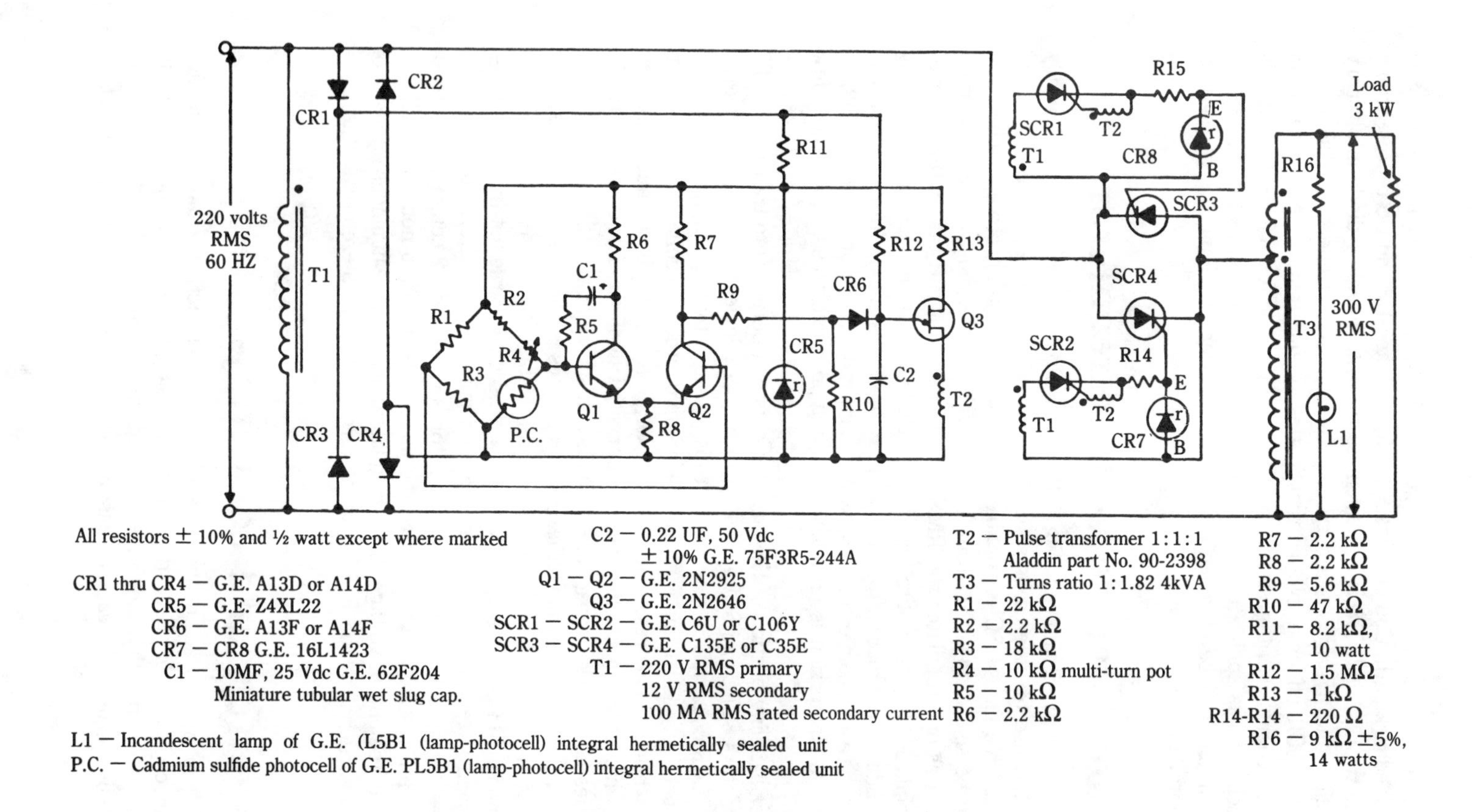

5-31 A phase-controlled voltage regulator for ac. Lamp L1 develops about 1.95 V across its terminals during steady-state operation. General Electric Semiconductor Products Dept.

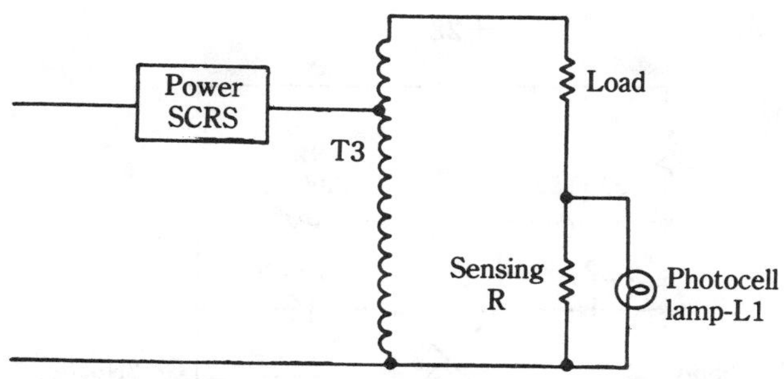

5-32 A slight lamp-circuit change produces an RMS load-current regulator. As in the ac voltage regulator, about 1.95 V across the lamp yields proper steady-state operation. General Electric Semiconductor Products Dept.

power levels most generally found in such power supplies. Therefore, the current through the device must be momentarily interrupted. This is usually accomplished by delivering the stored charge in a commutating capacitor to the anode-cathode circuit just prior to, or simultaneously with the turn-on trigger applied to the alternate SCR. Numerous commutating schemes have been devised, but not all are reliable under all load conditions.

The General Electric Company has pioneered the use of SCRs in inverters and converters. An early circuit is shown in Fig. 5-33. Although improved devices are now available, the basic configuration has served as a prototype for numerous and diversely rated inverters and converters utilizing SCRs. The 1-μF capacitor connected from anode to anode provides the commutation. When one SCR is triggered into its conductive state, a reverse-polarity pulse is delivered by the commutating capacitor to the anode-cathode circuit of the alternate SCR, thereby turning it off. In this type of inverter, the commutating capacitor must be optimized for a certain load, and additional commutating problems arise from inductive loads. But, when used within its load-variation limits, clean operation is obtained and the circuit is both reliable and efficient.

In SCR inverters of this type, much has been written about the matter of commutation. In my experience, it is necessary to use a good-quality commutating capacitor, preferably one made for this specific task. And, bargain basement SCRs should be avoided. Often, what appears to be commutating misbehavior is, in reality, inadequacy of gate drive; this is even more likely if the poor performance tends to show up before junction temperatures have had time to rise appreciably. Figure 5-34 depicts required peak gate amperage in a typical SCR as a function of the duration of the gate pulse. Notice that when the pulse is very narrow, much higher levels of peak gate current must be delivered from the trigger source in order to turn the SCR on. Even worse, this situation becomes further aggravated at higher temperatures. Because the SCR in a push-pull inverter circuit *both* supplies power and commutates the opposite SCR, triggering difficulties can readily manifest themselves as a commutation problem.

While on this subject, it is apropos to point out that it is common practice for SCR manufacturers to omit the circuit details of trigger sources—especially in polyphase systems. It is as if any old pulse circuit will do. This is unfortunate, for the triggering of the SCRs in inverter applications is far from being a trivial matter. In this case, the

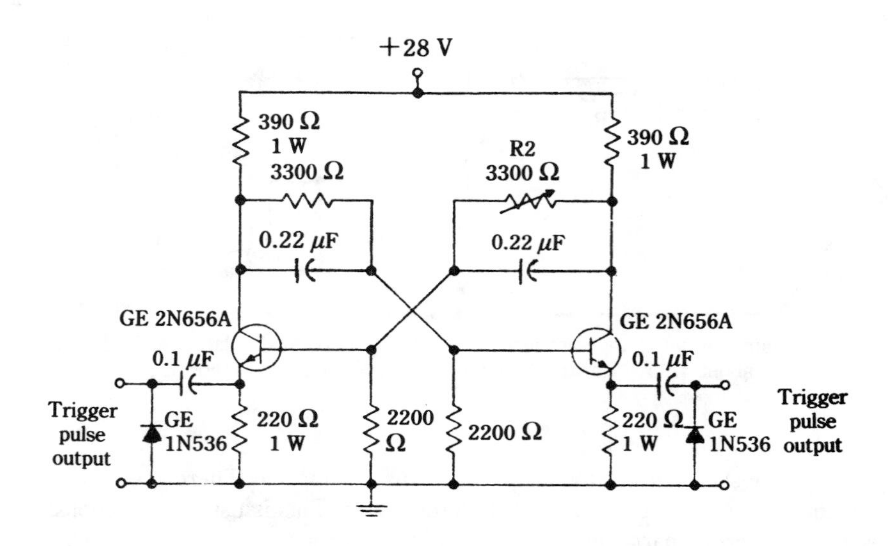

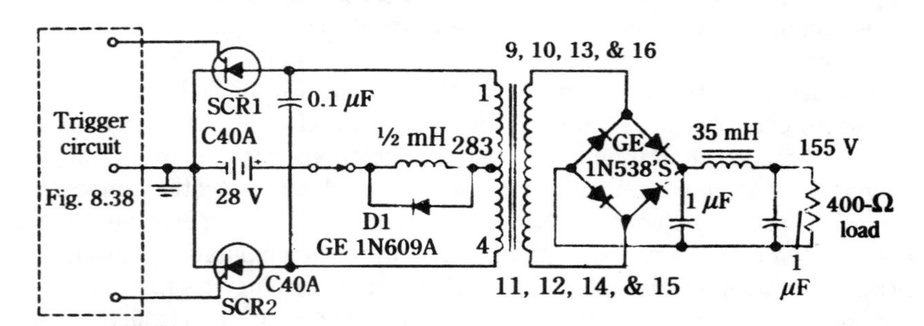

5-33 An scr converter and its trigger circuit. These basic circuits are basic to the extensive use of SCRs in many inverter and converter applications. General Electric Semiconductor Products Dept.

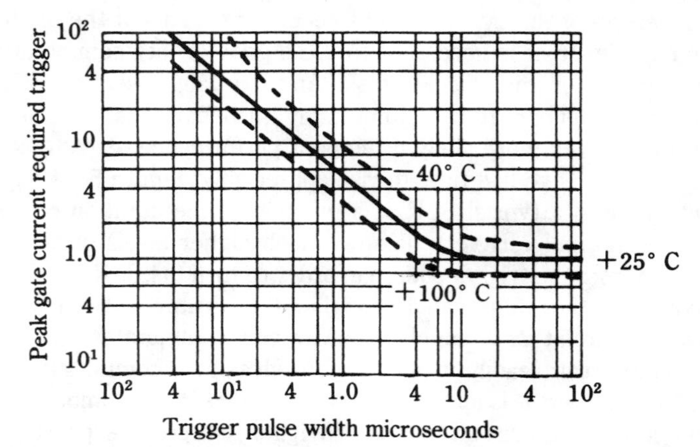

5-34 Peak amplitude of gate trigger pulse vs. pulse width. Skinny trigger-pulses require higher peak gate-current. The level of minimum triggering current is temperature dependent.

trigger circuit is shown—a multivibrator arranged to deliver sharply differentiated pulses, displaced 180° in phase, and of sufficient strength to satisfy all load and temperature conditions.

SCR inverter with wide-range load-handling adaptability

The inverter circuit depicted in Fig. 5-35 resembles the previous SCR circuit, but it incorporates a different commutation technique. In this scheme, the turn-on of one SCR shock-excites L1 and C5 into series resonance. The oscillatory voltage developed across the alternate SCR quickly turns it off. Oscillation is highly damped by resistances R_{10}, R_{11}, and their associated diodes. The diodes maintain the square waveshape of the output waveform by absorbing excess energy. Considerable voltage stress is thus diverted from the SCRs, for the peak SCR voltage could easily rise to several times the supply voltage without such damping.

What has been previously stated concerning the unique triggering requirements of various thyristor circuits emphatically applies here, for the trigger circuit must be considered as part of the inverter. Merely supplying turn-on pulses to this inverter will not enable it to commutate properly when delivering power into inductive loads. What is required is a source of precisely formed square waves. An ordinary free-running multivibrator is too dependent upon its RC timing elements to reliably maintain a 50 percent duty cycle, particularly when its frequency is varied. The UJT driven multivibrator employed in the square-wave trigger circuit, however, generates complementary square waves from two outputs, each having 50 percent duty cycles. The UJT pulses are used to synchronize the multivibrator switching transitions. This trigger circuit is also remarkably free of the "jitter" often attending multivibrator operation.

Inductor L1 tends to be small because a high-Q resonance is not sought in this technique. The use of a small inductor is abbetted by the selection of SCRs with a fast turn-off time. Commutation design procedure first involves the selection of the minimum size capacitor as follows:

$$C = \frac{t \times 1}{2E}$$

Where:

$C =$ commutating capacitor, in farads
$t =$ turn-off time of the SCR, expressed in seconds
$I =$ the current handled by the SCR just prior to commutation
$E =$ the dc supply voltage

After the minimum commutation capacitance has been calculated, the actual selection dictates a capacitor somewhat larger—up to several times if permissible by cost and space considerations. Inductor L1 is chosen to resonate with the commutating capacitor at a frequency corresponding to the half-period, which equals or exceeds the turn-off time of the SCR. Although not shown in the diagram, a large capacitor is also desirable across the dc input terminals, but would be unnecessary if it is already incorporated in the dc source and the connecting leads are not too long.

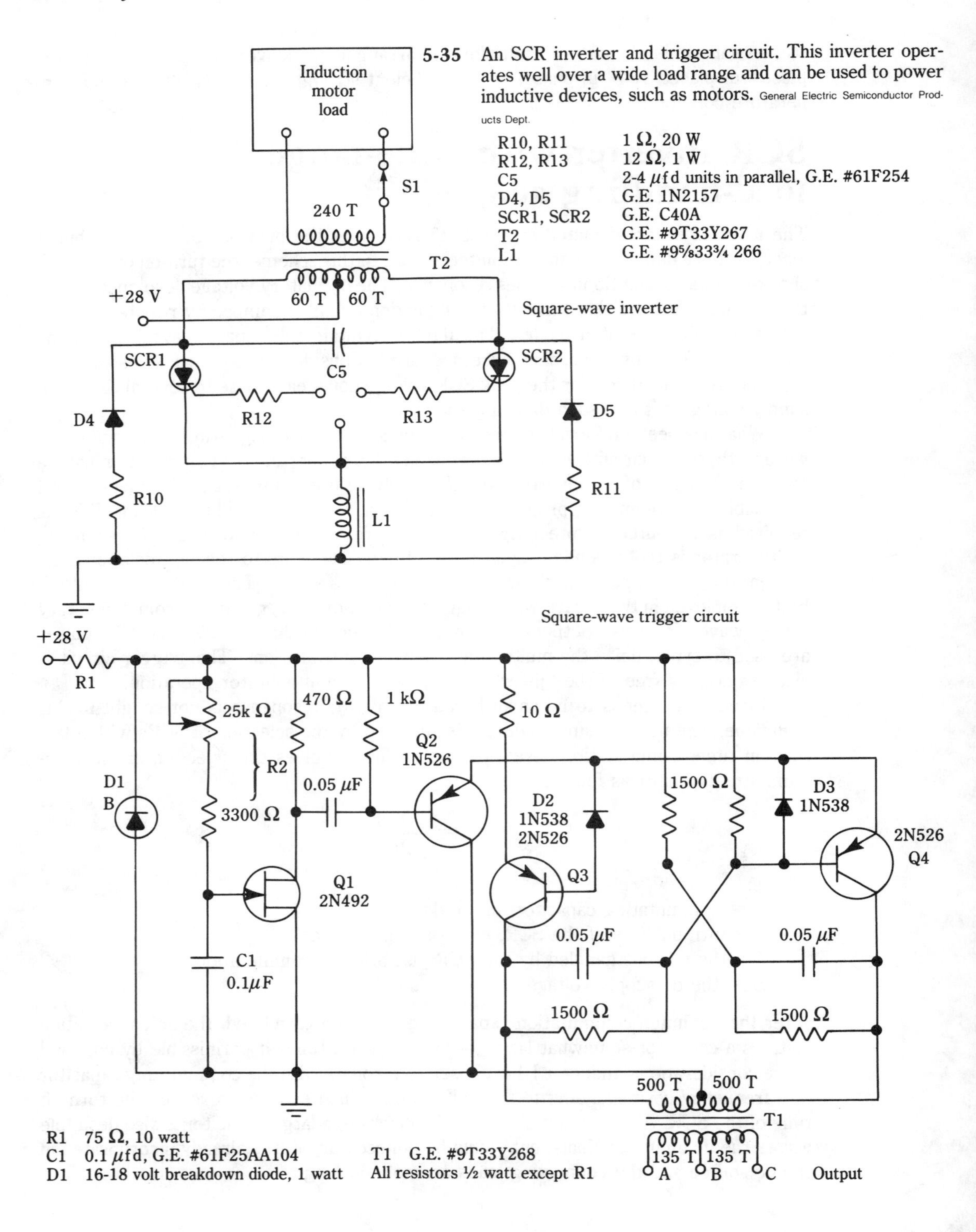

5-35 An SCR inverter and trigger circuit. This inverter operates well over a wide load range and can be used to power inductive devices, such as motors. General Electric Semiconductor Products Dept.

R10, R11	1 Ω, 20 W
R12, R13	12 Ω, 1 W
C5	2-4 μfd units in parallel, G.E. #61F254
D4, D5	G.E. 1N2157
SCR1, SCR2	G.E. C40A
T2	G.E. #9T33Y267
L1	G.E. #9⅝33¾ 266

R1 75 Ω, 10 watt
C1 0.1 μfd, G.E. #61F25AA104
D1 16-18 volt breakdown diode, 1 watt

T1 G.E. #9T33Y268
All resistors ½ watt except R1

The triac in inverter systems

Some triac circuit applications are closely associated with inverters, converters and power supplies; they therefore merit mention. For example, the designer or experimenter, could conceivably use the basic light-dimmer circuit in conjunction with a feedback loop to pre-regulate the ac input to a power supply. Such a scheme could replace the heavy and expensive ferroresonant transformer sometimes used for this purpose.

The triac solid-state relay shown in Fig. 5-36 is an excellent adjunct to certain inverter and power-supply applications. Here, the triac is either in its off or fully on state; thus it substitutes for the conventional mechanical switch, relay, or contactor, eliminating problems associated with wear-out and with arcing and pitting. But there is more than might initially meet the eye. The ZVS is a special RCA op amp known as a *zero-voltage switch.* As its name implies, this integrated circuit triggers the triac when the ac voltage applied to it goes through its zero-crossing. Because of this gate-signal timing, minimal electrical noise is produced. Moreover, the load is spared from the effects of certain kinds of destructive transients and current inrushes. Notice that the photo-coupled isolator not only provides isolation of the power circuitry, but it facilitates remote control and enables actuation from logic circuits. Essentially, a sine wave is delivered to the load, which from our point of interest, can be an inverter or power supply.

This type of solid-state relay cannot be used to switch direct current. However, it will be seen that the ZVS gating circuit needs no dc source for its operation.

Another useful triac circuit for the inverter and power-supply technician or engineer is shown in Fig. 5-37. This is the basic lamp-dimmer. Instead of the lamp, other loads, such as inverters and power supplies can be controlled, providing they are not too inductive. Essentially, the triac circuit substitutes for an expensive and bulky variable

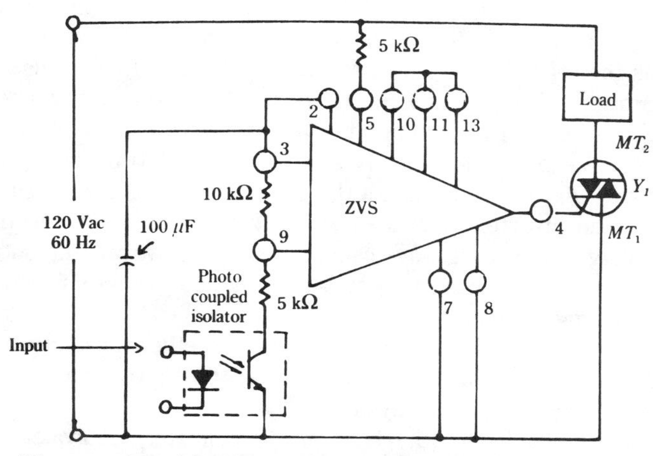

5-36 A triac solid-state relay with zero-crossing control and input isolation. This circuit is well adapted to controlling the application of ac power to inverters and to power supplies. RCA Solid-State Div.

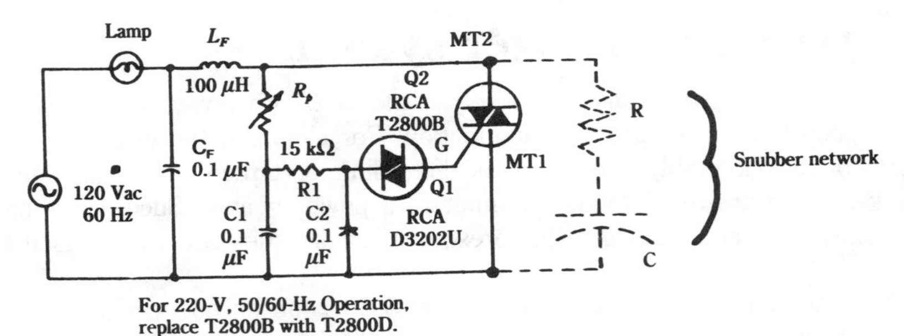

5-37 A triac lamp-dimmer circuit. Full-wave control of the RMS value of the ac load-voltage is provided. For inductive loads, an RC snubber is usually needed. RCA Solid-State Div.

autotransformer—these are not always readily available in the laboratory. The experimentally inclined can incorporate various electronic modifications to regulate the output, or to achieve remote control. A disadvantage compared to the autotransformer is that the controlled ac voltage is not sinusoidal (except at full output). The nature of the application will determine whether this is an adverse feature.

In order to make the circuit of Fig. 5-37 self-regulating, timing capacitor C1, could be shunted by a high-resistance photo-conductive element, illuminated by a small lamp deriving its operating current from the load terminals. Any tendency for an increase of output voltage would then be counteracted by an increase in the net time constant of the gate circuit, and, therefore, in the increased delay in triggering. This would occur because more intense illumination of the photo-conductive element would lower its resistance, thereby delaying the charging cycle of C1. Because such a scheme is somewhat tangential to the central themes of this book, the exact circuitry techniques will not be detailed. Suffice it to say that such an electronic voltage controller lends itself to control techniques that are not readily attainable with a variable autotransformer.

Two things should be kept in mind when it is desired to use a phase-controlled circuit (such as that of Fig. 5-37) with an inverter. If the circuit is used ahead of the inverter it is probable that the "snubber network" depicted in dotted lines in Fig. 5-37 will be necessary for proper commutation. This is because most inverters will appear as a somewhat inductive load. Resistance R is generally about 100 Ω and capacitance C is commonly 0.1 μF. On the other hand, if the phase-control dimmer circuit is to be used at the output of an inverter, it is mandatory that the inverter deliver a sinusoidal waveform of voltage—a square wave cannot be used. However, the output of square-wave inverters can be converted to a satisfactorily approximate sine wave by an LC low-pass filter in some cases.

Most triacs are intended for use at 50/60 Hz. Whether these can perform well at 400 Hz is not readily predictable. However, special 400-Hz triacs are available and find considerable deployment in avionic systems.

From a circuit standpoint, the triac is approximately equivalent to two SCRs connected in parallel, but oppositely polarized. Thus, both alternations of the alternating current cycle can be controlled symmetrically. But even better, a single gate suffices for such control. As already shown, control is accomplished by varying the time of firing, as in an ordinary SCR. Although the triac has become popular for use in solid-state relays, lamp dimmers, and electric motor control, direct implementation as an inverter or converter is minimal. If nothing else, the parallel, oppositely polarized arrangement together with the single control gate does not lend itself conveniently for use in inverter circuitry. Nor have these devices experienced developmental impetus for high-frequency operation. Their power-handling capability and current rating is, however, constantly being upgraded and they remain useful adjuncts to systems incorporating inverters, converters, and power supplies.

Alternative technique for regulating ac voltage

Another approach to the regulation of ac voltage is shown in Fig. 5-38. This circuit does not make use of an obvious feedback loop and is probably best described as an open-loop regulator. (A rigorous mathematical-analysis might deal with the concept of an "internal feedback" in the same sense that a two-terminal negative-resistance oscillator, such as a tunnel-diode, might be said to behave as if there was an internal feedback path.) In any event, regulator circuits (such as this one) generally don't attain the tight regulation readily feasible with conventional feedback and sensing arrangements. Nonetheless, the ac voltage stabilization is often satisfactory for practical purposes.

The circuit of Fig. 5-38 is, however, superior to the voltage regulator of Fig. 5-31 in one important respect—it has a faster response because slow-acting components (such as filamentary lamps and photo-conductive cells) are not used. In a rough comparison, the ac voltage-regulator of Fig. 5-38 might be considered an electronic substitute for a constant-voltage ferroresonant transformer, in which case there is a worthwhile saving in bulk, weight, and cost. Although it applies only 90-V to the load, the experimenter can increase this to the 120-V level by means of a conventional transformer, or an auto-transformer (such "linear" transformers are much smaller, lighter, and less expensive than the alluded-to ferroresonant type).

The first thing to understand about the regulator circuit of Fig. 5-38 is that there can be no voltage applied to the load, or current through it, unless there is some kind of a current path between the plus and minus junctions of the bridge rectifier (D1). Keep in

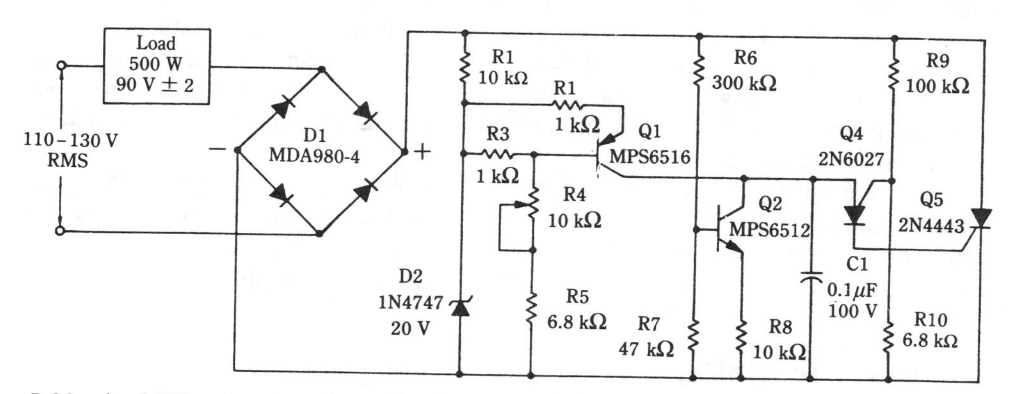

5-38 An RMS voltage regulator for ac power systems. Electronic regulation of ac voltage merits consideration when it is desired to dispense with the bulk, weight, and cost of a ferroresonant transformer. Motorola Semiconductor Products, Inc.

mind that the dc produced by this bridge rectifier, being unfiltered, is actually a wavetrain of uni-directional half-sinusoids. This being the case, it is possible to control the load voltage by varying the portions of these waves allowed to pass. Such variation is accomplished by the circuitry associated with Q1, Q2, and Q3. In essence, there is full-wave phase control similar in nature to that used in light dimmers and in motor-control circuits—even though the method of doing this is somewhat off the beaten path. Part of the reason for the unique approach is that an SCR (Q4) is used instead of a triac as the full-wave, phase-controlling power device (it is actually the rectifying-bridge that enables operation over both halves of the ac cycle).

Here is how phase-control and regulation are achieved: Q3 is a PUT (programmable unijunction transistor) that provides turn-on trigger pulses for the load-control SCR, Q5. The timing of these trigger-pulses is a function of how fast capacitor C1 is allowed to charge to the triggering level of the PUT. Actually, two circuit conditions govern the delay in production of trigger-pulses by the PUT. First, assume that the ac line-voltage increases. This raises the voltage across R10, which, in turn, forces timing capacitor C1 to have to charge to a higher voltage to cause PUT Q3 to generate a trigger for the SCR. The delay thereby imposed reduces the duration of the half-cycles of voltage applied to the load so that its RMS voltage is decreased. This constitutes phase-controlled regulation of the load voltage, with respect to ac line voltage. In the event of a lowering of the ac line voltage, the opposite sequence of events takes place in order to maintain constant load voltage.

There is a hitch in the above-described circuit action; unless some preventative means is taken, the PUT would fire at the very beginning of each half cycle and there would be a condition of latch-up. What is needed is a hold-off scheme to deprive the PUT of anode voltage during the initial rise of half-sine pulses from the rectifier bridge. This function is provided by Q1 and its associated circuitry. What happens is that Q1 remains in its nonconductive state until Zener diode D2 conducts; at that time, the base of Q1 becomes forward-biased and Q1 is able to apply operating voltage to the PUT (during holdoff-time, timing capacitor C1 is prevented from charging).

In addition to the fixed-voltage delay imposed by the Q1/D2 circuitry, there in a second means of producing the variable phase delay needed for regulation. This is provided by Q2 and its associated circuitry. Q2 simply acts as a line-voltage actuated shunting resistance across timing capacitor C1. So, if the ac line-voltage increases, Q2 conducts harder, delaying the time needed for the voltage charging C1 to attain triggering level of the PUT, and in turn, of the SCR. Thus, the tendency for load voltage to rise is counteracted.

It is evident that these regulation schemes stabilize the load voltage in the face of changing ac line voltage. It is intended for essentially fixed loads because there is no provision for regulating load voltage against load current. Figure 5-39 displays the performance of this ac voltage-regulator. A fixed 500-W (at 90-V RMS) load was used.

Ac power control by burst modulation in anti-parallel SCR circuit

In the mid-1970s, dc power supplies and inverters appeared to be a fairly mature technology. During the subsequent decade, however, it is clear that phenomenal ad-

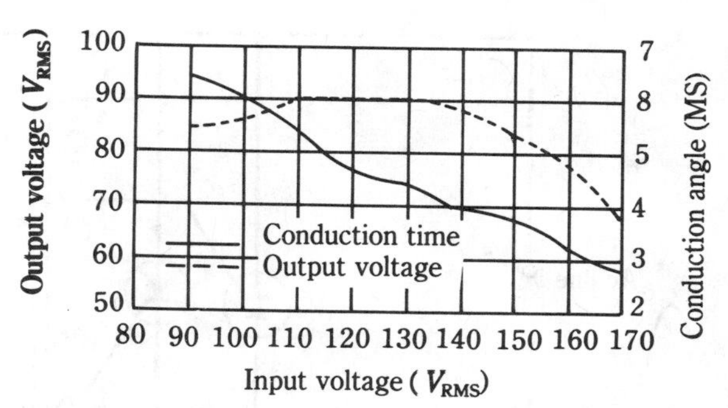

5-39 A line-voltage regulation band of the ac voltage-regulator of Fig. 5-38.

vancements took place. Lagging behind was parallel progress for the control and regulation of ac power—especially at high levels and at moderately high frequencies. Applications such as induction heating, ultra-sonics, welding, and integral horsepower motor control would benefit from devices (such as the triac), but capable of full-wave control of power and frequency required a great deal higher than that forthcoming from triacs. Triacs seem limited to about 1 kHz in the frequency domain, and manufacturers' catalogs do not list current capabilities much beyond several tens of amperes, or voltage ratings much greater than about 1000 V.

In contrast, SCRs (inverter-type) can operate efficiently to about 30-kHz and are available with kA and several kV ratings. In order to obtain triac-like full-wave control with SCRs, it has been customary to operate pairs in anti-parallel or back-to-back arrangements. This can work out very well, but because of isolation and phasing requirements of the two gates, the circuitry tends to be complex when control or regulation is also implemented. Because SCRs are such efficient and rugged devices, a suitable control technique long remained a tantalizing quest. Phase control (such as is used at low and moderate powers in light dimmers, small motor control, and for heaters of not-too-large capacity), is not usually the best way of controlling high-power systems. For one thing, the RFI and EMI can easily become unbearable. And as mentioned, phase control of two SCRs in a full-wave anti-parallel scheme can involve circuit and operational difficulties. The conventional phase-control circuit for controlling load power from anti-parallel SCRs is shown in Fig. 5-40.

Fortunately, Motorola has developed a device that can truly allow you to "have the cake and eat it, too." The MOC 3031 Thyristor Driver provides reliable triggering to anti-parallel SCRs together with electrical isolation. Based on the principle of the opto-isolator, this device is much more than just a substitute for transformers, for it also provides a unique method of power control, superior in certain applications to phase control (Fig. 5-41).

This control technique is known as *burst modulation* or as *integral-cycle control.* As the terms infer, load power is controlled by parcelling out integral (full-cycle) ac power. This greatly reduces RFI and EMI because the technique involves zero-voltage switching. The SCRs are maximally utilized, and problems of hysteresis and device tracking vanish. Obviously, this is not an instantaneous control method, so the nature and requisites of the load must be considered. It is ideal for heater loads, but it can also be satisfactory for uninterrupted power supplies (UPS). This appears a fertile field for experimentation. The basic circuit for such a burst modulation as supply is shown in Fig. 5-42.

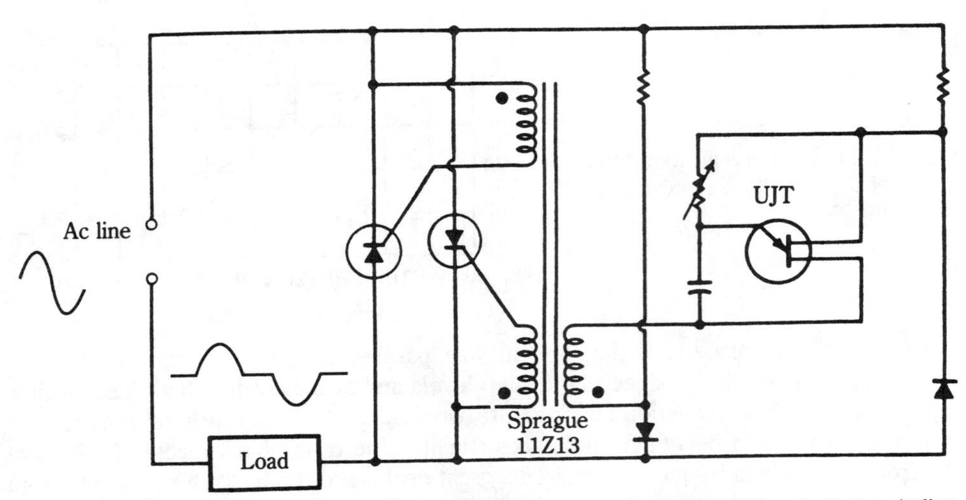

5-40 A conventional control circuit for ac supply using anti-parallel SCRs. A three-winding transformer is needed for this phase-control technique.

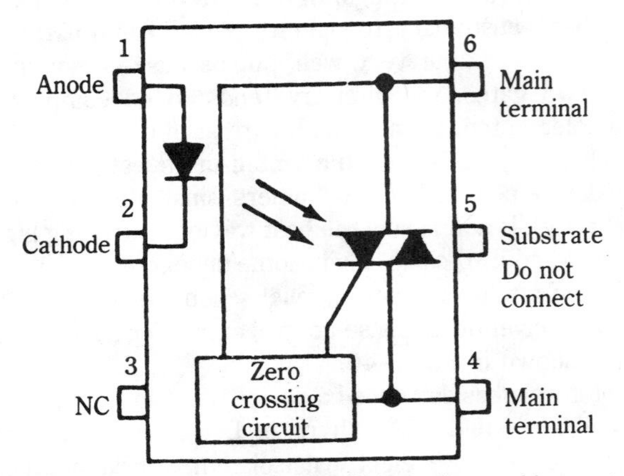

5-41 The Motorola MOC-3031 thyristor driver. This device combines optoelectronic isolation with a zero-crossing circuit that supplies properly phased pulses for triggering back-to-back anti-parallel SCRs. Control is by burst modulation. Motorola Semiconductor Products, Inc.

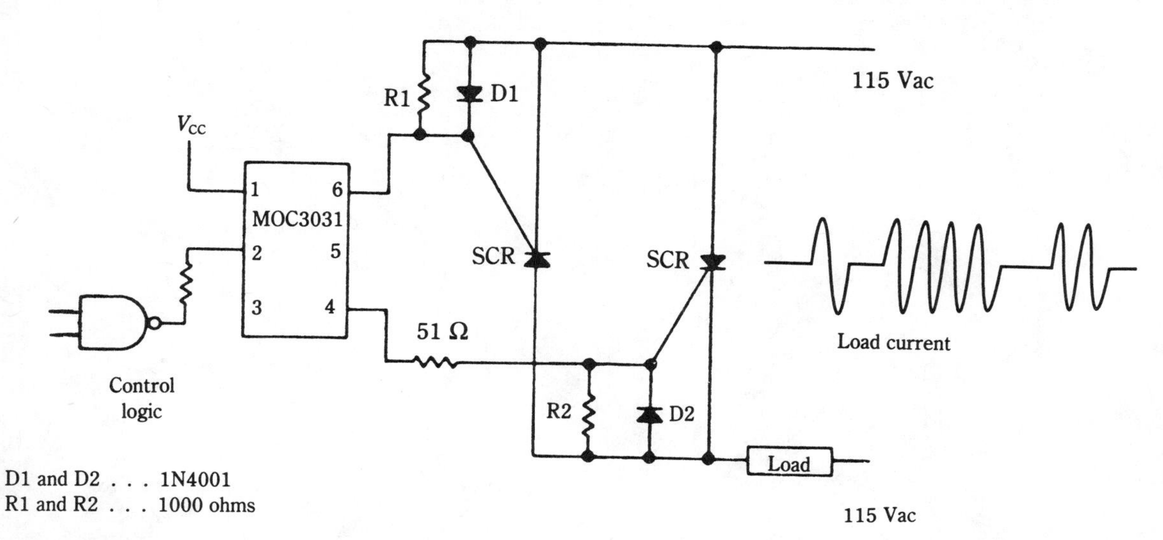

5-42 A burst-modulation ac supply using the MOC-3031 IC and anti-parallel SCRs. This control technique need not be restricted to 60 Hz. Depending upon the SCRs, operation at audio and higher frequencies can be realized. Motorola Semiconductor Products, Inc.

2 PART

Switching-type power supplies

In a treatise dealing with inverters and converters, it is only natural to discuss switching-type regulated power supplies as well. All of these systems components are "kissing cousins" — most generally, involvement with one leads to, or accompanies, involvement with at least one of the others. The last 14 chapters examine their theory, design, and application.

6
CHAPTER
An overview

MODERN SWITCHING-TYPE POWER SUPPLIES ARE CHARACTERIZED BY THE USE OF semiconductor devices that actually switch or interrupt the flow of current within the supply. Although this might sound like a rather odd mode of behavior, there are many benefits to be derived from such operation that cannot be obtained by traditional methods. These major benefits include high efficiency, small size, and the inherent capability to operate from a much wider range of input voltages. Additionally, as the cost of power semiconductors and integrated circuits continues to decrease, switching-type power supplies might also gain a very significant economic advantage.

Before embarking upon a detailed investigation of switching-type power supplies, it would be wise to become acquainted with the subject of power supplies in general. What, for example, is a power supply? What are you really dealing with? It is essential to pin down a few definitions at the very beginning to avoid later confusion. Even the nomenclature is cause for confusion, for a *power supply* does not usually generate or produce the power in itself, but rather it obtains the electrical power from a utility company located miles away. Rarely, when referring to a power supply, are we ever speaking of a battery or generator as the prime source of power—and even when we are, we most often exclude the battery or generator from our thoughts and concentrate solely upon the electrical circuitry that is connected to it.

Definition of a power supply

Electronic power supplies (or power supplies used in electronics) are most often charged with the task of altering, controlling, or regulating electrical power. The word *control* is quite inclusive: whether you rectify, invert, regulate, or change the voltage or current levels, some control technique must be involved. Such control of power delivered to a load can be achieved only by absorbing surplus power in the control device. To be energy conscious, the very concept of surplus or throwaway power deserves close scrutiny.

On the other hand, there are techniques that allow the control of load power without power dissipation in the controlling device. The basic methods of controlling electrical power to a load are shown in Fig. 6-1. Perhaps others might suggest themselves, but they will most likely be recognized as versións of those depicted. For example, there are other ways of using saturable-core devices, or magnetic amplifiers (usually shortened to *magamps*), and the rheostat is representative of many other devices, such as transistors, tubes, and thermistors.

Definition of a switching-type power supply

It is difficult to give a quick and concise definition that would distinguish switching-type power supplies from all other types. But it would be appropriate to say that a switching-type power supply is one in which the main flow of electrical power is generated, controlled, or regulated by means of switching devices. Most often, the switching-type power supply has an electronically regulated output. Switching-type supplies are usually more costly, and are thus expected to justify their performance, reliability, and cost on a comparative basis with conventional linear or dissipative-type power supplies.

Some power supplies involve a switching process, but are not ordinarily referred to as *switching-type supplies.* The reason for this amounts to nothing more than custom or tradition. A good example is the rectifying circuit shown in Fig. 6-2A, which from an academic standpoint would technically qualify as a switching-type power supply. How else can rectification be accomplished except through switching?

On the other hand, the mere fact that a power supply incorporates a switching process does not necessarily qualify it as a switching-type supply. For example, there are many new *programmable power supplies* of the linear type that utilize switching elements to establish the output voltage of the supply by switching resistances to alter the internal reference voltage (Fig. 6-2B). Other supplies, such as the old-fashioned vibrator type, generated and often rectified voltage using elements that are most definitely identifiable as switches.

The last three power supplies illustrated in Fig. 6-2 are more readily accepted as switching-type supplies by modern-day standards. In the popular converter, shown in Fig. 6-2D, the switching process is clearly evident. Switching in such supplies is usually performed by transistors or thyristors, often in conjunction with a saturable-core transformer that performs the principal switching task. Switching regulators often take the form illustrated in Fig. 6-2E, in which the transistor, diode, and inductor perform the major switching and regulating functions. For ac voltage control, regulation is accomplished by a thyristor switch, as in Fig. 6-2F.

Figure 6-3 illustrates a switching-type regulator in somewhat more detail. Such regulators are powered by an unregulated dc source. The switching process is used to perform the regulatory function by interrupting the flow of current from the unregulated supply. The operation of the switching device is controlled by an error amplifier or comparator, a feedback loop in the system, which continuously compares the output

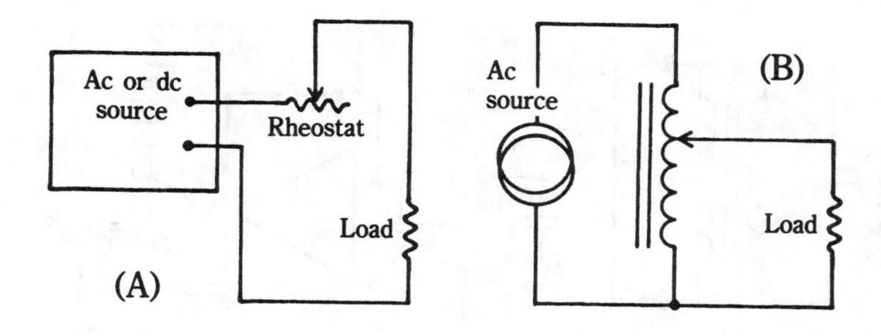

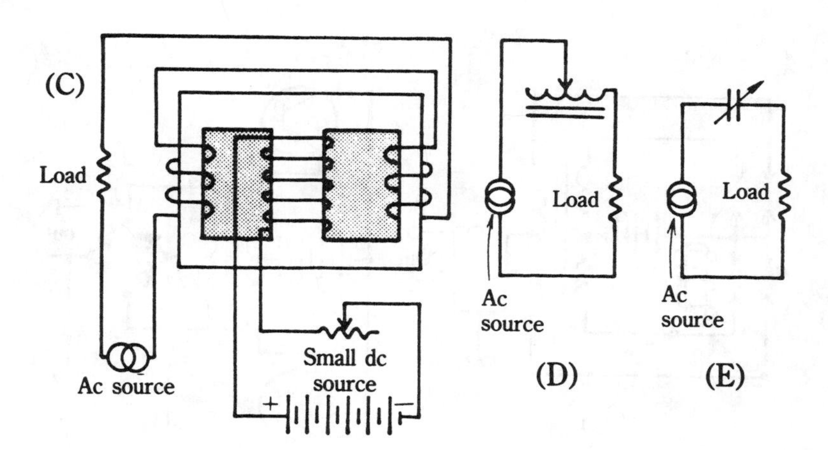

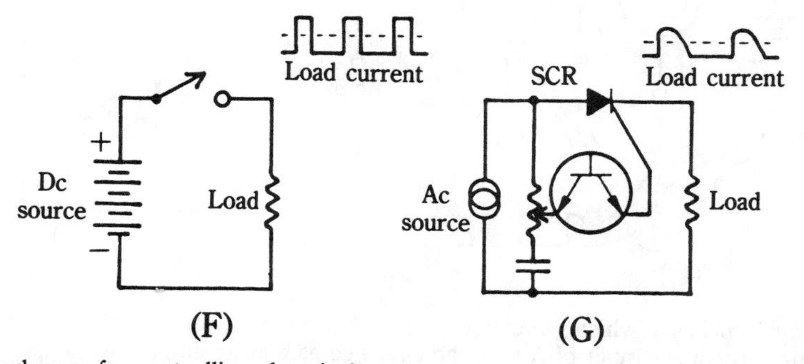

6-1 Basic schemes for controlling electrical power supplied to a load. Method A is probably the most common technique, but it is inherently the least efficient. All the other methods are, at least in their ideal form, nondissipative. Although the switching technique depicted in F appears to be primitive, it is the basis for modern high-efficiency regulating power supplies.

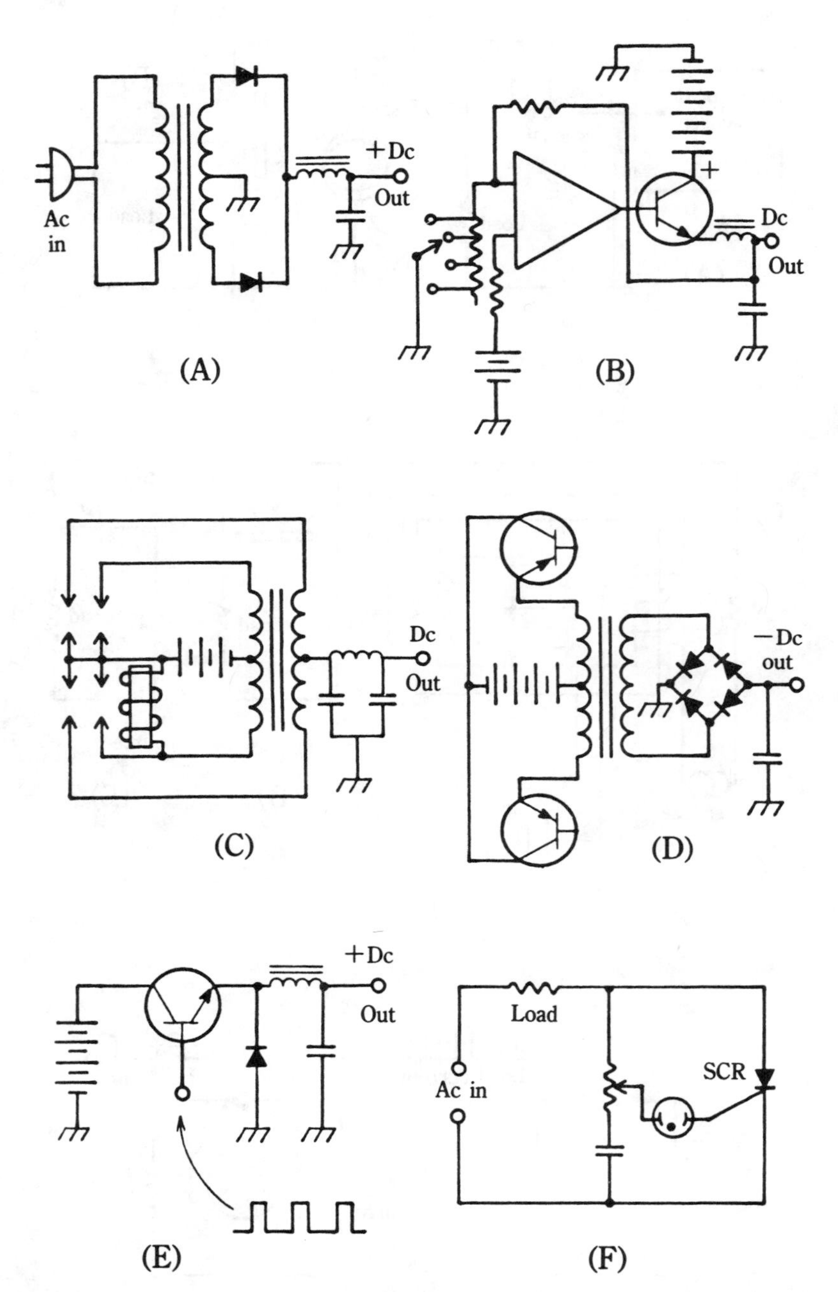

6-2 Various supplies in which switching techniques are used. From an academic viewpoint, the ordinary rectifier circuit of A must be classified as a switching power supply. In practice, however, rectifier circuits are not so thought of. On the other hand, the vibrator supply of C and its modern version, the saturable-core converter of C, definitely qualify as switching-type supplies. This is true also of the switching-transistor circuit of E, and the phase-control thyristor supply of F.

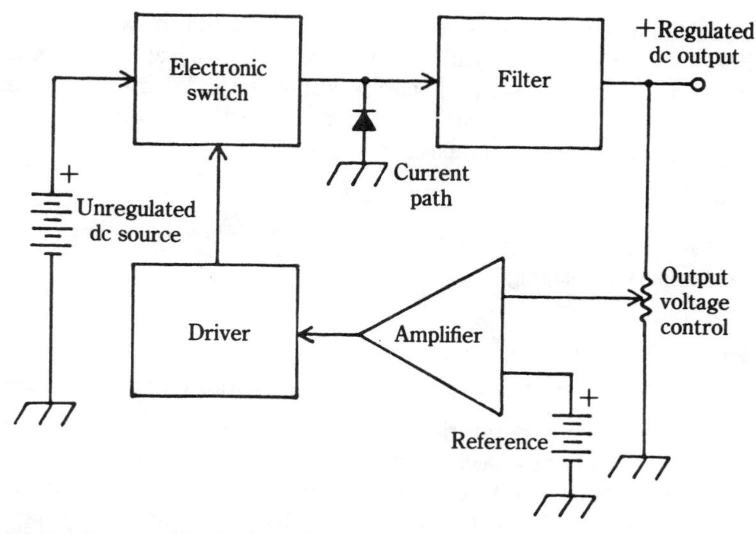

6-3 The block diagram of a basic switching-type power supply. The electronic switch can be a transistor or a thyristor, or a pair of such devices. Basically, this is an inverter with feedback to produce regulation. The filter and diode in the output more descriptively comprise an energy-storage system.

voltage to a reference voltage and automatically adjusts the switching operation to obtain the desired output voltage.

A switch is better than a rheostat

In *dissipative* regulation of voltage or current, power is deliberately thrown away. The dissipative element, often a power transistor, is imposed with the task of "soaking up" the excess power, and this naturally gives rise to heat-removal problems. It does not matter how sophisticated the control electronics are in such a power supply, the fact still remains that the dissipative element functions as a rheostat—a relatively crude method of regulating power.

In the switching-type power supply, a switching device is substituted for the dissipative device. Control or regulation of power is then achieved by varying the *duty cycle* or *repetition rate* of the switch, rather than its resistance. It is certainly valid to wonder what is so wonderful about this technique, considering that it must obviously be more complicated. The answer is that an ideal switch does not absorb or dissipate power—it is completely on or completely off, with no intermediate resistive state to dissipate power. As a result, the overall efficiency of a switching-type power supply is usually higher—much higher—than conventional dissipative-type power supplies. This factor is becoming an overwhelming point in favor of switching-type supplies because it results in significant advantages in size, weight, energy conservation, and cool operation. When these advantages are examined in relation to the requirements of the overall system, they often produce a significant economic advantage as well, particularly for high power levels.

There are a number of advantages that can be attributed to switching-type power supplies. But to be truly meaningful, the switcher must always be compared to a nonswitching, linear supply having similar power ratings. When this comparison is made, it will usually be found that the switcher has certain natural advantages over dissipative types, which can be summarized as follows:

- Higher electrical efficiency
- Lower operating temperature and relaxed heat-removal problems
- More compact packaging
- Lighter weight
- Wider range of input voltage
- Better "coasting" ability for momentary interruptions of the ac power line

These are the basic superiorities of the switching process. There are other parameters and characteristics (such as cost, reliability, and output purity), which will be discussed later. And, as might be suspected, there are also a few shortcomings. As with other circuits and systems, the switcher displays beautiful performance in one domain at the expense of poor behavior in others. There should be no delusion that the switching supply presents a panacea for all electronic ills. It does happen to be true, however, that its virtues often override its flaws, and this fact has resulted in the displacement of the dissipative supply by the switching type in an ever increasing number of applications. The switching supply has been a theoretical possibility for a long time, but it could not initially blossom into practical hardware. This has now been brought about by a combination of factors: semiconductor evolution, improved components, and advanced circuit techniques. The data presented in Table 6-1 is not all-inclusive, but it does provide initial guidance for the comparison of linear and switching-type supplies.

The principle of energy transformation

The goal of a near-dissipationless power supply is easily attainable if the requirement for regulation is removed, for then a simple transformer would be adequate for ac voltages, and an inverter or converter would be adequate for dc voltages. A transformer actually consumes very little power when properly designed, and its output voltage can be easily established by merely varying the ratio of turns between its primary and secondary windings. Similarly, an inverter or converter utilizes a high-frequency transformer whose turns ratio can also be varied.

The requirement for ***regulation*** of the output voltage or current, despite variations of the input voltage, greatly complicates matters. Regulation normally involves an error-correction process—circuits that can monitor the output voltage and make the necessary corrections to keep that output voltage within specified limits. A transformer does not lend itself to this type of operation because it is difficult to vary the turns ratio of the windings; there are mechanical methods of doing this, as in a variac or variable autotransformer, but fast, precise regulation demands a purely electronic process. In effect, what is needed is an all-electronic process for converting energy at one voltage

Table 6-1 A generalized comparison between switched and linear dc regulators.

Feature	Switch types	Linear types
Efficiency	65% to 85% is common.	25% to 50% is common.
Temperature rise	20°C to 40°C is readily achieved.	50°C to 100°C is not uncommon; depends greatly upon heat-removal techniques.
Ripple	20 to 50 mV peak-to-peak is often encountered. Smaller ripple voltage is usually difficult to achieve.	5 mV peak-to-peak is not difficult to attain and lower values can be had at greater cost.
Overall regulation	0.3% is common specification. Tighter regulation is usually difficult to achieve.	0.1% is commonplace and much tighter regulation is available at greater cost.
Power density	2.5 to 4 or 5 W per cubic inch for 20- to 50-kHz switchers. Higher switching rates can yield 25 to 75 watts per cubic inch.	0.3 to 1.0 W per cubic inch depending very much on power level, input voltage range, and heat removal hardware.
Isolation from line transients	Very good, often greater than 60 dB.	Generally inferior to switching types. Noisy line often plagues load.
RFI and EMI	Can be troublesome. Requires attention to shielding, suppression and filtering.	Less likely to be adverse factor.
Magnetics	Some designs can dispense altogether with bulky 60-Hz magnetics.	Bulky and expensive 60-Hz magnetics is evident in moderate and larger ratings.
Reliability	More parts, but recent designs capitalize on ICs. Enhanced reliability obtained from cooler operation.	Higher operating temperature often degrades reliability.
Cost	Cost dramatically decreases with higher switch rates. There is general tendency for decreasing cost as new semiconductors evolve. Cost crossover keeps going down; it might now be in 20-watt region.	Small linear types enjoy a cost advantage. However, with all the factors in the overall system considered, other cost factors become very significant in larger ratings. 60-Hz magnetics and thermal hardware can be high-cost items.

to energy at another. Ideally, this process should accomplish this energy transformation without introducing an appreciable amount of energy dissipation.

The modern switching regulator fulfills this demand through a unique switching process that provides for the temporary storage of energy in an inductive element, like that shown in Fig. 6-2E. In operation, the inductor is made to store energy during the time that the series switch is on, and to release that stored energy during the time that the switch is off. The output voltage or current of the switching regulator is then made controllable by simply varying the duty cycle or frequency of the switching process—a relatively easy task to implement electronically.

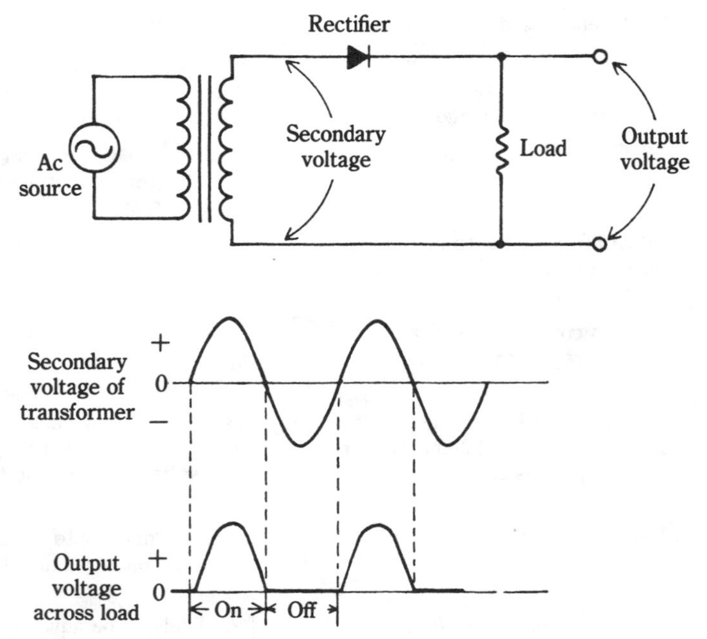

6-4 An instructive look at switching action in a simple rectifier circuit. The switching action of the half-wave rectifier is synchronized to the zero crossings of the ac secondary voltage. Assume that the diode is ideal with zero contact potential, zero forward resistance, and infinite reverse resistance.

This energy-transformation principle can be illustrated by first examining the operation of the simple half-wave rectifier circuit in Fig. 6-4. The rectifier diode actually performs a switching operation in that its function is identical to that of a switch synchronized to turn on when the energy to the load consists of voltage pulses having positive polarity.

If an inductor is now inserted in series with the load, as in Fig. 6-5, the energy pulses delivered to the load are stretched out so that they persist for more than a half-cycle of the applied ac waveform. With the switch in the circuit open, the pulses can never overlap to form a continuous voltage to the load. But with the switch closed, the energy stored in the magnetic field of the inductor is released into the load during the interval that the diode does not conduct. By appropriate selection of inductance and resistance values, the energy delivered to the load can be made continuous, with very little ripple.

The size or value of inductance is inversely dependent upon the operating frequency, or to be more precise, the pulse repetition rate. When driven by the 60-Hz power line, a half-wave rectifier circuit produces 60 pulses per second, whereas a full-wave or bridge circuit produces 120 pulses per second, two times the input frequency (this is a major reason for the preference of full-wave and bridge circuits over the simple half-wave circuit). The decrease in inductance becomes quite significant as the pulse rate is raised from 60 Hz to (for example) 60 kHz—for then the inductance value is also accompanied by a reduction in size and cost, though not by as large a factor.

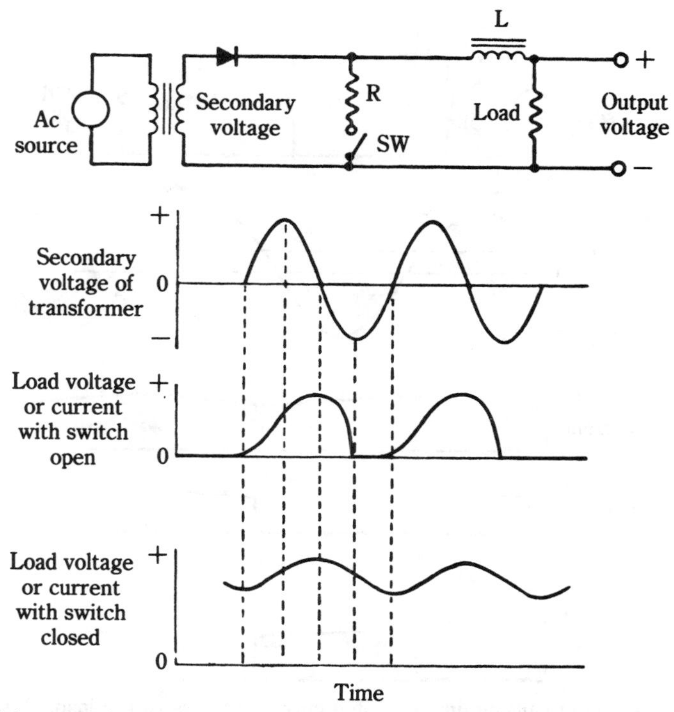

6-5 An inefficient, but instructive method of producing continuous current. Despite the half-cycle pulsing action of the rectifier diode, uninterrupted current passes through the inductor to the load.

This still serves to illustrate a significant advantage of switching-type power supplies that operate at high switching frequencies.

Returning to Fig. 6-5, the problem with using a simple resistor is that it dissipates energy and so introduces a form of energy loss that defeats our goal of a near-dissipationless power supply. Although the presence of the resistance can be used to prolong the release of energy stored in the inductor, it does so only by dissipating some of that energy. Fortunately, this problem is easily overcome.

If the resistor is replaced by another rectifier, the circuit in Fig. 6-6 results. Assuming that the diodes are ideal, the switching operation is entirely dissipationless; when a diode is conducting, its forward resistance is virtually zero, and when it is not conducting, its reverse resistance is high enough to be considered infinite. Rectifier D1 imparts energy to the inductor when it conducts, and it only conducts during the time that the ac input voltage exceeds the load voltage. Rectifier D2 provides a current path for the inductor during the interval that D1 is not conducting, as is shown by the waveforms in Fig. 6-6. Although the indicated waveforms show that the switching of the current through the rectifiers occurs instantaneously, this would not be so in a circuit using real diodes, for they would exhibit a forward resistance when conducting, as well as a forward voltage drop, and this would tend to slow the switching transition.

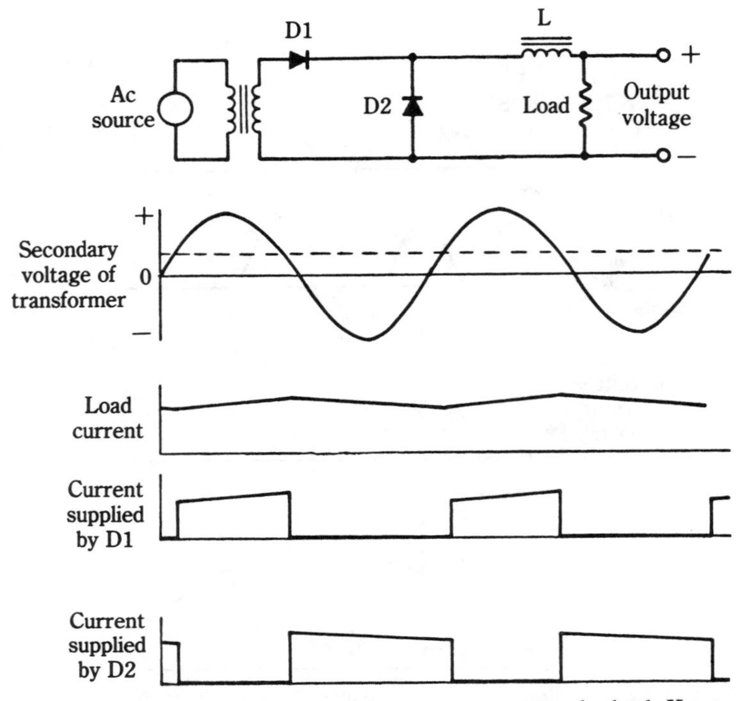

6-6 A better method of causing uninterrupted current to pass to the load. Here, an assumed nondissipative diode is substituted for the resistance in the circuit of Fig. 6-5. This "free-wheeling" or "catch" diode gates energy stored in the inductor to the load, but is inactive during forward conduction of rectifier-diode D1.

The circuit in Fig. 6-6 is representative of both ac and dc switching regulators. If rectifier D1 is replaced by a controllable switching element, and a dc input voltage is provided, the switching regulator in Fig. 6-2E results. For ac switching applications, rectifier D1 could be replaced by a thyristor, and output-voltage control could be obtained by varying the phase angle of conduction, in a manner similar to that shown in Fig. 6-2F. In either case, the output voltage is made controllable by varying the conduction time of the switching element. Therefore, it becomes possible to regulate the output voltage or current through a near-dissipationless switching process.

Regulating the switching-type supply

There are a number of circuit configurations, operating modes, and switching devices used in switching-type supplies. It is surprising how many different approaches can be used to attain similar objectives. All of the diverse methods rely upon the all-on and all-off conductive states in whatever switching device is used. That is, all switchers avoid partially conductive states. Indeed, this is a good definition of how they differ from linear dissipative regulating supplies. Even with the singular stipulation that only the two extreme states of conduction are allowed, there is considerable variation in the way that switching can be implemented. Some of these are listed as follows:

- Variation of both frequency and on time, either randomly or with constant off time
- Constant frequency with variable on time
- Constant on time (or pulse width) with variable frequency
- Half-wave, duty-cycle modulation of a sine wave (as with SCRs)
- Full-wave, duty-cycle modulation of a sine wave (as with triacs)

Moreover, the switching process can be implemented with the switching device inserted either in series with the load or essentially in parallel with it. The switching can be done at the frequency of the power line, or at thousands or even at millions of times per second. As might be anticipated, each approach offers unique performance features, tradeoffs, and economic variables. On the other hand, the state of this art is very fluid—the various design techniques compete with one another. The ultimate desirability of one or another approach is often dependent upon the particular development state of certain components. For example, if today one technique appears excessively costly or unreliable because of an inordinately high number of active devices needed, it is quite likely that an integrated circuit will make its debut tomorrow for the express intent of reducing device count to an inconsequential few. The art and technology of switching-type supplies is dynamic and fast moving, so one of the first things you should do is familiarize yourself with the various approaches found in these interesting power supplies.

7
CHAPTER
Theory

ONE OF THE MOST IMPORTANT FUNCTIONS OF SWITCHING-TYPE POWER SUPPLIES is the control or regulation of the supply voltage used to power electronic circuitry. Such supplies are normally required to maintain a fairly constant output voltage, despite widely varying input voltages. The regulated output voltage can be either higher or lower than the unregulated input voltage, but in most applications, the output voltage will be lower than the input voltage. The most common type of switching regulator is a three-terminal system that accepts an unregulated dc input voltage and provides a regulated dc output voltage. This type of regulator is examined first because it has more universal applications.

Switching regulators used to convert ac line voltage into a dc output voltage are normally implemented by the addition of a conventional rectifier-type power supply to first change the ac into unregulated dc. Switching-type power supplies used to provide a regulated ac output voltage are most often referred to as *inverter-type power supplies*; these are examined in the next chapter.

A simple switching regulator

The simplified diagram of the switching-type regulator shown in Fig. 7-1 will provide basic insights into the various techniques for approaching the goal of dissipationless regulation. Obviously, the emphasis in power supply design is primarily on voltage, though sometimes current regulation is also needed. In either case, the operation of the regulator depends upon setting up a feedback loop to monitor the output voltage or current and to provide a correction or error signal that is used to automatically adjust the operation of the switching device. This type of feedback operation is able to stabilize the regulator's output, despite variations either in the unregulated input voltage or in output load parameters. Any regulated supply can, for special applications, be divested

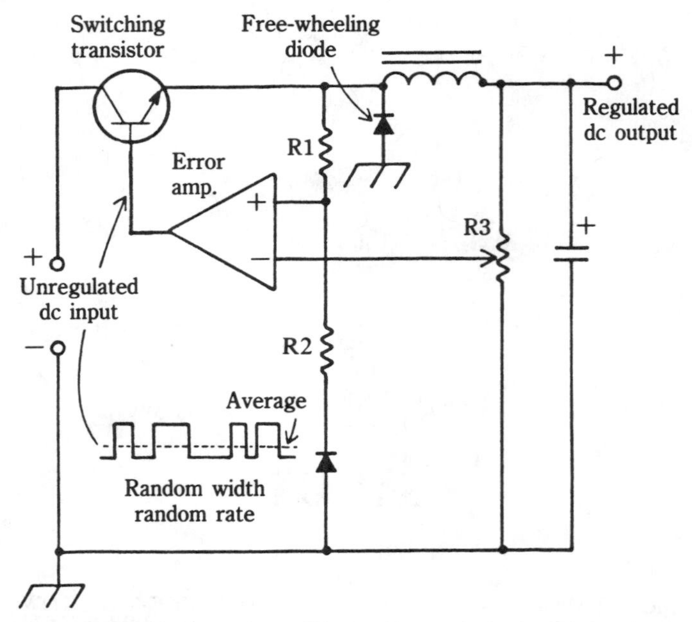

7-1 The self-excited switching regulator. The oscillatory behavior of the circuit is affected by input and output conditions, causing variations in both pulse width and pulse rate. Experimentation is usually needed for best performance.

of its automatic stabilizing operation by simply severing the feedback loop. This can be desirable, for example, in motor-control applications, where manual control of motor speed is sometimes required. However, most electronics are best served by automatic, closed-loop operation; thus, the following discussions will focus upon these topics.

In Fig. 7-1, the basic configuration and main elements of a simple voltage-regulated switcher are shown. In order to center attention on the switching circuitry, the unregulated power supply is not shown. An interesting aspect of this arrangement is that it is *self-oscillatory.* The oscillation frequency or switching rate is primarily determined by the inductance and capacitance of the output filter, but it is also affected by other factors. The output voltage level is proportional to the average value of the on times during successive pulses. Thus, the function of the filter is to average out the amplitudes of the switching pulses to produce a constant output voltage whose amplitude is equal to the arithmetic average of the pulses delivered by the transistor switch. Notice that the switching waveform is "wild" in that both pulse width and repetition rate are subject to variation. This imposes no great problem, however, for the action of the feedback loop is such that regulation is implemented by variations of pulse width, pulse rate, or of both of these parameters.

In this supply, and in most others, the so-called "free-wheeling" or "catch" diode will be seen. From the previous discussions of the half-wave rectifier, you will recall that a resistance connected in this circuit position enabled the current flow in the load to be more sustained than otherwise, but that the resistance had the objectionable feature of dissipating power. The free-wheeling diode extends the current-sustaining phenomenon into the realm of usefulness, for the diode provides high conduction when it is needed, and negligible dissipation otherwise.

The sense of simplicity that is derived from inspecting this self-oscillatory regulation can be somewhat deceptive. You might well ask, for example, why such a simple arrangement was not popular long ago. Part of the answer resides in the circuitry, implied but not shown, within the symbol of the error amplifier. Whether obvious or not, a lot is demanded from this functional element. If you want a lot of performance out of the switching regulator, and at the same time wish to avoid physically massive filter components, a nominal switching rate is needed that is many times the power line frequency. For reasons to be covered later, this tends to minimize the operation in the vicinity of 20 kHz. A good error amplifier (or comparator) was, at first, an elusive object.

Additionally, suitable switching transistors, free-wheeling diodes, and even electrolytic filter capacitors were only marginal, insofar as their suitability for high-frequency switching supplies was concerned. The cost of acceptable components was often prohibitive. The liberties you could take with design simplifications were also limited because competition with linear regulators, where superb regulation and noise-free operation were obtained relatively easily, made the cost and effort involved in producing a good switching regulator unjustifiable.

One of the really significant turning points in both the technological evolution and the economics of the switching-type supply was the advent of inexpensive, highly reliable integrated circuits. Thus, for a dollar or two you can now acquire an IC operational amplifier (op amp) for use in the schematic position, depicted by the triangle in Fig. 7-1. Such an op amp might comprise the equivalent of a dozen or more discrete elements, and these are direct-coupled for dc operation. Even more compelling to the designer than the op amp is the IC linear regulator, for these contain not only an op amp, but an excellent self-contained voltage reference. The thermal stability and electrical performance of this voltage reference exceeds that which is generally attainable from simple zener diodes. On a cost basis, it is competitive with voltage regulators. Although it was originally intended for linear operation, it is also quite suitable as a building block in switching regulators. In fact, most manufacturers of linear voltage regulators will gladly provide application notes showing how to use their regulators to make switchers (even better is the present availability of many dedicated control ICs specifically intended for use in switchmode power supplies).

The best of two worlds

The combination of switching and linear techniques has led to levels of performance not readily attainable using one technique alone. Although switching-type supplies are often simple configurations involving an electronic switching or chopping element, the trend of development increasingly favors a system approach, wherein other functional blocks, such as linear regulators, are also involved in the overall scheme.

Consider, for example, a regulated power supply consisting of a switching-type regulator followed by a linear regulator. In such a setup, the switching circuitry can be considered a *preregulator* for the linear output regulator. This is representative of the simplest of system approaches, but even this system has compelling features. The linear regulator is superior to the switching type in ripple attenuation, in the development of low output impedance, and in obtaining regulation within tight boundaries. But, for many applications, the low operating efficiency of the linear regulator presents severe thermal, electrical, and packaging problems. The low efficiency of the linear regulator is greatly aggravated with increased input voltage from the unregulated power supply. In order to take care of fluctuations in the ac power line, as well as worst-case conditions of load and temperature and a modest margin of safety, the dc input voltage often must be much higher than is compatible with reasonable operating efficiency. Unfortunately, the dissipation in a series-pass transistor increases in proportion to the voltage it must drop to produce its regulated output voltage.

It should be evident that a preregulator would relieve the series-pass transistor of a considerable burden, and that the overall efficiency could be increased if the preregulator itself was very efficient. With such a combination, line voltage could increase greatly with negligible worsening of the overall efficiency. The preregulating switcher might operate at an efficiency of 85 percent; the linear regulator might operate at a nominal efficiency of 50 percent, and the switcher would prevent it from dipping below this figure at high line voltage. At high power levels, an improvement of 10 to 15 percent in operating efficiency can be quite meaningful. Later, a number of other switching-type power supplies that use the system approach are investigated.

The self-excited switching regulator

The self-oscillating switching regulator in Fig. 7-1 excels in its utter simplicity. Although the operation of the switcher is determined in part by the inductance and capacitance of the output filter circuit, its operation is externally affected by both the input supply and output load. As a result, the pulse width and pulse rate of the self-excited regulator might vary considerably because the output filter performs the primary task of filtering instead of tuning.

In the self-excited switcher, the relationship between pulse width and pulse rate is not entirely random because the switcher is required to deliver a specific output current and voltage to the load. In general, however, the switcher tends to vary its pulse rate in response to changes in load current and to vary its pulse width in response to changes in unregulated dc input voltage. This is a generalization and is not meant to imply that the two actions are entirely interdependent, but rather that these are the dominant effects.

To a certain extent, the design of a self-excited switching regulator is greatly complicated by the fact that both pulse rate and pulse width are permitted to vary. As a result, the behavior of the self-excited switcher is difficult to predict. In practice, the initial design must be set up using very loose approximations, sufficient to obtain a working circuit; subsequent refinements are then made during actual operation, at which time changes in component values of up to 100 percent are not at all uncommon.

The operation of the self-excited switcher in Fig. 7-1 is easily explained in general terms. The sense amplifier performs a comparator function; that is, it compares the output voltage of the switcher (fed back via resistor R3) to a reference voltage, which in this case is provided by a zener diode. When the output voltage of the switcher drops too low, the sense amplifier turns on the switching transistor; when the output voltage rises too high, the switching transistor is turned off. The only problem encountered in this simple scheme is in making the switcher oscillate reliably under all conditions of dc input voltage and output load current.

To ensure reliable operation, it is necessary to add a certain amount of hysteresis (provided by R2 and R3) into the feedback loop to introduce sufficient positive feedback for guaranteed oscillation and fast switching of the transistor. But the presence of these resistors in the feedback loop also affects the pulse width, pulse rate, output ac ripple voltage, and average dc output. Consequently, although the self-excited switcher can boast of having few components, the design of the switcher is actually complicated by the fact that there are not enough components to isolate the controlling parameters that govern its operation. Still, the self-excited switcher remains quite popular in less-demanding applications, where its low cost and few parts can be compelling features.

The externally excited switcher

In the self-excited switcher of Fig. 7-1, it was observed that both the pulse width and pulse rate varied freely, and that such freedom introduced design problems in guaranteeing reliable oscillation. At the sacrifice of a slightly more complex circuit, it is possible to establish two possible modes of operation, simply by making either the pulse width or pulse rate a constant parameter. These two modes can be summarized as follows:

- Regulation is obtained by holding the pulse rate (frequency) of the switcher constant and permitting only the pulse width (the on time) to vary.
- Regulation is obtained by holding the pulse width constant and permitting only the pulse rate to vary.

In the self-excited switching regulator, the division of labor between varying the frequency and varying the pulse width is more or less capricious. No doubt, one parameter or the other could be made relatively constant, thereby shifting the corrective action to the parameter with greater freedom of deviation. These objectives are more easily achieved in switching circuits, which do not rely upon self-oscillation for their basic operation, but instead use separate oscillators to produce the waveform, which turns the switching transistor on or off. It then becomes easier to independently modulate or control either rate or duration without affecting the other.

It is often desirable that circuit function blocks be isolated to permit them to be optimized for one exclusive function. Otherwise, you find yourself in essentially the same position as the early proponents of "reflex" radio sets. In these now obsolete circuits, one stage or channel might handle RF, IF, and perhaps audio frequencies as well. Although the idea is intriguing, there are practical drawbacks in attempting to optimize all of these operating modes at the same time. In the case of the regulator, more reliable performance and better operating characteristics are generally achieved by allowing the switching transistor to be simply a voltage-controlled switch. You can also separate the functions of positive and negative feedback and isolate the reference voltage at the sense amplifier. By dispensing with the need for self-oscillation, the positive feedback path is entirely eliminated.

With separate oscillators, you are free from such malfunctions as failure to start, for example, at low temperatures and with a heavy load, and from certain transient disturbances, which tend to "shock" the LC filter into oscillation at a "natural" frequency unsuitable for regulation. Certain changes, such as simultaneous line and load changes, can also drive the pulse rate and pulse width in antagonistic directions, thereby upsetting the regulation. Finally, as a designer, you can more accurately specify the values of the filter components if only one switching parameter is allowed to vary. Switching regulators of this type are easier to mass produce within given specification boundaries. The alternative—overdesign or the "brute-force" approach—can easily negate the apparent low-cost feature of the self-excited switcher.

The fixed-frequency, variable pulse-width regulator

The operation of the system in Fig. 7-2 is similar to that of the self-excited type, but the switching rate is now fixed at the frequency of the separate oscillator. From a practical point of view, the operation of the supply can now be optimized by merely shifting the frequency until optimum results are obtained. This is often desirable because the initial design cannot take into account such variables and unknowns as the operating inductance and Q of the choke, the inductance and effective series resistance of the filter capacitor, and the specifications of the switching transistor and the free-wheeling diode. It is also nice to have independent control of the frequency from the viewpoint of the powered equipment; switching regulators generally produce greater ripple than their linear counterparts, and it is often profitable to have the option of adjusting the ripple frequency to minimize load disturbances.

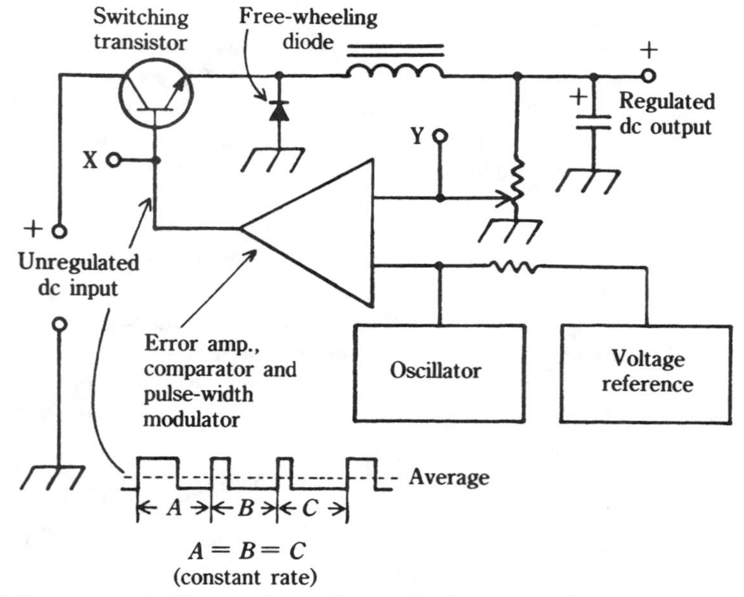

7-2 Fixed-frequency switching regulator. Here, the switching rate is established by a separate oscillator. Regulation is accomplished by variation of on-time pulse width. Design is more predictable than the self-excited circuit of Fig. 7-1.

Constant pulse width, variable-frequency regulator

The alternate fixed parameter scheme is shown in Fig. 7-3. Here, the pulse rate of the monostable or "single-shot" multivibrator is varied by trigger pulses obtained from a voltage-controlled oscillator. The duration of the on pulses applied to the switching transistor are determined by the time constants within the multivibrator, and remain fixed. This is a rather elegant scheme that allows independent adjustment of both the pulse width and the pulse rate. It is not easy to say that either this method or the one previously described using fixed frequency is better, although it is possible to cite certain advantages for one or the other. Any possible superiority is a function of the design method, so it is probably best to say that although one designer could attain superb results with one approach, another designer might achieve equally favorable results with another. There are fine points of difference having to do with output ripple, allowable adjustment range of the dc output voltage, and isolation from line transients, but both fixed-parameter techniques have good performance capabilities (the fixed-width technique works exceptionally well in resonant-mode supplies, to be described later).

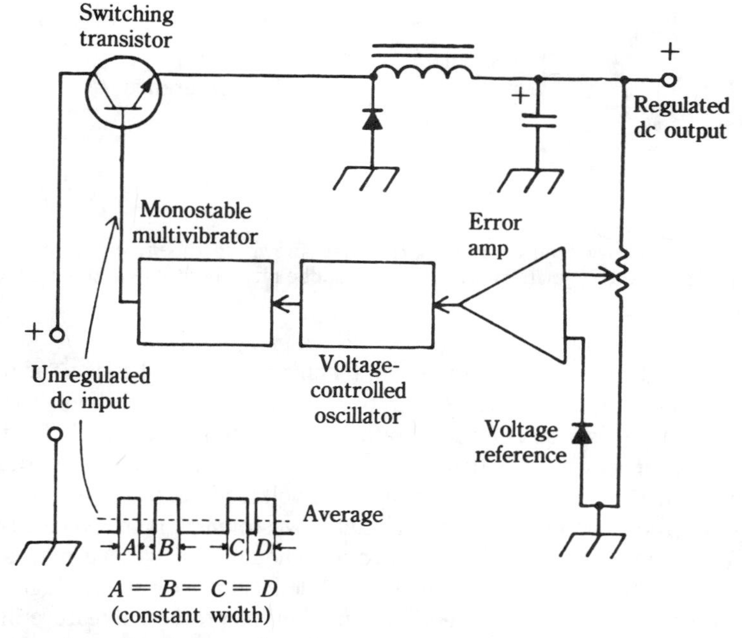

7-3 Alternate scheme for the operation of a switching regulator. The monostable multivibrator generates on pulses of fixed duration. The pulse rate, however, is governed by the voltage-controlled oscillator.

The shunt switcher

Thus far, all of the switching regulators covered thus far have utilized a series switching element. We know from linear regulators that shunt elements are sometimes advantageously used, so it would be only natural to use a shunt-connected switching element to accomplish regulation. It so happens that the shunt switcher is practical, and would assume the form depicted in Fig. 7-4. This particular arrangement uses a constant pulse-width technique, but fixed-frequency schemes can also be implemented. The shunt switcher develops output voltage levels that are higher than that of the unregulated input supply.

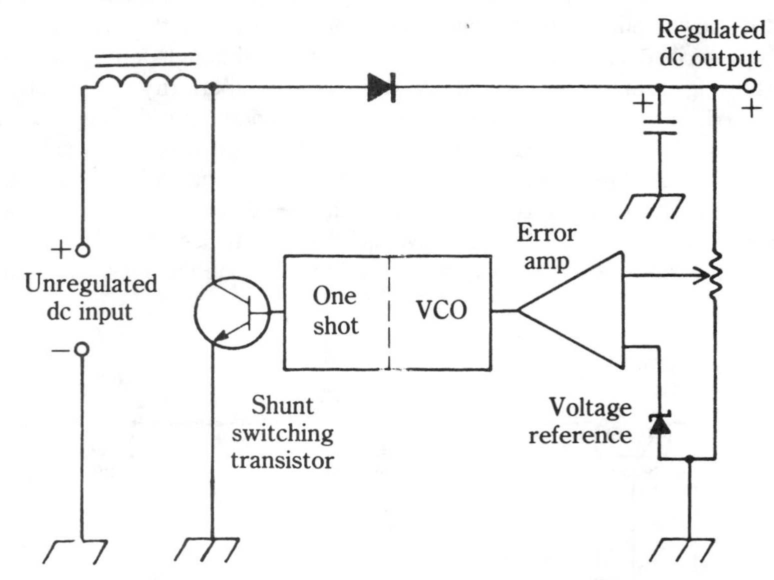

7-4 The basic shunt-switching circuit. Otherwise known as the *flyback circuit,* the salient feature is that the regulated output voltage can be much higher than the unregulated input voltage.

A possible advantage of this switching technique is that it handles radio frequency and electromagnetic interference easier because the current through the switching transistor following turn on is a ramp, rather than a step, function. A disadvantage of the shunt switcher, however, is that it has a lower efficiency than the a series switcher; this results primarily from the fact that the input voltage is less than the output voltage, so voltage drops occurring in the input circuitry waste much more power. But the fact that the shunt switcher is able to produce a higher output voltage makes the shunt switcher very useful in converting low-voltage dc battery power into higher voltages.

The shunt-switching technique offers an inherent protective feature in the event that the switching transistor locks up in either an open or closed state: in either case, the load would be spared the likelihood of destruction. If the switching transistor stays in its on state, it would act as a "crowbar" short across the supply, and with appropriate design could interrupt a fuse or circuit breaker—even destruction of the switching transistor is generally preferable to damaging costly equipment. If the switching transistor remains off, there would be a decrease in the output voltage, and the load would again be protected from damage because the output voltage of the regulator would then be no greater than the dc input voltage.

The operation of the shunt switch is analogous to that of other circuits using the "flyback" principle. Included among these are automobile ignition systems, electronic switching supplies for geiger counters, and high-voltage supplies in television receivers. The difference in the shunt switching regulator rests in the use of a feedback arrangement to stabilize the output; otherwise, all flyback circuits use the counter electromotive force developed by either an inductor or the primary winding of a transformer.

Obtaining variable pulse widths

The block diagram of Fig. 7-5 represents one method of achieving the constant-frequency mode of operation depicted in Fig. 7-2. Only the control circuitry is shown, but its association with the switching transistor and accompanying components is indicated by reference to points X and Y in both figures.

At first glance, the arrangement of Fig. 7-5 might appear rather complicated; however, the operating principle is straightforward, and the use of integrated circuits makes for a low component count. This is an excellent example of translating concept into hardware, a task that would be severely hampered if each functional block were constructed from discrete devices. Here, just a few inexpensive ICs together with some external components produce a complete pulse-width modulator.

The square-wave oscillator does not have its frequency affected directly, so it is possible to vary the frequency manually to optimize the operation of the circuit. A simple square-wave generator could be constructed using a multivibrator because the accuracy of the frequency is often critical. The integrator converts the square wave into a triangular wave, which is then summed with the output voltage from the sense amplifier. The Schmitt trigger is a level-sensitive circuit that produces a rectangular output pulse, as illustrated in Fig. 7-5. As the output voltage of the sense amplifier varies in response to the output of the switching regulator, the triangular waveform is shifted up and down, varying the trip point of the Schmitt trigger. This produces an output pulse of varying duration that is then applied to the switching transistor. A driver block is shown in Fig. 7-5, representing an extra stage of amplification, but this is frequently omitted in switching regulators of 1-A capacity or less.

There are, of course, many other ways to implement the same functional mode of operation, including purely digital techniques. For precise frequency control, an external frequency source could be used. And there is no requirement that the waveform be square or even triangular; many systems incorporate ramp-type waveforms and sine waves. An example of the linear-ramp technique is illustrated in the block diagram of Fig. 7-6, and it typifies many such designs.

The functional-box approach of the switching regulator system in Fig. 7-6 uses a block-marked *analog-to-digital converter.* Such boxes already do exist in integrated circuit and modular form, but the important fact is that the box generalizes the basic function to be achieved in obtaining a variable pulse-width output. Although the schemes illustrated in Figs. 7-5 and 7-6 provide useful insights, practical implementations increasingly use designated control ICs, in which the PWM function, together with many "bells and whistles" are built into the chip. Reduced to its simplest form, the switching regulator no longer represents a stumbling block to the designer.

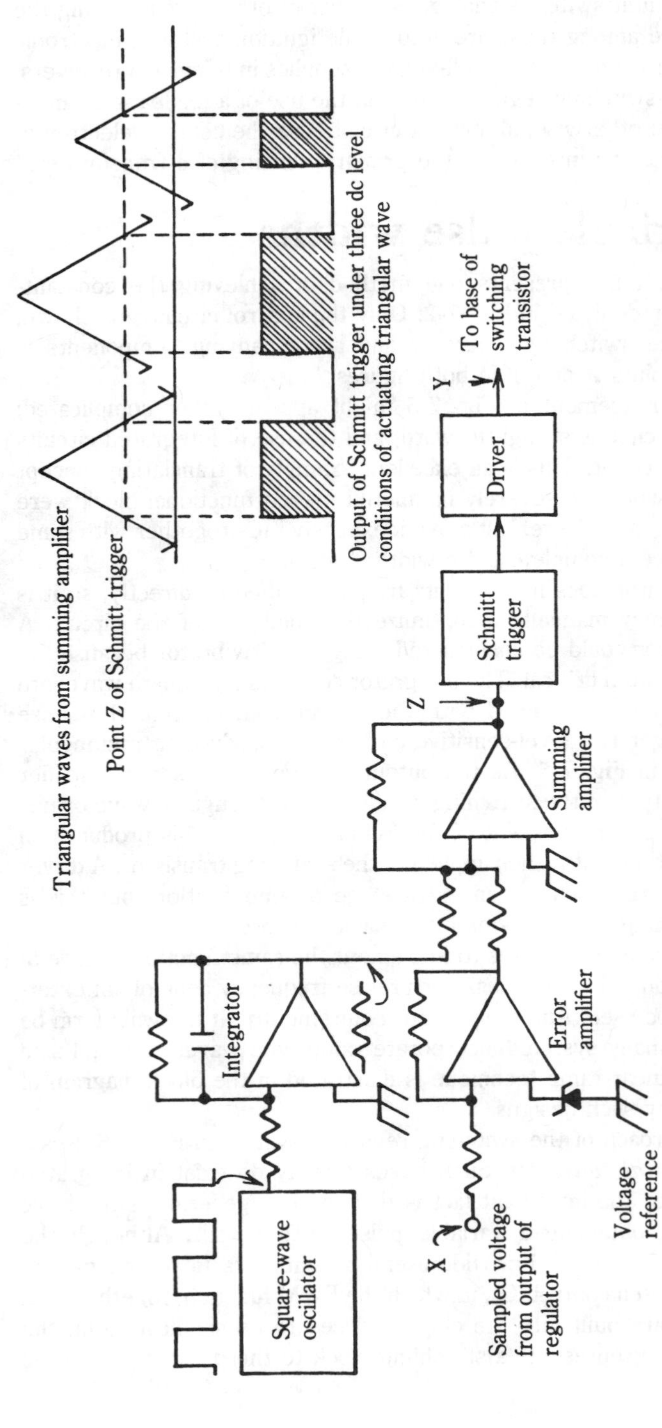

7-5 A block diagram showing basic concept of a pulse-width modulation system. In practice, the Schmitt trigger incorporates a small amount of hysteresis so that its turn-on voltage is slightly higher than its turn-off voltage.

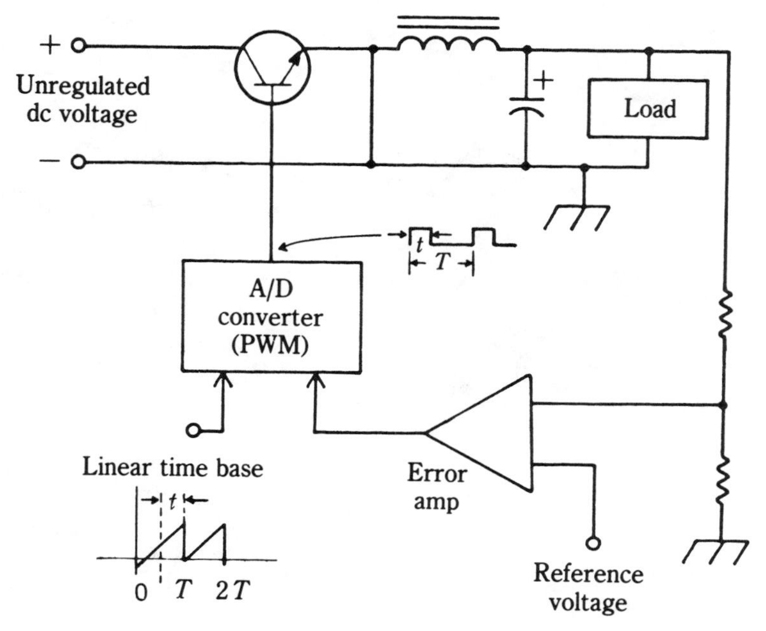

7-6 Another view of the PWM switching technique. Here, the pulse-width modulation takes place by means of an analog-to digital converter. This is one of the several schemes incorporated in many of the PWM-control ICs that are commercially available.

8
CHAPTER

Switching-type power supplies

THE SIMPLE SWITCHING REGULATOR PRESENTED IN CHAPTER 7 REPRESENTS one approach to efficient power-supply design. In this chapter, several other power-supply systems will be presented, many of which use the simple switching regulator as a part of their complete system. Because the simple switcher has already been covered, its function in a larger system will be readily understood as will variations of the simple switcher, using SCRs and triacs instead of transistor switching elements.

Ac-powered switchers

Switching-type power supplies that use arrangements, such as those shown in Fig. 8-1 represent a quantum jump in power supply advancement. These regulators are, in essence, systems of functional blocks. The major feature of these regulators is the elimination of physically massive power transformers and other power-line magnetics. The net result is a smaller, lighter package and reduced manufacturing cost, resulting primarily from the elimination of the 60-Hz components.

In both systems in Fig. 8-1, the incoming ac power is immediately rectified and filtered. This unregulated dc power is then converted to a frequency many times that of the power line, at least in the 20-kHz range. Then, by one or another type of regulator, the high-frequency power is changed to regulated direct current. Because of the high switching frequency, magnetic core and filter components are minuscule compared to their 60-Hz counterparts. Moreover, the 60-Hz filter at the front end of the system can be quite simple, since subsequent regulation provides electronic filtering of power-line ripple frequencies.

Schemes of this kind represent a bold new concept in power-supply design. There are actually many versions, all incorporating the basic idea of ***internal frequency step-up***. The generation of the high-frequency power can be performed by a saturable-core inverter, although other nonsaturating types are also used. A large-scale imple-

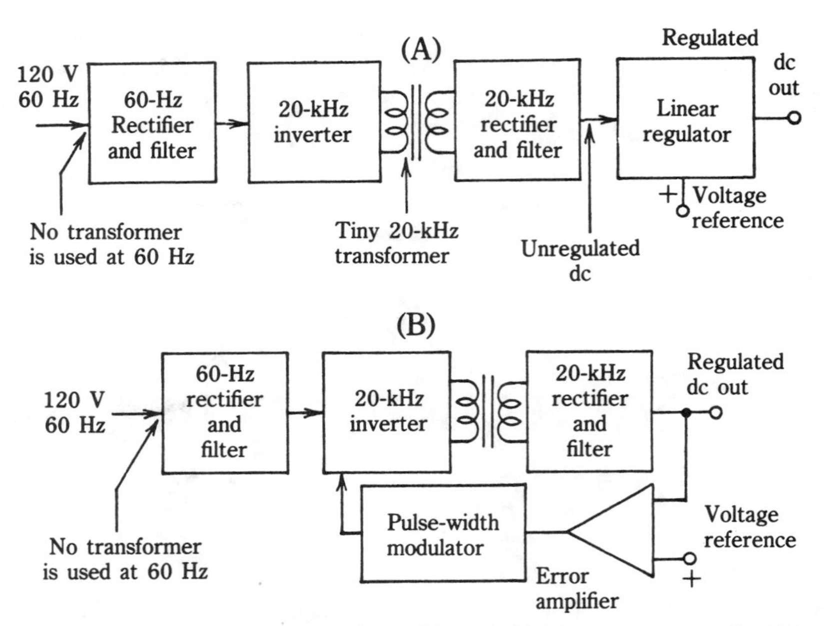

8-1 Two topographies that are commonly used in switching-type power supplies. The hybrid approach in A combines high efficiency of the inverter with precise voltage control of a linear regulator. Arrangement B provides even greater efficiency by using only switching techniques for regulation. Ripple and noise are, however, higher than in A.

mentation of this design philosophy was carried out by Tektronix Inc. when the firm introduced a new line of oscilloscopes, which, though brimming with the latest test functions, were dramatically more compact than previous instruments. The use of the sophisticated switching supplies enabled the oscilloscopes to operate at reduced temperature and consume appreciably less power than their predecessors.

Voltage-stabilized inverter

Figure 8-2 provides a more detailed inspection of one way in which the frequency step-up can be accomplished in the inverter block of Fig. 8-1B. The inverter depicted is not the self-excited type common to such applications as the generation of ac from automotive batteries; rather, this inverter is driven from a logic-controlled source, starting with a square-wave oscillator. The two switching transistors function essentially as a push-pull amplifier: each transistor is alternately driven full on and then off, and a rectangular wave is developed across the primary of transformer T1. Unlike self-excited inverters, the operating frequency of this inverter is not dependent upon saturation of the core of T1.

The prime operating feature of this arrangement is that although the square waves originate from the oscillator, across transformer T1 the duration of the rectangular wave is varied by the logic circuitry in the *pulse-width control* block. This variation occurs in response to the amplified error signal in such a way as to regulate the level of

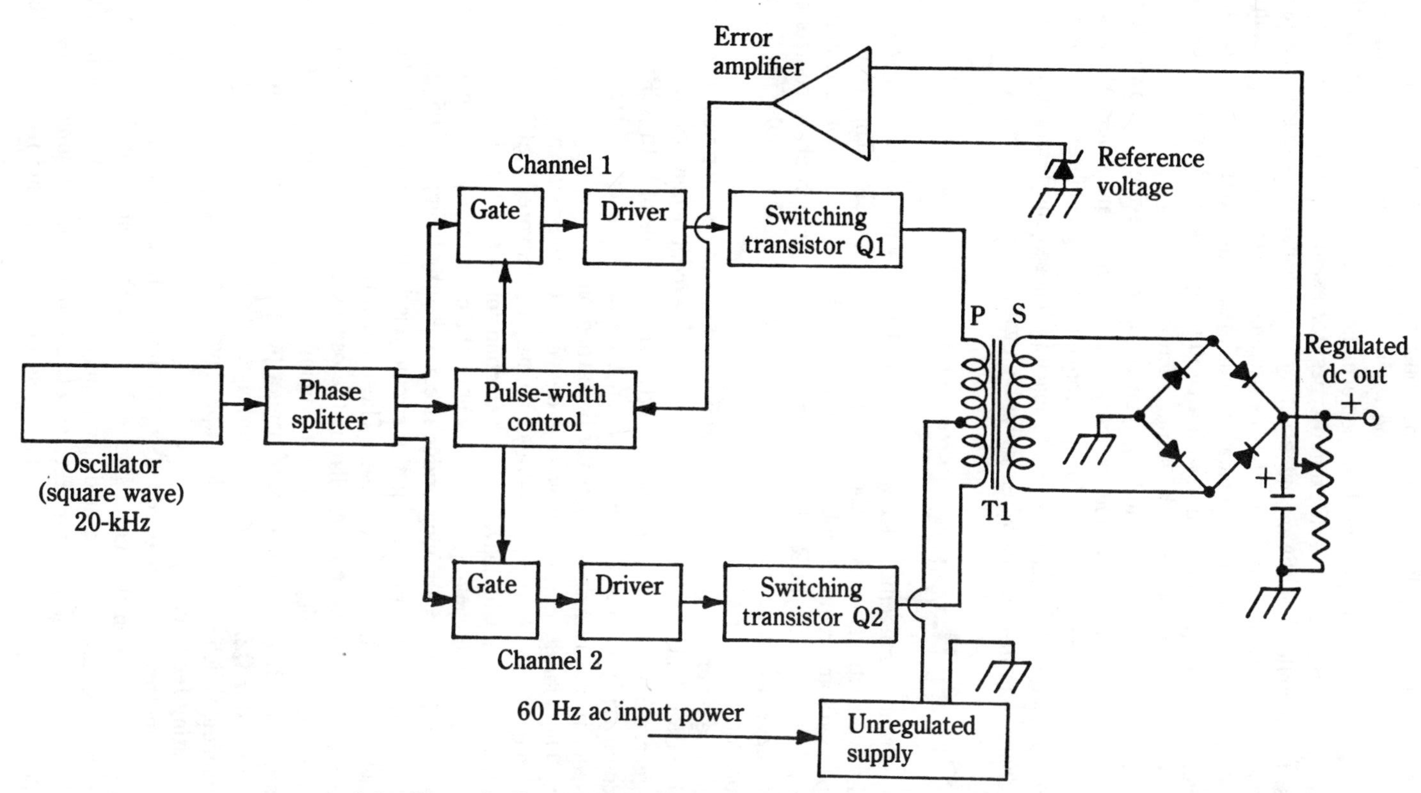

8-2 The major functional blocks of a voltage-stabilized dc-to-dc converter.

the dc output voltage. When the output must be increased, the pulses of the rectangular wave are longer; the converse is true when the output voltage must be decreased. An inverter of this type has excellent performance and is virtually devoid of the malfunctions which often plague self-excited types; for example, there are no starting difficulties, the likelihood of the circuit bursting into oscillation is greatly reduced, and the pulse-width control ensures that both switching transistors will never be simultaneously *on*. A *phase splitter* provides a balanced push-pull signal to the two processing channels; it can be simply a transformer with a centertapped secondary, or it can be an active circuit designed to perform the same function.

Unlike most of the switching-type regulators previously shown, neither a filter choke nor a free-wheeling diode is necessary with inverters. One of the objectives in the design is to use such a high switching frequency that good filtering of the output can be accomplished with a relatively small capacitor. These considerations also tend to reduce feedback loop instabilities when using high-gain error amplifiers.

SCR full-wave regulator

The circuit arrangement shown in Fig. 8-3 represents a practical way of combining full-wave rectification with the desirable features of SCRs. This technique is easy to implement, and requires neither a centertapped transformer nor critical firing circuitry for the gates of the SCRs. Regulation is achieved through phase control of the SCRs, but there are some subtle aspects to this circuitry that might not be immediately apparent.

The main rectifying bridge is comprised of two conventional rectifying diodes and two SCRs. This bridge provides no output, except when one or the other SCR is triggered into its conductive state. Although the gates of both SCRs receive the same turn-on trigger, only that SCR, which then has its anode polarized positively, with respect to its cathode will actually be turned on.

The average current or voltage delivered by the rectifying bridge is governed by the time delay in a given half-cycle until the "favored" SCR turns on; thus, this rectifying bridge will deliver various fractions of half-cycles. The greater the fractional part of a full half-cycle, the greater will be the current or voltage delivered to the filter. Because the main rectifying bridge is controlled by timing the triggers to the SCRs, it follows that the remainder of the circuitry is largely devoted to the production and timing of the SCR trigger pulses. These triggers are generated by UJT Q1, which is connected as a relaxation oscillator; the oscillation frequency is, among other factors, determined by timing capacitance C, timing resistance R, and the effective resistance of transistor Q2. The occurrence of trigger pulses is controlled by varying the conduction of transistor Q2. In turn, Q2 is controlled by the error signal obtained from the error amplifier, which samples the dc output level of the regulator, as is the case with most regulating feedback networks.

A method is also needed for synchronizing the UJT oscillator to each zero crossing of the ac wave on the power line. Otherwise, the control of the SCRs will occur in a hit-and-miss fashion. The auxiliary supply is not filtered; there is no filter capacitor following its bridge rectifier. Because of this omission, the ripple in the output causes a momentary dip to zero, twice per cycle of the power-line frequency (the ripple fre-

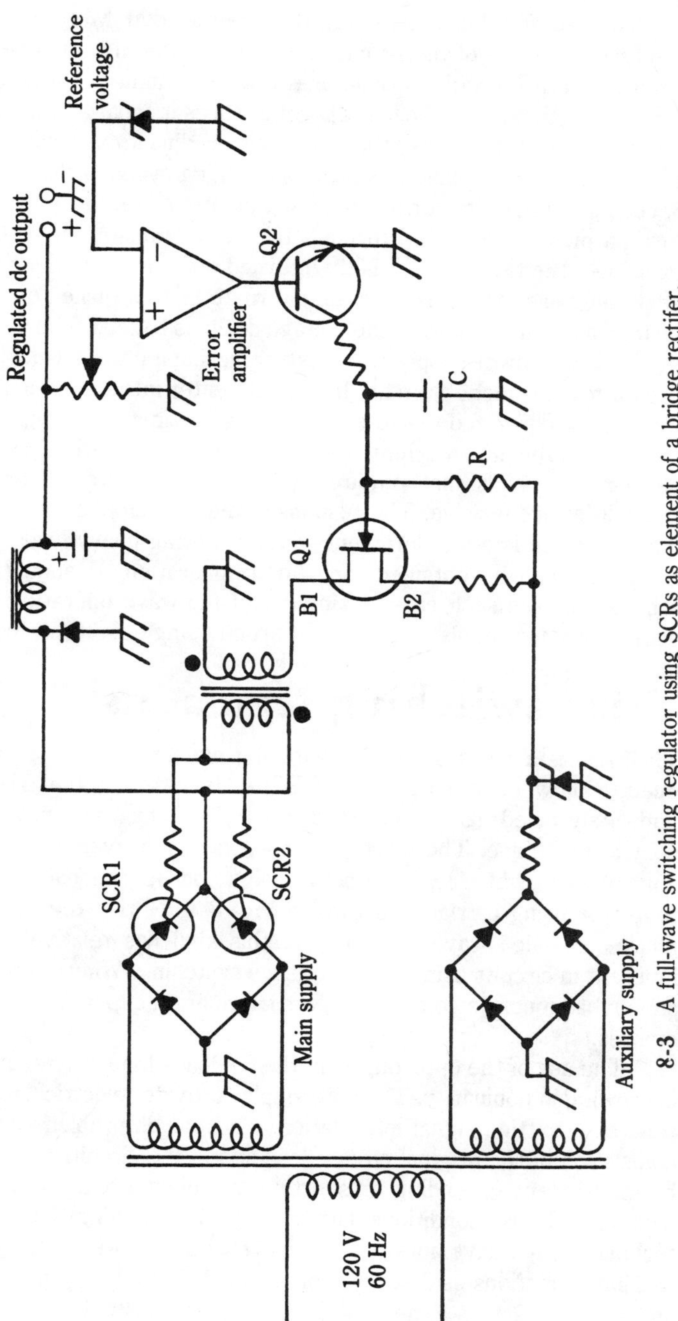

8-3 A full-wave switching regulator using SCRs as element of a bridge rectifer.

quency is double the frequency of the rectified ac wave). In this case, the output of the auxiliary rectifier dips to zero 120 times per second. Moreover, these dips take place at the zero crossings of the ac line. These dips interrupt the operation of the unijunction circuit, so the relaxation oscillator is not free-running, but is forcibly restarted at each zero crossing. Timing capacitor C commences its charging cycle each time the ac wave crosses zero; each of these charging cycles begins its rise from a near-zero voltage level. When the timing capacitor begins its charging cycle, nothing happens until the voltage developed across its terminals is sufficient to fire the UJT. When the UJT fires, a turn-on pulse is delivered through the pulse transformer to the SCRs. As previously explained, the favored SCR is then turned on. The same thing then happens during the next half-cycle, only this time the alternate SCR is turned on. The SCRs automatically turn off as their anode/cathode voltage declines to zero. The zener diode associated with the auxiliary power supply provides a waveshaping function, making the ripple easier to use for the unijunction timing. It does not enter into the basic philosophy of the system; the line-synchronized dip to zero remains the important aspect of operation.

This type of switching regulator generally merits serious consideration when dealing with loads that require tens or hundreds of amperes; even higher current capabilities are feasible. The response time of such line-frequency switchers is limited, but that often is not a deterrent when powering heavy loads. Better performance and economics can be obtained for 400-Hz operation. It should be mentioned that this scheme enables such easy attainment of full-wave operation that there is hardly any justification for regulators designed around single SCRs in half-wave configurations.

Triac switching regulators

The triac is a more desirable switching element, if considerations of power and frequency allow its use in place of SCRs. In essence, the triac functions as a pair of oppositely poled SCRs connected in parallel. Thus, you have a push-pull or full-wave control technique. The analogy is not exact, however, because the triac has a single control gate, which leads to both design and operational simplicity. A switching-type regulator using a triac is shown in Fig. 8-4. Notice the symmetry of the illustrated phase-controlled wave. A supply of this kind has relatively slow response, which is inherent in circuits using line-frequency switching. Where this is not a deterrent, such a supply has much to commend it, for triacs are inexpensive and very efficient switching devices.

The use of the optocoupler in the feedback loop represents a circuit technique that is growing in popularity. The optocoupler provides electrical isolation, while at the same time transmitting signal information. This avoids problems having to do with dc level translation in the original circuit design, and the circuitry is safer from damage caused by inadvertent grounding or shorting, as might occur from alignment procedures or from certain load conditions. Otherwise, this switching-type regulator involves no new techniques that have not been extensively used in previous thyristor control. The error amplifier functions as a straightforward differential amplifier; it provides linear control and does not saturate, as in some of the previously described regulators. There is no reason to manufacture a switching waveform here because it already exists in the power line.

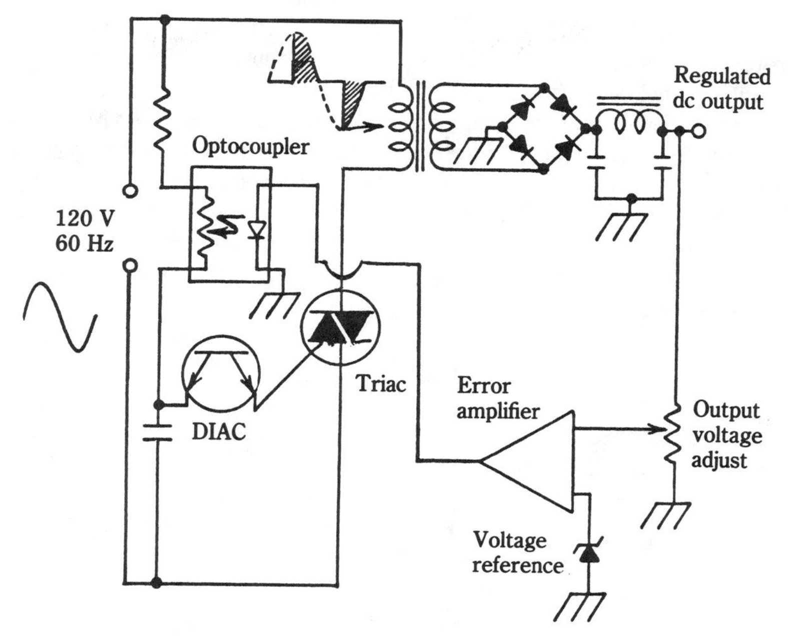

8-4 A full-wave switching regulator using a triac. The optocoupler acts as a variable resistance for phase control of the gate circuit of the triac. This is a desirable arrangement because of the electrical isolation between the ac line and the control electronics.

Where 400-Hz power is available, this type of regulating supply presents some compelling features. The transformer and filter components would be appreciably reduced in size from their 60-Hz counterparts, and newer triacs can perform reasonably well at this frequency. At present, SCRs are strides ahead of triacs in both power and frequency capability. On the other hand, most electronic applications do not require the high power levels required by traction motors or industrial machinery.

Making switchers out of linear regulators

Those who have constructed linear regulators will recall that one of the most commonly encountered malfunctions in these circuits is self-oscillation. This is particularly true when, for the sake of tight regulation, the error amplifier has very high gain. The oscillation can be superimposed upon the dc output level so that it might not actually interfere with the operation of either the power supply or the load. Generally, however, such oscillation produces all manners of troubles and component failures. Indeed, the oscillation is frequently in the form of a square wave because the error amplifier alternately saturates at one extreme then the other. At the same time, the "linear" series-pass transistor is being switched between cutoff and saturation.

The operational mode of this malfunctioning linear regulator simulates, in some respects, the intentional operation of the self-excited switcher. In fact, the transformation from linear to switching is often as simple as adding a few extra components.

Manufacturers of on-the-chip linear regulators can also provide application notes on switching-type regulators. This is true even if the switching regulator was far from their thoughts during the development of the IC linear regulators; IC linear regulators suitable for use in switching-type regulators are the 723, LM104, LM105, and 550.

The close, but long-elusive, relationship between a linear and a switching-type regulator is clearly illustrated by the two discrete power supplies shown in Fig. 8-5. In this case, a deliberate conversion of the linear circuit was made in order to obtain switching-type regulation. The change is effected by the addition of inductor L1, the 1N3491 free-wheeling diode, and capacitor C2 (which converts stages Q2 and Q3 into a variable-duration multivibrator).

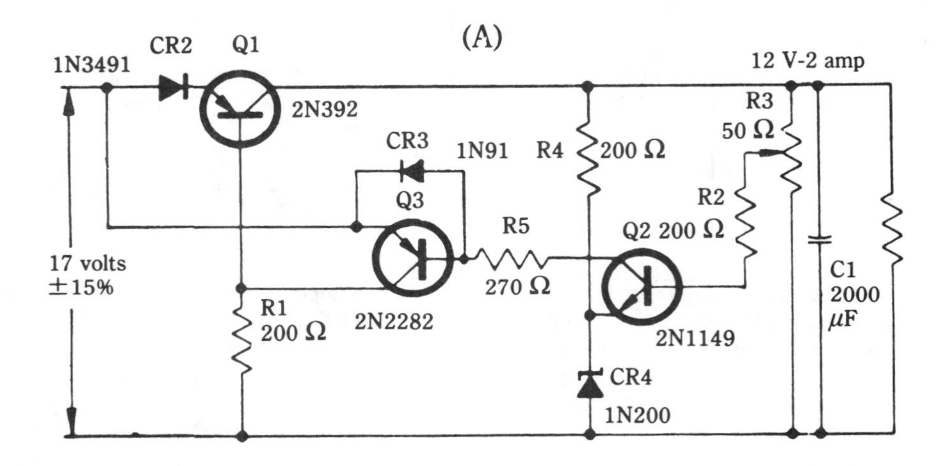

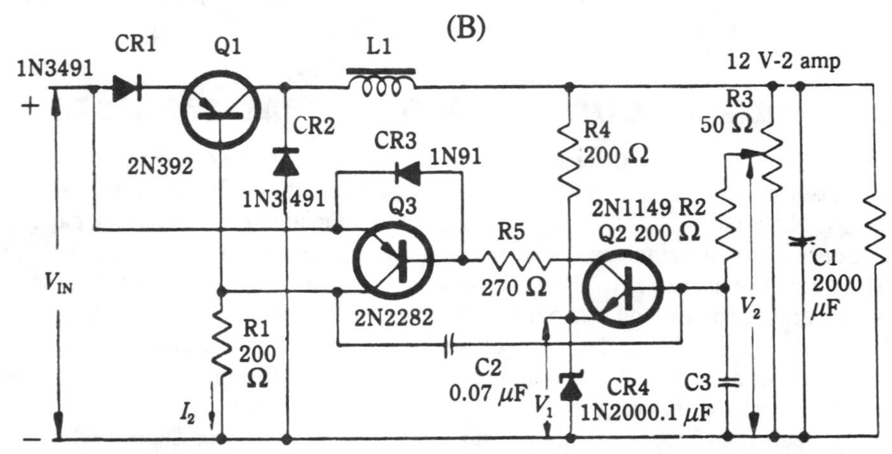

8-5 A simple conversion of dissipative linear supply into high-efficiency switching supply. (A) Original linear regulating circuit. (B) Modified circuit for switching operation.

Constant-current regulation

Although the application demand for voltage-regulated switchers has predominated, it should be appreciated that *constant current* switchers are also feasible. The modifications necessary to accomplish this mode of operation are analogous to the conversion of a linear voltage regulator to a current regulator. In both the linear and the switching-type regulator, current regulation is implemented by sensing the voltage drop developed across a tiny resistance inserted in series with the load, rather than by sampling the voltage produced across the load, as in voltage regulators. When a voltage regulator maintains constant voltage across this series resistance, it thereby regulates the current

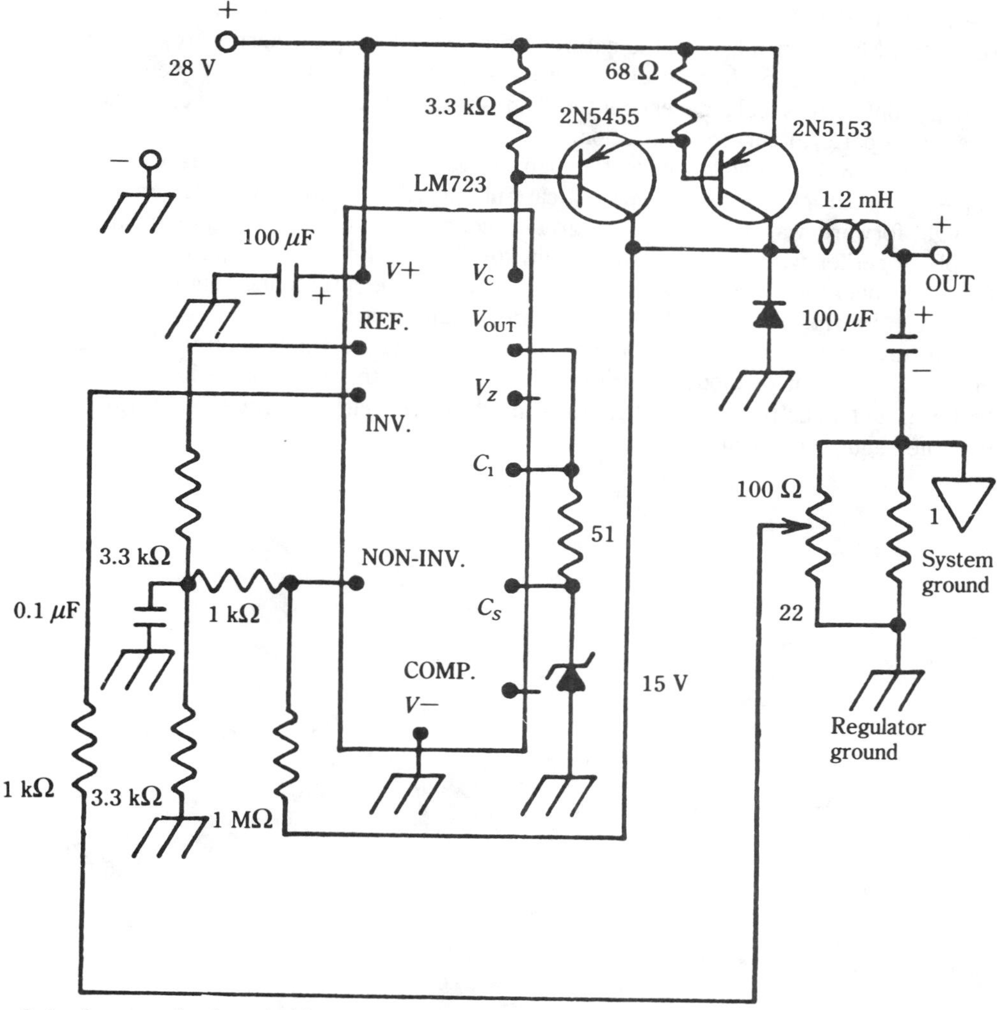

8-6 An example of a switching-type current-regulating supply. Notice that the regulator and the external load do not share the same grounding system.

flowing through the load. Thus, the basic action of the two modes of regulation is quite similar.

From a dynamic standpoint, there are some important differences. In voltage regulation, one strives for as low an output impedance as possible; the opposite is true for current regulation. Ideally, the equivalent circuit of a perfect current regulator incorporates a generator with infinite impedance, but in a practical current regulator, a high impedance will suffice, provided that it rises with frequency. Thus, the residual inductance in the filter capacitor and its leads is not as detrimental here as in voltage regulators. As a matter of fact, it is often desirable to greatly reduce the value of the filter capacitor in current regulators to achieve better transient response. This might cause instability in the feedback loop, but corrective measures can usually be applied elsewhere in the circuit; for example, at the frequency compensation terminal of the IC regulator. Much depends upon the nature of the application of the current regulator; for battery charging and electrochemical processing, rapid response is not needed, and minimal difficulty should be encountered in altering conventional voltage-regulating circuits to perform current regulation.

In Fig. 8-6 an adaptation of the popular LM723 is shown. Just as with linear current regulators, you must be very careful regarding grounding arrangements. A straight-forward way to circumvent ground conflicts is to allow the unregulated dc supply, together with the regulator circuit, to "float." The standard ground symbol in Fig. 8-6 would then signify nothing more than a common interconnecting bus, but one isolated from the true ground at the powered equipment.

A further means of protection for the constant-current supply is often obtained by inserting a resistor or diode in series with the load. In this way, active loads (such as batteries and inductive windings) are prevented from unleashing damaging energy back into the regulator circuitry.

9 CHAPTER

Noise considerations

NOT TOO MANY YEARS AGO, ENGINEERS DECLARED THAT THE SWITCHER WAS doomed to a shadowy and nebulous existence in electronic systems. The dire forecast was predicated upon the "fact" that switching supplies were inherently noise generators. Overlooked by these prophets of doom were matters such as the following:

- Any time a rectifying circuit is associated with a large input capacitor, highly nonsinusoidal current waveforms are produced. Because of the power levels often involved, strong harmonics are generated over an extremely wide frequency spectrum. Thus, the average linear regulator is not so innocent when it comes to noise generation. Typical current waveforms in capacitor-input rectifier circuits are shown in Fig. 9-1.
- The noise production from a switching-type power supply should always be compared with that of a linear supply having the same power rating, rather than to an ideal, noiseless power source. When such a comparison is made, it will often happen that the switcher is actually the smallest noise contributor.
- It often happens that digital loads generate more noise than switching-type supply itself.
- Many powered systems utilize CMOS or other logic with high noise immunity.
- Considerable understanding has been attained with regard to noise filtering and suppression techniques. Even though the switching process may be prolific in the generation of harmonic and spurious frequency components, this energy can easily be contained within the physical confines of the power supply.
- There is often the option of choosing the switching frequency so that the residual noise imposes minimal interference with the particular equipment being powered.

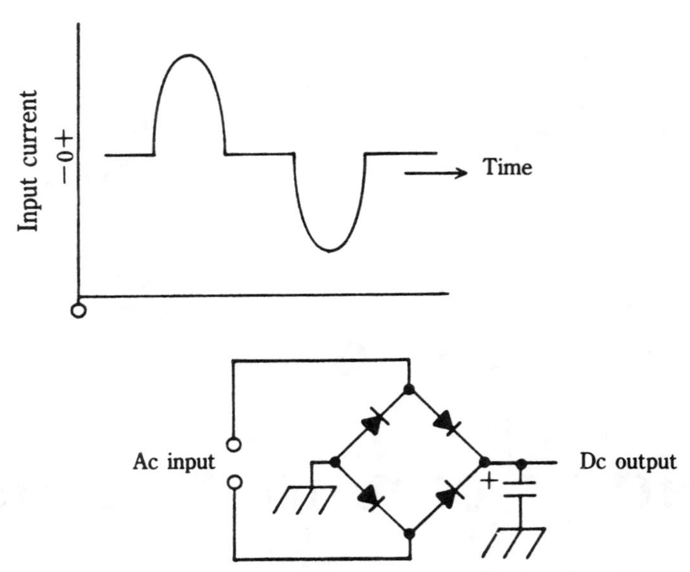

9-1 A harmonic-rich waveform at input of capacitor-loaded rectifier. The pulse-like wave contains harmonic energy. Although the line frequency is 60-Hz, odd-order harmonics extending into the megahertz region can be generated in certain instances.

- Some switchers can provide greater rejection of power-line transients than linear regulators.
- Because of new ICs, fast-recovery diodes, superior capacitors, and other improved components, the overall operation of switching regulators is better than ever. This enables more effective tradeoffs to be made for the sake of noise reduction.

So much for the defense of switchers. The fact remains that they are potentially offensive, with regard to noise production. Admittedly, there are some applications where it would not be wise to consider the use of these supplies. For example, a sensitive radio receiver is best powered by linear-type supplies. Similarly, instrumentation that is predicated upon a high signal-to-noise ratio could have its performance degraded by a switching-type power supply—particularly when the signals being monitored or detected are extremely weak.

For a great many applications, success will depend much upon design and installation techniques. Although you must be guided by logic and by available facts in the implementation of shielding and grounding methods, the final optimization is invariably empirical. You must be willing to experiment, to observe, and to try various combinations and permutations. The reward is often a surprising reduction in conducted or radiated noise. Unlike many electronic products, the switcher experimentation does not end with the breadboard, but is continuous in the initial prototype models. Finally, after various modifications, noise will have been reduced to the point of diminishing returns, with respect to the expenditure of man-hours and money. At this stage, reasons can probably be found to account for some of the strange noise-reduction methods.

Noise-reduction techniques

Figure 9-2 shows a number of commonly used noise-reduction techniques. Most of these adversely affect efficiency, regulation, or other performance parameters, so they must be implemented judiciously, with due regard for permissible tradeoffs.

- Output filter *A* can be used to attenuate high-frequency noise in the RF region, as well as ripple at the switching rate. This filter is outside of the feedback loop; therefore, the dc resistance of the inductor will degrade the voltage regulation. As shown, this filter is a feedthrough type. The "inductor" is simply a small segment of the conductor surrounded by a ferrite bead, and the "capacitors" consist of the built-in ceramic insulation. Feedthrough filters are excellent for removing high-frequency noise, but are limited by their relatively low inductance for attenuation of lower-frequency noise components. Conventional filters are usually necessary for filtering out switching frequencies and lower-frequency transients. In order to recover the lost regulation caused by the additional output filter, attempts are sometimes made to include it within the feedback loop. If this can be done without upsetting the loop stability of the regulator, it is desirable, but more often than not, this will lead to erratic operation.
- Capacitor *B* is often a tenth to a hundredth the value of the output capacitor to which it is connected in parallel. This added capacitor provides bypass action for the higher frequencies, where, because of internal resistance and inductance, the main filtering capacitor is no longer effective. Sometimes another even smaller capacitor is added. Thus, the main filter capacitor might be a 100-μF aluminum electrolytic type, the second capacitor could be a 10-μF solid-state tantalum type, and the third capacitor could then be a 0.1-μF ceramic type.
- Components *C* and *D* are intended to slow the rise and fall of the switching waveform. Although this gets to the heart of noise production, you must be prepared for a sacrifice in the efficiency of the switching regulator. In practice, it is often found that a very small smoothing of the abrupt switching transitions goes a long way in noise reduction. In general, the less reliance upon this method, the better. The improvement in switcher performance attained in recent years has been attributable to faster response in the switching transistor and free-wheeling diode.
- Capacitor *E* and low-pass filter *F* serve similar purposes in reducing noise feedback into the unregulated supply. Because much of this noise is actually caused by the switching transistor, capacitor *E* acts as an energy reservoir and carries a high ripple current. Its grounding, together with that of the free-wheeling diode, is generally made separate from all other components; with poor grounding, long leads, or bad physical placement, this capacitor can be the source of much trouble, because an inductive loop can be formed to feed noise into both the input and output of the regulator. Such an inductive loop can also be responsible for erratic switching behavior.
- Filter *F* is usually used to attenuate higher frequencies, particularly when there is an appreciable distance between the unregulated supply and the switching

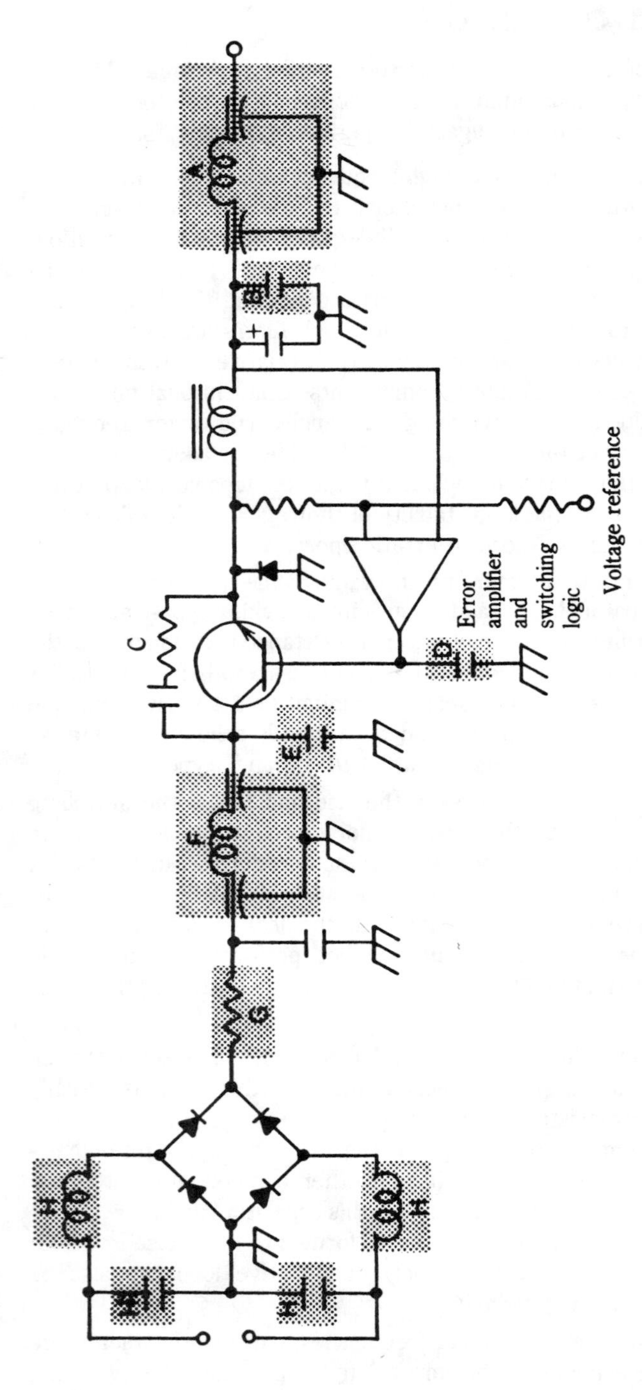

9-2 Techniques for reducing electrical nosie in a switching-type power supply. The shaded components provide slowdown, filtering, and isolation functions, as described in the text.

transistor. A feedthrough type is often used. Excessive inductance is dangerous in this circuit position because of the negative-resistance characteristic of the switching transistor; oscillation and other instabilities occur all too easily.

- Resistance *G* tends to make the input current to the rectifier more sinusoidal rather than pulse-like. This measure lowers the efficiency of the rectification system.
- Another source of rectification noise is caused by the reverse-recovery characteristic of the rectifying diodes. For 60 Hz, it might prove unwise to use diodes classified as fast-recovery types. In more complex switching-type regulators, where fast-recovery diodes must be used to achieve high rectification efficiency, you have a choice of "soft" or "abrupt" recovery types. Where the Schottky diode is applicable, you can obtain virtual freedom from this source of noise.
- Power-line filter elements *H* are used to prevent switcher noise from entering the power line. This is highly desirable, for many noise problems stem from conduction and radiation of noise through the power line. Such noise is often routed from one part of a system to another, and defies remedies generally successful with more direct paths.
- The physical placement and orientation of the switching supply, as well as the connection to the common system ground merits investigation. Supplies that incorporate toroidal magnetic components are advantageous when used in close proximity to sensitive circuits. Even ventilation holes and apertures in the power supply's cover might permit leakage of high-frequency noise (even the shielding capability of coaxial cable varies widely, depending upon the mesh, thickness, and conductivity of the outer sheath).

Noise generation in thyristor circuits

The abrupt turn on of thyristors generates harmonics, which have sufficient energy to interfere with many types of systems. This energy extends well into the megahertz range—even when the switching rate is only 60 Hz. Because the amplitudes of the higher harmonics approach zero asymptotically, interference can occur in radio and TV receivers. Turn off also contributes to the electrical noise generated by phase-controlled thyristor circuits. The noise, however produced, gets into equipment by both conduction and radiation. It is particularly desirable to keep the noise from entering the ac power lines, for it then tends to radiate, and find its way into sensitive circuitry.

Figure 9-3 shows how a low-pass filter can be used in a triac circuit to prevent the noise both from being passed into the load and from entering the power line. The configuration of the filter is primarily for attenuation of outgoing energy, but such a filter also protects the triac from incoming noise. The performance of the filter is degraded by the connection method shown in Fig. 9-3A, because the relatively high capacitance at terminal 2 of the triac tends to bypass noise frequencies around the inductor. In B, however, this capacitance is advantageously used, for it now acts as the shunt arm of a pi filter.

The simple filter configuration of Fig. 9-3B often appears in phase-control circuits utilizing triacs and SCR. Not infrequently, the noise generation is actually increased.

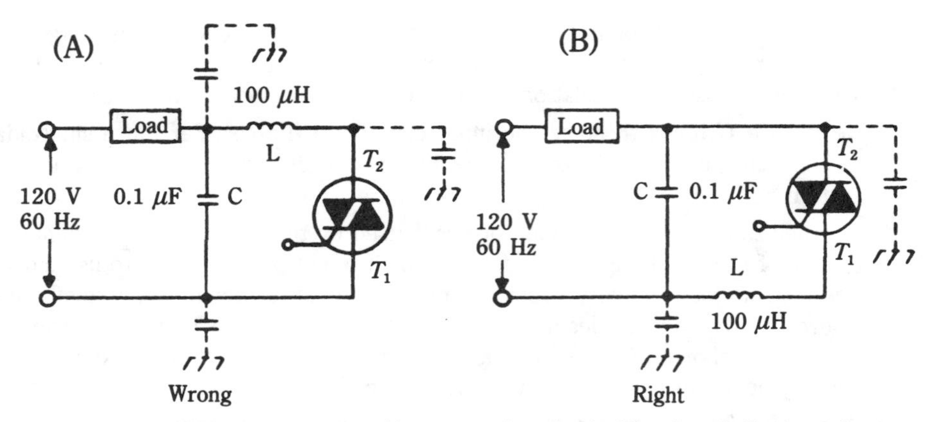

9-3 An application of a low-pass line filter to a triac circuit. The L and C values shown are typically used in the 100- to 1000-W range of power levels. The filter cut-off frequency is in the vicinity of 50 kHz.

What generally happens is that the filter is shock-excited, and the ringing that results momentarily turns the thyristor off. It quickly turns on again, and the load might not be affected by the momentary power lapse, but noise is generated by such erratic operation. In fact, the higher the Q of the filter, the more likely this might be to happen. Lowering the Q of the inductor can be helpful, but this technique is self-defeating in that it quickly slows down the attenuation of the filter in its rejection band. A better approach is shown in Figs. 9-4 and 9-5, for both triacs and SCR. An additional RC combination is added to the shunt arm of the filter; this produces the required damping, but is not nearly so detrimental to the performance of the filter as is the reduction in the Q of the inductor.

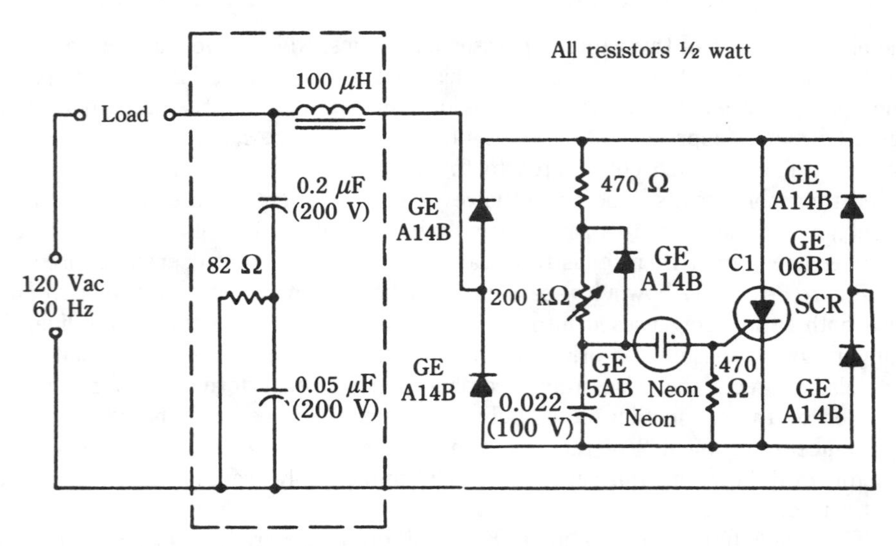

9-4 An RFI filter used in a full-wave SCR phase-control circuit. General Electric Semiconductor Products Dept.

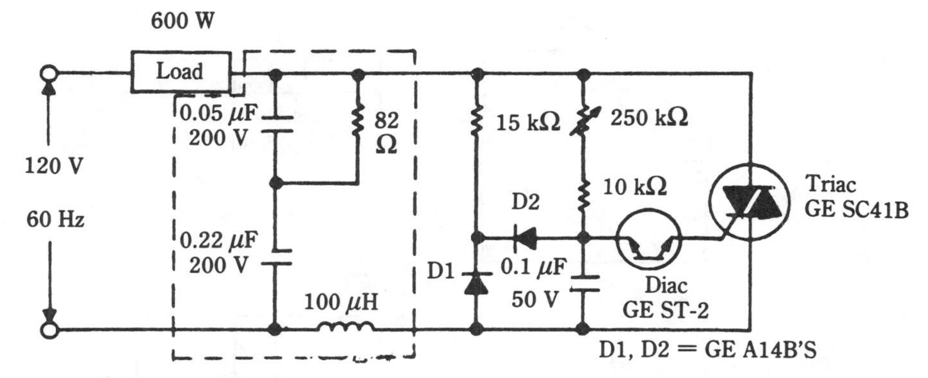

9-5 An RFI filter used in a full-wave triac phase-control circuit. General Electric Semiconductor Products Dept.

Electrical noise sources

There is more to switching than the mere making and breaking of a switch. So far, this chapter has covered the topic of noise in terms of its remedies, rather than its causes. Because noise originates as a direct consequence of the switching process, it is time to examine the manner in which noise is produced. Although this requires a certain amount of mathematical theory, this aspect of the discussion is kept at a minimum, for much more can be gained from a qualitative approach than from a purely mathematical expression.

First, consider the circuit in which the direct current to a resistive load can be made and broken by a single-pole, single-throw switch, as in Fig. 9-6. Suppose this switch can be operated at various rates, and that you shall not now concern yourself with such secondary effects as arcing, bouncing, or with finite contact resistance. You have then a "perfect" switch. When it is open, it offers infinite resistance to the flow of current; at closure, its ohmic resistance is zero. Just to make the situation ideally simple, assume that the transition between its two states is accomplished very, very quickly, compared to the on or off time—even at high switching rates.

If you observed the voltage waveform across the load resistance, you would see a nice, clean square or rectangular wave. But just how "clean" is this train of pulses? The number of pulses per second is exactly that corresponding to the number of times per second that the switch is placed in its on position. If the switching process is exceedingly well controlled, you could produce a train of uniformly wide pulses. Because the oscilloscope screen displays this repetition rate or frequency, and shows no transients or other frequencies, have you not demonstrated a "clean" switching process?

The answer is both yes and no. Yes, the switching operation is as clean as it can be, devoid of transients, spikes, bounce, arcing, wave distortion, etc. But no, there cannot be a nonsinusoidal wave that is mathematically or electronically clean; according to the important mathematical principle known as the *Fourier theorem*, all waveforms except the sine wave possess harmonics of the fundamental frequency. All of these harmonics are pure sine waves (or identically shaped cosine waves). Any waveshape can be resolved into a number of sine waves, each harmonic frequency having its unique

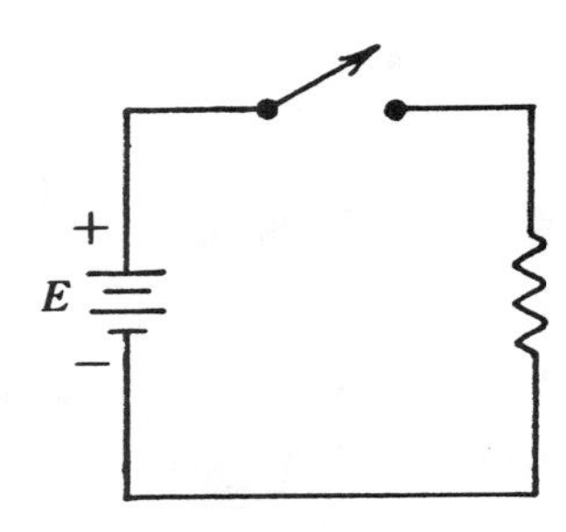

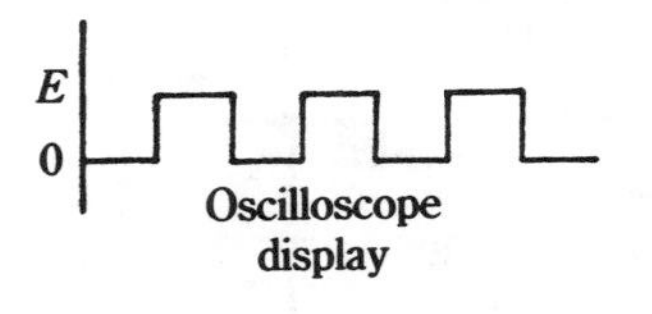

DC value, or average value of the pulse train when filtered.

First harmonic (f) is the fundamental frequency and is equal to the pulse repetition rate or switching frequency.

Second harmonic ($2f$)

Third harmonic ($3f$)

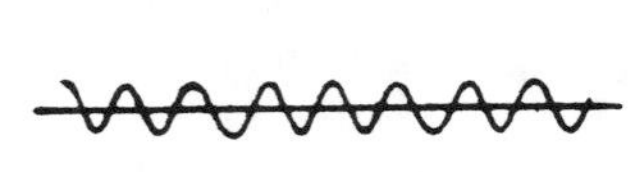

Higher-order harmonics are also produced, but not shown. These decrease in amplitude with increasing frequency, but can have significant amplitude-even at megahertz frequencies.

9-6 The apparently clean waveform that is produced by a switch is actually rich in harmonics. Only a pure sine or cosine wave can justifiably be called ***electronically clean.*** Much of the noise in electronic circuits is the consequence of non-sinusoidal waveforms.

amplitude and phase relationship. Conversely, you can synthesize any waveshape by appropriately combining such harmonically related sine waves. Rectangular switching waves in power supplies have many harmonics; the same is true of SCR and triac waveshapes.

Figure 9-7 provides extended insight into the consequences of making and breaking the circuit at a periodic rate. This is a universal graph, depicting harmonic production as a function of both pulse rate and pulse duration. You see that the spacing of the harmonics is governed by the pulse repetition rate; the actual harmonic frequencies are inversely proportional to the pulse duration, or width. Those groups of harmonics that are shown as being negative are simply 180° out of phase with the harmonics contained in the first group. But all harmonics have positive energy and manifest themselves as signals that radiate and cause interference to susceptible systems.

Notice that the successive groups of harmonics grow smaller as the harmonics they contain become higher in frequency, but they never become zero, except at periodic

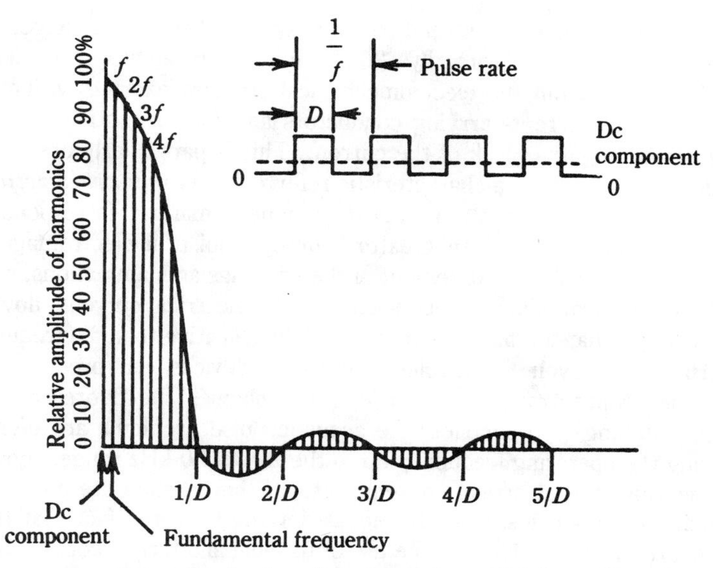

9-7 The frequency spectrum of a simple switched waveform. This is a universal curve that can be applied for any pulse repetition rate f and any pulse duration D rectangular pulses only. The narrower the on-time of the pulses, the greater the number of harmonics, and the closer is their spacing. Negative amplitudes are indicative of relative phase of the harmonics—they are just as "real" as those with positive amplitudes.

zero crossings, such as $3/D$, $4/D$, $5/D$, etc. In practice, substantially zero amplitudes could be achieved through the low-pass action of stray circuit capacitance and resistance. Nevertheless, interference at very high frequencies can be produced by such simple switches, particularly when they are used to interrupt high energy levels or voltages. It should also be pointed out that some complicated consequences result when the load is something other than pure ohmic resistance; indeed, the counter electromotive force associated with the interruption of current in an inductive circuit can produce a very substantial amount of noise.

In summary, it can be said that the switching or interruption of currents and voltages in a circuit always give rise to potential sources of objectionable noise. Nevertheless, the desirability of a near dissipationless power supply has stimulated the development of numerous—and often quite simple—techniques of noise reduction. As a result, switching-type power supplies often achieve much lower noise levels than comparable dissipative types using linear control methods. This accomplishment probably results from the fact that designers of switching-type supplies are much more aware of both the causes and cures of electrical noise, and so are not likely to fall into the traps that plague many dissipative designs.

Acoustic noise sources

Mention should also be made of acoustic noise in power supplies. Such noise usually emanates from electromagnetic devices, such as inductors and transformers, but can also be produced by capacitors and current-carrying conductors.

All electromagnetic devices tend to emit acoustical noise. In conventional 60-Hz supplies, the audible noise is primarily 120 Hz, double the line frequency. The source of the noise is derived from the electromechanical stresses exerted within the components, between the current-carrying conductors and the magnetic core materials, at each positive and negative peak of the current. This is particularly prevalent in nickel alloys, which tend to exhibit a characteristic referred to as ***magnetostriction***, resulting in periodic elongation and constriction of the magnetic material in response to varying magnetic flux. However, an even greater source of noise in electromagnetic components is caused by the shock excitement of the windings and laminations, which tend to ring or vibrate at much higher frequencies than the electrical currents flowing through them. Considering that the human ear is especially sensitive to audio frequencies in the 1- to 5-kHz range, even 60-Hz electromagnetic devices can produce considerable harmonic noise concentrations (a problem that plagues the fluorescent lighting industry). In switching-type supplies, the acoustic noise problems are often solved by simply raising the operating frequency up to the 20- to 30-kHz range, above the range of human hearing (the electrical consequences of this method are covered later). At lower switching frequencies, some benefit is derived from the fact that the magnetic components are much smaller, and the use of toroidal and ferrite cores is much quieter than laminated steel cores. And the lower dissipation of switchers means that shock mounts and standoffs can be used to prevent acoustical coupling because there is less need for a solid contact to the chassis for thermal heatsinking.

Other sources of noise generation, such as capacitors and conductors, also benefit from the smaller size requirements afforded by switching-type supplies. The use of aluminum chassis and structural parts also decreases the possibility of electromagnetically induced noise, as does the use of sound-deadening foams and potting compounds. As a result, the acoustic noise resulting in a switching-type power supply is no worse than in a dissipative supply—and at high frequencies, it might be considerably better.

10
CHAPTER
Switching ac voltages

MOST POWER SUPPLIES ARE LINE OPERATED. THE OUTPUT OF SUCH SUPPLIES CAN be either ac or dc, but the initial problem encountered is that of how to switch or control the ac voltage. In preceding chapters, we have introduced three basic switching devices that can be used with ac: the diode, the SCR, and the triac. This chapter examines these three switching elements with their associated circuitry.

The half-wave diode rectifier

It is inevitable that the half-wave rectifier be considered because both the circuit configuration and the switching process appear fairly simple. The instrumentation and interpretation of ac and dc readings, however, require an understanding of how waveshapes relate to the concepts of effective and average values. As shown in Fig. 10-1, the readings of the dc and ac meters are not what the unversed in ac theory would expect, but the measuring instruments, after all, do give reliable readings of the quantities they monitor. When these readings are correctly interpreted, design and servicing become a predictable cause-and-effect relationship. The fact remains, however, that the switching circuit in Fig. 10-1 generates a nonsinusoidal waveform because segments or fractions of complete sine waves are not considered to be sine waves at all. Therefore, the half-wave rectifier is a harmonic generator. In practice, harmonic generation in such a circuit can be more severe than would be anticipated by a Fourier analysis of sine pulses. The reason is that the dc component of this wave train can produce varying degrees of saturation in the core of the transformer. Considerable overdesign of the transformer is necessary, with respect to rectification, if you wish to avoid the effects of operation in the region of core saturation. Unfortunately, overdesign is also required to compensate the low utilization of the transformer when subjected to this waveform, for the harmonics increase the hysteresis and eddy-current losses in the transformer core.

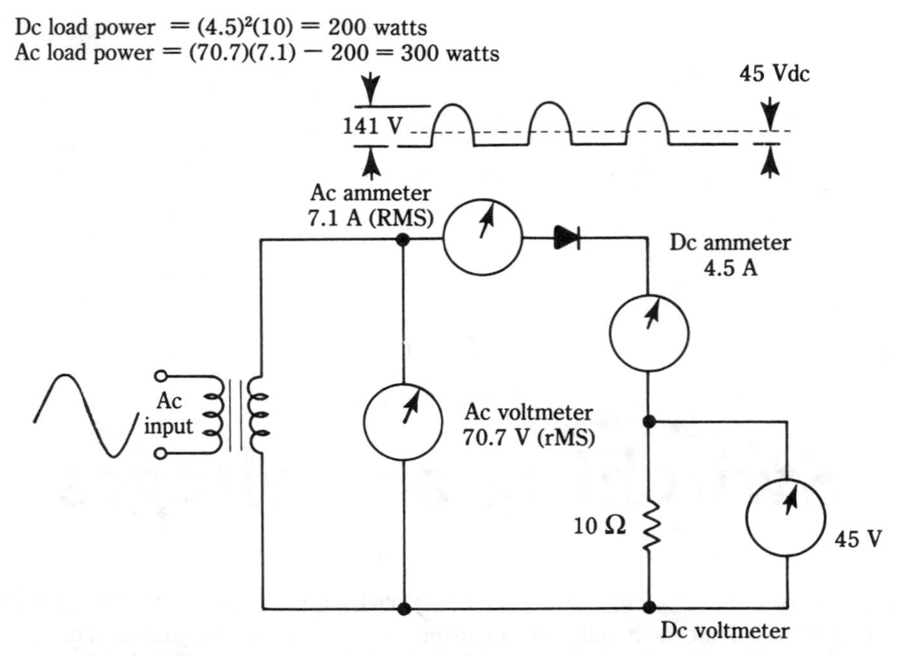

10-1 A low-efficiency energy conversion from a simple half-wave rectifier. As depicted above, the input power is 500 W. Assuming an ideal diode, 60% (i.e., 300 W) of the power in the load is ac fundamental and harmonics. This leaves 40% of the load power as the desired dc (i.e., only 200 W).

Regulation

An added disadvantage to the simple half-wave rectifier is that it is unable to regulate or control its output voltage; that is, the output voltage delivered to the load varies in relation to the unregulated ac input voltage. Even more complex rectification systems, such as full-wave rectifiers with complicated LC filters, are unable to provide significant regulation of the output voltage. (A possible exception in such systems is the use of voltage-regulating transformers that are able to exercise control over the output voltage through specially designed saturable cores and tuned windings.) Thus, although the half-wave rectifier proves to be a simple ac switch, it also proves to be an impractical switch, inasmuch as there is no way to control its operation to effect voltage regulation.

Conversion efficiency

Our half-wave rectifier discussion is, admittedly, an oversimplification, but it is quite useful in illustrating the relative inefficiency of primitive rectification systems. As Fig. 10-1 indicates, the half-wave rectifier exhibits an ac-to-dc conversion efficiency of about 40 percent. Full-wave rectifiers, with double the number of rectification pulses, have efficiencies of about 80 percent, a substantial increase. The addition of capacitive and inductive filtering components can further increase this efficiency. The fact remains, however, that an increase in efficiency is accomplished only with an increase in cost and complexity. Overall power conversion efficiency is further degraded by the regulation process-primarily by dissipative (linear-type) regulators.

The significance of this efficiency factor is depicted in the waveform of Fig. 10-2, which is of the type obtained from a full-wave rectifier system that has a capacitive-input filter for temporary energy storage between rectification pulses. In electronic systems, which are by far the most common, the efficiency of the power supply is affected primarily by the portion of the unregulated input power that is wasted in the regulation process. Thus, in Fig. 10-2, the amount of usable power is contained only in the lower rectangular portion of the waveform, which represents the regulated dc output voltage (V_{OUT}) of the power supply. The upper portion of the waveform, representing the ac ripple voltage (V_R) and voltage margin (V_M), is wasted energy that greatly reduces the efficiency of dissipative-type linear power supplies. The ripple voltage can only be reduced by increasing the amount of filtering, and hence the cost of the power supply. The voltage margin is a function of the variations in ac input voltage and the minimum voltage drop required across the series-pass regulating element, representing a worst-case design constraint.

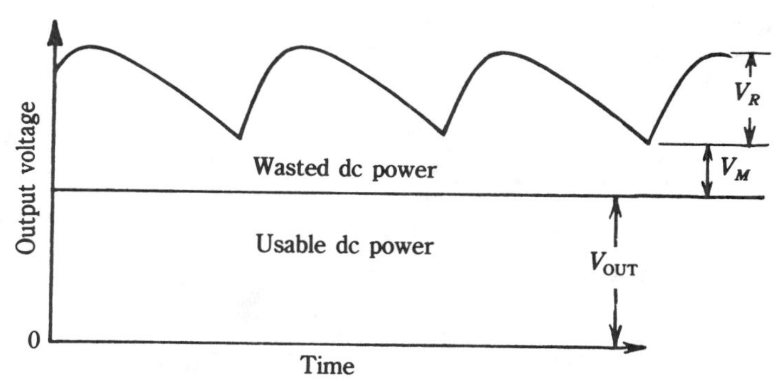

10-2 A typical power dissipation situation in linear. The wavy line represents the unregulated input-voltage from a full-wave rectifier filtering. Only *part* of this impressed power will appear as output voltage from the linear regulator. The remainder is wasted in the series-pass element of the linear regulator. In contrast, a switching regulator can utilize nearly the entire input-voltage amplitude.

Switching-type power supplies use the entire input waveform depicted in Fig. 10-2 because all of the waveform represents "usable" power. A switching regulator, for example, would compensate for the ripple-voltage component of the waveform—not by dissipating it, but by varying the pulse rate or pulse duration of its switching process to convert the ripple voltage into usable output power. Similarly, the voltage margin, which varies with fluctuations in ac line voltage, presents no problem to the switching regulator, for the switching regulator also varies its pulse rate or pulse duration to compensate for such fluctuations. As a result, switching-type power supplies generally place lesser demands upon their input rectification systems than would be the case with dissipative supplies, and this tends to reduce the size of the power supply and compensate for the cost of the switching components.

The SCR switch

The SCR switch resembles the half-wave rectifier, but has the advantage that it is controllable. The circuit, shown in Fig. 10-3, also contains instrumentation. This switch is not only a generator of prolific harmonics but also causes the power factor of its ac feed line to be low, despite the fact that the load may be pure resistance. (The SCR circuit behaves as an inductive reactance in this respect.) In Fig. 10-3, the power factor becomes progressively worse (lower) as the conduction angle is made smaller.

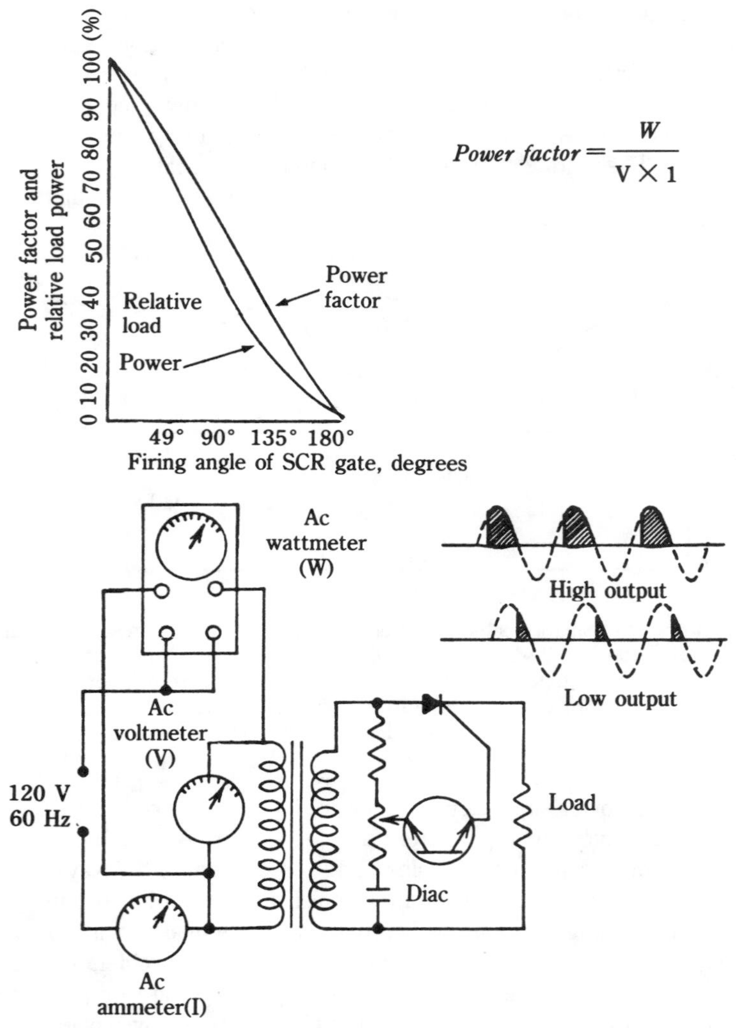

10-3 The phase control of power in the load also varies the power factor in the ac line. This might be surprising because the power factor in the resistive load always remains at unity. Another accompaniment to phase-controlled power is added noise-generation from the non-sinusoidal waveshapes.

Incidentally, the use of a conventional power-factor meter provides yet another example of complexity arising from apparent simplicity. It so happens that this instrument responds to the phase difference between the fundamental voltage and current. But when appreciable harmonic energy is present, as with SCR switching, an erroneous reading is obtained—the departure from accuracy can exceed 100 percent.

When the SCR switch is delivering low output (corresponding to a small angle of conduction), it becomes somewhat nebulous to refer to the switching process as one of high efficiency. It might be true that relatively little dissipative power is involved in either the on or off positions of the switch, but the high current consumed through the power line because of the low power factor can itself be the mechanism through which appreciable power is lost.

The phase-controlled switching method used in such circuits creates very abrupt transitions in the output waveform, and this greatly increases the magnitude of noise-producing harmonics. The dc conversion efficiency of the SCR system is no better than that of the half-wave rectifier; if anything, it is worse. Thus, although the SCR circuit makes a good ac switch, inasmuch as it is able to control the output power level, it is a bad switch with regard to power factor, conversion efficiency, and noise generation.

The triac switch

The triac switching process is a push-pull or full-wave version of the SCR switching mode (Fig. 10-4). Its propensity for generating noise is similar to that of the SCR, for its harmonic spectrum is also extensive. If the two sections of the triac are balanced with respect to firing potential, there will be a tendency toward cancellation of the dc component flowing through the supply transformer. However, the low power-factor, which accompanies small angles of conduction is not overcome by use of this thyristor.

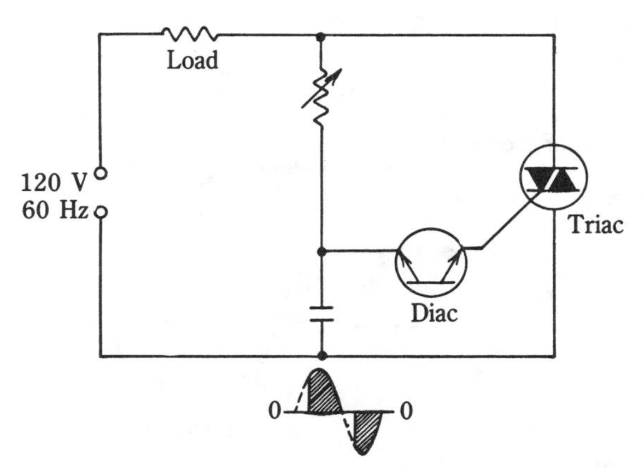

10-4 Fullwave, phase-controlled triac. Notice the symmetrical switching waveform. This is the basic circuit used in light dimmers. Commonly available triacs do not operate well much beyond 400 Hz, and therefore have not been prominent in dc supplies or regulators. A rudimentary input filter is usually called for to prevent switching noise from backing up into the ac power line.

The zero-crossing switch

The block of Fig. 10-5 constitutes a switch, which, by virtue of its switching process, produces few operational side effects. Such a switch makes and breaks the circuit without generating significant harmonics, altering power factor, or saturating the supply transformer (if one is used). This switch does not produce an erratically varying inrush current, as ordinarily results when a transformer is suddenly energized from an ac source. At the same time, this switching process utilizes the dissipationless principle of switch control to full advantage.

The circuitry illustrated in Fig. 10-5 is used to switch the triac on and off at the precise instant that the ac input waveform crosses through its zero-voltage point. The advantage of this approach is derived from the Fourier postulate that the sine wave is the major building block of all waveshapes. By switching at the zero crossing of the input sine wave, abrupt switching transitions are prevented; only the fundamental sine wave passes through to the load. In truth, a small transition does occur in going from the off

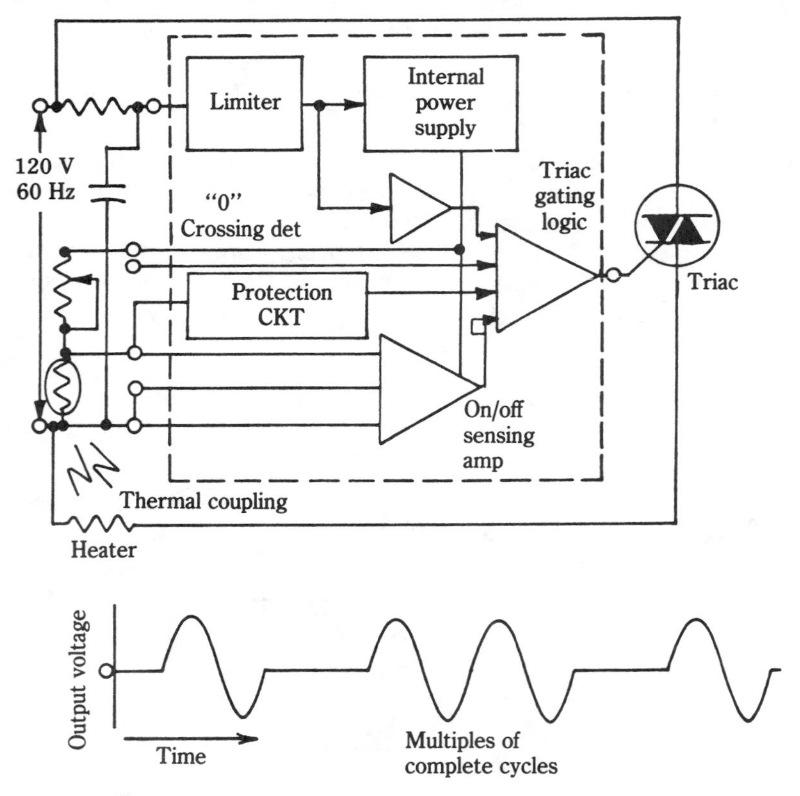

10-5 A zero-crossing switch for ac loads not requiring fast response-times. Operation is limited to delivery of one or more complete sine waves. Such integral-cycle behavior (also known as *burst-modulation*) greatly reduces thyristor switching noise. This control technique is excellent for heaters, provides smooth control of motors, and extends the life-span of filamentary lamps.

state to the initial ramp of the sine wave, but the noise produced at this instant is negligible when compared to the sharp transition encountered when attempting to switch the waveform at any other time during the cycle.

The arrangement illustrated in Fig. 10-5 actually comprises a switching regulator —although not the type generally used for powering electronic equipment because it does not convert the ac input voltage into a dc output. Still, this type proves to be extremely useful in regulating many ac loads, such as heating elements in temperature-controlled ovens. In this latter example, a thermistor might be used as the temperature sensor for activating the triac circuit. Temperature regulation would occur through the action of the triac circuit in delivering multiples of complete sine waves to the heating element in the oven. The response time of such a regulator is relatively slow, but the requirements for this and similar applications rarely require high-speed regulation. The advantages of such a system are high efficiency, relative simplicity, and low noise generation, which permits close proximity to sensitive electronic equipment.

Although the zero-crossing mode of switching, when applied to sine waves, proves to be the "cleanest" technique, it does not enjoy widespread use in actual switching regulators. Such switching slows down the response time, which already suffers poorly in comparison with that of linear regulators. If a 20-kHz sine wave is applied to a load in groups of complete cycles averaging, for example, 10 cycles per group, it is as though we were dealing with a 2-kHz wave insofar as response time is concerned—the response time cannot be faster than the duration of one group of cycles. Although a much higher frequency version of this type might be implemented in the future, at present, slow response has been an obstacle in using this technique for supplying power to electronic loads.

11 CHAPTER

Switching dc voltages

ONE OF THE MOST FAVORED TECHNIQUES FOR SWITCHING DC VOLTAGES involves an electronic equivalent of the single-pole/single-throw mechanical switch. Interruption of the current flowing through this type of device is responsible for much noise generation, but such effects can be suppressed with a wide variety of relatively simple filters. The switching circuit shown in Fig. 11-1A produces a pulsed voltage across the load, and though it is possible to define an average value for the pulse train, it is no more than a concept. Except for a few nondemanding applications, such as charging storage batteries, the equivalence between the pulsed power and the average level is more academic than practical.

By means of a capacitor, a considerable improvement is brought about in Fig. 11-1B. The sawtooth wave comprises segments of exponential charges and discharges of the filter capacitor. If the capacitor is large enough, the minimum value of the wave never dips to zero, and a truly direct current passes through the load. This simple scheme has much to recommend it, but it leaves much to be desired. For example, if you wish to diminish the ac ripple voltage superimposed upon the dc level, an impractically oversized capacitor is required. Aside from the expense and massive proportions, such a large capacitor would have an inordinate slowdown effect on the response time of the regulator. Of course, the switching rate could be speeded up, also causing the ripple amplitude to decrease, but only so much enhancement of filtering performance can be secured in this manner, for ultimately the dissipative losses in the switch itself begin to mount up. Instantaneous currents into the capacitor tend to be high, and these are similar to the surge current in rectifiers that flow when a power supply is first turned on.

LC filters and free-wheeling diodes

It is only natural to think of adding an inductor. Unfortunately, an inductor by itself is useless for chopped currents because the opening of the switch must release the energy

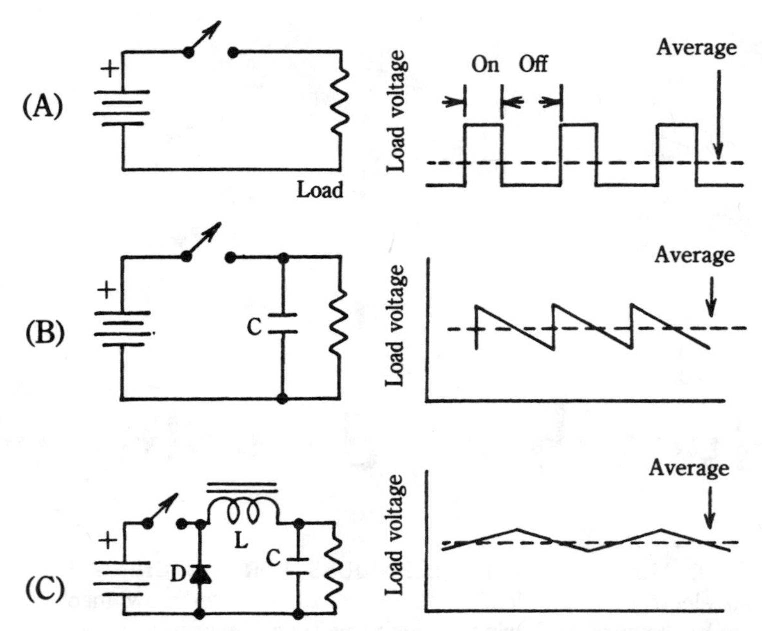

11-1 The evolution of dc switching circuits. (A) This waveform is too intermittent; "average" is not practically realized. (B) A fairly sustained average voltage can be obtained because of the capacitor. (C) Best results are provided by this scheme. The diode enables near-continuous load current throughout the switching cycles. The diode also suppresses inductive arcing when the switch is opened.

previously stored in the magnetic field of the inductor, which causes arcing and other manifestations of high voltage. A mechanical switch might have its contacts burned or welded together, and it is needless to detail the damage likely to overtake a semiconductor switching element.

Suppose that, for the purpose of arc suppression, a diode is connected as shown in Fig. 11-1C. The diode does not interfere with the injection of pulses into the inductor, and is so polarized as to absorb the high-voltage transient generated when the circuit is interrupted. Although such applications of diodes are not new, in this particular circuit it turns out that the diode also helps bring about a steady delivery of power to the load. When the switch is open, the load current is then supplied by the stored energy in the magnetic field of the inductor, with the diode making a complete path for the current to flow.

A more relevant way to view the diode is a means whereby the energy in the magnetic field of the inductor (stored during the previous *on* time of the switch) is converted into useful current flow in the load. Without synchronizing techniques, or any additional circuitry, the diode provides for this current flow at the right time—when the switch is in its off position. The diode enables the LC filter to perform in an acceptable manner; and because of the diode, you should become accustomed to thinking of the LC filter as an energy reservoir, rather than as merely a frequency-selective network. Various names for this diode are *catcher diode, coasting diode, bypass diode, flywheel diode, free-wheeling diode,* and *commutating diode.*

Switching and voltage regulation

With a few simple components—a switch, an inductor, a capacitor, and a diode—the electrical power available from a dc source can be interrupted, smoothed again, and then applied to a load in the form of essentially pure dc power. Thus far, the duty cycle and frequency of the switching process were nonvarying. You know from chapter 7 that this is not so in actual regulating supplies, for the effect of varying the pulse width and repetition rate is what permits us to control the output voltage of the regulator. This is illustrated in Fig. 11-2. When such control is made automatic by means of an appropriate sensing technique and feedback loop, the result is regulation of the dc voltage applied to the load.

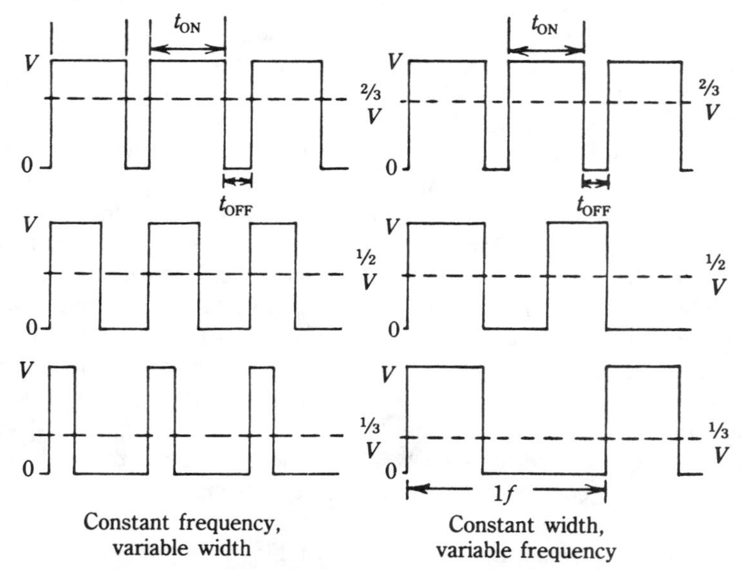

11-2 Two basic methods of pulse modulation for controlling the dc output voltage. The left-hand waveforms exemplify control with constant-frequency, variable-width pulses. The right-hand waveforms exemplify control with constant-width, variable-frequency pulses. In both instances, the storage elements in an LC filter would produce the average values, which correspond to the dashed lines.

The two methods of control have much in common. The output voltage decreases in situations when the ratio of on time to off time is decreased; this can be brought about by decreasing either frequency or pulse width (on time). Thus, the two sets of switching waveforms in Fig. 11-2 represent equivalent methods of bringing about a decrease in output voltage—even though different approaches are used. This can be described by the equation:

$$V_{OUT} = \left(\frac{t_{ON}}{t_{ON} + t_{OFF}}\right) V_{IN}$$

Notice that the denominator of this equation represents the time for one full switching cycle. If the switch remained closed, the output voltage would equal the input voltage; if the switch remained open, the output voltage would be zero. Obviously, actual output voltages will be somewhere between these two extremes. It is not easy to say that better results should be forthcoming from either the constant-width or the constant-frequency technique; excellent performance has been obtained with both methods, and much depends upon the skill with which the designer implements the control function (the constant-width, variable-frequency method is favored for the resonant-mode regulated supplies).

In self-excited switching regulators, both frequency and on time are allowed to vary. Which parameter is more influential in bringing about regulation depends upon many factors. Although such random variation of the switching waveform leads to entirely satisfactory regulation, many designers feel it is better engineering practice to operate with either the frequency or the on time constant. Some restrictions on these parameters are justified, from a design standpoint, in order to place safe limits on the operation of the regulator. In many circuits, if the on time is too short or the frequency too high, the series-pass transistor begins to dissipate excessive energy; and if the on time is too long, excessive currents can damage the transistor or saturate the inductor.

The possibility of using a constant off switching mode has not, thus far, been mentioned. This mode can be viewed as a degradation of the constant-frequency mode. In both modes, the variable parameter is the pulse width, or on time. In the constant-frequency mode, the denominator of the fraction is held constant, whereas in the constant off mode, that part of the fraction is allowed to change as on time is varied. The net result is that with variations in on time, the value of the whole fraction does not change as fast in the constant off mode, as it does in the constant on mode. In order to produce maximum regulating effect, the value of the fraction should be sensitive to changes occurring in the variable parameter, and the constant off mode is definitely not as sensitive. But a boost for the constant off mode involves the fact that digital waveforms from logic circuits can sometimes be conveniently implemented in this fashion.

Other things being equal, the higher the switching rate the better because inductors, capacitors, and transformers. However, other things do not remain equal—switching losses go up with faster switching rates because of the greater time consumed in achieving turn-on and turn-off, electrolytic capacitors become less effective, and losses tend to mount in the free-wheeling diode. For many years, frequencies in the 20- to 30-kHz region were favored because this region is well above audible frequencies, but still represent acceptable tradeoff between efficiency and economics. Presently, switching rates of several-hundred kHz and greater can be used very efficiently. On the other hand, where audible noise does not tend to be objectionable, reasonable overall performance is often obtained at several kilohertz. The penalties for low-frequency switching rates are weight, size, and of course, cost.

Inductor current

A key to the operation of the basic switching circuit shown in Fig. 11-3 is what happens to the current passing through the inductor. Because the inductor is subjected to

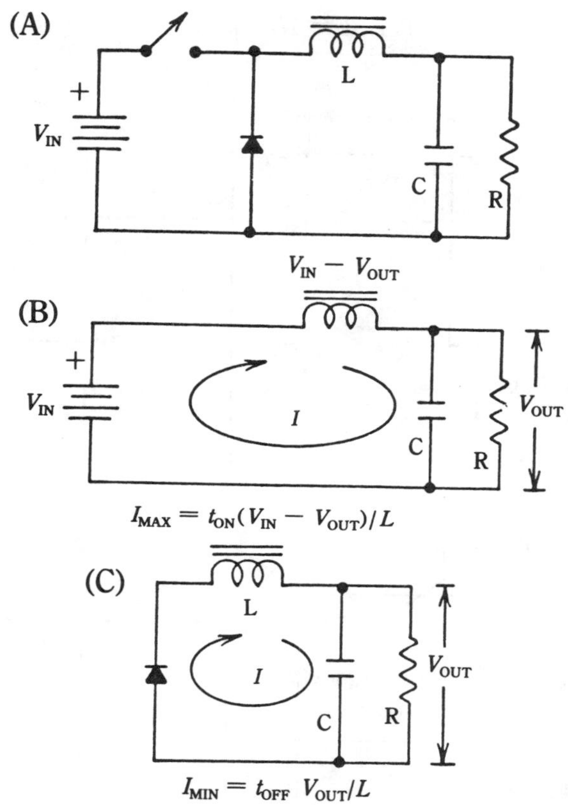

11-3 The current peaks in the inductor are functions of on and off times. (A) The basic switching circuit. (B) Current path during on time. (C) The current path during off time. To use the equations, express current in amperes, time in seconds, and inductance in henrys.

step-function voltages, it is easy to determine the current through it. For small changes in V_{OUT}, the current through the inductor is just a waveform made up of ramps or linear slopes. The peak amplitudes of such a wave can be calculated from purely geometric considerations. Referring to both Figs. 11-3 and 11-4:

$$I_{MAX} = t_{ON}\,(V_{IN} - V_{OUT})/L$$

In so many words, the factor $(V_{IN} - V_{OUT})/L$ is the slope of the on ramp and, when multiplied by the elapsed on time (t_{ON}), produces the highest current (I_{MAX}). A similar procedure is used for the calculation of I_{MIN}. In this case, the off slope is represented by the factor E/L.

The important thing about the current through the inductor is that it is essentially of a sustained nature, despite its sawtooth waveform. And unlike the inductor current in a simple half-wave rectifier circuit, there is no dip to zero. In the switching circuit, the inductor and the free-wheeling diode work together to produce a near constant current through the load. In Fig. 11-3C, direction of current flow prevailing after the switch

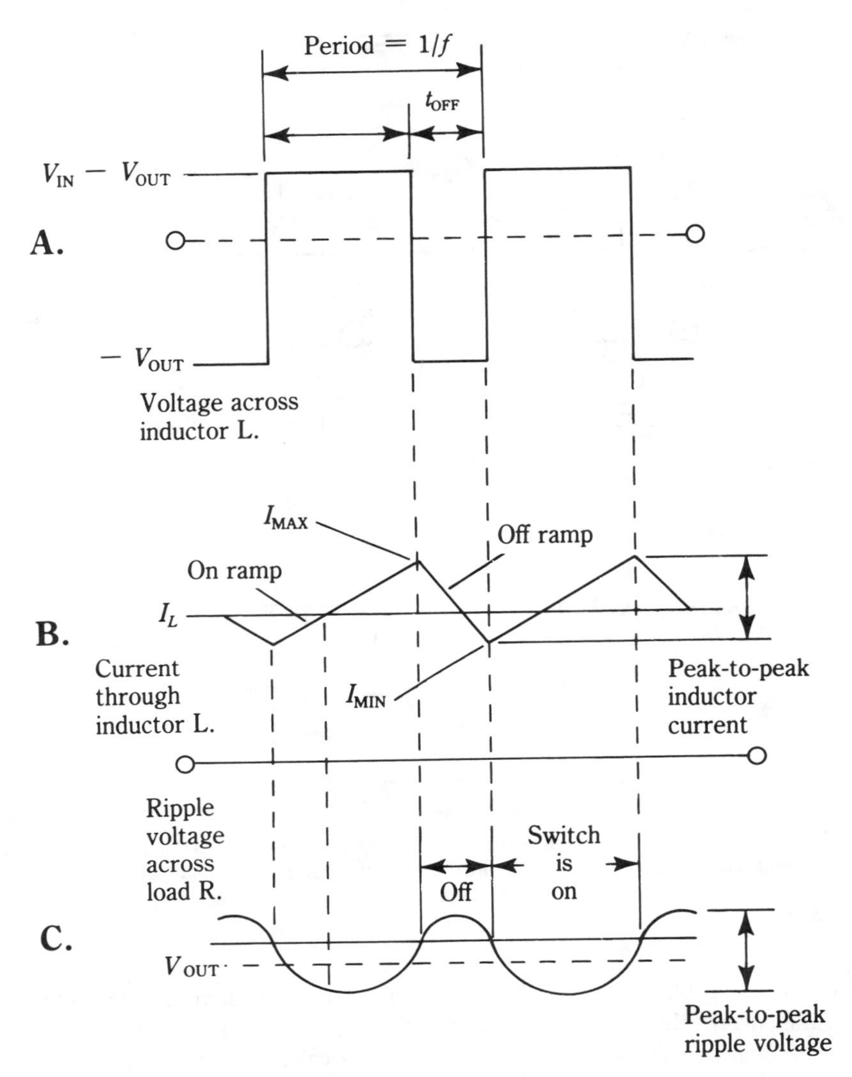

11-4 The waveforms associated with the switching circuit of Fig. 11-3. (A) The average voltage across the inductor is zero. (B) The average of the ramp-current through the inductor is equal to the load current. (C) The ripple is caused by the operation of the switch. Notice its phase relationship to the inductor current ramp. Largely because of equivalent series resistance in the capacitor, ripple is non-sinusoidal.

turns off is the same as during the time the switch remains turned on—even though the voltage across the inductor has reversed.

Expected waveform of the ripple voltage is not always in evidence when the output of a switcher is observed because the electrolytic filter capacitors have appreciable series resistance. The waveform is therefore a composite of the voltages developed across internal resistances and capacitances. However, this section deals only with idealized components to establish basic operating mechanisms, although the major challenge in the development of switching-type regulators has been in the manufacture of ideal components. The theory of the switching-type supply has been around for a long time, but the continuing refinement of these supplies is primarily contingent upon still better switching transistors, filter capacitors, free-wheeling diodes, and other critical components.

Switch or transformer?

An important aspect of previous discussions pertaining to the basic dc switching circuit is the matter of control. Specifically, the dc output voltage is governed by the duty cycle of the switching wave. This is true whether the switching speed or the on or off times of the switching cycle is varied. It is both interesting and instructive to look at this control function. The output voltage is determined only by the switching parameters and the input voltage. Specifically, the important relationship:

$$V_{OUT} = \left(\frac{t_{ON}}{t_{ON} + t_{OFF}}\right) \times V_{IN}$$

Where:

t_{ON} is on time, and t_{OFF} is off time of the switch. Inasmuch as $t_{ON} + t_{OFF}$ denotes the period of one switching cycle, and period is the reciprocal of frequency or switching rate f, the equation can also be written:

$$V_{OUT} = ft_{ON}\ V_{IN}$$

To further simplify the form of this basic relationship, we can let the factor ft_{ON} be represented by the letter k. Then the expression for output voltage becomes simply, $V_{OUT} = kV_{IN}$. It is also true that input current I_{IN} and output current I_{OUT} are related by k; that is, $I_{IN} = kI_{OUT}$. I_{IN} is the current that would be measured by inserting a dc ammeter in one of the battery leads of the circuit in Fig. 11-1; and I_{OUT} is the same quantity as previously designated I_L, the inductor current, also equal to the load current.

Although this section has dealt with the current in the inductor, this is the first mention of the currents flowing into and out of the switching circuit. It might appear somewhat unnatural that these two currents are not the same, but this probably stems from our tendency to think of the switching transistor in much the same fashion as a series-pass transistor in a linear, dissipative regulator. The series-pass transistor in the linear regulator is a "rheostat" and the dc currents flowing into it from the dc source out of it into the load are almost identical. Not so with the switching circuit!

Rewrite the voltage and current expressions in terms of their common factor, k; you obtain $k = V_{OUT}/V_{IN}$ and $k = I_{IN}/I_{OUT}$. Because quantities equal to a common quantity are equal to each other, you can write, $V_{OUT}/V_{IN} = I_{IN}/I_{OUT}$, which can finally be written:

$$I_{IN}V_{IN} = V_{OUT}I_{OUT} \text{ or } P_{IN} = P_{OUT}$$

which says that the input power and output power of the switching regulator are identical—and if there were no dissipative losses in the regulator components, the switching-type power supply would have an efficiency of 100%, the same theoretical efficiency as a transformer!

In Fig. 11-5, these relationships are summarized and compared to the input/output relationships in a simple ac transformer circuit. Here, you can see the justification of describing the basic dc switcher as a "dc transformer." The fundamental feature of ideal transformer action is the ability to change either voltage or current with no loss of power. Other methods of changing dc voltage or current are accompanied by power loss. Typical examples are the use of the rheostat or analogous devices, such as the series-pass transistor in linear regulators, the potentiometer, and the level-changing sources, such as the zener diode. In all such methods, a power loss is absorbed by the

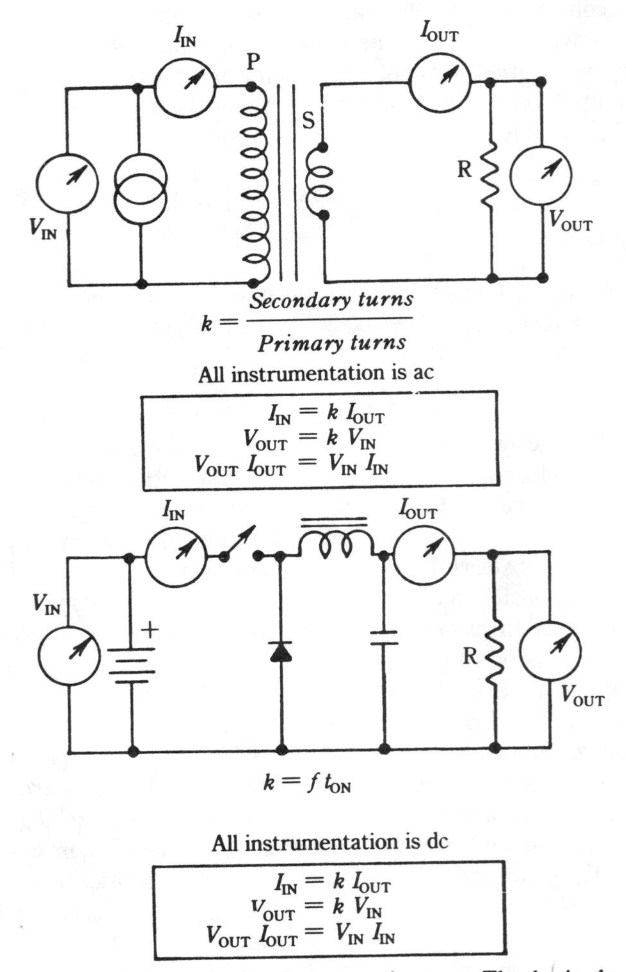

11-5 An interesting comparison of ac and dc transformers. The basic dc switching circuit functions as a true transformer. The equations show the basic similarities in these transformers. What is called a *dc-dc converter* wouldn't be misnamed by referring to it as a *dc transformer.*

inserted element, and in none of these examples can more current be extracted than that which enters the circuit. Unlike these conventional ways of changing dc voltage level, the basic dc switching circuit has, like the ac transformer, the property of delivering a greater current to the load than the current initially injected into the circuit. This is the most important feature of the basic dc switching circuit, and it is directly responsible for the important advantages that switching regulators have over dissipative types.

Although the switch undoubtedly makes it possible to construct a dc transformer, the switch itself is not solely responsible, but rather it works in conjunction with the energy-storing inductor. The role of the inductor in the circuit is analogous to that of a variac or variable-voltage auto-transformer, except that the inductor is not tapped and cannot, by itself, vary the output voltage. Instead, the switch is the controlling element that determines both how much energy is to be stored in the magnetic field of the inductor and how that energy is to be released to obtain the desired output voltage. Of course, the switch requires additional circuitry to sense the output voltage and control its operation, but the analogy is essentially correct. And, similar to the autotransformer, switching circuits exist that are able to both step up and step down the input voltage so that output voltages can be either higher or lower than the input.

A common pitfall in the use of driver transformers

It appears natural enough to use a transformer as the coupling medium between a PWM voltage source and an output stage. You might think that this is a fail-safe technique for driving the output stage. After all, transformers have long been used for similar purposes in audio amplifiers, where it is well-known that a properly designed transformer can faithfully reproduce complex waveforms. It also is a verifiable fact that many switchmode supplies use transformers in their driver circuits. Not withstanding this, such transformer coupling often looms up as a problem in getting an initial design to work properly. The trouble is often traceable to a characteristic of transformers that is often overlooked. It turns out that the faithful reproduction might not be forthcoming, regardless of the way the transformer is designed.

Assume that the primary winding of a transformer is driven by a unipolar PWM wave, such as a control IC might produce. Let the duty cycle initially be 50%. The situation is depicted in Fig. 11-6. Most electronics practitioners would realize, either from theory or from experience, that the secondary wave would be a true ac wave with perfect symmetry, and with a peak-to-peak amplitude of 20 V (assuming the transformer-ratio to be 1:1). But, what happens when the duty cycle is other than 50%? It is here where the men get separated from the boys!

Assume that we are interested in driving an N-channel power MOSFET that needs 10 V at its gate to fully turn the MOSFET on. With the 50% duty cycle, these requirements are met by use of the 10-V positive excursions of the secondary waveform. Now, look what happens when the duty cycle is 75%. The peak-to-peak amplitude of the secondary waveform, it is true, remains 20 V. But the amplitude of the positive excursion that drives the MOSFET gate is now only 5 V. Only the transformer remains

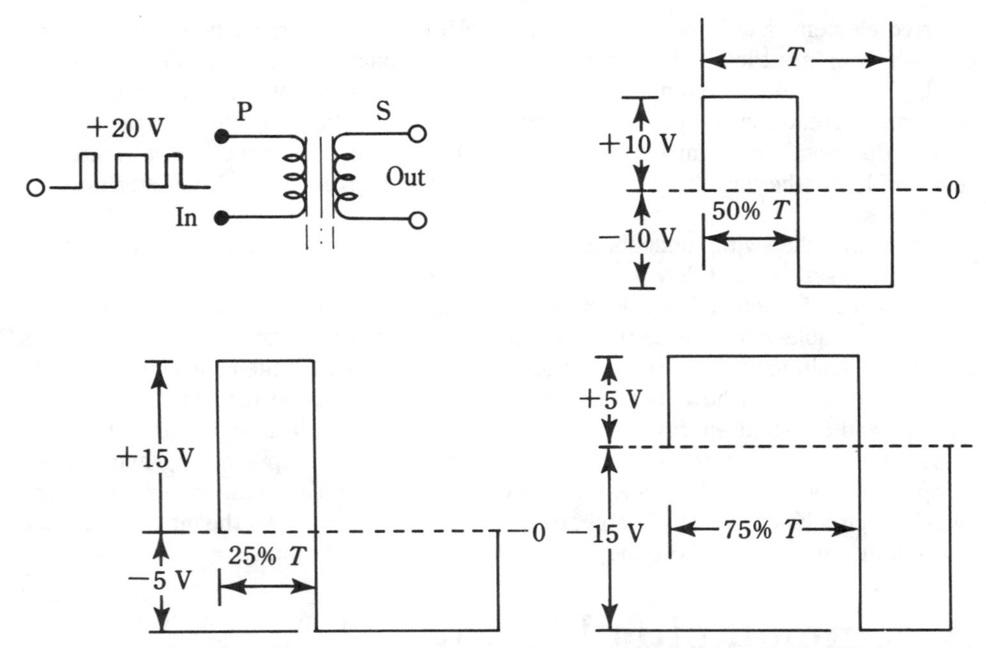

11-6 The secondary waveforms of a transformer with unipolar PWM voltage applied to the primary. The situations for three duty-cycles are shown. Notice the amplitude variations in the secondary. This behavior exists for an ideal transformer and can cause "mysterious" troubles in driving a switching stage.

"happy"—its law of operation states that equal volt-seconds have to be generated in the positive and negative excursions of its secondary waveform. This translates to equal areas of the alternations as depicted graphically. This can be seen in the third case where the duty cycle is 25%. Here is an abundance of gate drive. The net valve of volt-seconds has remained zero in all-three duty cycles, but the amplitude distribution has certainly changed.

How can this behavior be reconciled with the fact that transformers are often used for coupling unipolar PWM waves? The answer is that this can work if the duty cycle range is not too great. Thus, it could conceivably be acceptable to let the gate drive-voltage vary between 10 and 15 V. However, to design for such a compromise, you must be aware of this characteristic of the transformer, or else be very lucky.

The arrangement shown in Fig. 11-7 overcomes the duty-cycle dependent effect in the transformer 25 just described. Here, logic gates are used to convert the unipolar PWM pulse to bipolar format which represents a true ac-waveform. Because the primary of the transformer is no longer exposed to a dc component, there is no watt-second distribution problem; the induced pulses in the secondary remain symmetrical about their zero-axis, regardless of duty cycle. The output of this circuit would ordinarily connect to the primary of an output transformer to be followed, in turn, by a full-wave rectifier and filter. Although the driver transformer will be a small component, the output transformer tends to be physically large because it handles a higher power-

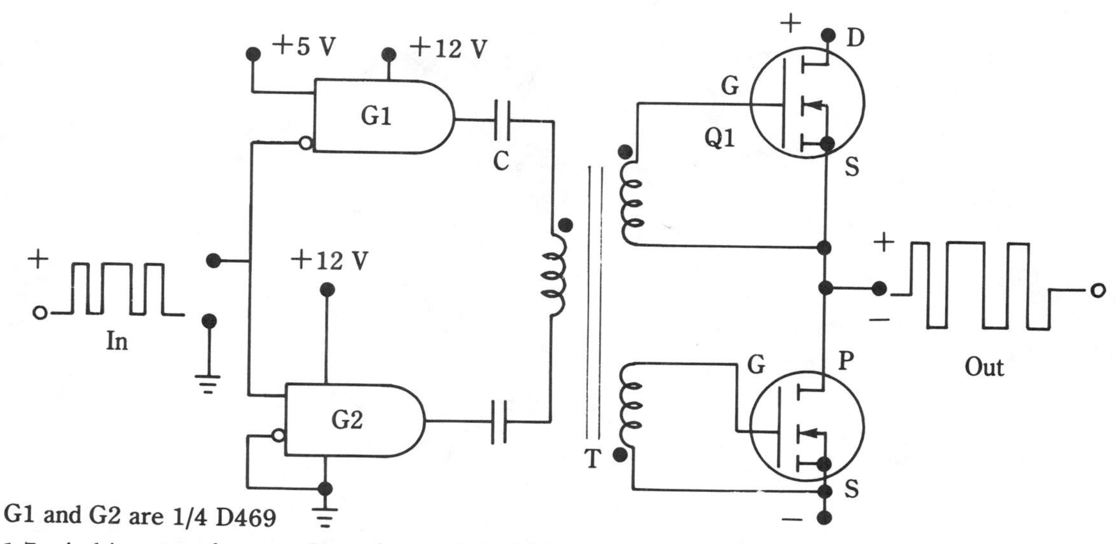

11-7 A driver transformer scheme for properly driving output switching stage. The primary of the drive transformer is presented with a true ac waveform so that there is no dc component to affect induced voltage in the secondary. The output transistors are properly driven regardless of the duty cycle.

level. Both transformers, in any event, must operate in their linear regions. Inadvertent saturation of core components is a common cause of malperformance, being somewhat elusive to discover because it tends to be duty cycle sensitive.

As might be suspected, it is not necessary to use two capacitors as depicted in Fig. 11-8; one will suffice, but it should be large enough to preserve the flat-tops of the PWM pulses. Notice that this scheme can be used to drive single-ended output stages by dispensing with the lower secondary-winding and Q2. For such a single-ended stage, the output transformer will be inserted in the drain lead. Also, by appropriate scaling of the driver transformer ratio, these techniques are applicable to bipolar transistors. It should be pointed out that many control ICs can be connected to provide a complementary driving waveform, making it necessary to use external gates, such as shown in Fig. 11-9. Also, the drive transformer, itself, is not always needed; much depends upon the isolation desired, and the drive characteristics of the output transistors.

12
CHAPTER

Passive switching components

A DELIGHTFUL ASPECT OF DESIGNING, CONSTRUCTING, OR SERVICING SIMPLE unregulated power supplies is the freedom you have to use a wide variety of parts and components. The flexibility extends to the physical layout, and virtually nothing in these primitive supplies is critical once the basic requirements dictated by voltage breakdown and power-handling ratings have been complied with. Often, tradeoffs can be made without incurring significant degradation of performance; for example, if a certain inductance is not available for the filter choke, a considerably smaller or larger one might suffice with only a compensatory increase or decrease in the filter capacitance.

Then came the modern, linear voltage regulator. Because of the more stringent needs of complex circuits and systems, the linear regulator, with its high-gain feedback loops and much tighter specifications on performance parameters, required greater care in the choice of components and in the layout of parts and required consideration of new factors in the operating environment. Deviation from recommended components quickly yielded such plagues as oscillation, latchup, and poor regulation. So, the linear regulator required more experience, more know-how, and more insight into application requirements than did its predecessor, the simple, unregulated power supply.

When the modern, switching-type regulating power supplies came into being, they were accompanied by a great many superior components that took some of the magic out of their design. As a result, even though switching-type supplies are generally more complicated than linear supplies, the use of integrated circuits and special switching components has made switching-type supplies much easier to work with. But to take advantage of these new components, the essential characteristics of these devices must be thoroughly understood.

The output filter capacitor

A good place to start is with the *output filter capacitor.* In the first place, the basic design equations of the switching-type supply clearly reveal that you cannot proceed on the premise, "if a little is good, more is better." Switching frequency, duty cycle, choke inductance, and input and output voltages are all interrelated, so any brute-force approach must be ruled out. It is also important to realize that you get more than we bargained for in electrolytic filter capacitors. That such capacitors also contain resistance and inductance should not come as a surprise because stray parameters are encountered all the time in electronic designs and parts, and they are frequently ignored. But the stray inductance and internal resistance of electrolytic capacitors are of prime importance in the operation of switching-type power supplies, and therefore they cannot be ignored.

Figure 12-1 depicts the impedance of an aluminum electrolytic capacitor as a function of frequency. Notice that the plot is made on log/log paper to linearize the reciprocal relationship between frequency and capacitive reactance. This relationship is described by:

$$X_C = \frac{1}{2\pi fC}$$

Where:

X_C = the capacitive reactance in ohms
f = the sine-wave frequency in hertz
C = the capacitance in farads

The capacitor obeys the capacitive-reactance equation up to about 10 kHz, for up to that frequency, the plot is a straight line. The slope, incidentally, is the same for all ideal capacitors and for actual capacitors at low frequencies. Figure 12- 1 shows that

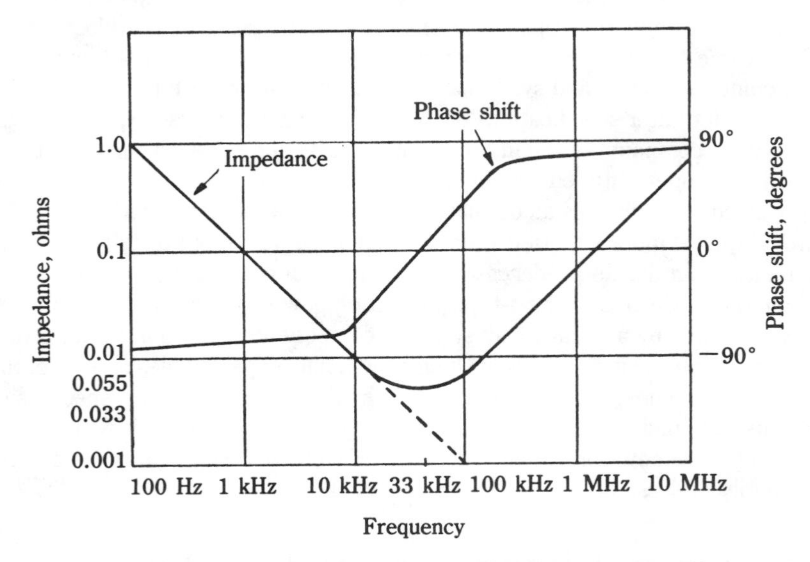

12-1 The impedance and phase characteristics of a 1500 μF electrolytic capacitor. The initial straight portion of the impedance plot represents pure capacitive reactance. The dashed projection would be that of an ideal capacitor zero resistance and zero inductance. The upswing of the impedance curve above 33 kHz is because of internal inductance.

this slope implies a decrease in reactance by a factor of 10 for each tenfold increase in frequency, which can also be expressed by stating that the capacitive reactance decreases 6 dB per octave of frequency (20 dB per decade of frequency).

Equivalent circuit of the capacitor

You could interpret the circuit characteristic depicted in Fig. 12-1 as a low-Q resonance produced by a series arrangement of capacitance, inductance, and resistance. (See also Fig. 12-2.) Because these curves were plotted from measurements obtained with an electrolytic capacitor, you can conclude that, in addition to capacitance, the capacitor does contain inductance and resistance. And as the frequency is raised, the inductive reactance of the capacitor's foil and connecting wires eventually dominates. Therefore, the series circuit of Fig. 12-2 is a very down-to-earth representation of the component that is referred to as an electrolytic filter capacitor. Most other components can be similarly represented; for example, a grid-dip meter will demonstrate that small ceramic capacitors with their leads shorted together behave as resonant LC circuits, with the inductance in the leads, and the sharpness of resonance governed by internal resistance.

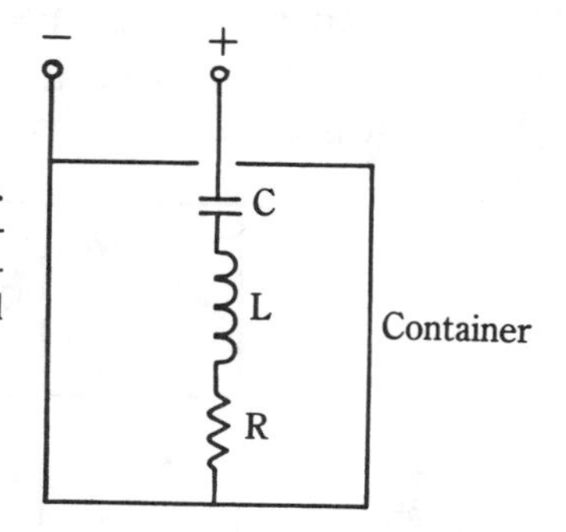

12-2 The equivalent circuit of an electrolytic filter capacitor. With appropriate element values, this circuit will reproduce the curves shown in Fig. 12-1. In this equivalent circuit, L is equivalent series inductance ESL and R is equivalent series resistance ESR.

The equivalent series R and L are responsible for the following undesirable effects in switching-type power supplies:

- Both the shape and the magnitude of the ripple voltage deviate from that which would be obtained from pure capacitance.
- The design of the switcher is more difficult than would otherwise be the case because the output ripple of these supplies is one of the design parameters in optimizing the switching frequency, the duty cycle range, and the inductor.
- The phase/gain margin that governs the stability of the feedback loop can turn out to be inadequate because of the uncapacitor-like phase behavior at higher frequencies.
- Although much is made of the dc regulating ability of power supplies, a generally more compelling reason to use regulated power sources is to present a low ac impedance to the load over a wide frequency range. The presence of equivalent series R and L tends to defeat this objective.
- The R value is not easy to pin down, and its variation can be considerable with

respect to different brands, and to aging effects. This also makes it difficult to manufacture capacitors with compliance to tight specifications.

- Much of the cost disadvantages attributed to the switching-type supply stem from the need for capacitors with very low *R* and *L* values.
- The equivalent series *R* is quite temperature-dependent, further complicating design and application.

The best expedient is to order special capacitor types made specifically for switching-type power supplies, and only from reputable vendors. Various manufacturing techniques and processes are used to reduce both R and L; for example, there are "stacked" units having a rectangular packaging configuration and four (rather than two) leads. There are also proprietory fabrication methods which are never detailed in manufacturers' literature, and all such methods usually lead to higher costs. Makers of switching-type supplies have, of course, tried to circumvent the higher-cost specialty capacitors, and one widely used technique is to parallel several smaller capacitors, instead of using a single large unit, the results of which are shown in Fig. 12-3.

Interpreting impedance and phase

In Fig. 12-1, a dip occurs in the impedance at approximately 33 kHz, at which point the phase shift is zero. This is no accident, for in every series-resonant circuit there is a

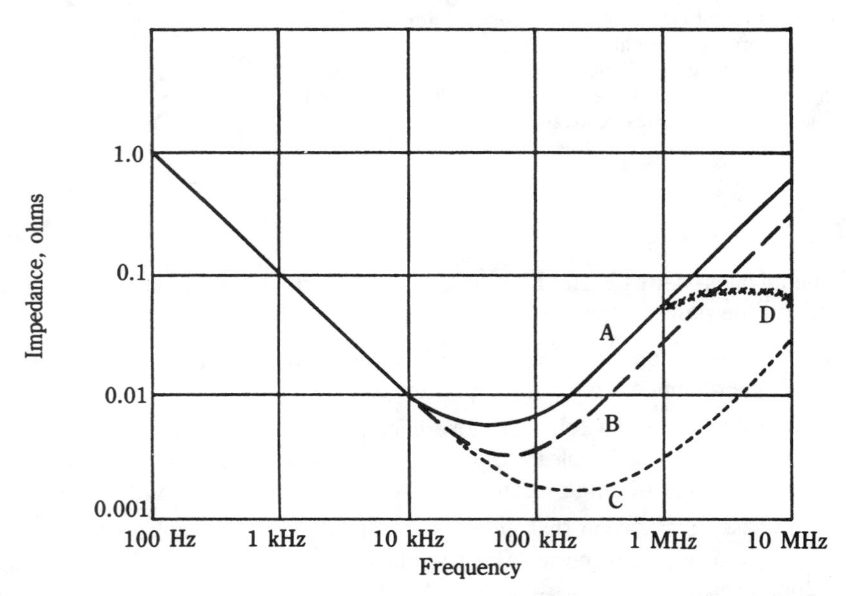

12-3 Remedial measures for lowering ESR and ESL in output filter capacitors. (A) An impedance plot of nominally 1500 μF aluminum electrolytic capacitor. (B) Improvement from paralleling several equal-value capacitors to total 1500 μF. (C) An improvement from substituting a four-lead "stacked"-construction capacitor. (D) General nature of improvement when a relatively small tantalum solid capacitor is shunted across the 1500-μF unit. Mylar and ceramic capacitors are also used as shunts for decreasing ESL in the megahertz region.

resonant frequency defined by this condition. The impedance at resonance is called the *equivalent series resistance (ESR)* of the electrolytic capacitor. For the capacitor of Fig. 12-1, the ESR is about 0.055 Ω. As a rule, the lower the ESR, the better suited the capacitor is for the output filter of a switching-type power supply.

The capacitor's impedance is essentially capacitive reactance up to frequencies of about 10 kHz. Then, resistance becomes increasingly predominant, manifesting itself in pure form at the resonant frequency, in the vicinity of 33 kHz. Thereafter, inductive reactance of the leads and capacitor structure begins to dominate at frequencies above 300 kHz.

Although such a way of evaluating electrolytic capacitors is not new, it did not have as much significance for the linear-regulating supply as it does for the switcher. Not only are the characteristics depicted in Fig. 12-1 typical, but the region where the inflection of the impedance curve occurs—in the 10- to 100-kHz range—is of prime importance for switching supplies. Suppose, for example, that this capacitor is used as an output filter for a switcher operating at 20 kHz. For the numerous harmonics of 20 kHz, which are generated by the switching process, the effectiveness of this capacitor is greatly diminished, as its ability to attenuate these harmonics depends upon its ability to bypass or short them to ground. This situation is particularly unfortunate, because the switcher tends to be a prolific generator of high-frequency harmonics, and these noisy harmonics in the output not only plague external equipment, but also affect the internal feedback loop used to regulate the output voltage.

The ESR rating

Because the equivalent series resistance (ESR) exerts a major influence on the impedance of the capacitor in the 10- to 100-kHz range—the region of optimum efficiency and performance for many switchers—this parameter should be determined as soon as it is known that the voltage rating and capacitance value are suitable. It is becoming increasingly common for manufacturers to supply this information in their technical literature, but it is not, however, always readily available. A much used method of estimating the ESR is by means of the equation: $ESR = P_D/2\pi fC$ where P_D is the dissipation factor, f is the measurement frequency in hertz, and C is the capacitance in farads. The dissipation factor has long been used as a figure of merit for capacitors, and was typically measured at 60 or 120 Hz for filter capacitors, or at 1000 Hz for coupling capacitors (the measurement frequency f must be known before you can calculate the ESR). Although the P_D is useful in estimating the ESR of the capacitor, it is useless in estimating the ESL, which may be the dominating factor at switching frequencies.

The ESR actually decreases as temperature is increased. This temperature dependency is particularly strong in the below-zero temperature range, as is shown in Fig. 12-4A. This, together with decreasing capacitance (Fig. 12-4B), has upset the predicted performance of many a switcher that performed well in the environment of the laboratory, but was later expected to duplicate results in more rugged environs.

The ripple-current rating

All too often ignored is the ripple-current rating of the output capacitor. The probable reason this rating is overlooked is because it has not been of great importance in the

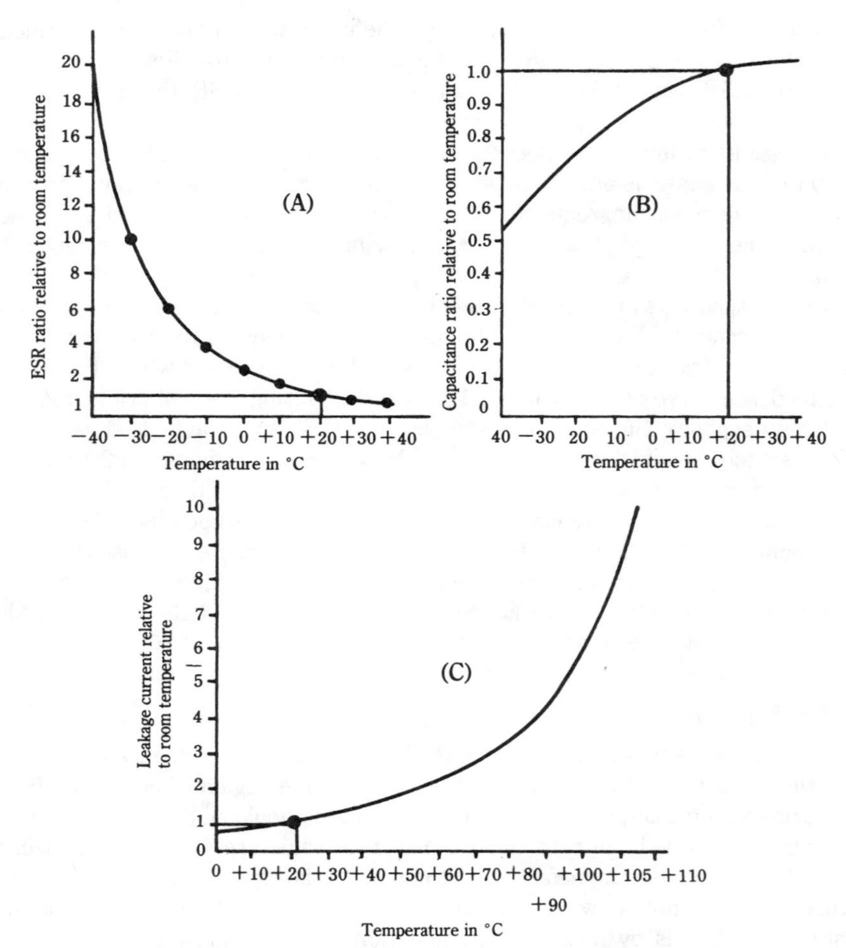

12-4 Typical temperature dependencies in electrolytic filter capacitors. (A) ESR as a function of temperature. Normalized for unity at room temperature. (B) Capacitance as a function of temperature. Unity at room temperature. (C) Dc leakage-current as function of temperature. Unity at room temperature. The capacitor in all the above situations is a 1500-μF "computer-grade" electrolytic type.

linear regulator. In a switching-type power supply, however, the lifespan of the output capacitor is bound to be adversely affected if its ripple-current capability is not adequate for the job. Fortunately, all major capacitor makers can provide ripple-current ratings. Your task is to determine the rating required, and there is a general rule of thumb method that has proved to be quite reliable. Although somewhat empirical, this method is predicated on sound logic. The following assumptions apply:

- The waveshape of the current variation in the inductor is a sawtooth.
- This current constitutes the only ripple to which the output capacitor is subjected.

- A typical peak-to-peak value of this sawtooth current is 20 percent maximum direct current (I_L) delivered to the load. This is based upon the generally encountered guidance to design, which suggests making the inductor current increase and decrease 10 percent, with respect to the dc component of the inductor. As a consequence of the foregoing, the ripple current (I_R) in the output capacitor can be computed from the formula:

$$X_C = \frac{1}{2\pi f C}$$

The factor in the denominator is introduced to convert the peak-to-peak current wave into its rms value. The relationship can be further simplified to $I_R = 0.058 I_L$. Generally, it is wise to incorporate a margin of safety after this computation has been made, to ascertain that the manufacturer's rating corresponds to the appropriate ripple frequency of the switching frequency of the power supply.

The ripple current generates internal heat in the capacitor, with the attendant changes in temperature-dependent parameters, such as those shown in Fig. 12-4. Not shown, because of its somewhat nebulous nature, is the relationship of life to internal temperature, as elevated temperatures greatly reduce the life expectancy of any electrochemical component. It has often been difficult to ensure a moderate ambient temperature for capacitors, much less to aggravate the situation by permitting excessive ripple currents. Even with an appropriately rated capacitor, dangerous internal temperatures can develop when there is no provision for heat removal from the external surface of the case.

Selecting a capacitor

There are two basic families of electrolytic capacitors from which to choose: aluminum and tantalum. The aluminum types are available in many quality grades and fabrication techniques. The tantalum types comprise distinct subtypes: foil, solid, and wet slug. At first glance, these might present a bewildering array with regard to selection. The old philosophy from the days of unregulated supplies, and carried over in linear regulators, was to search out capacitors having the highest figure of merit, defined by (capacitance × voltage)/cost. Obviously, this resulted in getting the most for the least.

However, the concept of getting the highest rated voltage and capacitance for the least cost can be very misleading when you are working with switching-type power supplies. As has been discussed, the impedance characteristics, primarily ESR, must be taken into account if the full potential of the switcher is to be realized. With line-frequency, thyristor-type supplies, cheap aluminum electrolytics of dubious quality can often be used, but all too often this leads to noise generation, poor reliability, and generalized criticisms of all switching-type power supplies. Good-quality aluminum electrolytics probably provide the best compromise, if one must be made, between cost and performance. These comprise "computer-grade" types and other types made especially for switchers. Some of the specialized constructions, however, have very compelling features, such as very low ESR and ESL, rather than low cost.

The outstanding feature of all tantalum types is their high volumetric efficiency; which is to say that a lot of capacitance can be stuffed into a given volume. The wet-slug tantalum has been the undisputed champion in the capacitance-versus-volume contest;

and in applications where size and weight assume great importance, such tantalums naturally merit attention. Solid tantalums appeal where there is great emphasis on longevity—both shelf life and operating life—but they have not been generally available in either large capacitance or high voltage values, as have other types. The foil-type tantalum is a good capacitor for switchers, but is not cost-competitive with aluminum types.

It is true that good capacitors have played a big role in making the switcher a technological and commercial success. But it is also true that the present success of switching-type supplies has inspired the capacitor makers to devise even better-suited types. Because of this impetus, it is difficult to say that one type is inherently better than another. At one point in time, and for a given application, one type might offer the optimum mix of parameters per dollar.

Where switching rates are high and current demand is low, nonelectrolytic capacitors will more frequently be found as adjuncts to the previously discussed electrolytic types, providing bypassing at high frequencies where the ESL of electrolytics becomes too high. The present trend for high switching-rate supplies is to dispense with electrolytic capacitors, using in their stead poly-propylene, or ceramic types. These are often packaged for surface-mounting to keep ESL low.

Whatever type of capacitor is selected, successful use depends greatly upon packaging and wiring techniques. For example, the ESL requirements of a capacitor can be reduced through careful wiring procedures. Short, wide straps are the best, and even these can be paralleled to further reduce self-inductance. It would be self-defeating to buy a four-lead capacitor, with elegant impedance characteristics, and then connect it into the circuit using a foot or so of conventional wire. More and more, the art of designing switching-type supplies is incorporating radio-frequency techniques.

The inductor

Trying to make an ideal component, such as a perfect solid-state switch, is an unattainable goal, but it focuses your efforts in the right direction. Thus, in the case of the switching transistor, you know that you must emulate as closely as possible a switch that is a perfect conductor during on time, a perfect insulator during off time, and one that changes its conductive state instantaneously. Pursuing a similar philosophy, what would your idealized inductor be like? The needs of the switching-type power supply indicate that such an inductor would be magnetically unsaturable, would have zero dc resistance, and would have no stray capacitance in its windings. This immediately suggests an air-core inductor, but for higher inductance values this would require too many turns of fine wire, which would also increase interwinding capacitance. The capacitance is undesirable because it promotes both series and parallel type resonances that degrade circuit operation. At the very least, you would hope that the first resonant frequency of your inductor would be much higher than the switching frequency.

The only way to reduce the resistance of an air-core inductor is to use large wire, which leads to an oversized monster. In order to cut down the number of turns needed, you must include a ferromagnetic core. And, to avoid the effects of premature and especially abrupt saturation of the core, an air gap must be provided. The optimum, but practical, inductor is then a product that results from tradeoffs involving wire size, number of turns, effective magnetic-core permeability, and air gap size. Because all of these quantities are interrelated, the design of a suitable inductor is as much an art as a science. And, as might be anticipated in such circumstances, more than one approach can be used. The magnetization and loss curves of Fig. 12-5 show the nature of the conflicting parameters involved in the design of a practical inductor.

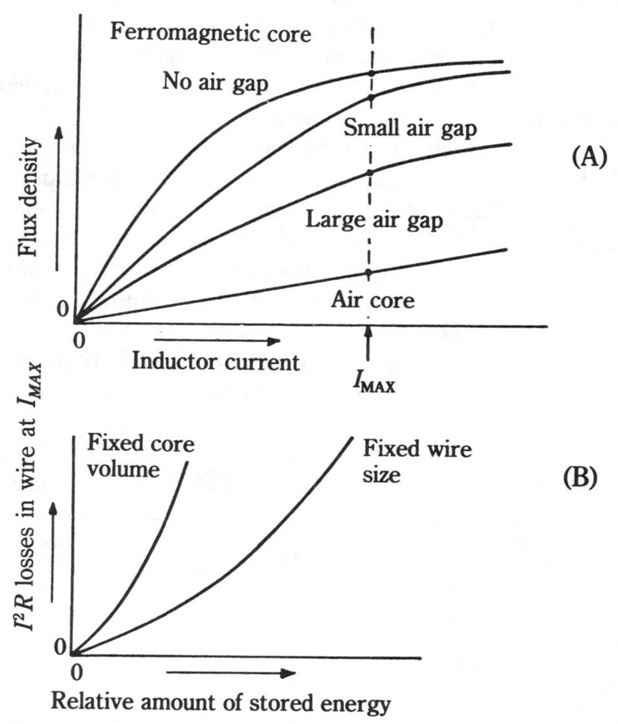

12-5 Many trade-offs are involved in the design of a good inductor. (A) Linear magnetization curve all desirable. Found only in air-core inductors, but may be approached with ferromagnetic core with suitable air gap. (B) Too large an air gap requires too many turns of wire and increases I^2R losses. Rule of thumb: Limit I^2R loss to about one percent of supply output power.

Core materials and structures

It is convenient to use *ferrite cup cores* because the wire can then be easily wound on a bobbin and inserted into the core. Additionally, it is not difficult to establish a suitable air gap empirically in order to linearize the magnetization curve. Another benefit provided by the air gap is a relatively gentle passage into saturation when large currents are passed through the windings.

Molybdenum Permalloy toroids (ring-shaped cores) also make excellent cores for these inductors. Of course, there is some sacrifice in winding convenience compared to bobbin windings, but it fortunately happens that inductors in switching-type power supplies, and especially those operating at the higher frequencies, require relatively few turns of heavy wire, compared to the monstrous "chokes" used in other supplies. The necessary air gap is actually present in this material, but it is effectively distributed throughout its volume in ferromagnetic particles that are separated by nonmagnetic binder material. As a result, the manufacturer has good control over the effective air gap in toroidal cores, and they are available in various A_L values (A_L being the inductance in millihenrys for 1000 turns of wire). Cores with relatively low A_L values tend to make optimum inductors for switching-type supplies because a low A_L corresponds to a large effective air gap. Powered iron toroids have overlapping characteristics with Permalloy toroids with regard to the requirements of switching-type supplies, and thereby merit competitive consideration.

Ferrite toroids are a possibility, but the saturation characteristics are not as easily controlled, and this leads to design headaches. The emphasis in ferrites has been primarily directed to making higher A_L values, rather than lower ones. The manufacturers of ferrite material have, however, made available an intriguing selection of core structures, which appear to be application-oriented toward the needs of switchers because their two-piece construction permits the inserting of air gaps.

Other core types have been more or less successfully used. See Table 12-1. These include laminated structures and air cores specialized high-frequency designs. Laminated cores suffer from relatively high hysteresis and eddy-current losses at ultrasonic switching rates.

Table 12-1 Saturation characteristics of inverter core materials.

CORE MATERIAL	SATURATION FLUX DENSITY IN KILOGAUSS (1000 LINES PER SQ CM)
60 Hz Power Transformer Steel	16 – 20
Corosil, Hipersil, Silectron, Tranco	19.6
Deltamax, Orthonol, Permenorm	15.5
Permalloy	13.7
Molypermalloy	8.7
Mumetal	6.6
Ferroxcube 3E2A	~ 3.5

The least that can be expected from an inductor that enters saturation too rapidly is a change in the switching rate of a self-oscillating regulator. To a greater or lesser degree, depending upon other factors, this can increase ripple voltage and worsen dc regulation. A manifestation not quite so benign is the catastrophic destruction of the switching transistor, or the free-wheeling diode. After all, when the energy stored in the inductor is required to abruptly change, the deficit or excess energy must be either quickly supplied or absorbed which tends to produce high, often destructive, peak currents in the switching transistor and free-wheeling diode. Consequently, it rarely pays to skimp on the design or cost of an inductor when you consider the cost and reliability of other associated components that it affects.

13 CHAPTER

Traditional semiconductor switching components

THE SEMICONDUCTOR COMPONENTS THAT ARE COVERED IN THIS CHAPTER are the long-used discrete devices that have reliably performed the switching operation in the power supply. Switching transistors and diodes generally perform the dc voltage switching in switching regulators, inverters, and converters. Triacs and SCRs perform a similar function in switching ac voltages. There are, however, interesting marriages of these components in many systems so that it cannot be said that any one device is limited to a specific switching operation. But it can be said that a knowledge of these devices is essential because their optimum operation and survival in a switching-type supply places great emphasis on their ability to function as a truly ideal switch.

Switching transistors

There are many types of switching transistors available, and these run the gamut from modernized germaniums to sophisticated monolithic Darlingtons. Often, more than one type will suffice for the job at hand. In general, the choice is narrowed by such considerations as whether an npn or pnp type is required, and whether strong consideration must be given to high voltage, high switching speed, high current, or cost.

Initially, one has to be aware of the basic differences between an ideal switch and an actual switching transistor. In Fig. 13-1, a segment of an ideal switching waveform is illustrated; together with it is the actual waveform that a real switching transistor would produce. The perfect rectangular wave is a fantasy—one to be approached, but never attained. The beauty of such a perfect switching waveform is that losses are zero all the time. The ideal switch is either all the way on or all the way off, and the transition between the two states is instantaneous. When this ideal switch is on, it delivers full current with no voltage drop across its terminals; when this ideal switch is off, it withstands the full voltage across its terminals and passes no current. The trapezoidal

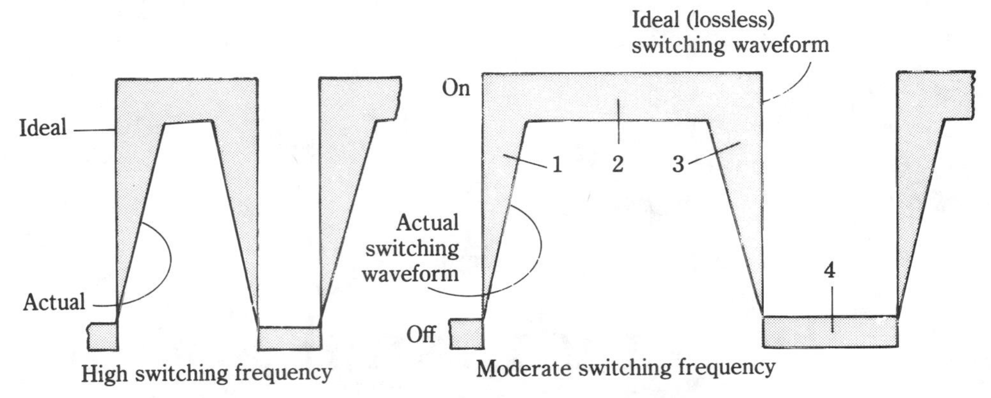

13-1 Power is dissipated in four regions of a real transistor when switching. In contrast, an ideal switching transistor would not dissipate any power. (1) Power dissipated during turn-on. Switching transition. (2) Power dissipated during on-time. (3) Power dissipated during turn-off switching transition. (4) Power dissipated during off time.

wave, the actual switching waveform, is something that must be tolerated in a real transistor, but much time and effort is spent urging it to become more like the ideal.

The actual switching waveform has four regions of power loss. When this real switch is off, it is not entirely off; it allows a small leakage current to flow to the load. Because the emitter-collector voltage is highest during the off interval, this loss, though often low, can still be significant. When the real switch is in its on state, it again is not entirely on; the collector-emitter saturation voltage, $V_{CE(sat)}$, is sustained across its collector-emitter terminals. Although this voltage might be relatively low, it occurs just when the collector current is highest, and therefore this loss is also very significant.

The shaded areas of Fig. 13-1 reveal yet another deviation from the ideal. The actual switch requires time to make on and off transitions. During such transition times, the switch is operating as a linear element, a class-A amplifier; it is dissipating power because it has appreciable voltage across it, as well as appreciable current through it. Fortunately, the switch usually spends only a relatively short time in these transition regions, but if the switching frequency becomes too high, the transition power losses become dominant. Power dissipated during these transitions also contributes greatly to transistor failures, and for this reason manufacturers often provide graphs showing various combinations of voltage and current that constitute "safe operating areas" during switching transitions.

General transistor characteristics

Table 13-1 provides generalized guidance with regard to the commonly used transistor types and fabrications. Because of economics, availability, second-source requirements, and ignorance, strange transistors are sometimes found in strange places. Excellent monolithic, triple-diffused Darlingtons are now in a state of accelerated technological development; these structures are not inherently relegated to their former low place in the matter of frequency capability. There is also considerable performance overlap among the various devices.

Table 13-1 Guide to application of power transistors.

	Germanium (pnp)	Silicon single diffusion (npn)	Silicon epitaxial base (npn/pnp)	Silicon double diffusion (npn/pnp)
Features	Low cost. High current capability. Low collector-emitter saturation voltage.	Electrically rugged. Superior thermal characteristics often make this silicon device a better choice than germanium.	Practical combination of ruggedness, frequency capability, and voltage rating. A useful workhorse for many power-supply applications.	Best bipolar process for high-frequency capability and high switching speed. Current rating is good.
Shortcomings	Limited frequency capability. Effective heat removal needed (junction temperature is limited to vicinity of 110 °C). Collector voltage is generally rated below 100 V. No npn.	Limited frequency capability. Moderate voltage ratings. Not commonly available in pnp.	Possibly marginal performance in high-frequency switching circuits.	If too much emphasis is placed on frequency capability; electrical ruggedness suffers. Generally, the SOA is low.
Uses	Series pass elements in linear regulators. Inverters, converters, and switching regulators operating at low and medium frequencies.	General-purpose power device in linear regulators and in switching applications up to medium frequencies.	General-purpose device with parameter blend which is often easier to apply than those of the single diffusion transistor.	High-speed switching applications utilizing good design. Protective techniques are often used.

	Silicon triple diffusion (npn)	Silicon Darlington (npn/pnp)	Power MOSFET (n-channel)	Synthesized transistor (simulated npn)
Features	Second-best bipolar device for frequency capability and switching speed. Has best voltage rating.	High current gain. High input impedance. Favorable cost and production aspects. Parameter combo is suited for many uses. Electrically rugged.	No thermal runaway. No secondary breakdown. Very high switching speed. Very high input impedance. Internally protected gate. Enhancement mode. Easy paralleling.	Complete overload protection. Base drive as high as 40 volts with no damage. 0.5 μsec switching time. 3 μA base current. Interfaces CMOS or TTL. Low cost.
Shortcomings	Current rating is generally inferior to other silicon transistors. Cost tends to exceed "workhorse" types.	Some loss of circuit flexibility. Trade-offs in frequency, voltage, and current capabilities must be carefully evaluated. Cannot be driven to saturation. Therefore this device has a high voltage drop.	In some circuits, the internal (body) diode can cause trouble. ON resistance might cause excessive voltage loss and degradation of efficiency.	Limited current and voltage ratings. (Approx. 2A and 40V.) No pnp version is presently available. Vulnerable to self-oscillation if care is not exercised in layout and by-passing.
Uses	High-voltage regulators, inverters, and converters. Generally used in switching applications where both high frequency and high voltage exist.	Series-pass elements. Drivers. Moderate speed switching. Useful in reducing parts count and manufacturing costs. Motor control.	Appears optimally suited for linear and switching supplies involving wide power-range. Merits consideration for high switching rates. Excellent for motor control.	High-reliability, low-power output element in both switching and linear supplies. Especially useful in cost-effective designs because protective circuitry is not needed.

	Power MOSFET (p-channel)	Depletion-Mode power-MOSFET	SENSEFET power MOSFET	Insulated Gate Bipolar Transistor (IGBT)
Features	Many are marketed with very similar characteristics to popular n-channel types. Simplifies circuit topography in certain applications.	Device normally ON. Gate bias can be applied to either cause harder turn-ON or to turn the device OFF. Development has focused on n-channel types, p-channel types are possible.	Has extra element that provides a tiny fraction of actual load current for control purposes.	Exhibits high input impedance for a power MOSFET and the low voltage drop of a bipolar transistor can have good reverse-blocking ability. No body-diode effect.

Table 13-1. Continued

Power MOSFET (p-channel)	Depletion-Mode power-MOSFET	SENSEFET power MOSFET	Insulated Gate Bipolar Transistor (IGBT)
Shortcomings			
Availability not as extensive as n-channel MOSFETS and cost tends to be higher for similar ratings.	Not widely available compared to enhancement-mode MOSFETS. Commonly used circuit techniques are geared to enhancement devices. Limited current and voltage ratings.	Not widely available in extensive range of ratings.	This is a high-voltage and high-current device that is best-suited for low and moderate frequencies. Best IGBTs are limited to 5-kHz, 20-kHz and 50-kHz types might be useful.
Uses			
Most applications are similar to those in which n-channel types have been used. These included switch-mode power supplies and motor control. The p-channel MOSFET is particularly well-suited for use in bridge-type synchronous rectifiers.	Well-suited for certain constant-current circuits. Can be used in linear regulators. Simulates normally ON mode of certain relays.	Excellent for use in current-mode switching regulators. When so deployed, power-wasting sampling resistance is not needed. (The resistance that is required has negligible power dissipation.)	Excellent for switch-mode supplies operating at 600 volts (and higher) and at tens of amperes, but at 20 kHz or preferably lower rates. Device enables very efficient motor control.

Motorola

Germanium transistors formerly made by Motorola, Delco, Sylvania, etc. are now provided by Germanium Power Devices, Andover, MA.

Manufacturers indicate frequency capability in two ways. A common method is to designate f_T, the frequency at which the current transfer ratio of the common-emitter configuration becomes unity; f_T is also known as the *gain-bandwidth product* or *beta-cutoff frequency.* The gain declines with respect to increasing frequency at 6 dB per octave; knowing this, you can work backwards and establish the approximate beta cutoff frequency, defined as being 3 dB below the current gain (β) at a low frequency, such as 1000 Hz. The beta cutoff frequency should be at least 10 times the switching frequency; this is predicated upon the assumption that it takes 10 harmonics to construct a fairly rectangular waveshape.

The preceding frequency evaluation might not be valid for switching rates above several kilohertz because storage phenomena, incurred in the on or saturation mode, tend to predominate. Therefore, transistor manufacturers often give data more relevant to operation in the switching mode, such as the delay, rise, storage, and fall times under specified operating conditions. From such information, you can plan the switching rate and duty cycle in a realistic way.

The triple-diffusion structure can be subdivided into the *annular* and the *etch-cut* variations. The annular structure tends to display voltage ratings up to 400 or 500 V, and the f_T can be 30 MHz or so. The etch-cut provides a classic example of tradeoff, for its voltage capability approaches 1000 V, and its f_T may be on the order of 10 MHz. The safe operating area reveals more electrical ruggedness for the high-voltage than for the high-frequency types.

The loss mechanisms depicted in Fig. 13-1 do not tell the whole story about dissipation in the switching transistor. As you might suspect, power is also dissipated in driving the transistor, because a strong drive is required to force the transistor well into collector saturation during the on period. However, this also tends to keep power dissipation from $V_{CE(sat)}$> low. Therefore, it would appear that selection would favor transistors with high current gain and low emitter saturation voltage.

To a considerable extent, every time you optimize a parameter in a transistor, you pay for it by the degradation of another parameter. If you postulate an overall figure of merit for transistors by multiplying together the numbers representing all of the significant parameters, this product would tend to remain approximately constant; for one parameter is increased at the expense of a proportionate decrease in some other parameter. For example, voltage capability tends to deprive the transistor of frequency capability. From such a vantage point, advances in transistor technology can be defined as processing or fabricating techniques, which result in improvement of all or most of the significant performance characteristics.

Switching considerations

Driving a switching transistor is not just a matter of producing saturation during the on time. In Fig. 13-2, it is seen that the base current does not cease immediately when the

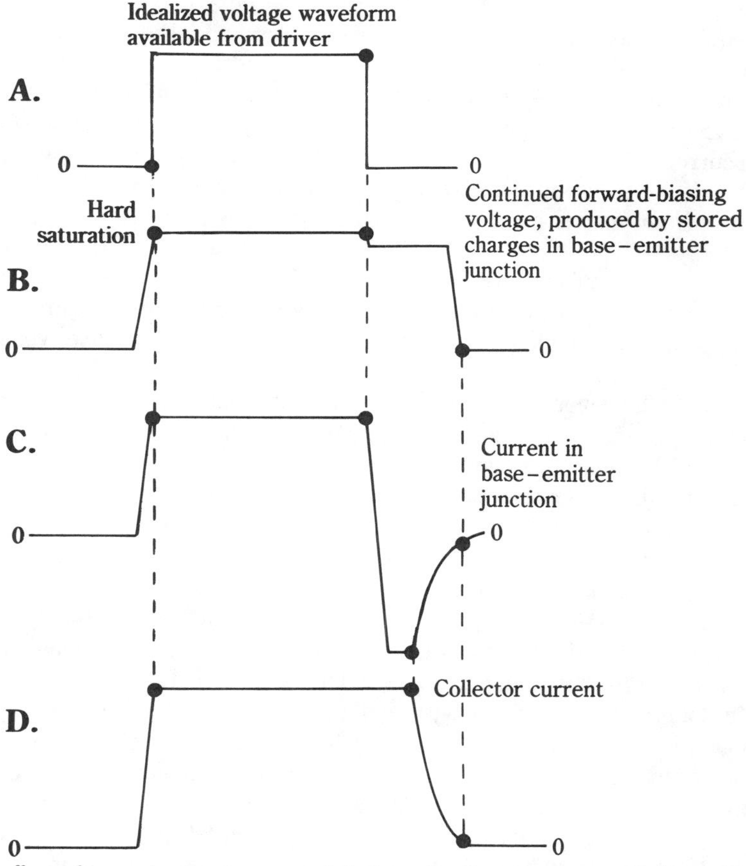

13-2 An effect of base-charge storage on fall time of collector current. (A) Ideal drive voltage delivered to base of switching transistor. (B) Voltage waveform actually monitored at the base. (C) Current in base-emitter junction. (D) Current wave at the collector: this is the current available to the external circuit inductor during on-time.

transistor is supposed to be off. Because of charge storage in the base-emitter junction, some time is required for the base current to decay to zero. During this time the collector current continues to flow as if the transistor were still driven from an external current source. This phenomenon largely governs the fall time of the collector current. On the other hand, the rise time of the collector current improves with increasing base drive current. Thus, the losses in the turn-on and turn-off regions of Fig. 13-1 are affected oppositely by base drive. Because the base drive is intimately connected with the loss in the on region, it is readily seen that there must be an optimum value of turn-on base drive. This is further complicated by the fact that the base is often made negative during the off period in order to decrease losses at this time and to reduce turn-off time. As might be expected, optimum base drive is frequently determined empirically on a prototype model in an endeavor to obtain the highest efficiency possible.

Because of parameter tradeoffs in transistors, there is no universally accepted type. A germanium transistor complies with the requirements of one switcher, whereas another requires a sophisticated, triple-diffused, silicon type. But there is one criterion all designers must abide by: the concept of *safe operating areas (SOA)*. The SOA imposes bounds on transistor operation in addition to that which would prevail from the standpoint of power dissipation alone. Stated simply, you cannot simultaneously apply maximum current, voltage, and power ratings. The concept of SOA is shown graphically in Fig. 13-3. If the transistor was a simple passive device, such as a 100-W resistor, it generally would not matter if the 100 W resulted from 10 V and 10 A from a dc source, or 50 V pulses at 10 A from a switched source with a 20 percent duty cycle. Not so with the transistor, for it is subject to collector current limitation, to secondary breakdown at much lower power dissipation than would result from mere extension of the curve depicting maximum allowable power dissipation, and to collector-emitter voltage limitation because of avalanche breakdown.

Figure 13-3, however, is not complete. With real transistors, a power/temperature derating curve generally must also be applied. And it turns out that there is an allowable relaxation of the boundaries of the SOA for switching waveforms. As might be expected, a more favorable use of the transistor's ratings becomes possible with shorter on times. Therefore, a more realistic approach to SOA is that shown in Fig. 13-4, from which allowance can be made not only for the basic bounds illustrated in Fig. 13-3, but for the way in which these bounds change with respect to duty cycle and to temperature. Thus, you need a family of SOA curves with on time as a parameter, and a power/temperature derating curve. Such information is usually provided by the major manufacturers of switching transistors, but it might not encompass all conditions of secondary breakdown; if in doubt, go to the manufacturer for more complete information.

Although you can estimate whether a switching transistor is "fast" or "slow" just from considerations of gain-bandwidth product f_T, a reliable indication of exact switching performance is not readily obtained in this manner. Therefore, transistor manufacturers also provide curves that depict delay, rise, storage, and fall times of those transistors likely to be used for switching. A typical set of such curves is shown in Fig. 13-5. By summing up delay, rise, storage, and fall times, you can more easily come to meaningful conclusions concerning allowable switching rates and duty cycles. The curves also enable one to estimate the losses likely to be incurred during rise and fall times. Other

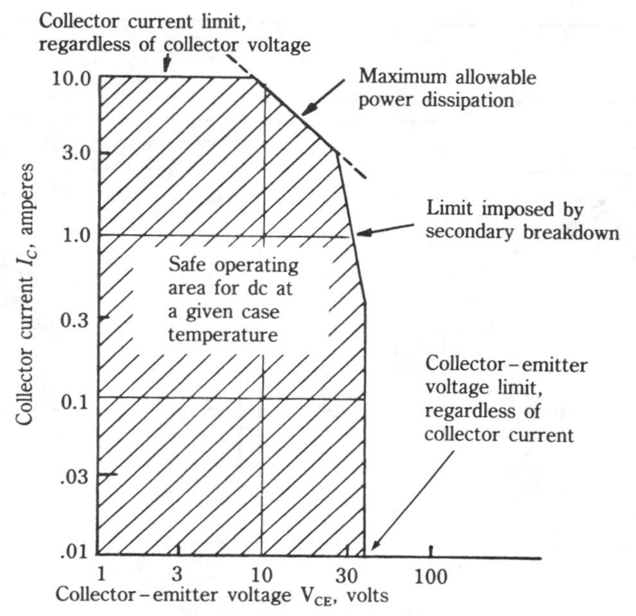

13-3 An illustration of basic concept of safe operating area of a power transistor. Just any combination of collector voltage and current is not permitted. Allowable voltage/current are limited by secondary breakdown, by maximum-allowable collector current, and by maximum allowable collector voltage. Even though these limitations are not violated, you must *also* abide by the maximum-allowable power dissipation.

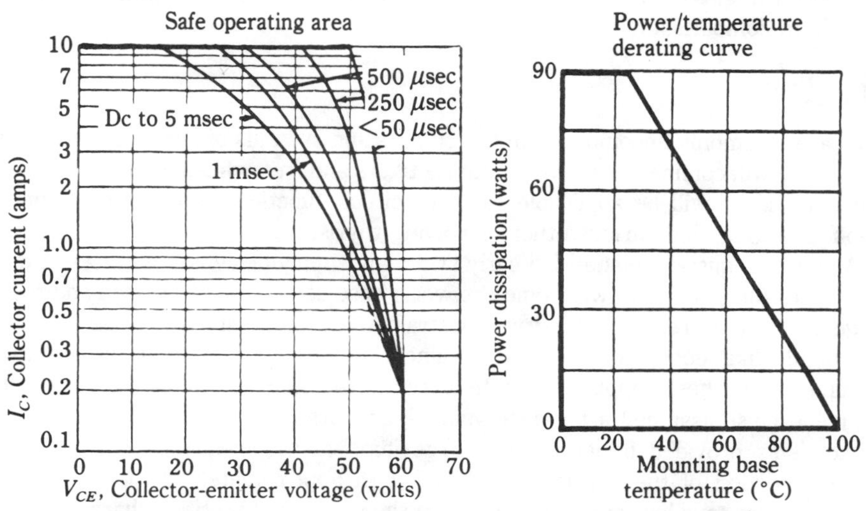

13-4 A typical set of curves showing SOA and power/temperature derating information. When the mounting base temperature exceeds 25°C, the allowable operating conditions revealed from the SOA curves have to be modified. If the power/temperature derating curve is plotted in terms of junction temperature, a formula will be provided to enable calculation of mounting-base temperature. In principle, junction temperature is the more-fundamental parameter. Motorola Semiconductor Products, Inc.

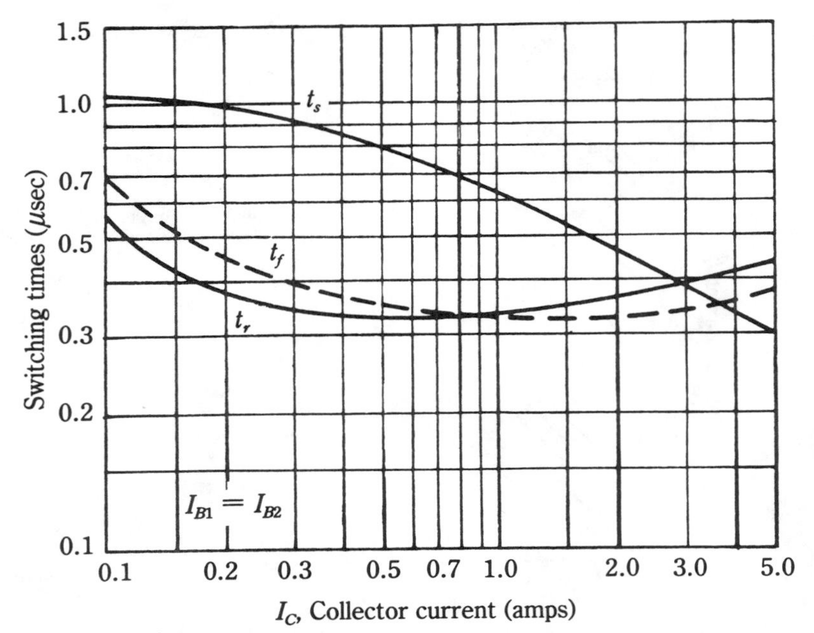

13-5 A typical set of curves showing times governing switching speed. t_r Rise time. t_s Storage time. t_f Fall time. A fourth curve, turn-on time delay t_d is sometimes depicted. Its value is usually small compared to the other times. Motorola Semiconductor Products, Inc.

things permitting, the on time should be great enough to render rise time and fall time negligible in comparison.

Diodes

Diodes are commonly encountered in the circuits and systems of switching-type power supplies. We will confine our discussion to rectifiers and free-wheeling diodes; however, the general ideas will be applicable to other circuit functions, such as commutation, hold-off, charging, and various others involving thyristors.

A common mistake made with early switching-supply designs was to select diodes according to their ability to withstand the worst-case conditions of forward and reverse voltage, forward current, temperature constraints, and surge and short-circuit current. This sounds like good, conservative, engineering design; and so it was, until the switching rate of these supplies began to increase. After several kilohertz was reached, the audible noise assumed a nuisance value—or worse, as higher power levels were reached. This provided impetus for a large jump in the frequency domain, to the ultrasonic portion of the spectrum, starting with approximately 20 kHz. It was no longer sufficient to select diodes just because they had good rectifier characteristics at 60 Hz; you now have to be concerned with the frequency capability of diodes.

Real diode behavior

You have been deluged with the near-ideal characteristics of diffused-junction diodes, and such diodes are indeed quite remarkable. Their forward voltage drop is reasonably low for many applications, and little contention can be offered with regard to their very low leakage currents—even at high voltage levels. Nor can they be faulted on such matters as reliability or cost. In Fig. 13-6A, the half-wave rectification behavior of a large, diffused-junction silicon diode is about as ideal as you can expect at 1 kHz (of course, you do not see the losses from less-than-perfect conduction).

If, however, the frequency is increased to 50 kHz, the presence of the reverse-conduction region serves to drastically lower the rectification efficiency. The rectifier runs hotter and the overall efficiency of the power supply is reduced. As with switching transistors, the inability of the diode to faithfully turn off is caused by stored charges. The storage takes place during forward conduction; both the junction and the bulk material are affected and must be depleted of these charges. This does occur, but during the time required for charge "clean-out" by the reverse-voltage portion of the sine wave, excess power dissipation is taking place in the diode. And at the same time, the output voltage is lowered and the filter capacitors are subjected to increased ripple current. The rate of recovery from this postconduction behavior can be manipulated; best results are attained when the amount of absorbed charge is reduced as a consequence of processing and fabrication of the diode. Quick recovery, however, can generate other problems, such as RFI and circuit disturbances from ringing phenomena. Two diode manufacturing techniques have resulted in satisfactory high-frequency rectifying diodes.

In one technique, *fast-recovery* characteristics are produced, usually by gold doping the silicon semiconductor material. A significant improvement occurs, and substantially clean rectification is obtained at higher frequencies. As might be anticipated, such success is a tradeoff—adversely affected are the forward voltage drop and the reverse-leakage characteristics. In the overall picture of diode action, the tradeoff is a favorable one for high-frequency operation because it is particularly desirable to eliminate or greatly attenuate reverse conduction stemming from charge storage.

The other method of dealing with charge storage is one of circumvention: a diode is made that does not make use of minority charge carriers, as there is virtually no charge

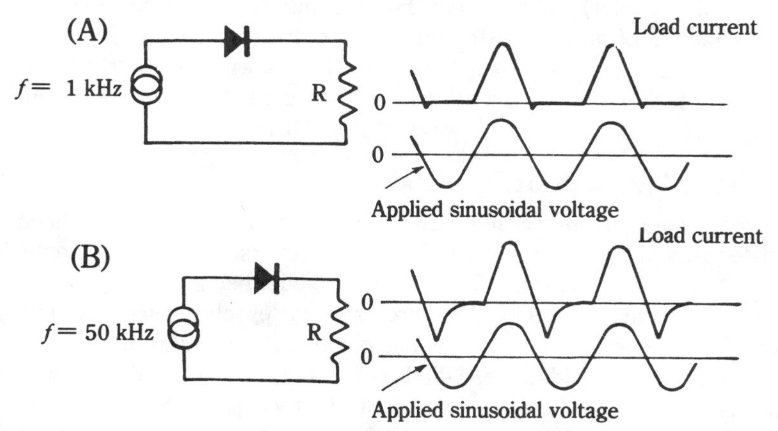

13-6 Rectifying behavior of a typical diffused silicon diode. (A) At 1 kHz, reverse recovery characteristics are barely perceptible. (B) At 50 kHz, reverse recovery is major feature of rectified current.

to become involved in storage. Such a diode is called a Schottky barrier diode, otherwise known as a *hot-carrier diode.* Unlike a silicon junction diode, the Schottky fabrication utilizes a contact between a metal and semiconductor material, usually n-type. With appropriate processing, a unilateral conduction, rather than an "ohmic" contact, is produced. For practical purposes, this diode is free of storage phenomena, and its high-frequency performance is affected only by its internal capacitance, which is small enough to permit good, clean rectification beyond 100 kHz.

If the worst that could be anticipated from ordinary rectifier diodes in high-frequency circuits was the requirement for more effective heat removal from the diodes or somewhat lowered efficiency, perhaps there would have been less impetus for the development of high-frequency diodes. But, because circuit processes are interrelated, the poor performance of one component can adversely affect the operation of other components.

Switching considerations

In Fig. 13-7A is shown a simplified high-frequency (50 kHz) converter. You can suppose that this is part of a switching-type regulated supply with a feedback path from the dc output to the inverter driver, which would control switching transistors Q1 and Q2 to vary the duty cycle of the inverter. But for the purpose at hand, consider only the 50-kHz rectangular wave that is being generated and rectified. Figures 13-7B and C then display the switching waveforms of good-quality diffused diodes and gold-doped fast-recovery diodes. It is evident that departure from ideal behavior is much greater with the conventional diodes than with the fast-recovery types.

Closer inspection reveals significant aspects of the distorted waveforms in Fig. 13-7B. In Fig. 13-7B, the diode current waveform shows considerable interference with the process of rectification. Among other things, much excess heat must be removed—this can be inferred by the large area of the wave beneath the zero axis. The painful consequences of the bad rectification appear in Fig. 13-7D, in which the switching transistors are subjected to a very high current peak—one enduring more than long enough to make its energy impact felt. These transistors are also forced to spend an appreciable fraction of on time in the unsaturated or linear region. It's a safe bet that the SOA will be exceeded by such behavior. Not shown are other possible adverse effects, such as higher ripple current through the output filter capacitor and increased output ripple. Notice the cleaner I_c wave of Fig. 13-7E.

Gold-diffused vs. Schottky diodes

Fast-recovery diodes are processed with the aid of gold doping of the silicon semiconductor material, and by this technique, very good reverse recovery characteristics are obtained at 100 kHz and beyond. The Schottky diode, being a majority-carrier device and inherently blessed with minimal charge-storage problems, also exhibits good reverse recovery well beyond 100 kHz. Is one diode better than the other? The excellent reverse recovery of the gold-diffused diode is obtained as a tradeoff with other parameters. One of these is the forward voltage drop. It happens that conventional silicon diodes are already in trouble when applied in low-voltage rectification systems. Because of computer technology, there is great emphasis on 5- or 6-V supplies, often with considerable current capability. The forward voltage drop of the rectifiers constitutes a

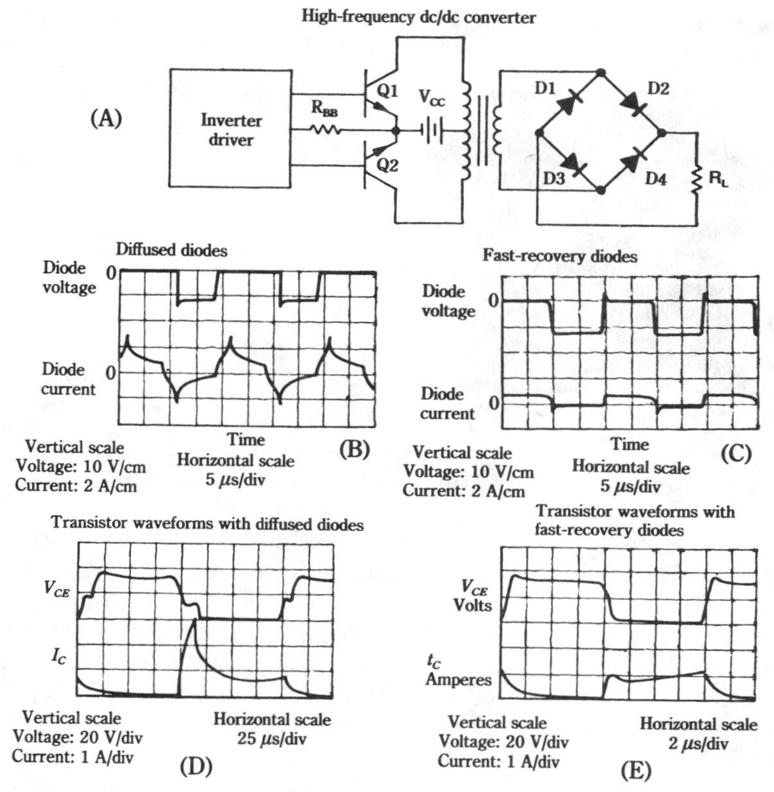

13-7 The comparison of diffused and fast-recovery diodes in a 50-kHz dc-dc converter. Motorola Semiconductor Products, Inc.

serious power loss at relatively low voltages. One helpful technique is to use a center-tapped full-wave rectifier circuit, rather than the bridge (which imposes the voltage drop of two diodes, worsening an already undesirable situation). Of course, a rectifier diode with lower forward voltage drop than silicon junction types must necessarily merit consideration, and this is where the Schottky diode shines.

Figure 13-8A shows the way forward voltage drops compare in Schottky and conventional silicon diodes of similar current capability. Not only is the Schottky diode a better choice for low-voltage, high-frequency rectification, but it remains the choice for low-voltage, low-frequency supplies because of its low forward drop. Its temperature capability, however, is generally inferior to silicon junction devices, but it is nevertheless superior to that of older germanium rectifier diodes. As a matter of fact, the Schottky diode is in its element in low-voltage applications because its reverse leakage current increases quite rapidly with higher reverse voltages. What technology will bring tomorrow is not too clear, but hither-to-used Schottky diodes approached excessive dissipation in the reverse region when the reverse voltage across it reached about 30 V, as shown in Fig. 13-8B; and the situation has been more adverse at high operating temperatures. Much development work is being expended to extend this temperature

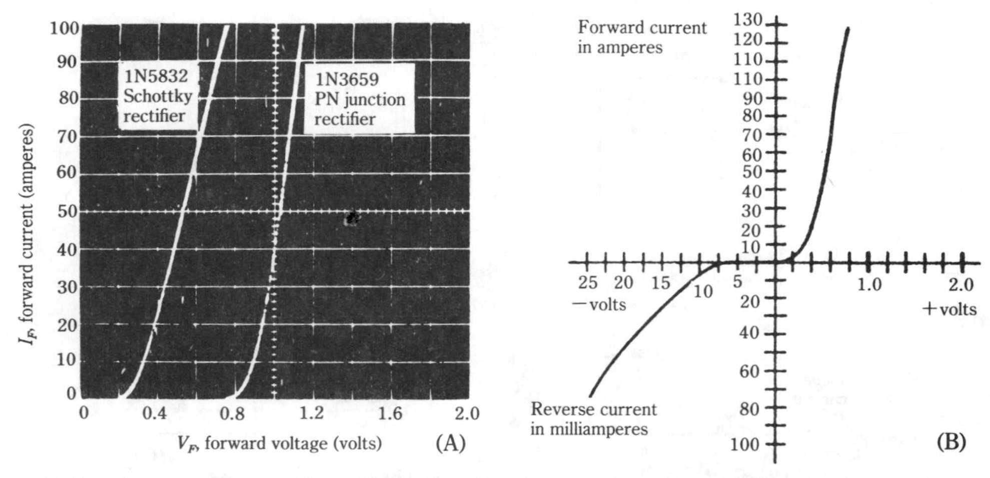

13-8 The comparison of forward voltage drop in Schottky and conventional silicon diodes.

limitation. Until recently, problems arose when the "junction" temperature hit 100° C, but currently available units are able to operate satisfactorily and reliably to 125°C, and beyond.

Fast-recovery, gold-diffused junction diodes become increasingly competitive as forward voltage drop diminishes in importance—that is, for the rectification of higher voltage, for example, about 10 V to 15 V. Where forward drop is not very important and the voltage capability of the Schottky does not suffice, it turns out that the gold-diffused diode might be a better choice.

Silicon-controlled rectifiers

SCRs, like switching transistors, are available in a wide variety of types. Indeed, the relevant characteristics for certain switching functions are not always easy to decipher from the specification sheets. The SCR is a device that is dependent upon *regenerative* circuit action. Once triggered, the gate is deprived of control and a number of malfunctions can occur. A simple SCR phase-control circuit can be used as a preregulator and operated from the 60-Hz power line. The troubles likely to be encountered in such a circuit tend to be minimal. However, in higher-frequency applications and in circuits such as inverters, where turn off is brought about by commutating techniques, it is possible to rapidly accumulate a box of burned-out thyristors without half trying!

Inexpensive SCRs are generally intended for the consumer market—for applications where the basic accomplishment is simple power control. These units are often made by the alloy-diffused process so that gate-triggering voltage and current, anode-cathode blocking voltage, and holding current can have sloppy tolerances. It would be wrong to say that these relatively inexpensive devices cannot be used in switching-type power supplies, but to use them under the illusion that all SCRs are the same is an open invitation to trouble. Some of the malfunctions common to SCR switching circuits are:

Failure to turn off This can result from excessive junction temperature, inadequate blocking-voltage rating, transients in the main supply, or a poor trigger circuit. A likely cause with "bargain" devices is a low *dv/dt* rating; the rate of voltage change caused by transients can turn such a device on, even though other turn-on mechanisms are absent — catastrophic destruction is likely.

Spurious turn on This is caused by basically the same reasons as above. Again, inadequate *dv/dt* figures prominently in this malfunction, tending to damage the SCR.

Good operation, but short life This is probable indication of inadequate *di/dt* capability. High initial current densities cause localized heating, which gradually brings about deterioration and ultimate destruction. Thermal fatigue of solder-bonded pellet is also a possible cause of malfunction.

Latchup and commutation difficulties at high switching rates Charge storage imposes time delays in SCRs, just as it does in transistors. The ratings of many SCRs, even quality units, apply for 60- or 400-Hz operation. Units for higher-frequency applications must be so specified.

For switching-type supplies, all-diffused SCRs (including epitaxial types) tend to be superior to alloy-diffused devices. An additional consideration involves the matter of gate/cathode geometry. The edge-fired geometry has been extensively used, but it often leads to unsatisfactory *dv/dt* and *di/dt* capability, and so it is not suitable for high-frequency operation. It is true that the edge-fired geometry requires relatively low gating current, but it is doubtful that this is a compelling feature in the overall picture. The initial cost advantage of edge-fired geometries has narrowed as more development has been directed toward the superior center-fired geometry. There are many versions of this new geometry, but all derive their superiority over the edge-fired technique from the fact that more cathode area is in conduction per unit time.

SCRs, in which the pellet is either soft- or hard-soldered, can have poor reliability in switching-type power supplies, particularly in those with marginal heat removal. A much better manufacturing technique is *compression-bonded encapsulation*, in which a spring system replaces the solder bond. For large supplies, controlling currents in the tens and hundreds of amperes, the *disc package* provides excellent thermal characteristics because the pellet is heatsinked from both sides.

Switching considerations

Thyristors enable the control of load power by interrupting the flow of load current, so they are similar in many respects to the basic switching circuit of Fig. 13-9. The thyristors used most often in switching-type power supplies are the SCR and the triac, although other *pnpn* devices are also important for triggering and timing. The transistor operates as a driven switch — with its conductivity at all times dependent upon actuating drive. The thyristor operates as a regenerative switch triggered from its off to its on state by a short-duration pulse. Once triggered, the thyristor supplies its own gate drive and very rapidly goes into saturation, thereafter, the gate loses its control entirely. It does not matter if subsequent trigger pulses are delivered to the gate or if the gate is disconnected from the circuit. Once the thyristor is in its on state, it is latched as a closed switch.

The thyristor remains on until the current through it is either interrupted or reduced below a relatively low value — the so-called *holding current.* The thyristor then reverts to its nonconductive or off state. Having thus been reset, it is again receptive to a gate trigger pulse. Because the simple thyristor circuit is fed from an ac source, load current is determined by timing the occurrence of gate triggers. In this way, more or less of each half-wave cycle is allowed to pass to the load. This method is

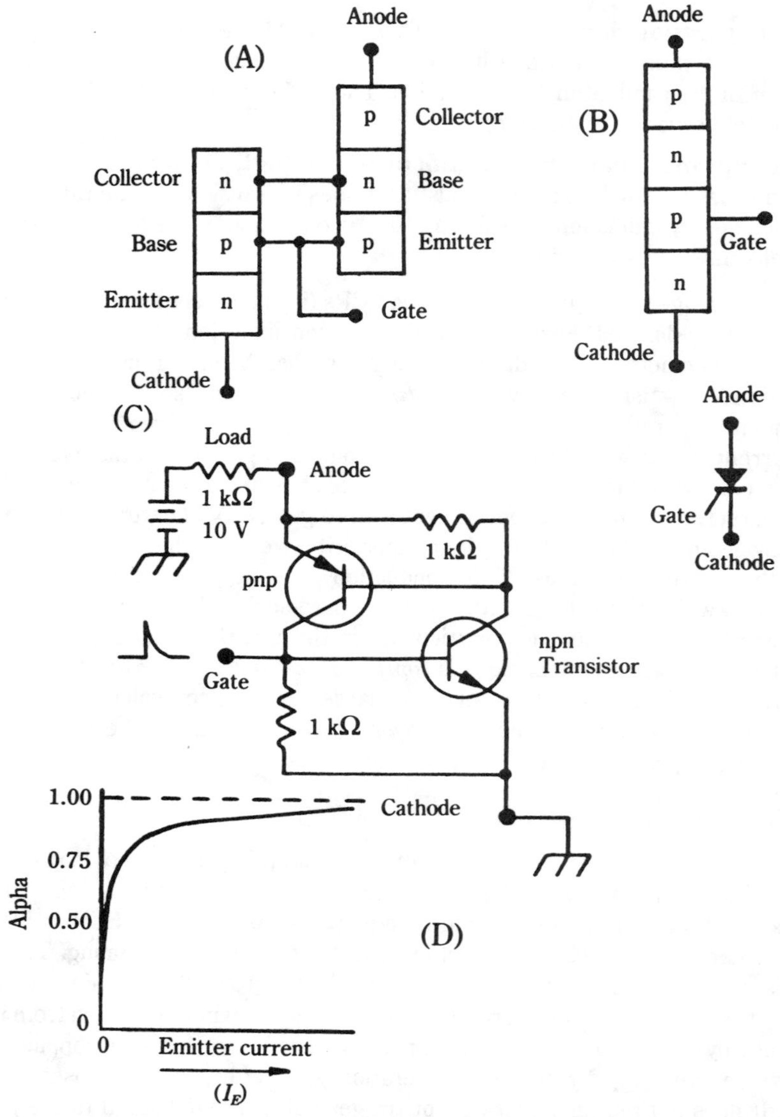

13-9 Model and equivalent circuit of the silicon-controlled rectifier SCR. (A) A physical model, made up of two transistor-like semiconductor devices. (B) A merger of the two structures in A. This is similar to actual SCR. (C) A circuit of a regenerative switch that acts in analogous fashion to an SCR. (D) Current transfer ratio alpha is strongly dependent on emitter current in the low-current region.

known as *phase control,* but the effect is similar to that of duty-cycle or pulse-width modulation.

Figure 13-9 shows physical and circuit models of the silicon controlled rectifier. The two-transistor configuration closely duplicates the actual operation of the SCR. In essence, this is a latching binary, or flip-flop, in which one state change (off to on) is provoked by a gate trigger pulse, and the alternate change of state (on to off) is brought about by the next zero crossing of the ac input wave (there are other modes of thyristor operation, but the described sequence of events is typical).

The useful behavior of the SCR stems directly from the fact that alpha (α)—the current transfer "figure of merit" in transistors—is a function of emitter current. Alpha is simply the ratio of output (collector) current to input (emitter) current with the transistor in the common-base circuit. The concept of alpha can be used in equations describing relationships in transistor circuits other than the common-base configuration, as in the dual-transistor circuit of Fig. 13-9. Because neither transistor has any forward-biasing arrangement, the stage is in its nonconductive state. Although a leakage current (I_{CO}) provides some forward-bias current to the base-emitter junction of the npn transistor, the α developed by this transistor at such low currents is too low to start any chain of events in the overall circuitry. Thus, this transistor, and the pnp transistor also, is "dead."

The injection of a short-duration positive trigger into the "gate" terminal provokes a drastic change in the circuit. The npn transistor has its α momentarily increased and thereby its collector current. This also serves to increase the forward-bias current of the pnp transistor, bringing it to life. The resultant collector current of the pnp transistor reinforces the forward base-emitter bias of the npn transistor, whether or not the gating pulse still exists. Once so initiated, both transistors are speedily transferred into their conductive states. Because one transistor reinforces the on state of the other, the entire circuit latches in the on state, allowing full current to be delivered to the load. In this latched state, trigger pulses have no further effect, and even a negative pulse applied to the gate will usually have no effect. To restore the circuit to its off state, the current must be momentarily interrupted either at the cathode or anode.

There are thyristors (e.g., triacs) with gate turn-off capability, but conventional SCRs are most often used as phase-controlled switches in power supplies, regulators, motor controllers, and other applications where power levels exceed several watts. Even at much lower powers, the conventional SCR is generally used.

The equation describing current flow through the transistors is:

$$I = \frac{I_{CO}}{1 - (\alpha_1 + \alpha_2)}$$

Where:

I = current flowing through the load

I_{CO} = leakage current through the circuit where the transistors are in the off states of conduction

α_1 = alpha of pnp transistor

α_2 = alpha of npn transistor

This equation states that the load current increases substantially as the sum of the two individual αs approaches unity, at which point the load current is determined entirely by the circuit's own resistance. The transistor circuit then behaves as a closed switch. Of course, neither this circuit nor the thyristor structure it simulates has zero resistance or zero voltage drop in its on state, but because both transistors are in hard saturation, the voltage drop across the anode/cathode terminals of the device is relatively low. Active or "hot" transistors are those in which a closely approaches unity. But the foregoing equation shows that even relatively poor junction devices, those with alphas considerably less than unity, can participate in the regenerative process, which requires only that the sum of the current transfer ratios attain unity.

Voltage/current characteristics

From what has been said so far about the SCR, it might well be supposed that the consequence of applying an ac voltage to it would be essentially nothing. The rationale could be that, providing no gate signal was applied, the two-transistor model would not permit anything greater than leakage current to pass. When the polarity of the anode-cathode voltage was proper for normal conduction, no conduction would occur because of the lack of emitter-base bias on both transistors. For reverse polarity of the applied voltage, you would expect no action anyway. This reasoning is valid, but only if the applied ac voltage is kept within bounds. One of these bounds is familiar to us, for if the reverse voltage is high enough, avalanche breakdown will occur. In Fig. 13-10, reverse breakdown occurs at point A, and is designated as BV_R. When the SCR is in a phase-controlled switching circuit, it is important to avoid reverse breakdown because this results in loss of both rectification and load control.

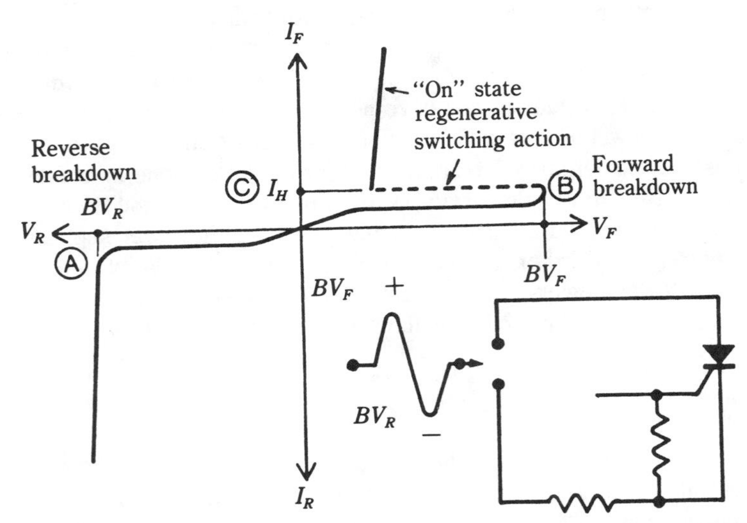

13-10 A voltage-current characteristic of the SCR with zero gate voltage. (A) This region is similar to avalanche or zoner breakdown in silicon diodes. (B) This breakdown region results from excessive anode-cathode voltage. (C) This region shows the holding current, L_H, and is the minimum anode current needed to sustain the SCR on its state of conduction. Note: Regions A and B must be avoided because the device is intended to be triggered into conduction only by a gate signal.

Figure 13-10 shows another breakdown at point B, designated as BV_F. It is very important to understand that this breakdown, for forward polarity of the applied ac wave, occurs without any gate signal. This might appear strange because the phenomenon resembles the action that you would expect from an initiating gate pulse—the SCR regeneratively switches to its on state and latches there. You are thus up against a basic limiting factor in circuit operation. What happens is that the leakage current increases enough to forward bias the emitter-base junction of the npn transistor. This increase is sufficient to induce both transistors into a state of saturation. Obviously, point B must be avoided if we wish to control the SCR only by the timing of gate pulses. In Fig. 13-10, because it was desired to display the total dynamic characteristics of the SCR, the impressed voltage was deliberately made higher than would be proper for actual circuit operation. Because of the effects of temperature, as well as the rather sloppy tolerances on many SCRs, it is wise to incorporate a healthy safety margin to assure operation that is free from either of these breakdown phenomena.

In Fig. 13-11, the applied voltage is well within bounds, so neither forward nor reverse breakdown occurs. A triggering network is normally provided so that gate-induced turn on can be adjusted. The triggering circuit could turn on the SCR immedi-

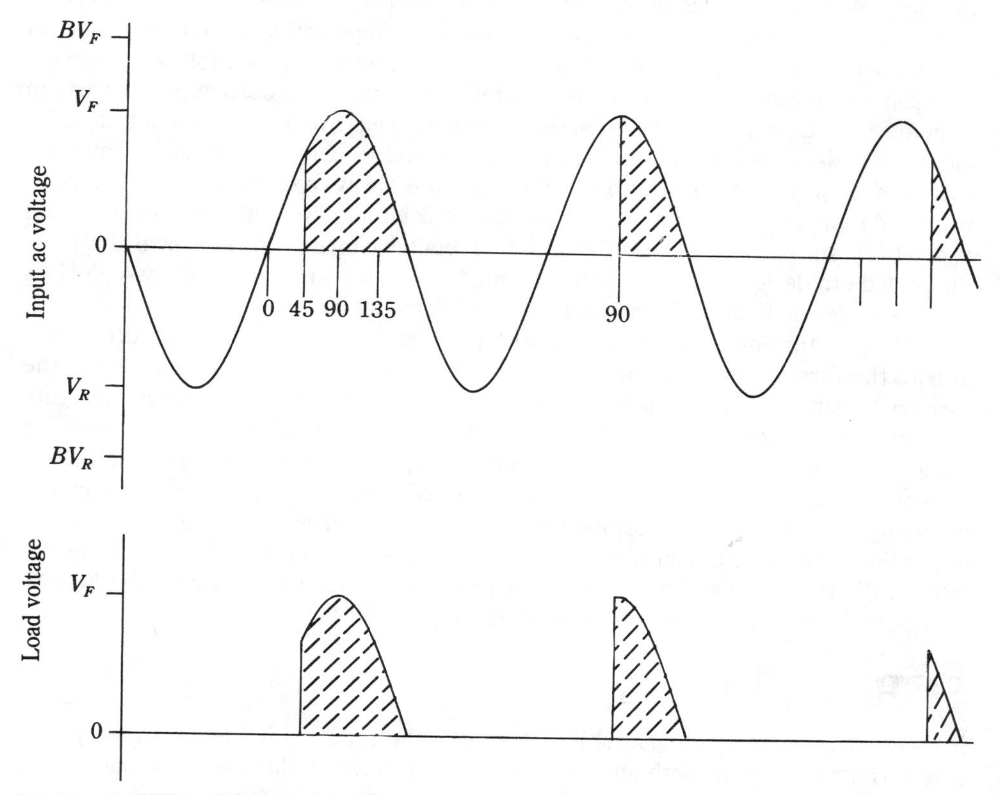

13-11 An SCR switching waveform. The three examples illustrate 45-, 90-, and 135-degree conduction-delay angles. The earlier the SCR is triggered into conduction, the more power is delivered to the load.

ately, at the positive-going transition of the input ac voltage, or the triggering could be delayed for a time. Figure 13-11 depicts several triggering delays. The shaded areas represent the portion of the cycle in which the SCR is in its on state, delivering power to the load. Simple triggering circuits merely sense the ac input voltage and trigger the SCR when some preset voltage level is attained; others utilize resistor-capacitor circuits, which permit delays to be greater than 90°; more sophisticated triggering circuits use feedback to regulate the output power.

Originally, SCRs were used almost exclusively for 60-Hz applications. Later, they performed efficiently in 400-Hz three-phase systems. At present, there are SCRs available for use in inverters that operate at frequencies beyond 25 kHz. The maximum frequency of the SCR is determined by the response characteristics of the semiconductor junctions. In a manner of speaking, the difficulties stem from excessive rather than attenuated response. As we increase the frequency of the applied voltage, we find an increased tendency toward turn on, irrespective of gate signal. The inherent capacitance existing across the junctions of the device can cause high-frequency turn on by allowing sufficient turn on current to be passed to the gate from the main anode voltage source. This is known as the *dv/dt* effect, being appropriately named because current through this capacitance is proportional to the frequency of the applied ac voltage.

Another performance parameter that must be kept within bounds is the *di/dt* value. This is the initial rate at which anode-cathode current increases following turn on. Lifespan and reliability are adversely affected if this rate is excessive, particularly in large devices. Often, *di/dt* can be slowed down by means of a small amount of series inductance. Because it might not be easy or practical to obtain timely and meaningful data, it is often wise to derate current capability. It is always a step in the right direction to provide for good heat removal. SCRs are available in different constructions, gate geometries, and packages. Consultation with a reputable manufacturer should always be the prelude to design, and even replacement, for the state-of-the-art for these devices progresses faster than their external appearance might suggest.

The gate turn-on characteristics of SCRs (and triacs as well) play an important role in both the turn-on time and the amount of dissipation that the SCRs incur during the turn-on transition. Manufacturers give both minimum and maximum values for gate turn-on voltages and currents. Although it might seem desirable to select gate-drive parameters that favor the minimum side of the specifications, this results in slower turn-on times, which inevitably lead to increased dissipation, particularly at higher operating frequencies. It is true that increased drive values also increase the amount of dissipation in the gate circuit of the SCR, but it is the total amount of power dissipated in the SCR that is most important. Consequently, it is actually easier on the SCR to provide near-maximum values of gate-triggering pulses.

SCRs and triacs

It is commonly stated in technical literature that the triac is a full-wave thyristor, one that permits passage of both alternations of the ac wave to the load. This sometimes leads to a misconception, for although the SCR operates as a half-wave rectifier, it does not follow that the triac functions as a full-wave rectifier. In most switching-type power supplies, the ultimate output is regulated dc. Of course, this does not necessarily exclude the triac, for rectification can always be imposed following the control function. A more natural preference for the SCR stems from its rectifying property, and the fact that triacs are more limited in voltage, current, and power ratings. From these considerations, there should be no surprise that the triac has not displaced the SCR in switching-type power supplies.

In Fig. 13-12 is shown the basic arrangement used for phase control of full-wave dc power with two SCRs operating in conjunction with a pair of conventional rectifying diodes. This basic technique is utilized in the switching supply illustrated in Fig. 8-3. An interesting method of accomplishing the same objective with a single SCR is depicted in Fig. 13-13. The gate-firing circuit uses a unijunction relaxation oscillator. For simplicity, the manually controlled version of this basic technique is presented—the power delivered to the load is adjustable by means of variable resistance *R* (a wider control range can be obtained if the UJT circuit is empowered from a small auxiliary dc supply, rather than from the main rectifying bridge, as shown). In a switching-type regulator, resistance *R* could be replaced by the collector-emitter circuit of a bipolar transistor; an output-derived dc error signal could then be applied to the base to vary the charging rate of capacitor C in such a way as to bring about regulation of the sampled output voltage.

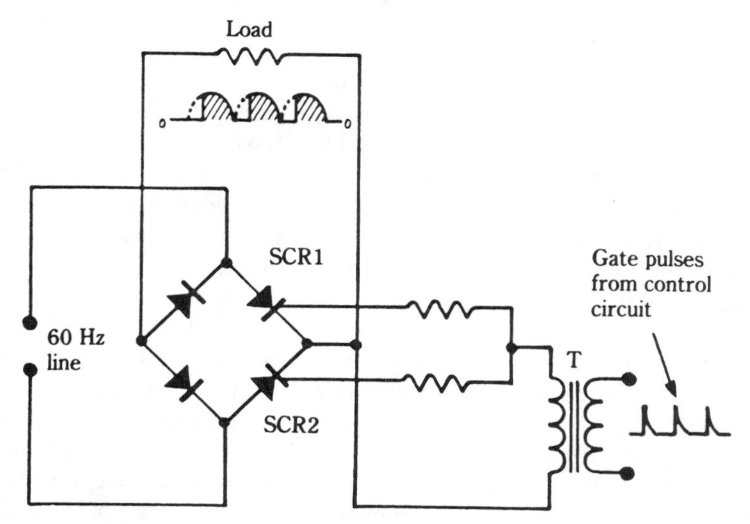

13-12 Switching circuit for full-wave power control by SCRs. The circuit is essentially a modified rectifying bridge in which two conventional rectifiers are displaced by SCRs.

Whether this method is used in manual or in automatic controlled systems, the UJT commences an oscillation cycle at each zero crossing of the 60-Hz wave. This constitutes synchronization with the power line and is an essential operating feature. This feature depends upon feeding the UJT with unfiltered dc. Therefore, if an auxiliary supply is used, as suggested, there should be no filter capacitor.

Triacs are, in general, limited to low-frequency applications, although future types might overcome this disadvantage. Phase-controlled triac circuits are similar to SCR

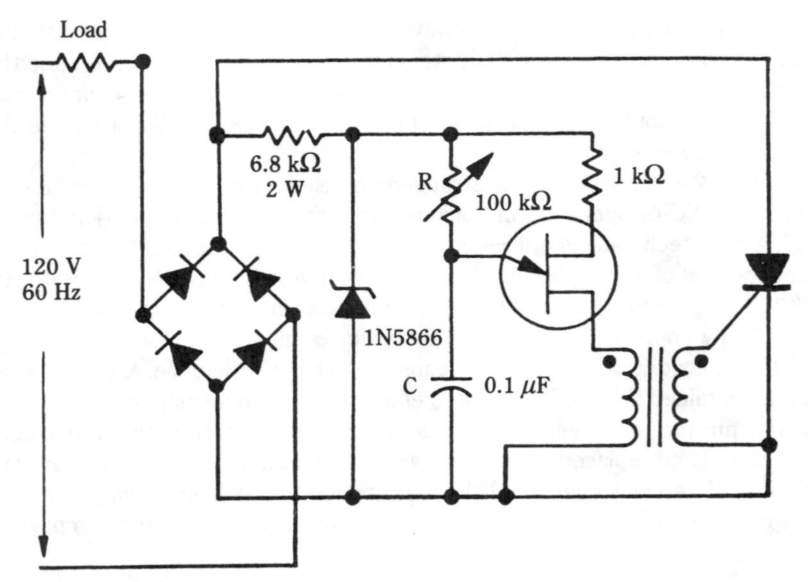

13-13 An alternate method of SCR full-wave control of power. In this arrangement, a single SCR produces a phase-controlled output for both half-cycles of the ac input voltage.

circuits, but they benefit greatly from the fact that the triac requires only one gate signal, whereas a full-wave SCR circuit generally requires two. The gate of the triac can be triggered using dc voltages of either polarity, although the gate tends to be somewhat more sensitive when the gate voltage is of the same polarity as the anode. Because triacs are normally used only for ac switching applications, the worst-case turn-on conditions must be accounted for in the design of the trigger circuits. As in the case of the SCR, it is always best to provide near-maximum gate drives, both to ensure fast turn on and to reduce device dissipation.

Figure 13-14 illustrates a triac and its switching characteristics. The triac also exhibits both forward and reverse breakdown voltage, BV_F and BV_R. The conduction characteristics of a triac are symmetrical, and a gate pulse can turn on the triac with anode-cathode voltages of either polarity. (Nomenclature such as anode and cathode are meaningless with triacs because the device is bidirectional in nature. Most manufacturers assign new terms for these electrodes—such as M1 and M2, for main polarity terminals. However, as long as you remember the bilateral nature of these electrodes, you can continue to use the anode and cathode terms, which are most familiar.) The triac is very much like two SCRs connected in parallel, but oppositely polarized, except for the triac's single gate terminal. The triac's behavior is otherwise equivalent to the SCR, and descriptions pertaining to the SCR are generally applicable to triacs as well.

The optoisolator

An interesting device, with considerable potential for use in switching-type power supplies, is the *optoisolator* (Fig. 13-15). Incorporating a light-emitting diode (LED) and a photodiode, phototransistor, or photo-Darlington light detector, this device provides

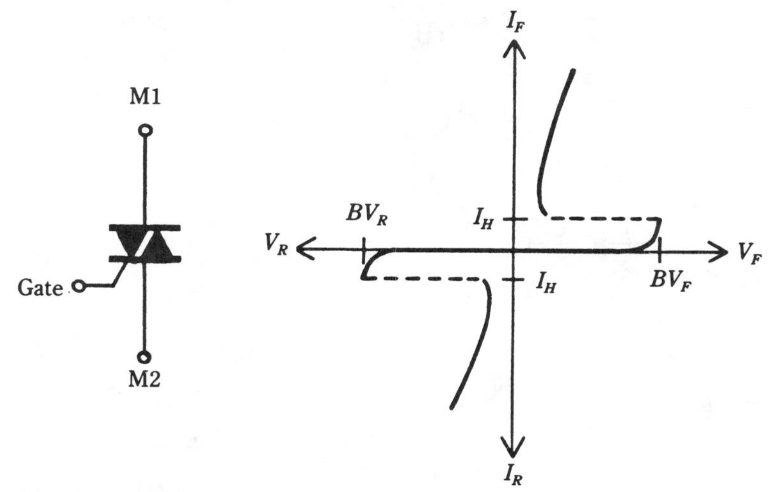

13-14 Although it has only one gate, the triac operates as a symmetrical ac switch. The load-current carrying terminals are better referred to as ***main terminals 1 and 2***, or ***M1 and M2***, rather than "anode" and "cathode."

the functions of level-shifting and electrical isolation. Both of these functions give designers many a headache in switchers without a 60-Hz power transformer, or in those in which a high-voltage output is developed. Of course, it can be argued that this device is not really new because packaged combinations of photoconductors, such as cadmium sulfide in conjunction with incandescent lamps, have been around for many years. The performance and reliability of some of these older versions were notoriously poor—most engineers looked upon them as novelties, and there was little attempt to make them commonplace items. However, the optoisolator is truly different, for both the transmitter and receiver of the optoisolators have frequency capabilities that are more than adequate for the most sophisticated ultrasonic switcher. This contrasts dramatically with the slow response of incandescent lamps, neon bulbs, and photo-conductive detectors made from various compounds of lead, selenium, cadmium, and sulfur.

Unfortunately, the predominant emphasis on the use of optoisolators has been for interfacing various logic circuits. The digital mode of operation will also find use in switching-type supplies, but the most immediate application appears to be in the switcher's feedback loop, where the optoisolator is used in its linear mode. The makers of these products have had some reservations about advocating such use, because the transfer function of the optoisolator is often not very linear. But the dynamic excursion of operating current need not extend into severely nonlinear regions, nor is the nonlinearity of great importance when the device is inserted in the negative feedback loop. The regulator derives its performance primarily from its error-signal canceling feature, and so is not adversely affected by nonlinearities, but it tends to compensate for changes occurring within the feedback path. In practical regulating power supplies, optoisolators are easily incorporated using standard design procedures to ensure stability in the feedback loop.

On the other hand, optoisolators tend to be quite temperature dependent, and their

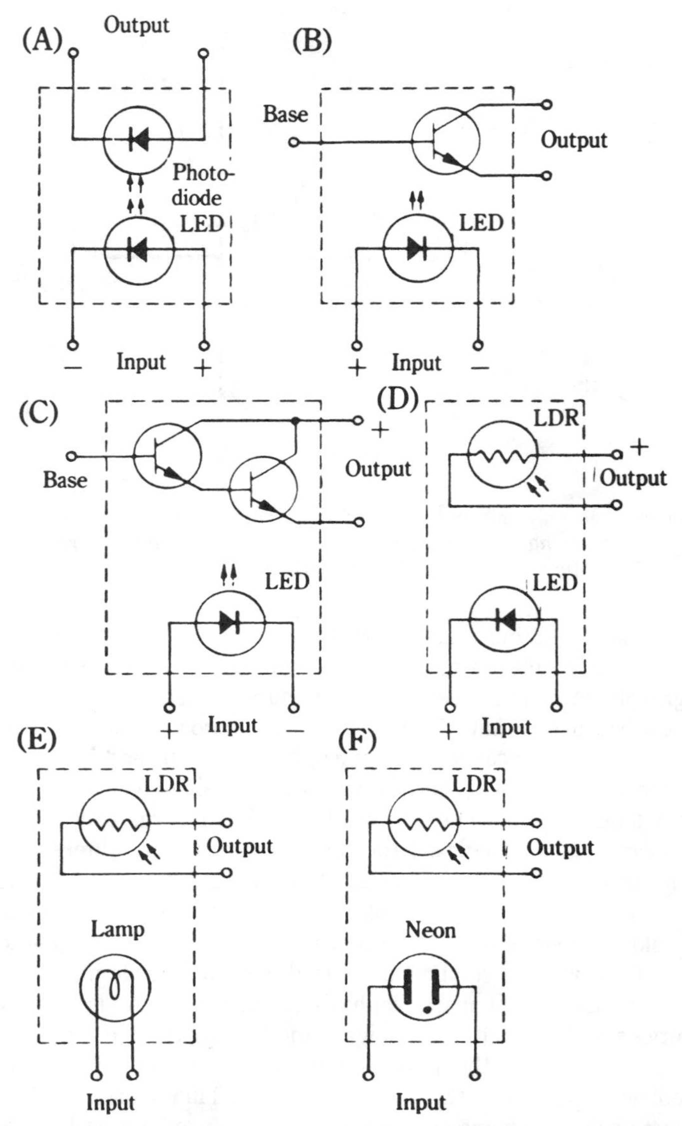

13-15 Optoisolators come in many formats. (A) LED in, photodiode out. (B) LED in, phototransistor out. (C) LED in, Darlington out. The devices depicted in D, E, and F, use photoconductive elements as detectors, which causes them to have sluggish response times. These devices are also known as *light-dependent resistors LDRs.*

current-transfer ratio varies greatly among the various types. Those with photodiode output devices have the best response characteristics, but the output current is, at best, a small fraction of the input current. An optoisolator with phototransistor output can develop current-transfer ratios in the vicinity of unity, and when a photo-Darlington output is used, the current-transfer ratio can range up to 10. The latter type is also the slowest, but units are available with responses adequate for switching regulators. In all of these types, the current-transfer ratio is greatly influenced by the forward input current—another way of describing their nonlinearity. The base leads of phototransistors can be left unused or can be used for inhibit, strobe, or other control purposes. Also, by connecting a resistance from base to emitter, higher frequency response can be attained at the expense of sensitivity. Figure 13-16 shows the current-transfer curves for a typical optoisolator with LED input and phototransistor output.

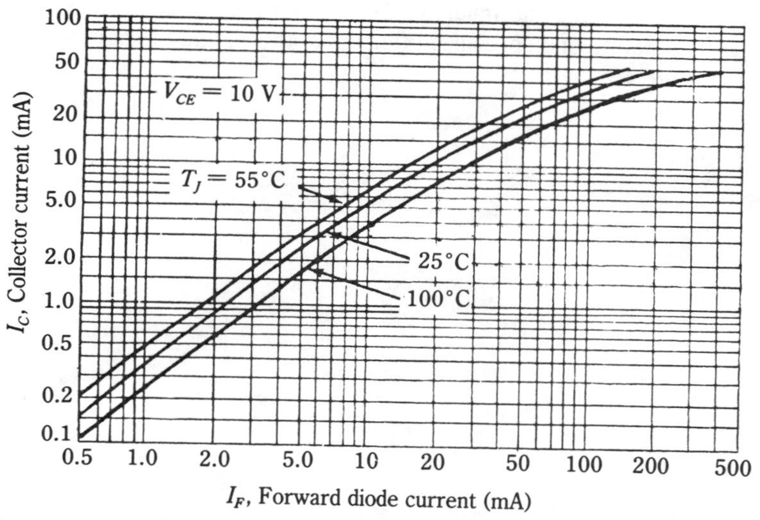

13-16 The current-transfer characteristics of the Motorola 4N25 optoisolator. This device uses an LED and a phototransistor. It attains a current-transfer ratio of about 0.5 for an input diode current of 10 mA. Motorola Semiconductor Products, Inc.

Some types use a photoconductive output cell. This can be cadmium sulfide, cadmium selenide, or a similar material. The response time of such a cell should be considered before the device is used in a feedback loop because photoconductive devices tend to be slow. LEDs are usually infrared types, but the LED in Fig. 13-15D is often peaked in the visible region to achieve better overall efficiency and speed of response. Photoconductive types also requires different circuit implementations because they generate no current or voltage, but rather a modulated resistance.

Optoisolators using incandescent or neon lamps are older versions. Long life spans can be obtained from incandescent lamps at very low currents. Telephone circuits have apparently achieved such reliability, but not all electronics designers have observed this precaution. An advantage of this type, as well as the neon-bulb version, is that an ac signal can be directly sensed. All optoisolators are enclosed in light-tight packages, and provide 1500- or 2000-V insulation between input and output. Special designs extend this isolation to well over 10 kV.

Besides the more common discrete power devices, there are also monolithic ICs in the form of "synthesized" power transistors (such as National Semiconductor LM195) and hybrid power op amps (such as RCA HC2000H). The LM195 can be driven from logic circuitry and can control loads of 25 W or so in switching circuits. Internal protection limits current and dissipation. Paralleling is easy—no ballast resistors are needed and usually the same drive suffices. Switching speed is about 500 ns.

The RCA power op amp can control 100 W in inverters and switching regulators. Its differential input provides useful design options. With its 7-A peak current rating, applications in motor control merit consideration.

Three interesting and useful devices of relatively recent availability are listed near the bottom of Table 13-1. These are the depletion-mode power MOSFET, the SENSE-FET power MOSFET, and the Insulated Gate Bipolar Transistor (IGBT). Because of their unique parameters and specialized functions, these, and other newer devices are covered in chapter 19.

14
CHAPTER

Voltage references and comparators

THE REGULATION CAPABILITIES OF A POWER SUPPLY DEPEND LARGELY UPON the stability of the internal voltage reference source and the associated circuitry that is combined with it. Although the ultimate accuracy of the output voltage depends upon the accuracy and temperature stability of the reference voltage, certain circuit functions—feedback-loop components, comparator or error amplifier characteristics—contribute to the overall stability and performance of the power supply. This is true of both linear and switching-type power supplies.

Voltage reference sources

The voltage reference source encountered most is the zener diode. Considerations of cost, reliability, and ease of implementation have been determining factors. Thus, in a large proportion of regulator schematics, you can quickly locate the zener diode reference source. Actually, the simple zener diode has shortcomings for applications where stability, temperature drift, and dynamic impedance are of importance. In the better quality linear regulators, these matters assume the utmost importance, and the voltage-reference source is often modified. In switching-type regulators, emphasis on tight regulation and temperature immunity has only recently received emphasis, because other inadequacies in the switching process did not make it worthwhile to pursue such goals. With modern components and circuit techniques, switchers are often designed to serve in applications that were previously the domain of the linear regulator. In such instances, it does not always suffice to simply throw in a cheap zener diode with a composition resistor.

As with the linear regulator, better results can be attained with a *voltage-reference diode.* In reality, this "diode" comprises two or more diodes—usually one is a reverse-breakdown (zener) diode and the others are normal forward-conducting diodes. The diodes are connected in series so that their opposite temperature coefficients tend to cancel; zener diodes above 5 V have positive temperature coefficients, and forward-biased junctions exhibit negative temperature coefficients of approximately −2 mV/° C. Although a 5.6-V zener diode might have a positive temperature coefficient of +2 mV/° C, a 7.5-V unit will tend to have a positive temperature coefficient of about 4 mV/°C. In the former case, a single forward-conducting diode would be used in an attempt to approach a zero temperature coefficient; in the latter case, two forward-conducting diodes would be necessary. This technique is not so simple as it might initially appear—either for the maker or the user. It is not at all easy to mass-produce diodes with the required uniformity to yield specific voltages and temperature coefficients when combined. The user must provide a constant-current drive, rather than a simple dropping resistor, if he or she is to advantageously use the device. Together, these result in high cost; but temperature coefficients in the vicinity of ±0.005 percent are attainable.

Actually, zener diodes were long ago misnamed. It so happens that the reverse-breakdown phenomenon derived from two mechanisms: field emission and avalanche. Field emission (zener breakdown) predominates at the lower end of the voltage range, whereas avalanching is the important breakdown mechanism near 10 V. Within this voltage range, both breakdown mechanisms are operative. Breakdown by field emission is actually a tunneling process, wherein charge carriers have a certain probability of jumping a forbidden zone. Such a breakdown mode has an exponentially described behavior, and its current-voltage characteristic is not very abrupt. Such a characteristic is undesirable because the actual breakdown voltage is not clearly defined. The dynamic impedance of such a gradual transition tends to be high—again undesirable in a voltage reference. A high dynamic impedance implies that a small change in current through the device will result in a relatively high change in voltage developed across its terminals; the lower the dynamic impedance, the better the bypass action for ac components. In this respect, a voltage reference with low dynamic impedance simulates a very large capacitor, but with the advantage that the impedance to ac is very low for all frequencies.

The avalanche-breakdown mechanism occurs when charge carriers in the junction are accelerated to a high velocity by an electric field. These carriers then impact valence-bound electrons, which impart their kinetic energy to other electrons. The phenomenon is cumulative, and is manifested by a very abrupt increase in current through the device (the action is similar to the ionization of a gaseous diode). Some time ago, voltage-breakdown diodes were introduced on the market in which breakdown in the 4- to 10-V range is predominantly by avalanching. These low-voltage avalanche (LVA) diodes display exceedingly sharp breakdown characteristics. Microamperes, rather than milliamperes, are required for their operation, thus greatly reducing self-heating. The tremendous improvement stems from processing techniques. The improved devices are, among other things, usually fabricated by the epitaxial, rather than the diffusion process. Unfortunately, these superior breakdown diodes are not depicted differently from ordinary types on schematics.

The reverse-breakdown characteristics for conventional and LVA 5.1-V diodes are compared in Fig. 14-1 and clearly show the difference in the transition region of the diodes. The almost vertical slope in the conductive region of the LVA diode is a superior

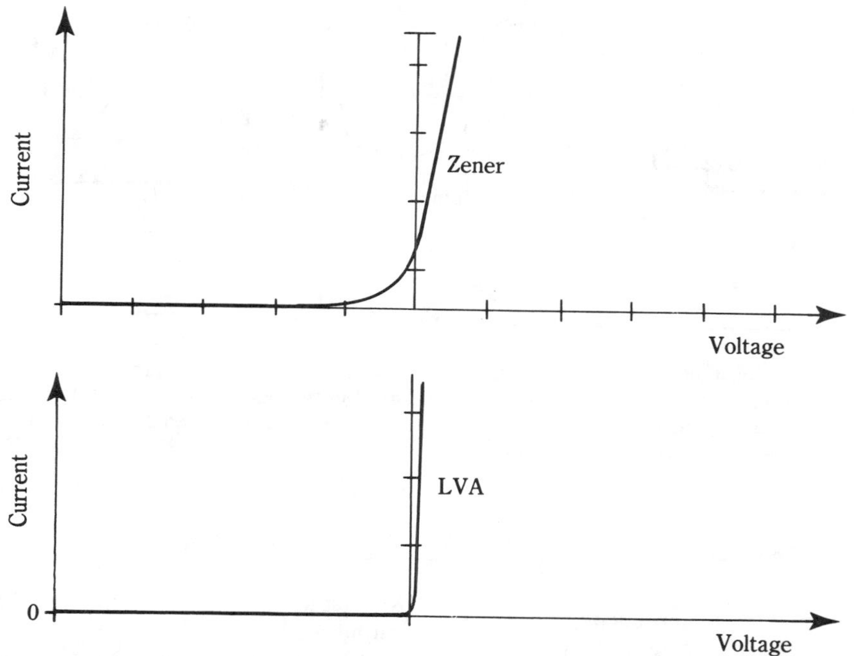

14-1 A comparison of conventional zener diode and low-voltage avalanche diode. The slow, upward bending of the conventional diode curve indicates high forward impedance; the sharp, more nearly vertical slope of the LVA diode indicates a much lower impedance.

voltage reference for regulators, but its abrupt turn-on characteristic and low-leakage current can enhance the performance of peripheral equipment as well. For example, SCR crowbar circuits are often used with both linear and switching- type power supplies to protect the load from overvoltage in the event of regulator malfunction. By using the LVA diode in the gate circuit of the SCR, the triggering can be made more precise than with conventional breakdown diodes. Not only can better protection be imparted to the load, but enhanced immunity to spurious operation from noise will generally result.

FET constant-current devices

The field-effect transistor is inherently a constant-current device, and its current-voltage characteristic is similar to the pentode electron tube. In the simplest situation, as depicted in Fig. 14-2, only two terminals are available to the external world, and the device is commonly referred to as a *diode.* The characteristics of this simulated diode are also shown in the figure, from which it is seen that a substantially constant current prevails over a large portion of the operating region. If a precision resistor is connected in series with one terminal of the device, as in Fig. 14-3A, a constant voltage will be developed across the resistor and can be used as a reference voltage with the provision that the sampling device has an input impedance that greatly exceeds the value of the series resistor. In regulated power supplies, this requisite is readily met by either a comparator or by a buffer stage.

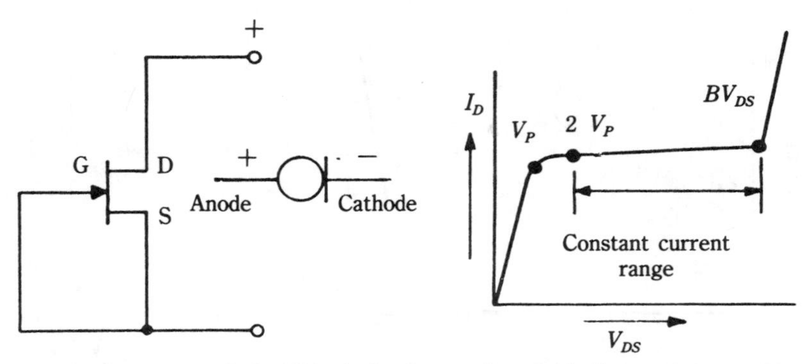

14-2 The constant-current diode. This device is actually a field-effect transistor with gate and source terminals connected together to form the cathode of the "diode." Most constant-current diodes are made from N-channel FETs. Specially designed devices of this kind are known as *current-regulating diodes (CRDs).*

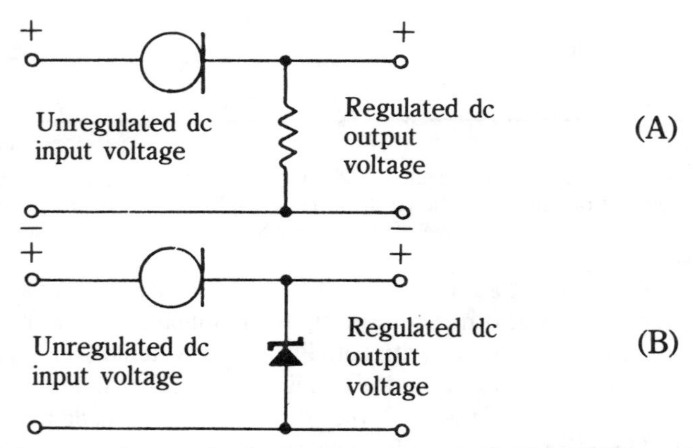

14-3 Two applications of the current-regulating diode (CRD). (A) An adjustable voltage reference can be had from this circuit by selecting an appropriate precision resistor. (B) This circuit provides a precise voltage reference by biasing the reference diode at an optimum current level for obtaining a low temperature coefficient.

Instead of using conventional FETs, better results are attainable from specially designed field-effect diodes that are processed to optimize the constant-current characteristic. These devices have only two terminals because there is an internal gate-source short. They are available in over 30 current increments from about 200 μsA to 5 mA. The minimum operating potential is very low—on the order of 1 to 3 V, and the forward-breakdown voltage is often as high as 100 V. Thus, these devices are adaptable to a wide variety of circuit situations. An advantage of using these specially made, current-regulating diodes (CRDs) is that a zero temperature coefficient can be achieved at the factory by associating the CRD with a resistor of equal, but opposite temperature coefficient. This usually prevails for units rated in the vicinity of 0.5 mA.

An excellent application for the CRD is in conjunction with a zener, LVA, or voltage-reference diode, as shown in Fig. 14-3B. A temperature coefficient of 0.001 percent over the 0° C to 100° C temperature range can be attained with this technique. For optimum results, both the CRD and the voltage-reference diode should be specified for zero temperature coefficient in the vicinity of 0.5 mA.

There is more than meets the eye in the combination of CRD and zener. The CRD in the constant-current source exhibits an extremely high impedance to ac. The contrary is true of a zener, or other voltage-reference source, in which an important figure of merit involves its very low ac impedance. When the two types of devices are combined in the manner shown in Fig. 14-3B, a unique low-pass filter is formed with a cut-off frequency near zero. Such a configuration theoretically offers very high attenuation to all ac frequencies. In practice, because of stray parameters, the circuit exhibits attenuation approaching 100 dB for frequencies up to several hundred kilohertz. Thus, most ripple and noise components riding on the unregulated supply line are effectively removed. In the simple circuit of Fig. 14-2, the output impedance is equal to $1/g_{oss}$ where g_{oss}, or an equivalent, is a commonly specified parameter. In this circuit, the constant current is equal to I_{DSS}, also a commonly specified parameter. The modified circuit of Fig. 14-4A provides any constant current value up to the value I_{DSS}. Additionally, output impedance increases as the regulated current is reduced by increasing R, because of the feedback action. The cascaded arrangement of Fig. 14-4B is capable of producing much tighter current regulation and much higher output impedance than is attainable from single-device circuits. For this circuit to operate properly, Q2 must have a higher I_{DSS} than Q1, preferably at least 10 times greater. And it is important that both FETs are provided with drain-source voltages at least twice their pinch-off voltage, V_p, also a commonly specified parameter of FETs. This criterion for V_p actually applies to the single-device circuits as well.

The CRD can be advantageously used as part of a voltage-divider network for sampling the output voltage of the switching regulator. Such a scheme is shown in Fig. 14-5. When so utilized, the error signal is not attenuated, as is the case with the

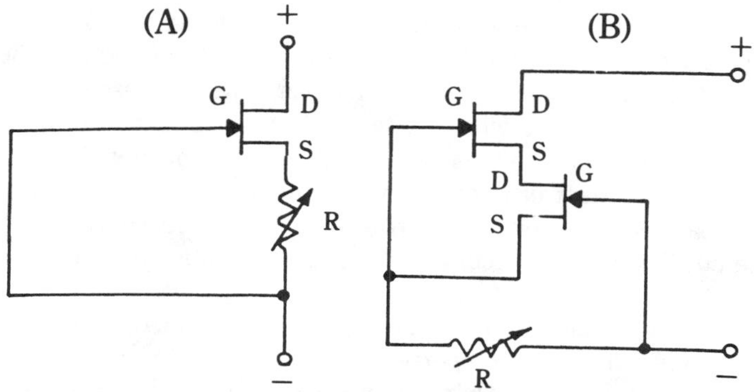

14-4 Constant-current sources with adjustable current levels. (A) A single FET circuit. Maximum current obtains for $R = 0$. (B) Cascaded arrangement of two FETs. This scheme yields better regulation and higher dynamic impedance than the single FET circuit. In both circuits, the manufacturer is able to manipulate the fabrication process to obtain zero-temperature coefficient for $R = 0$, or other values.

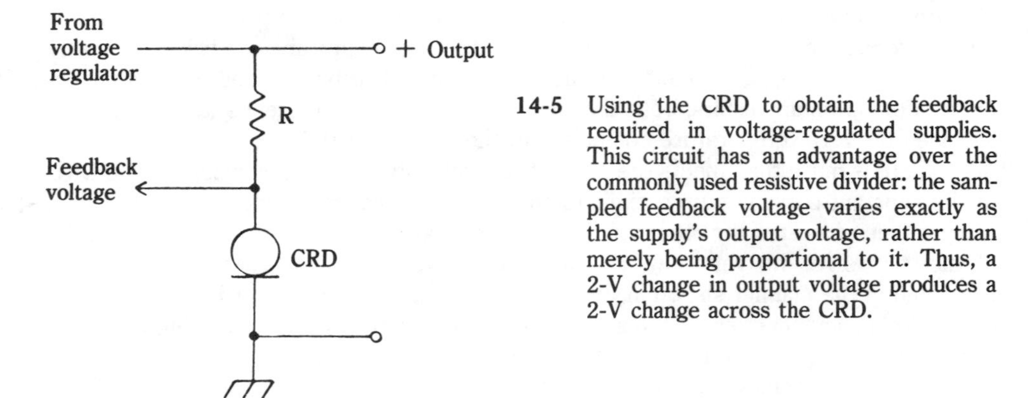

14-5 Using the CRD to obtain the feedback required in voltage-regulated supplies. This circuit has an advantage over the commonly used resistive divider: the sampled feedback voltage varies exactly as the supply's output voltage, rather than merely being proportional to it. Thus, a 2-V change in output voltage produces a 2-V change across the CRD.

conventional potentiometric sampling divider (an example of such an application can be found in the advanced switching regulator design of Fig. 17-15).

Synthesized low-voltage reference

One of the techniques sometimes used to obtain a reference voltage below that provided by zener diodes is to use ordinary silicon pn diodes in their forward-conductance mode. Approximately 0.6 V per diode is obtained in this manner. Thus, a 2.5-V reference results from connecting four diodes in series. This method is not very satisfactory because the regulation is quite sloppy, dynamic impedance is high, and reproducibility is not at all good. It might, therefore, be surprising that one of the finest low-voltage references is derived from the forward-conduction characteristics of pn junctions.

It has long been known that the relationship between the emitter-base voltage and collector current is highly predictable in transistors, and that two transistors in an unbalanced differential-amplifier circuit could be made to produce a current that was proportional to the absolute temperature (a useful property for electronic thermometers). A significant result occurs if the collector currents of the amplifier stages are maintained at a constant value, for it then happens that the base-emitter voltage of a third transistor (also operated with constant collector current) can be combined with the base-emitter difference voltage of the differential amplifier stages to produce a voltage that is independent of temperature. In other words, there is a certain way of summing the base-emitter voltage drops to obtain a reference voltage having a zero temperature coefficient. This occurs at the energy band-gap voltage of silicon, which is 1.205 V.

It should be appreciated that the attainment of the zero temperature coefficient, together with other desirable properties of a voltage reference source, was never practical with discrete devices. Only the monolithic fabrication process, with its tight thermal coupling between devices and its excellent reproducibility of device characteristics, could be used to successfully translate theory into hardware. The result is a new series of devices that offer exceptionally low temperature coefficients down to the 1-V reference level, which can be used to much advantage over the typical 6.2- to 8.5-V range of conventional zener diodes. And because no breakdown mechanism is involved, noise generation is very low.

The schematic diagram of the LM113 is shown in Fig. 14-6. Transistors Q1 and Q2 form an unbalanced differential amplifier. A voltage, proportional to the difference of their base-emitter voltages, appears across R4. To this differential base-emitter voltage is added the base-emitter voltage of Q4 in order to develop the zero temperature coefficient, as previously described. Transistors Q3, Q5, and Q7 provide a current-source load for Q4. The remaining transistors are primarily involved in buffering the output and in producing a low dynamic impedance.

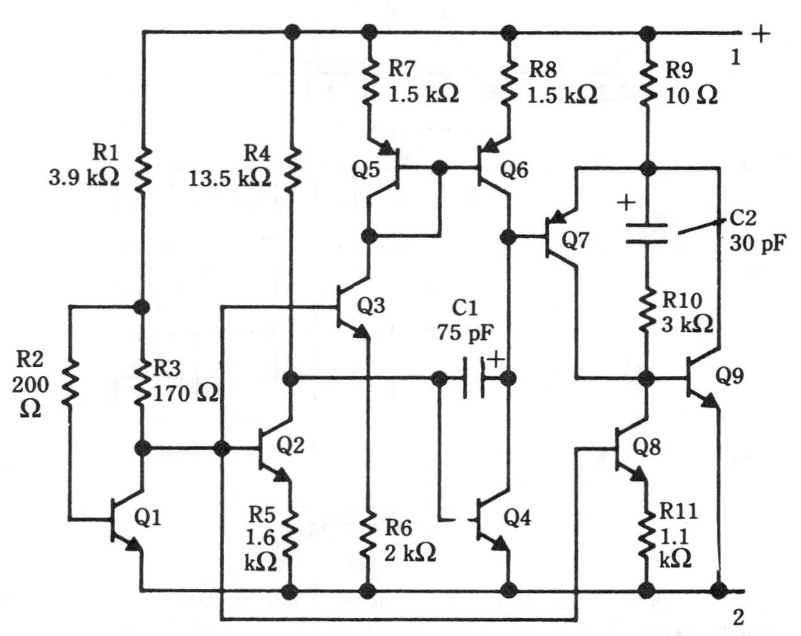

14-6 A schematic circuit of the LM113, a precision 1.2-volt reference source. Even though this is a relatively simple IC, it is obvious that it would be impractical to attempt to reproduce it with discrete devices. National Semiconductor Corp.

The characteristics of the LM113 are shown in Fig. 14-7. Notice the near-vertical slope of the "breakdown" region, inferring a low dynamic impedance. Although only the active-operating region of transistors are involved, the similarity to the characteristics associated with breakdown phenomena is quite remarkable. Indeed, the nearest conventional breakdown device—a 2.4-V zener diode—is inferior in all respects as a voltage reference. The temperature characteristic attests to the validity of using opposing temperature coefficients within the IC to establish a near-zero temperature coefficient over a wide temperature range.

A 2-V regulator using the LM113 "diode" is illustrated in Fig. 14-8. The LM113 is driven by an FET current source and serves as the voltage reference in a linear regulator involving an LM108 operational amplifier and a 2N2905 booster transistor. As shown, this circuit will operate on an input voltage as low as 3 V.

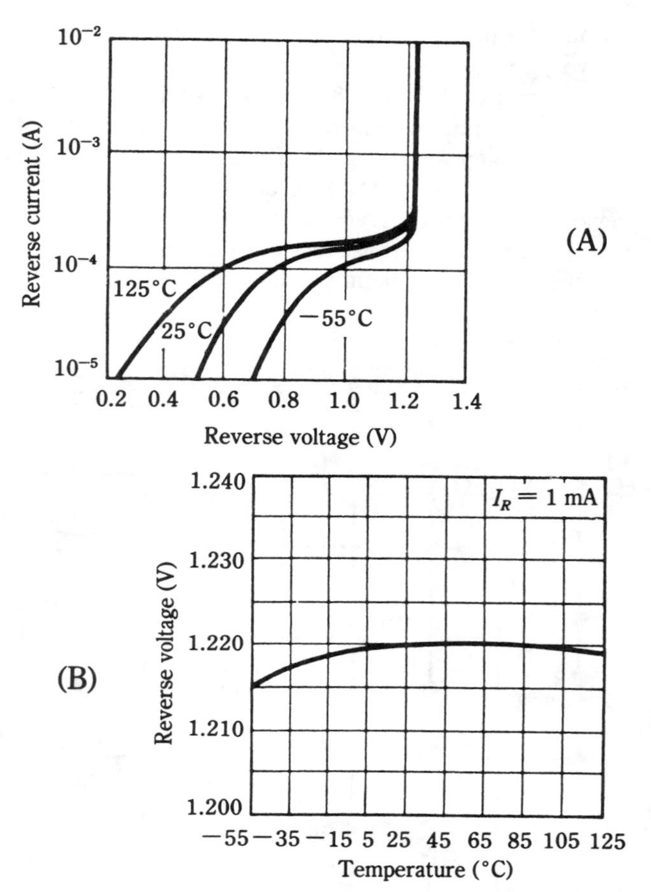

14-7 Characteristics of the LM113 voltage reference source. The output reference voltage becomes very stable at forward currents above 1 mA. Notice the expanded voltage scale used in depicting temperature effects. National Semiconductor Corp.

The comparator and error amplifier

This building block of regulatory circuits and systems has many names and more than one mode of operation. In the technical literature of servo systems, linear regulators, and switching-type supplies, you will find such terminology as error amplifier, error-signal amplifier, sensing amplifier, sensor, comparator, differential amplifier, and summing amplifier. All of these terms are not necessarily synonyms; they encompass more than a single kind of circuit behavior, but they do possess a common feature no matter how named, their circuit function is to generate output signals that distinguish between the input conditions of "too high" and "too low." Because these conditions pertain to the output voltage of the supply being regulated, the information contained in the response of the comparator (or its counterparts) can be utilized for correcting the very deviation responsible for such an error signal. Correction in the linear regulator occurs

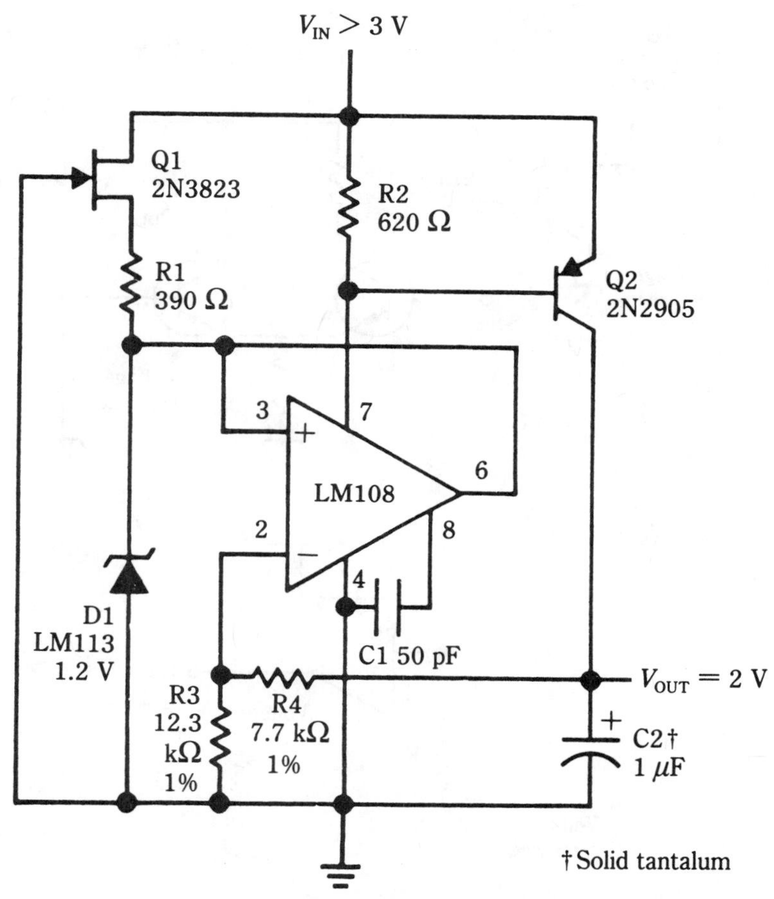

14-8 A 2-V regulator using the lm113 synthesized voltage-reference source. This reference source is especially applicable to low-voltage supplies, as covered in chapter 20. Although a complex monolithic array, this IC is generally depicted by the simple breakdown-diode symbol.

through the rheostat-like action of the pass transistor. In the switching-type supply, the correction results from an appropriate change in the duty cycle of the switching process.

In certain switchers, the comparator is very similar to those used in linear regulators; it might consist of a single transistor, a differential pair of transistors, a complex array of discrete devices, an operational amplifier, or an IC module specifically intended for such functions. Examples are shown in Fig. 14-9. The use of monolithic ICs as comparators reduces the burdensome tasks of design and manufacturing. Other problems associated with imbalance, offsets, temperature tracking, and device tolerances can be neatly circumvented with ICs. The physical layout is invariably more conducive to the requirements imposed by high frequency and high gain. Much evidence has been accumulated that points to improved reliability, with respect to the use of discrete

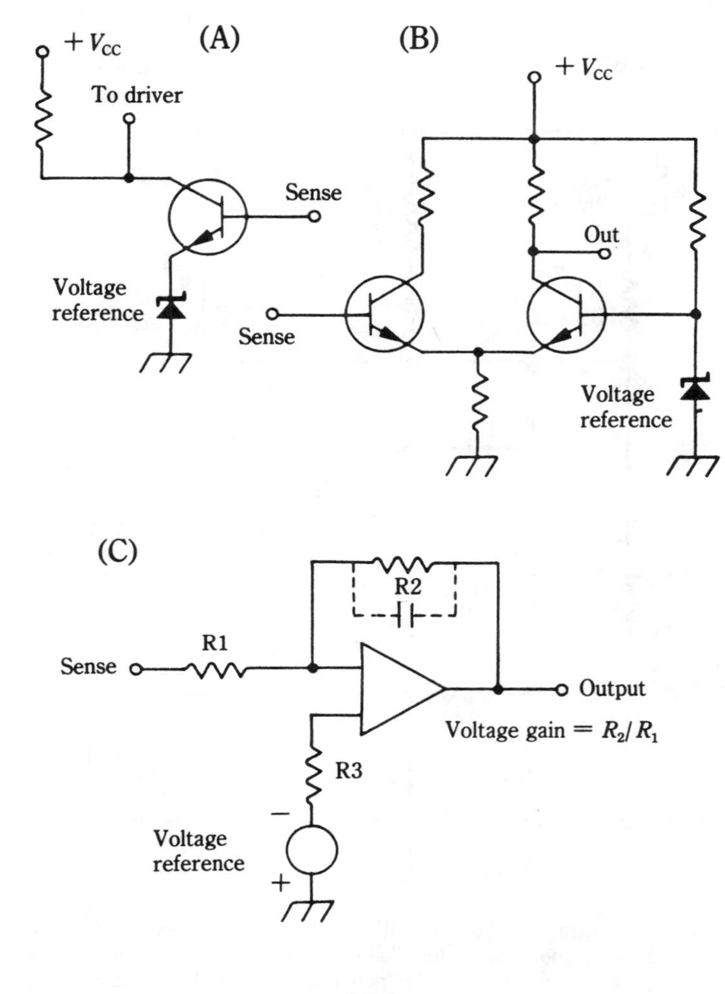

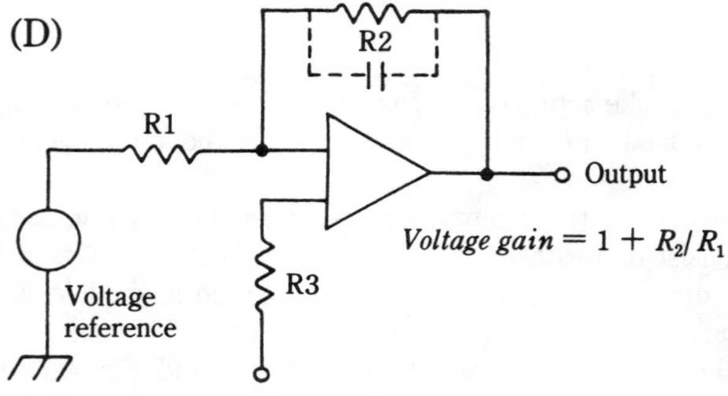

14-9 Simplified circuits of non-saturating voltage comparators. In op-amp circuits C and D, the capacitor placed across resistor R2 is used to roll off the high-frequency response, and so insure a stable phase-gain margin. Also, R3 should equal the parallel equivalent of R1 and R2 for optimum temperature-stability.

devices. Even an inexpensive 741 op amp can provide excellent results in a great many systems.

Two operational modes of comparators

In linear regulators, the comparator itself exhibits a linear transfer function. Although emphasis has rarely been focused upon linearity, the fact remains that such a comparator is essentially a class-A amplifier with dual inputs, it can be inverting or noninverting, and it can comprise one or more active devices. This type of comparator is also used in certain switching-type power supplies. The term *comparator* stems from the fact that the output voltage (or some portion of it) is sensed by one of the inputs and compared to a stable reference voltage applied to the other input. But this is not the only way of generating a usable response or using discrete devices or IC modules.

When you inspect the block diagrams and schematics of self-oscillatory switching supplies, it might not be immediately evident that the comparator is operating in a switching mode. Some comparators are three-state circuits—with no error being sensed, the output is zero; and with an input error in one direction, the output is saturated at one polarity; and the converse prevails when the input signal departs in the other direction from the reference voltage. In actual use, a two-state, on/off operation would more accurately describe the circuit events. Regulatory action occurs as the consequence of the relative time spent in the on and off response states. The switching transistor is controlled by the output of the comparator to correct the average dc output level of the supply. Regulation is then the result of appropriate variations in the duty cycle of the switching process. Despite the different operational methods of comparing, the switching regulator performs in very much the same way. (It should not be thought that the saturating-mode comparator is a flip-flop; it remains a linear amplifier, but one of great sensitivity that is deliberately overdriven.)

Considerations in the selection of the comparator

Whether the comparator in the switching-type power supply will comprise a single transistor, operational amplifier, or special IC module is a decision that must be deferred until the operational mode of the comparator is determined. In linear regulation schemes, the comparator usually provides amplification over a wide voltage range, and the demands imposed upon it can be quite different from the comparator, which responds to an error signal by saturating in one or the other polarity. In the first instance, you deal with a linear error amplifier, and you expect amplified versions of deviations from a reference level. This calls for a device capable of voltage or current gain, with the added stipulation that polarity sense must be conveyed, and that is not a difficult requirement to meet. It turns out that the amplification need not even be truly linear over its dynamic range—as is often the case with a single transistor. Such a sense or error amplifier must not be driven into saturation. Its frequency response must be sufficient to handle the response time of the overall regulator; it must be free from self-oscillation and from major contributions to feedback-loop instability. These are not formidable obstacles.

It is only natural to consider the operational amplifier as a nonsaturating, linear comparator. By virtue of its differential inputs, high open-loop voltage gain, low-impedance single-ended output, and other compelling features in the realm of cost and packaging convenience, these devices appear tailor-made to the needs of the switcher. Among those that have already proven their reliability in the field are National Semiconductor types, 709, LM101, LM741, and LM747. Some op amps have internal frequency compensation, a feature that often simplifies design and manufacture of equipment. On the other hand, an op amp with provision for external frequency compensation can usually provide wideband performance, and often permits an extra dimension of circuit flexibility. For example, it is sometimes advantageous to manipulate the phase gain characteristics of an op amp to compensate for deficiencies elsewhere in the overall feedback loop.

Another very useful technique for the linear sensing of the error signal involves the use of IC voltage regulators, such as the National Semiconductor LM723 family. Here, you benefit not only from the desirable features of operational amplifiers, but also from characteristics and modes of performance relating directly to regulating systems. For example, the LM723 contains its own reference voltage; not only is this desirable from the standpoints of convenience and economy, but you would be hard pressed to duplicate its temperature stability with ordinary zener-diode circuits. Moreover, this internal reference has sufficient current capability to provide a stable voltage level for external circuits, with negligible degradation of stability or change in absolute value. This and similar IC voltage regulators have provisions for overload protection, remote control, and programming. A maximum, rather than a bare minimum, of terminals is brought out to allow flexibility in circuit design. For example, the input op amp comparator can be used in either the inverting or noninverting mode.

The saturating-type comparator operates as a switch. The operation is usually open loop in order to advantageously use the tremendous gain attainable. Operational amplifiers have successfully fulfilled this function, but there have been certain obstacles to trap the unwary. At the very outset, the op amp used in this way must be able to perform without latching up. A latchup mechanism prevents a return from the saturated state. This is not as likely to be encountered as it once was; the problem has been solved in new designs. The response (slew rate) of the saturating comparator should be very fast—even faster than might be deemed necessary for driving the subsequent switching circuits. Fast response is not, as with a power transistor, required to hold down dissipation, but rather to prevent self-oscillation. Recall that the gain is "wide open" between the extremities of bipolar saturation. Often this problem will tax the ingenuity of the designer. A little forethought is generally the best remedy—you must simply be mindful of the routing of the input and output leads from the comparator; another reason why the designer should know not only dc and ac techniques, but also the art of high-frequency layout.

Some typical saturating-type comparators using op amps are shown in Fig. 14-10. In order to enhance stability and increase noise immunity, the saturating comparator often uses hysteresis so that its operation is somewhat like that of a Schmitt trigger. This is implemented by means of a positive feedback path, as can be seen in examples C and D. Such a technique decreases the sensitivity of the comparator, but it is usually found to constitute a desirable tradeoff in switching-type power supplies. The hysteresis characteristic actually provides one of the basic design parameters of such supplies, the peak-to-peak ripple voltage.

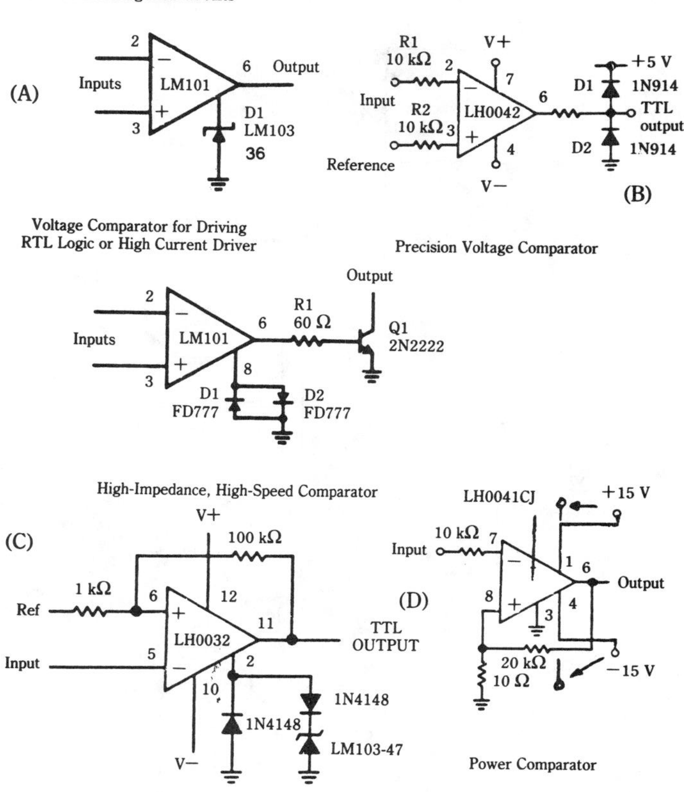

14-10 Saturating-type voltage comparators using op amps. (A) "Workhorse" voltage-comparator circuits. (B) Comparator with an FET op amp. (C) Comparator with input-voltage hysteresis. (D) Comparator with hysteresis and high-current drive capability.

A wide selection of op amps is available for use as a saturating comparator, and many circuit techniques can be applied to speed up, stabilize, or otherwise manipulate the performance. But, the use of specially designed modules should merit serious attention. It happens that much attention was given to the saturating comparator by the semiconductor processors in order to comply with the demands of the computer industry. The results of this concentrated effort are now available in the form of specialized ICs, which are optimally suited for use as saturating comparators—even more so than high-quality op amps. Some of this work has been pioneered by the National Semiconductor Corporation, and has resulted in many excellent IC comparators. Particularly suited to the needs of switching-type power supplies are the LM710 and LM711 families. The fact that these units are compatible with most logic levels is a significant factor in the light of ever-increasing use of gating and logic techniques in the regulating loop of switching-type supplies. The circuit of a monolithic voltage comparator is shown in Fig. 14-11.

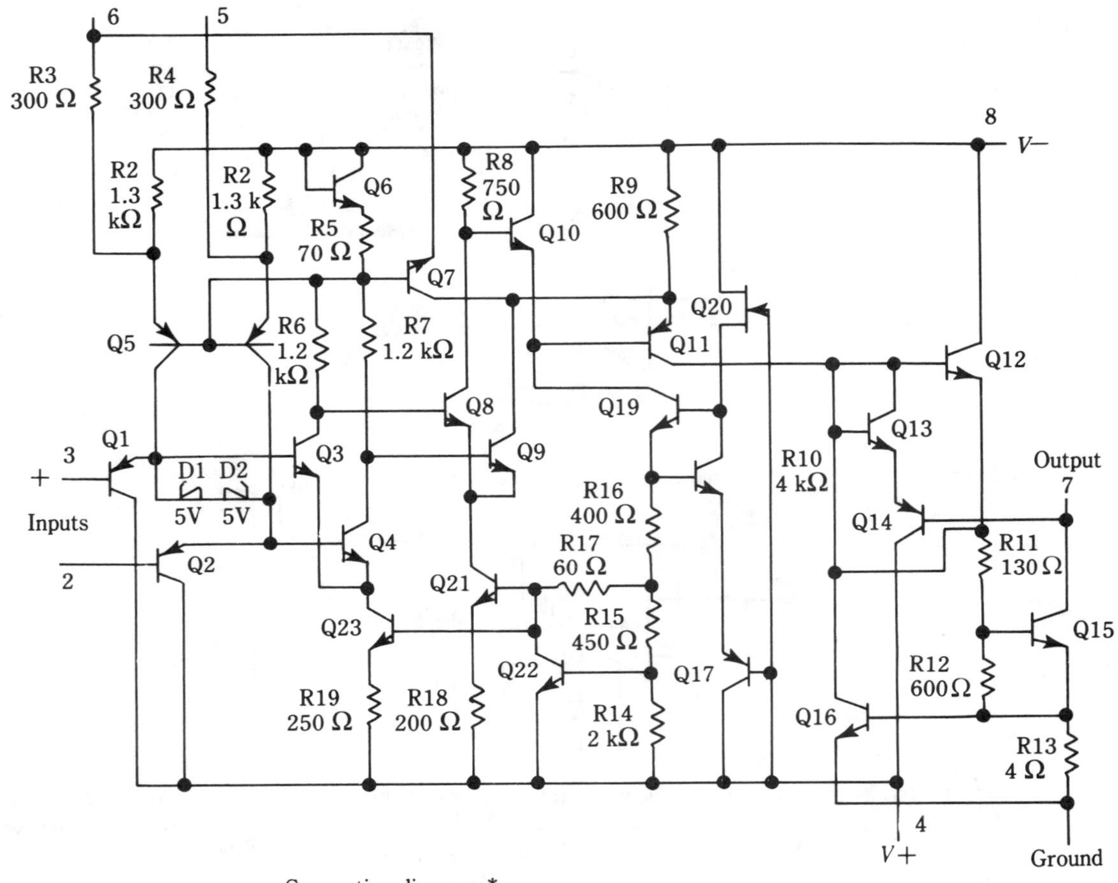

Connection diagrams*

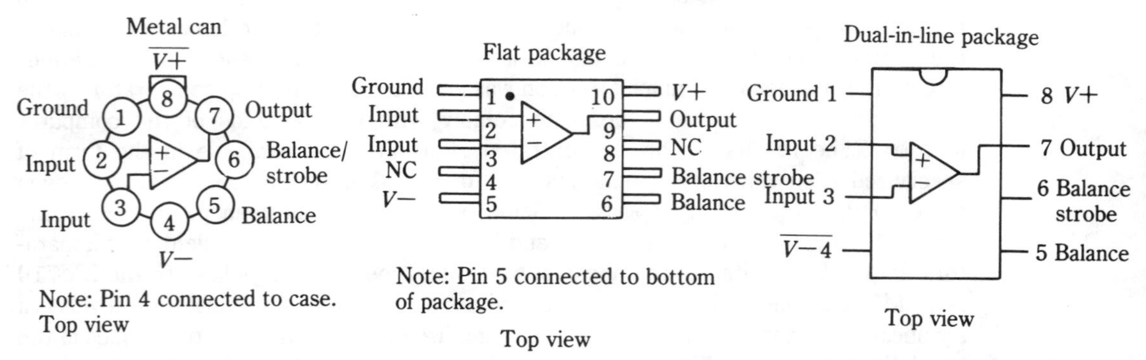

14-11 The internal circuitry of the LM311 voltage comparator. The use of such dedicated ICs generally yields better results than implementations using op amps. National Semiconductor Corp.

A particularly useful comparator is the LM2901, a quad op amp that operates from a single power supply. Not only are the four individual units useful as saturating comparators, but if only one is required, the others might be connected to provide other functional blocks in the switching-type power supply, such as a one-shot multivibrator, zero-crossing detector, square-wave generator (free-running multivibrator), binary stage (bistable multivibrator), OR gate, and AND gate.

Circuit building blocks

The versatility of the monolithic IC in general, and of the operational amplifier in particular, is virtually boundless. These devices facilitate the speedy and economical construction of efficient and cleanly performing switching regulators. Depicted in Figs. 14-12, 13, and 14 is a selection of circuits particularly suitable to the many needs of switchers. A number of different ICs are also shown; some types are interchangeable, but each type is endowed with its unique operational features.

The circuitries of Fig. 14-12 are linear circuits yielding proportionate response to input signals. In contrast, those of Fig. 14-13 saturate during their operational cycles, and thereby produce pulse-like waveforms. The group of circuits in Fig. 14-14 are combinations presenting six circuit functions that are potentially applicable to various tasks in the control section and feedback loop of switching-type power supplies. Many of these ICs were originally developed by the National Semiconductor Corporation, and numerous applications have been developed by this firm for switching-type supplies. It is suggested that such manufacturers be contacted for assistance in determining the best devices for use in switching-type supplies, or in any other application. New devices are being introduced at an amazing rate, and even the possibility that some new device might better meet your specific needs makes it well worth looking into.

The linear regulator

Although replacement of linear regulators by switching types is one of the basic themes of this book, it should not be inferred that linear or dissipative regulation of voltage or current is not useful. Beyond the simplest of switching-type supplies, you invariably find that linear regulators have important involvements in the overall switching system. There are domains of application where the linear regulator, despite its low operating efficiency, has compelling advantages over switching types. As has been already pointed out, the linear regulator tends to provide closer regulation, less ripple and noise, greater bandwidth, and in many cases, simpler design and implementation. In this book, interest in the linear-regulating technique stems from several considerations.

The building block of switching-type power supplies, such as those depicted in Figs. 14-12, 13, and 14, obviously require sources of dc operating power. Because of monolithic ICs, it is as easy and economical to provide regulated supplies as it is to provide unregulated supplies. The use of regulated operating power leads to better circuit performance because of isolation provided between circuit sections, more predictable performance inherent from stable voltage (or current) sources, and greater attenuation of power-line disturbances. There are other advantages too, such as circuit-

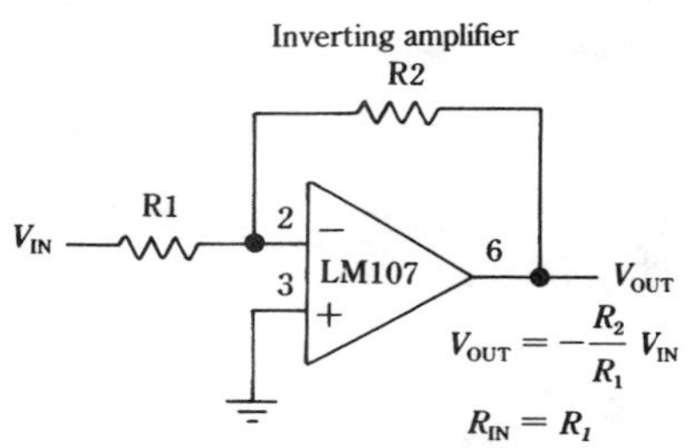

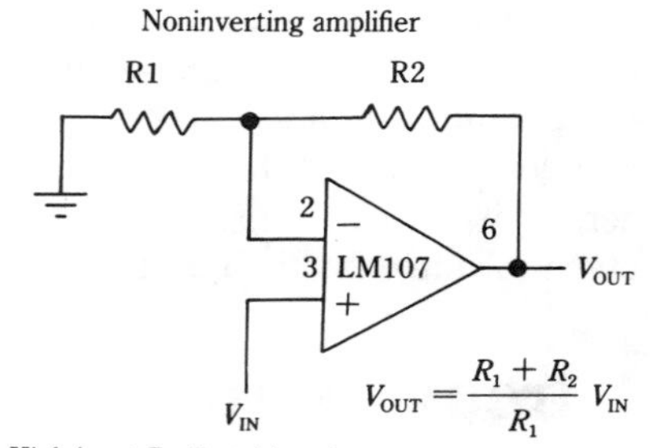

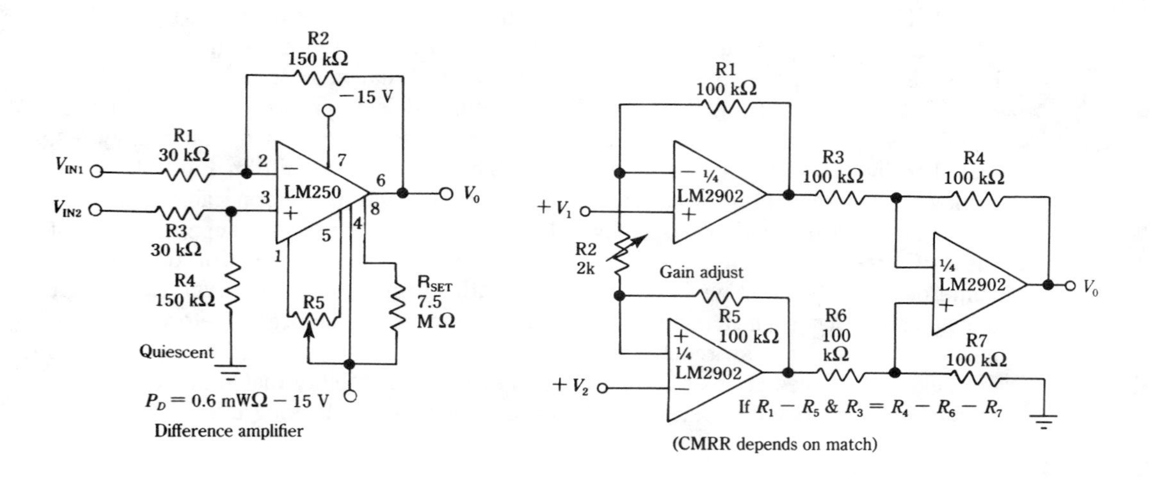

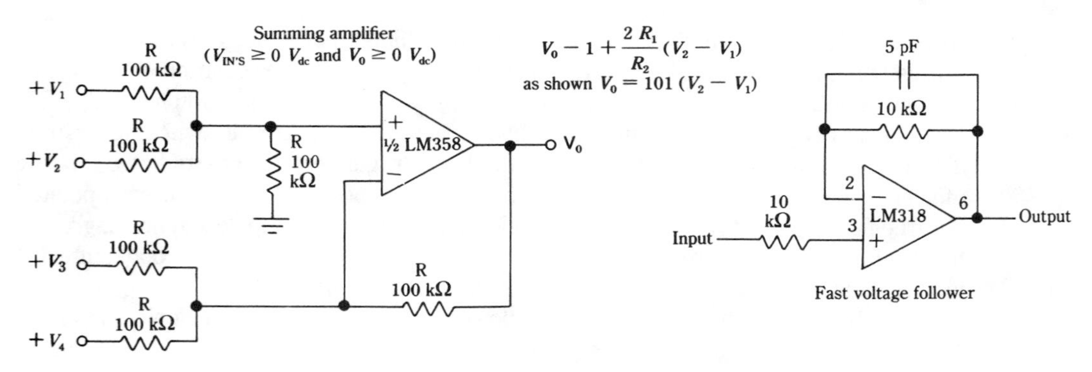

Where: $V_0 = V_1 + V_2\ V_3\ V_4$
$(V_1 + V_2) \geq (V_3 + V_4)$ to keep $V_0 > 0$ V_{dc}

14-12 Useful linear circuits configured around operational amplifiers. These circuit functions will be found in many applications involving the sensing, comparing, and amplifying of the error signal. National Semiconductor Corp.

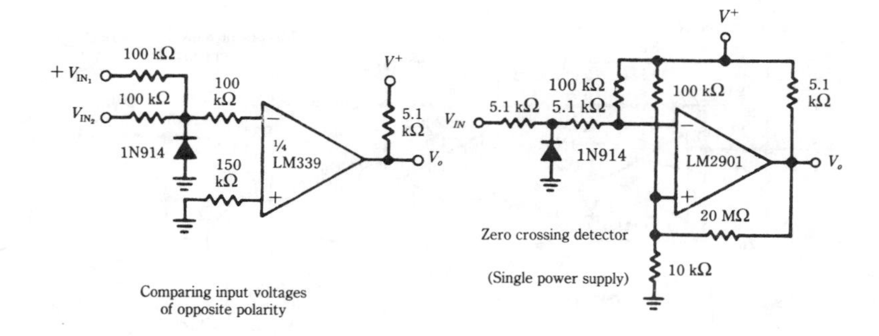

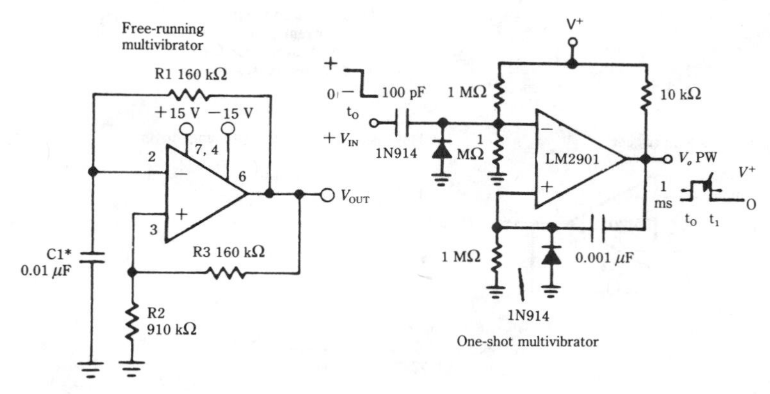

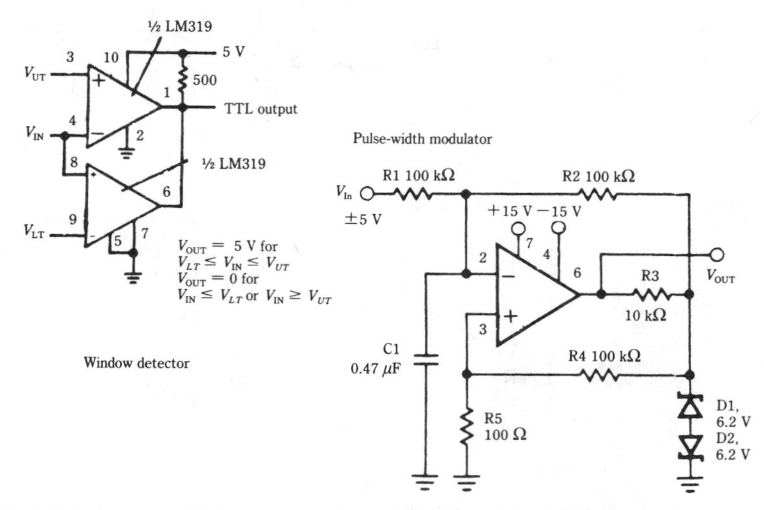

14-13 Useful circuits involving saturable operation. These and numerous derivatives are commonly encountered building blocks of the switching-type regulated power supply. National Semiconductor Corp.

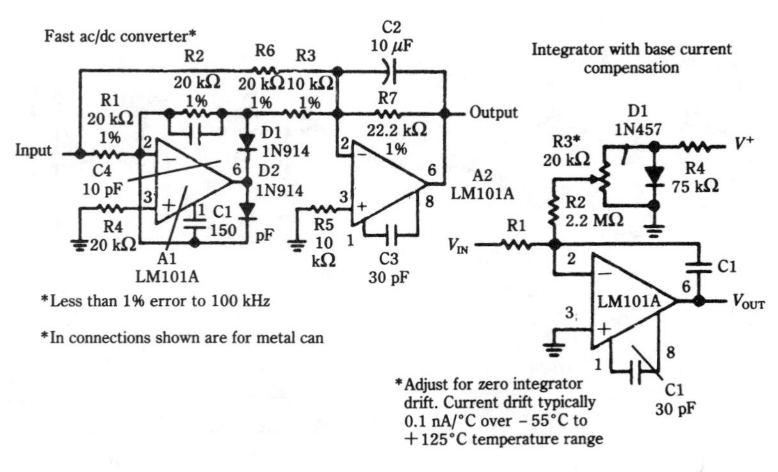

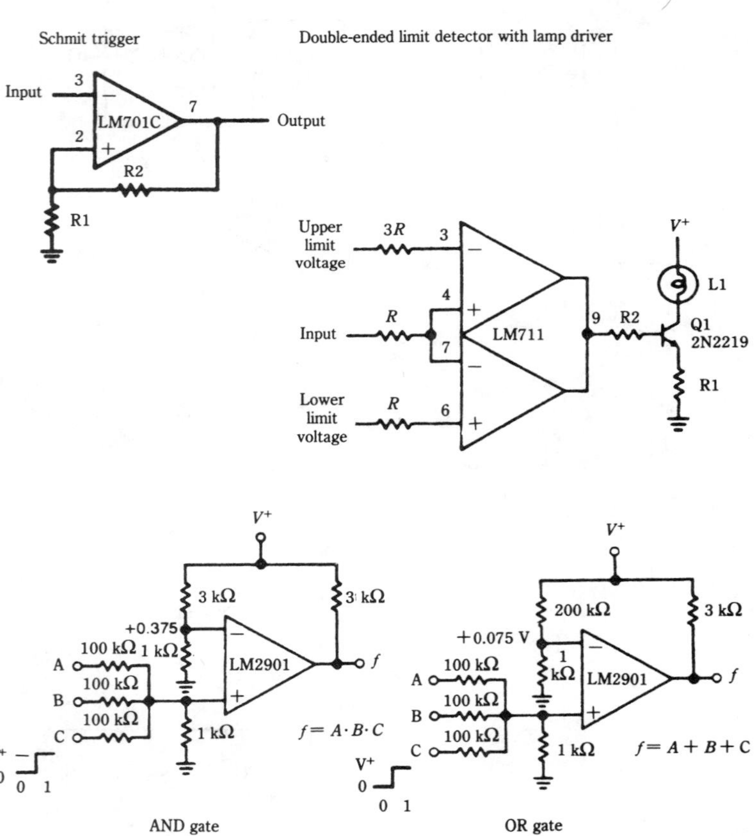

14-14 Additional circuits with application potential in switching-type supplies. These and similar configurations are often involved in the duty cycle processing portion of the switching-type power supply. National Semiconductor Corp.

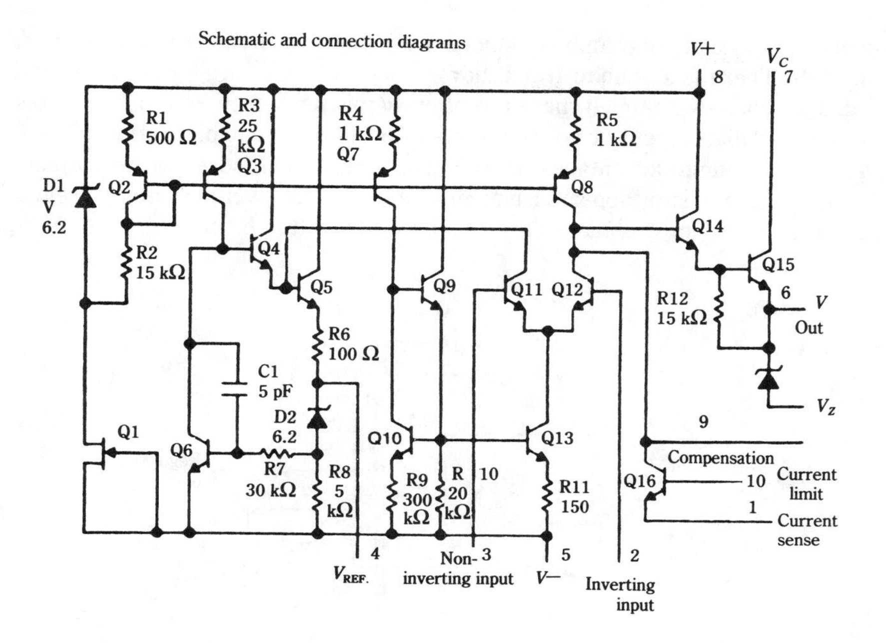

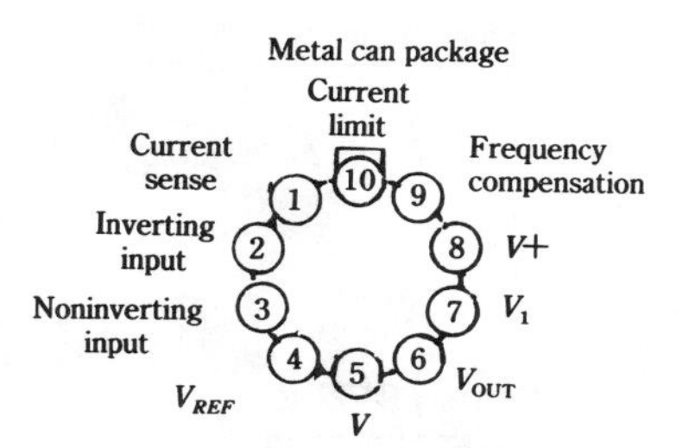

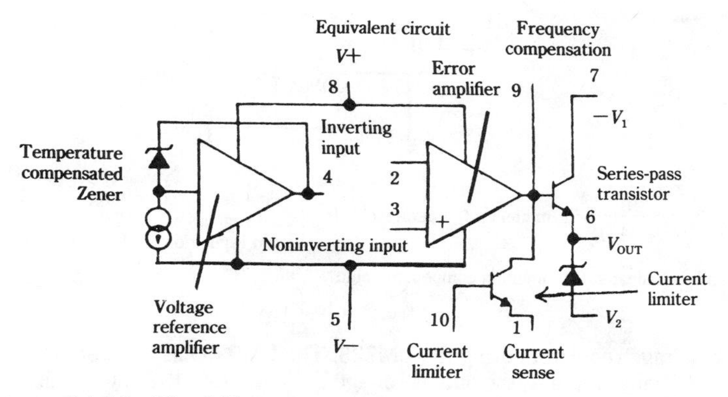

14-15 The LM723 linear regulator. National Semiconductor Corp.

control flexibility and programming options. Discrete devices can also be used, as shown in Fig. 8-5. There is a definite trend, however, to use monolithic modules specifically intended for such use, though these are often supplemented by external discretes that boost the available current or voltage or provide auxiliary control functions. The use of these ICs has definite advantages in packaging, production, cost, and performance.

The linear regulator appears frequently as part of the overall system of regulation, even though switching techniques are heavily involved (the block diagram of Fig. 8-1A

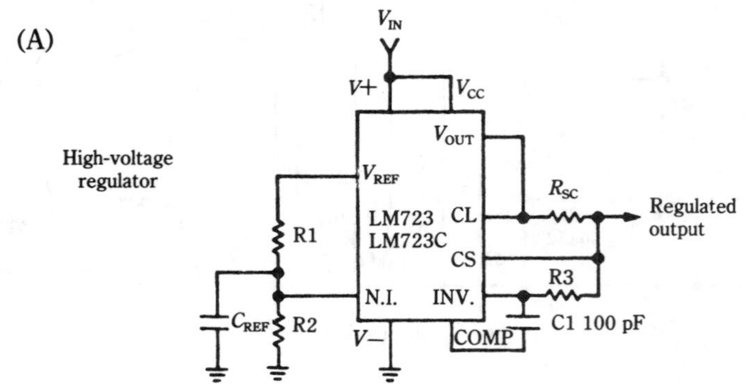

Note: $R_3 = \dfrac{R_1 R_2}{R_1 + R_2}$ for minimum temperature drift.

Typical performance	
Regulated output voltage	5 V
Line regulation (Δ V_{IN} = 3 V)	0.5 mV
Load regulation (Δ I_L = 50 mA)	1.5 mV

(V_{OUT} = 2 to 7 Volts)

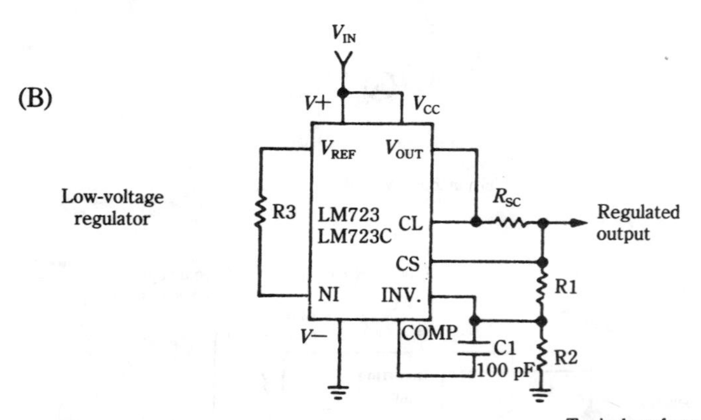

Note: $R_3 = \dfrac{R_1 R_2}{R_1 + R_2}$ for minimum temperature drift.

Typical performance	
Regulated output voltage	15 V
Line regulation (Δ V_{IN} = 3 V)	1.5 mV
Load regulation (Δ I_L = 50 mA)	4.5 mV

R_3 can be eliminated for minimum component count.

(V_{OUT} = 7 to 37 volts)

14-16 Basic voltage regulators using the LM723. The LM723 has an internal 7-volt reference source. Therefore, in A, the output voltage is $V_{REF}R_2/R_1 + R_2$, whereas in B the output voltage is $V_{REF}R_1 + R_2/R_2$. National Semiconductor Corp.

shows such an arrangement). And the linear regulator is sometimes advantageously used in the feedback loop of a regulating system.

One of the most versatile of voltage regulator ICs is the LM723 series. This module can provide excellent performance compared to discrete circuits and has few "frozen" design parameters. The manufacturer can illustrate numerous applications and many others can be implemented through the ingenuity of the user. This flexibility is largely achieved by bringing out as many circuitry junctions as is feasible in a standardized package. For example, the voltage reference and the comparator terminals are

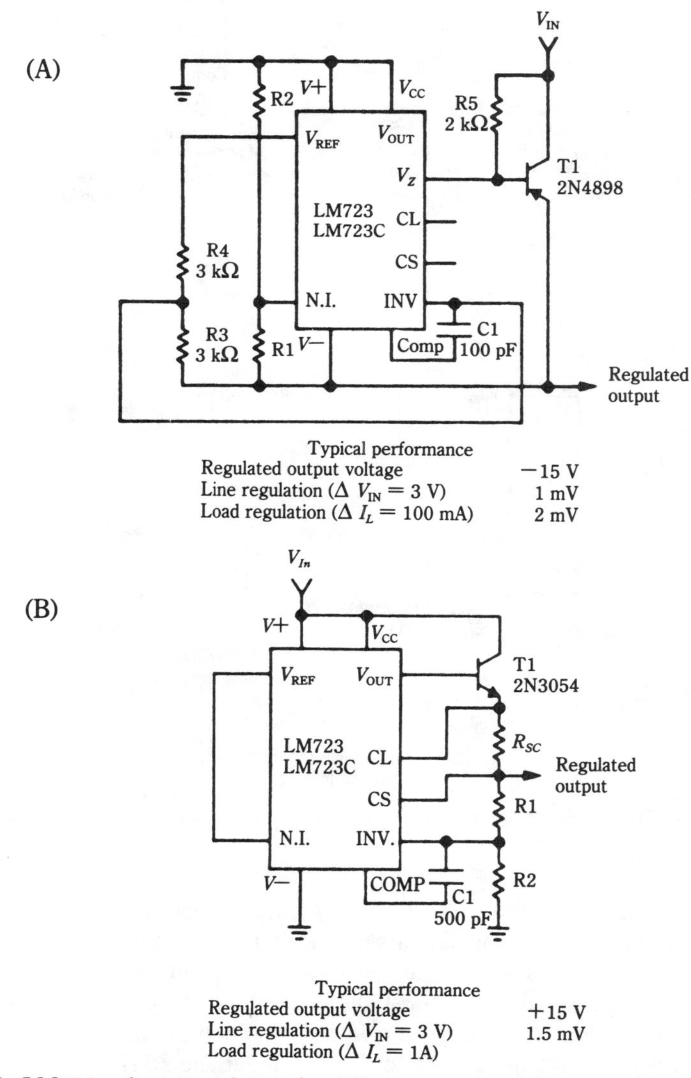

14-17 Basic LM723 voltage regulators for different polarity outputs. (A) Negative regulated output. (B) Positive regulated output. National Semiconductor Corp.

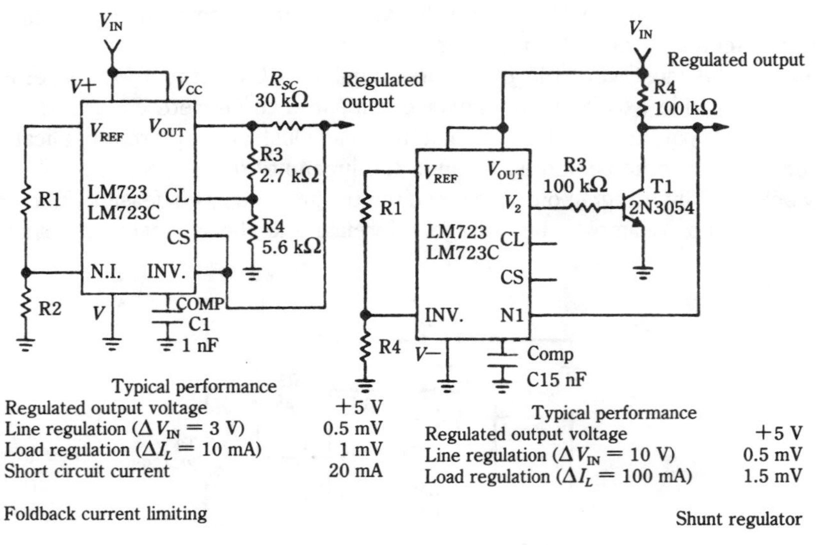

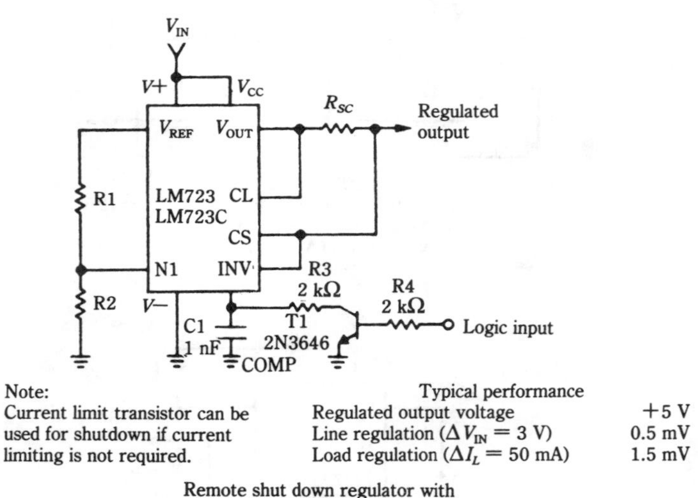

14-18 Additional applications of the LM723 linear voltage regulator. National Semiconductor Corp.

brought out separately on their own pins. The output impedance of the voltage reference source is very low—enabling its application for other circuit functions with negligible loading effect. The module itself can supply up to 150 mA, and external transistors can easily be added to boost the current capability to many amperes. The circuit is shown in Fig. 14-15, together with connection-pin information.

Basic voltage-regulating circuits configured about the LM723 are shown in Figs. 14-16 and 14-17. The reference voltage is nominally 7.15 V with a tolerance of approximately ±5 percent from unit to unit. Both positive and negative voltage regulators can be readily implemented. Additional applications of the LM723 are depicted in Fig. 14-18. Not only do foldback current limiting, current limiting at a constant value, and remote shutdown constitute protective techniques from the effects of over-voltage, over-current, and short circuits, but these circuit functions lend themselves to various other control objectives that occur in regulating systems. For example, the remote shutdown circuitry could be associated with a large capacitor in order to provide a "soft" startup. The shunt regulator of Fig. 14-18 has the inherent feature of being short-circuit proof. A shorted output circuit simply causes the regulator to shut down, and when the short is removed, the regulator automatically resumes normal operation —a feature not always easy to attain in the more commonly encountered series-pass type of regulator.

15

CHAPTER

Switching regulator design

CONSIDER FIRST THE SIMPLEST OF SWITCHERS—THE FREE-RUNNING OR self-oscillatory type, such as shown in Fig. 15-1. A designer with experience in linear regulators invariably encounters a few stumbling blocks here. For example, the common-sense approach to good filtering with an LC circuit would be to design a low-pass filter with considerable attenuation at the switching frequency. Although this would seem to make sense, it is not the proper starting point in the design sequence. A switching regulator must have a certain amount of output ripple for it to work at all. Indeed, the predominant parameter of the output "filter" capacitor is often its equivalent series resistance, rather than its capacitance. It is more essential that the required output ripple be supported, than for the LC circuit to function as a good low-pass filter. Of course, there is a conflict of interest—ideally, it would be desirable to make the ripple arbitrarily low because a low ESR capacitor would reduce capacitor heating and would improve the high-frequency characteristics of the regulator, but such freedom cannot be permitted by the fact that the output ripple voltage is intimately involved with other important design quantities.

Free-running switcher designs

The fact that output ripple voltage is a basic parameter in the operation of the self-oscillating regulator, rather than merely an incidental nuisance to be clobbered with a brute-force filter, can be appreciated from the following relationships:

$$t_{ON} = \sqrt{\frac{(2LC)(\Delta V_{OUT})}{V_{IN} - V_{OUT}}}$$

$$t_{OFF} = \sqrt{\frac{(2LC)(\Delta V_{OUT})}{V_{OUT}}}$$

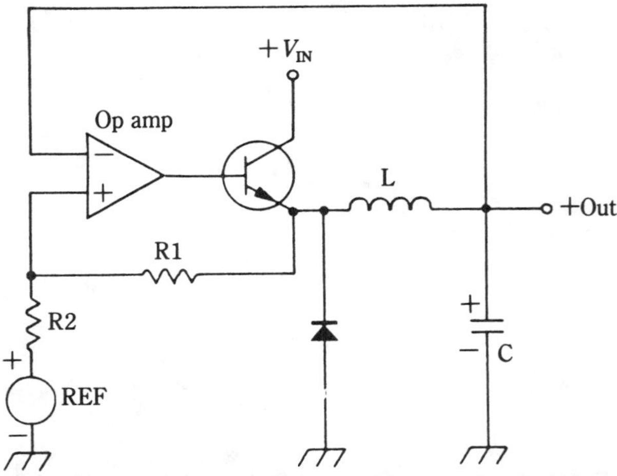

15-1 A simple self-oscillating switching regulator. A guide in making approximate calculations is to assume that the p-p output ripple voltage is equal to the hysteresis voltage of the op amp.

$$f = \frac{1}{t_{ON} + t_{OFF}}$$

Where:

$f\Delta V_{OUT}$ = the switching frequency

t_{ON} = the time the switch is closed

T_{off} = the time the switch is open

L = the inductance of the output choke

C = the capacitance of the output capacitor

V_{IN} = the input dc voltage of the regulator

V_{OUT} = the output dc voltage of the regulator

$\blacktriangle V_{OUT}$ = the peak-to-peak output ripple voltage of the regulator

A useful design equation can be derived from the above relationships, as follows:

$$C = \left[\frac{V_{IN} - V_{OUT}}{(2L)(\Delta V_{OUT})}\right]\left[\frac{V_{OUT}}{f V_{IN}}\right]^2$$

Thus, you have a means of computing the value of the output capacitor in terms of several quantities that are usually known because they are imposed by the conditions of application, or by cost or size motivated choices. The exception here is the inductor L—the value of L should not be determined by resonance formulas. It would not serve any useful purpose to express L in terms of C because L is determined from other considerations. The waveform diagram of Fig. 15-2 will provide useful insights into the role played by the inductor.

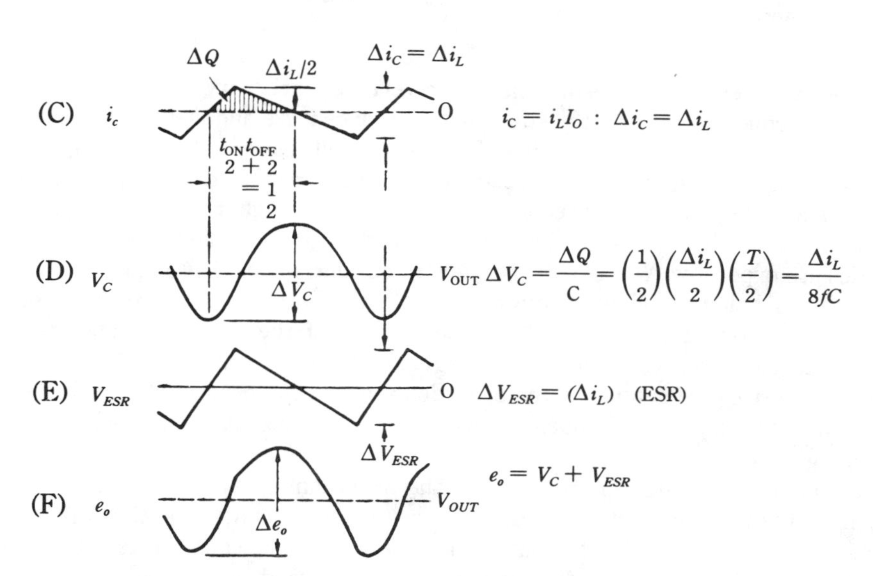

15-2 Some pertinent waveforms in the simple switching regulator. At high switching rates, ▲V_{ESR} tends to predominate over ▲V_C, especially when electrolytic filter capacitors are used.

The inductor must be large enough so that the peak current through it will not be greatly in excess of the maximum steady-state load current. This is to enable the switching transistor to stay within its safe-operating area, and similarly to hold down the peak currents in the free-wheeling diode. At the same time, a moderate peak current relaxes the design requirements on the inductor itself. The basic equation for the inductor is:

$$L = \frac{(V_{OUT})\ (t_{OFF})}{I_L}$$

Where:

I_L = the peak-to-peak ripple current through the inductor. Also, $I_L = (2\ I_o\ (n - 1)$, where n is the ratio of the peak inductor current to the maximum average dc output current, I_o.

Using actual numbers, the above procedure will become clear. Assume it has been decided to hold the maximum inductor current to 1.2 times the maximum dc load current. Then, $n = 1.2$, and $n - 1 = 0.2$. For this situation, the equation for L becomes:

$$L = \frac{(V_{OUT})\,(t_{OFF})}{(2)\,(0.2)\,(I_O)} = \frac{(V_{OUT})\,(t_{OFF})}{(0.4)\,(I_O)}$$

The value of t_{OFF} is derived from the relationship,

$$t_{OFF} = \left(\frac{1}{f}\right)\left(1 - \frac{V_{OUT}}{V_{IN}}\right)$$

The parameter that is established from relatively arbitrary considerations is the switching frequency, f. From the standpoint of reduced size and weight, you choose as high a switching frequency as possible. But, the overall cost and the speed capability of the switching transistor and the free-wheeling diode act as restraints. The output capacitor also begins to look predominantly inductive at higher frequencies.

It turns out that devices and components with really good high-frequency characteristics are expensive. Yet another frequency consideration is that of audible noise; so, other things being equal, it is often desirable to switch at least at an ultrasonic frequency. It has been found that a good balance of the numerous, and generally contradictory factors, occurs in the 20-kHz region. Here, the efficiency is still good with moderately priced elements, and the operation is above the perceptible sound frequencies. In many cases, higher frequencies actually yield diminishing returns with regard to packaging dimensions.

Summing up the design procedure of the free-running regulator, you should first determine the input and output voltage, the switching frequency, and the output ripple voltage. It is to be noted that the output ripple voltage is the consequence of the ripple current in the inductor. Therefore, it is easiest to proceed as follows:

1. Write down the values of V_{IN}, V_{OUT}, and f.
2. Choose a value for n, the ratio of peak inductor current to maximum load current. Nominal values range from 1.1 to 1.4.
3. Computer t_{OFF}.
4. With the information derived in steps 1 and 2, determine L.
5. Output ripple voltage $\blacktriangle V_{OUT}$ is approximately equal to the hysteresis of the comparator. $\blacktriangle V_{OUT}$ is also the voltage developed across the capacitor impedance because of the ripple current in the inductor. It is often true that the predominant parameter of the capacitor is not its reactance, but its ESR. If you know the ESR, the peak-to-peak ripple voltage across the capacitor is approximately the product of the peak-to-peak inductor current and the capacitor ESR. In Fig. 15-2, the hysteresis of the comparator is approximately $(V_{OUT}R_2)/(R_1 + R_2)$.
6. Calculate output capacitor, C. A little contemplation reveals that various LC combinations are possible. High C and low L leads to improved transient response and recovery time; however, too small an inductance leads to high-

peak current demands from the switching transistor at the time the regulator is first turned on, and under certain overload conditions.

The driven, synchronized regulator

When a switching regulator is actuated from a fixed-frequency source, its behavior is modified from that of the self-oscillating mode. Most significantly, the output ripple voltage is no longer a basic design parameter; rather, consideration must now be given to the amplitude of the drive voltage. The drive voltage is a triangular wave as it enters the comparator. The lower the drive voltage, the greater must be the loop gain of the regulator. But you must not go too far in this direction because noise and transients will ultimately override the error signal and produce erratic operation. Loop instability plagues any feedback system when the gain is too high. And the bandwidth of the regulator tends to decrease with increasing loop gain—producing poor transient response and recovery time. On the other hand, high loop gain does result in better regulation, improved ripple, and faster response to noise transients. The fact that the loop gain is a function of the amplitude of the triangular wave appearing at the comparator leads to a very convenient method of determining loop gain.

It is possible to attenuate the ripple voltage much more in a driven regulator than in a self-oscillating regulator. But, unbridled enthusiasm is out of place here because the ESR of the output filter capacitor imposes a limit on capacitor filtering ability. Rejection of input ripple and noise also tends to be inferior to that ordinarily attained from the self-oscillatory mode of operation. The important feature of both operational modes, however, is that inductor L can be considered as an energy reservoir—a source of continuing load current when the switching transistor is off. Accordingly, the inductor can be designed in the same way for both the driven and self-oscillating regulators.

The capacitor, however, need not be calculated, with regard to a required output ripple voltage. Rather, it is more convenient and relevant to use the resonance equation,

$$C = \frac{1}{4\pi^2 f_C^2 L}$$

in which f_C is a small fraction of the switching frequency, f. Frequency f_C nominally ranges from 1/20 to 1/50 of the switching frequency. Probably a maximum fractional value for f_C would be on the order of $f/5$, but the physically smaller filter components so obtained would constitute a tradeoff for a smaller margin of loop stability. The cutoff frequency of the LC filter commonly ranges between 150 and 500 Hz in regulators operating between several kilohertz and several tens of kilohertz. Therefore, the attenuation imparted to 60-, 120-, and 180-Hz input ripple is essentially "electronic" in nature.

Grounding techniques and component layout

A common failure in the construction of a switching-type power supply is to assume that because dc and audio frequencies are involved, that parts layout, lead length, and grounding techniques are only of secondary importance. Such an attitude can quickly lead to faulty operation, and it can be faulted for three reasons:

1. Much of the energy exchange occurring in the system occurs in the high-order harmonics. This makes stray parameters important, as well as proximity effects. Thus, you must often approach the matter of physical arrangement in a manner similar to that involved in the successful design of a TV tuner.
2. Very high peak currents can be involved. If these peak currents are coupled into sensitive portions of the circuit, faulty operation almost becomes a certainty.
3. High gain is involved. When a comparator is between its two saturation states, it behaves essentially as a linear amplifier. Unlike most op amp applications, the gain of the comparator is "wide open" during its brief transition intervals between saturation, so it is very vulnerable to picking up stray signals at its input terminals. The high loop gain existing in a high-performance regulator also poses a stability problem in itself. It should not be assumed that because there is a positive feedback path in a self-oscillatory regulator, that any further attention need not be paid to the occurrence of inadvertent positive feedback —nothing could be farther from the truth. There still remains the error-canceling negative feedback provision that must meet the stability criteria of any closed-loop system.

Figure 15-3 provides some guidance in parts location and grounding techniques. Of prime importance, the free-wheeling diode and the input bypass capacitor must return together to a separate ground, well away from the switching transistor base connection. These components should not be grounded at the same ground point used for the output filter capacitor because if this were done, all space between that ground point (or region) and the unregulated power supply would be embraced by strong circulating current, tending to create inductive pickup in various components and leads. The input bypass capacitor is not always depicted in schematic diagram, and ideally it should not be needed. But in practice, you can usually find that this component is needed, particularly when there is appreciable distance between the unregulated supply and the switching transistor.

Voltage stepup in the shunt switching circuit

The shunt-switching circuit of Fig. 15-4 simulates a voltage-stepup transformer. When the switch opens, the energy stored in the magnetic field of the inductor is rapidly released in the form of a large voltage transient. This is the "spike" caused by counter-electromotive force, which is advantageously used in ignition, TV flyback, and Geiger-counter power-supply circuits. It is also the source of so-called switching transients in many applications not intended to produce high voltages and the inductance is often of a parasitic nature. In any event, the voltage transient produced in this circuit is "dumped" into output capacitor C through diode CR. Because of the rectifying action of the diode, the charge accumulated in the capacitor is not discharged when the switch is

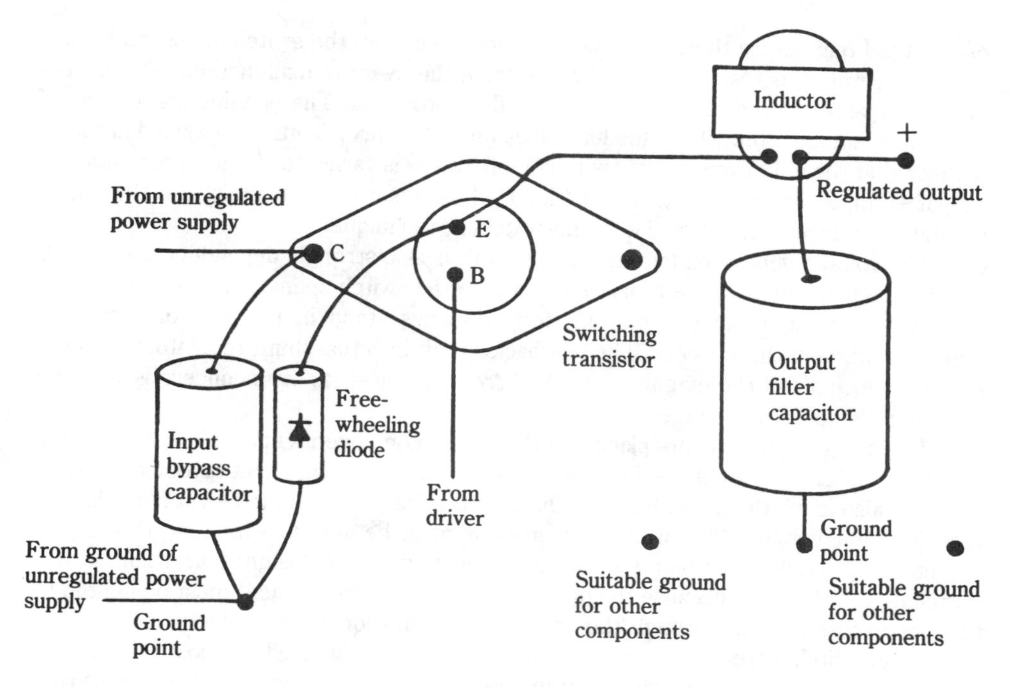

15-3 General guidance in parts layout and in location of ground points. The idea is to prevent high-intensity currents in the free-wheeling diode and in the input filter capacitor or bypass capacitor from forming an induction loop embracing the transistor base lead, or other system components.

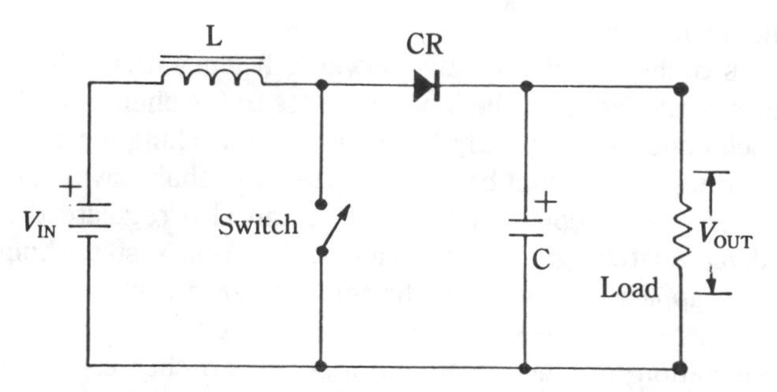

15-4 The basic concept of the shunt switching regulator. The salient feature of this switching technique is that the output voltage can be considerably higher than the input voltage. Operation is suggetive of a dc step-up transformer.

in its closed position. The capacitor behaves as a reservoir, giving up its energy to the load, but not more rapidly than its rate of charge replenishment.

Not only is the closed switch isolated from the charged capacitor, but it is also isolated from the unregulated dc source, V_{IN}. This is attained by virtue of *limited closure time.* Under such conditions, the current demand from source V_{IN} is in the form of a ramp. Long before the current can become excessive, the switch opens, and a new cycle of operation is ready to start. Of course, in the event of malfunction, in which the switch remained closed, a short circuit would be produced. The opening and closing of the switch is under control of a feedback loop and reference. Control is exerted in such a manner that the duty cycle of the switching operation is varied to maintain regulation of output voltage. The simple circuit of Fig. 15-4 serves only to demonstrate the fundamental ideas incorporated in the shunt-switching technique.

The voltage generated by this arrangement is also strongly dependent upon the Q of the inductor, and upon the rapidity with which the switch opens. It might appear that such a design might be difficult, but few obstacles stand in the way of successful implementation of this concept. The feedback circuit in actual shunt regulators does not care too much about the magnitude of the error voltage—its main concern is with the direction of the error voltage.

The design requirements placed on the circuit components in a shunt switcher are not much different than those for a series switcher, and component layout considerations are also much the same. Because the output voltage of the shunt switcher is higher than its input voltage, the switching transistor must be able to withstand this higher voltage. Self-oscillating shunt switchers are not as easily designed as their series-switching counterparts because the on time of the switching element must be limited to prevent excess current through the switch and saturation of the inductor.

If the switch were to remain open, the output voltage would be almost equal to the input voltage because the voltage drops across the diode and inductor are usually negligible. In addition, the output current under open-switch conditions is also equal to the input current—and none of this current flows through the switching device, which is quite opposite the behavior of the series switcher. Thus, in the shunt switcher, the switching operation is only needed to supplement the output power required by the load. In the series switcher, the switching operation was required to limit the power delivered to the load.

The shunt switcher tends to have lower efficiency than the series switcher, primarily because of the fact that the input currents to the shunt switcher are considerably larger, which contributes greatly to ohmic and switching losses. Nevertheless, in applications requiring power from battery supplies, the shunt switcher is not only able to generate higher power-supply voltage, but it can also regulate them—something that a simple dc/dc switching converter cannot do. As a result, shunt switchers are finding many new applications in small battery-operated systems, such as electronic calculators and watches, where small size and efficiency are extremely important.

The basic equation for the inductor in a shunt switcher can be quickly obtained using the equations of the series switcher, by observing the rearrangement of the circuit components in relation to the input and output voltages. Thus:

$$L = \frac{(V_{OUT} - V_{IN})\,(t_{OFF})}{I_L}$$

The on time is not usually dependent upon considerations of the LC circuit, but rather upon a maximum or worst-case time limit. Frequently, the on time is fixed at a constant value, determined by the equation:

$$t_{ON} = \frac{(L)\,(I_L)}{V_{IN}}$$

and the off time is varied to obtain the desired regulation. Such an approach has certain advantages, and it greatly simplifies the startup circuitry of the switcher because the on time does not immediately have any effect upon the output voltage. A disadvantage is that the output ripple voltage increases as the average load current decreases so that the regulation becomes poorer. An alternative is to choose a lower value for t_{ON}, corresponding to a lighter load condition, or to make provision for reducing or varying the on time after startup.

One problem with the shunt switcher is that of dc feed-through—there is no way of completely turning off the output voltage. In case the switching element fails by opening up, the input voltage passes directly through to the output devices being powered. The shunt-switcher circuit in Fig. 15-4 is also not directly applicable for obtaining output voltages that are less than the input voltage.

A modification of the shunt switcher is depicted in Fig. 15-5, in which the inductor element uses a secondary winding, both for isolation and voltage stepdown. Although the output is shown having a negative ground return, it is equally possible to use a positive ground. Notice the polarity markings on the inductor windings—the diode must be reverse biased during the on time of the switch for proper operation. The design of a good inductor also requires that the two windings be closely coupled magnetically, otherwise large voltage spikes will appear across the switching element, thereby shortening the life expectancy of that device.

Let's save these switching transistors

The idea of using switching action, rather than rheostat dissipation to control and regulate dc voltage and current is actually an old concept. However, popularity was slowed for a long time because of certain perceptions in the technical community. Chief among these was the notion that the switching technique was inherently unreliable in practice. It was also felt that any advantages over linear supplies was more than counterbalanced by the generation of electrical noise. Ultimately, enough insight was

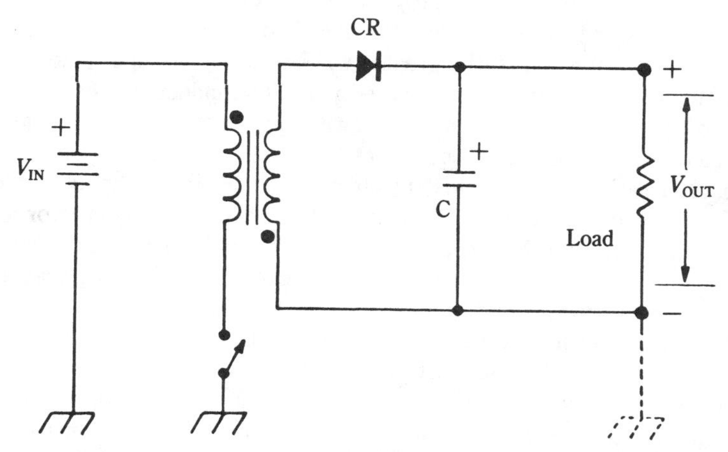

15-5 A shunt switching technique for elimination of dc feedthrough. In this scheme, the dc output voltage can be either positive or negative, as well as higher or lower, than the input voltage. Additionally, it allows galvanic isolation between input and output circuits.

attained so that reliability could be handled on an engineering basis. When control was finally gained over the destructive switching transients, it was found that, as if by magic, much of the noise problem was also alleviated. Once the mystery of blown transistors yielded to explanation, the semiconductor firms contributed to the art by producing transistors with greater electrical ruggedness.

Before the advent of power MOSFETs, it had been observed that the destruction of bipolar switching transistors could be associated with certain operational conditions. For example, a not uncommon cause-and-effect relationship was the destruction of the switching transistor at the moment the supply was energized from the ac-line. This was a current-inrush problem agitated by the vulnerability of the transistor to certain *combinations* of collector current and voltage, as depicted by the manufacturer's SOA curves. A two-fold solution has cured this problem. First, the use of thermistors in the ac line or rectifier circuit initially reduces the current surge following energization of the supply. The surge stems from uncharged capacitors and from momentary operation of the switching transistor in its linear mode. Two examples of the use of NTC thermistors are shown in Figs. 15-6 and 15-7.

In addition to, and sometimes in place of, the thermistors, most control ICs incorporate "soft-start" time-delay circuitry for holding off drive to the switching transistor for a fraction of a second, or so, after the supply is energized. This circuitry also guards against a related type of transistor destruction that can occur when the supply is shut down.

The elusive destroyer of the switching transistor turned out to be switching itself. During initiation and termination of the on period, the transistor experiences unfavorable combinations of collector voltage and current. These combinations are susceptible to the fatal secondary-breakdown phenomenon, wherein the destruction of the transistor occurs even though there is no detectable rise in its operating temperature prior to the failure. It also had to be discovered that the voltage and current combinations leading to forward-bias breakdown and reverse-bias breakdown were different and, generally speaking, had to be handled separately. "Reverse-bias" can simply be the zero-voltage applied to the base to turn the transistor off.

When a transistor turns off, the current fall-time tends to produce two side effects. The more abrupt the turn-off is, the greater will be the the voltage spike caused by any inductance this current change is associated with. The inductance can be the leakage inductance of a transformer, or, at high switching rates, can be the leads on the printed-circuit board. The other side effect becomes prominent when the turn-off is delayed, or is relatively slow. This manifests itself as increased power dissipation because of greater duration of simultaneous voltage across the transistor and current through the transistor. Both side effects increase switching losses, lower efficiency, increase noise, and worst of all, can destroy the transistor from penetration of its RBSOA (Reverse-bias operating area).

Ideally, you would like to see the transistor turn off instantaneously in a zero-inductance circuit. Both of these performance parameters are impossible in practice. The practical goal is rapid turn-off and minimal inductance. Often, there will be dangerous spike energy, despite all precautions that can be taken. Various clamping and "snubbing" techniques are used to absorb this transient energy. A number of passive networks have been devised for this purpose. The simplest, and perhaps the most

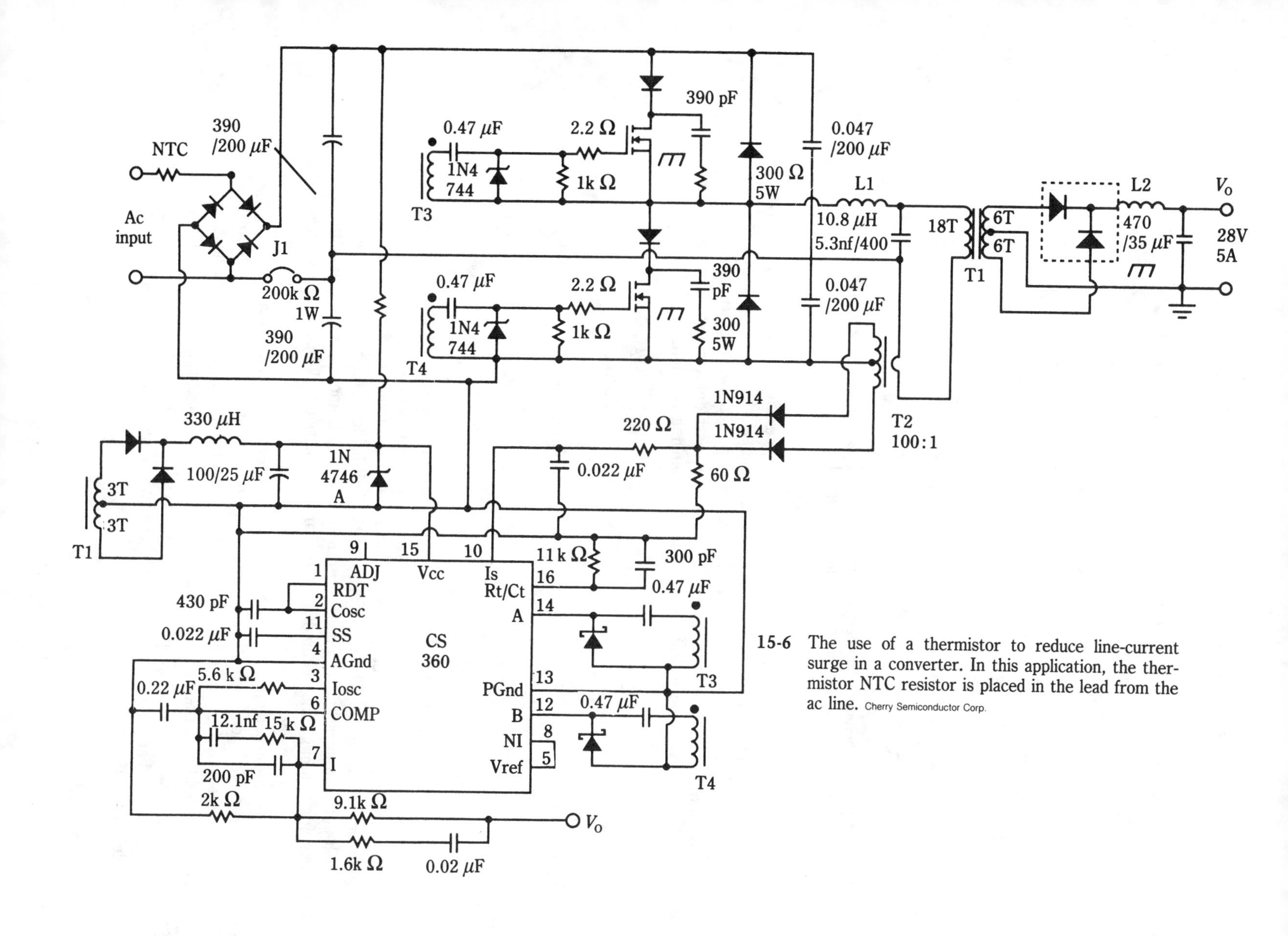

15-6 The use of a thermistor to reduce line-current surge in a converter. In this application, the thermistor NTC resistor is placed in the lead from the ac line. Cherry Semiconductor Corp.

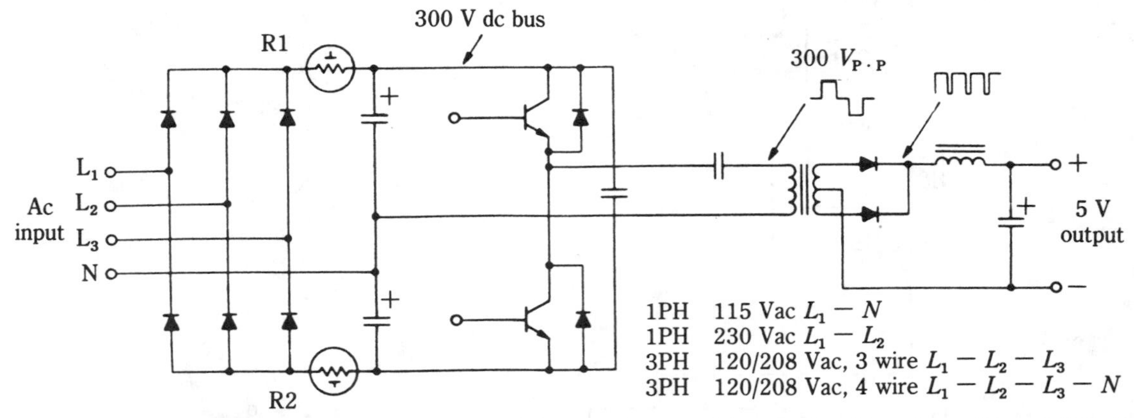

15-7 A partial schematic of an off-line converter that shows the use of thermistors. Thermistors R1 and R2 limit current inrush when converter is first energized.

commonly encountered of these is a simple RC series circuit connected from collector to emitter, as shown in Fig. 15-8. More complex networks might be called *load-line shapers,* but they too might be correctly referred to as *snubbing circuits.*

Optimum values for the simple RC snubber is best determined empirically. However, initial values can be calculated from the following relationships:

$$C = \frac{(I)\,(t_f)}{V}, \quad R = \frac{t_{on}}{C}, \text{ and } P = \frac{CV^2(f)}{2}$$

where

C = capacitance in Farads (multiply by 10^6 to obtain value in microfarads)
V = the peak switching voltage
R = resistance in ohms
P = power rating of the resistance in watts
f = switching rate in Hertz

Voltage, V, is actually the V_{CEX} value found in transistor spec sheets. The V_{CEX} value that must be withstood is either V_{IN}, the input voltage, or twice V_{IN}. The value that must be dealt with can be gleaned from inspection of the various switching circuits (shown in Fig. 15-9). The superiority of the half-bridge circuit over the conventional push-pull configuration is evident from their V_{CEX} requirements. Notice, also, the use of clamping windings and clamp diodes to channel inductive kick-back energy back to the input power supply.

The free-wheeling diode commonly encountered in switchmode supplies usually didn't receive undue attention as long as switching rates were in the 20- to 40-kHz region. All too often, one got away with the notion that this diode was just another rectifier, and that the main idea was to select a switching transistor with appropriate frequency capability. The basic circuitry involved was that of Fig. 15-10. The empirical observation that temperature rises in these two-semiconductor devices were "reasonable" sufficed.

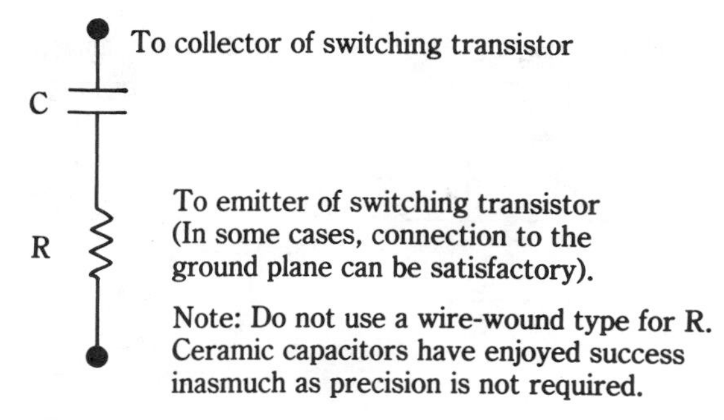

15-8 A simple RC snubber network for absorbing inductor kickback energy. The capacitor must have a low ESL; the resistor must be non-inductive.

As the art advanced and both higher switch rates and higher voltages were involved, "mysterious" failures of the switching transistor became more commonplace. Actually, this was often accompanied by other failed components and devices in such a manner that it wasn't always clear which element initiated the chain reaction. When it ultimately became evident that the switching transistor was, indeed, the initial failure, the natural response was to learn why. It shouldn't be too surprising that the switching transistor should be exceptionally vulnerable to failure. Inspection of the FBSOA curve in Fig. 15-11 shows one archiles heel of a typical power transistor. The other is secondary breakdown tending to occur from combinations of voltage and current with the transistor off.

There is an arbitrary current limit up to point "A". It can be assumed that this is largely caused by the internal bonding wires and connection of the transistor. From point A to point B, various combinations of collector current and voltage enable you to abide by a 500-W dissipation limit. In other words, the transistor behaves in this region much as a power-rated resistor that doesn't care too much about the combination of voltage and current just so its rated power capability is not exceeded.

Now, look what happens at point "C"; the allowable power dissipation is dramatically reduced to about 30 W! You can already begin to suspect that certain modes of misbehavior in the switching circuit could trigger secondary breakdown in the switching transistor—a usually destructive phenomenon. Acceptance of this idea then leads to recognition of inherent departures from ideal rectification characteristics in the free-wheeling diode. These are essentially, finite voltage-drop during forward conduction, a forward recovery delay, a reverse-recovery characteristic, and internal capacitance. The last-two behaviorisms—especially, readily cause transient effects resulting in penetration of the forward-biased SOA boundaries, leading to second-breakdown and destruction of the switching transistors. The free-wheeling diode itself, although the actual culprit, often survives this catastrophe!

What happens is that the free-wheeling diode, instead of turning off when the switching transistor turns on, remains conductive, thus subjecting the transistor to a transient, but heavy current drain. The delayed turn off of the diode stems from stored charge during its previous on time. This behavior is known as the diode's reverse-recovery characteristic. At high switching rates, the diode's self-capacitance also contributes to this transient loading of the switching transistor. The basic idea during design is to

1. Single-transistor flyback converter

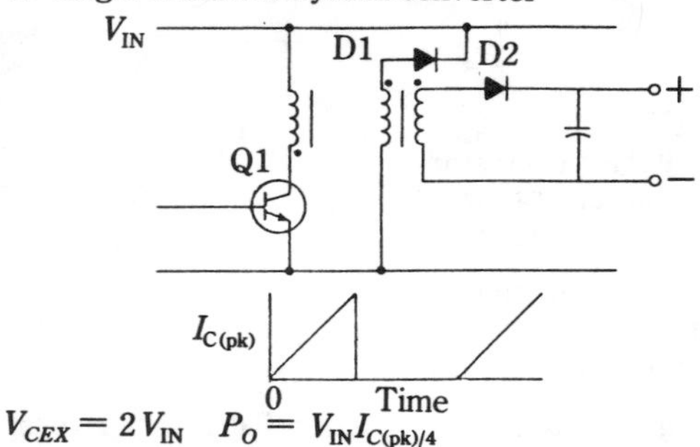

$V_{CEX} = 2V_{IN}$ $P_O = V_{IN}I_{C(pk)/4}$

2. Two-transistor flyback converter

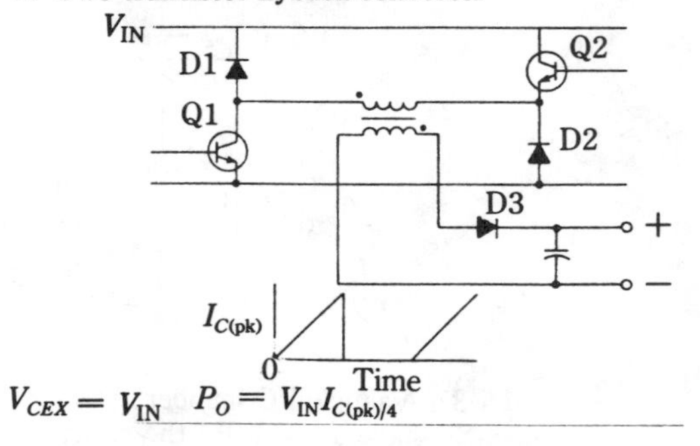

$V_{CEX} = V_{IN}$ $P_O = V_{IN}I_{C(pk)/4}$

3. Single-transistor forward converter

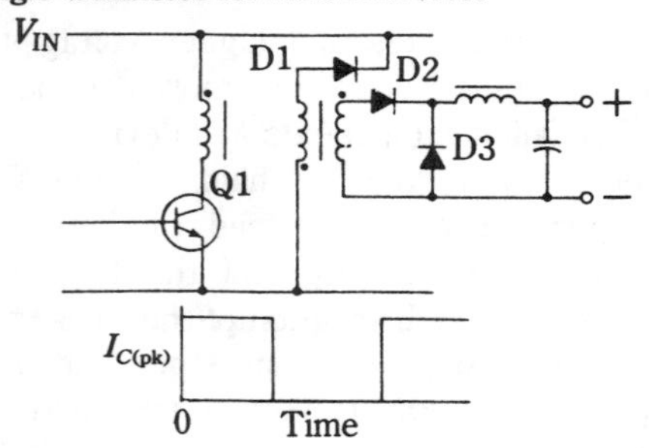

$V_{CEX} = 2V_{IN}$ $P_O = V_{IN}I_{C(pk)/2}$

4. Two-transistor forward converter

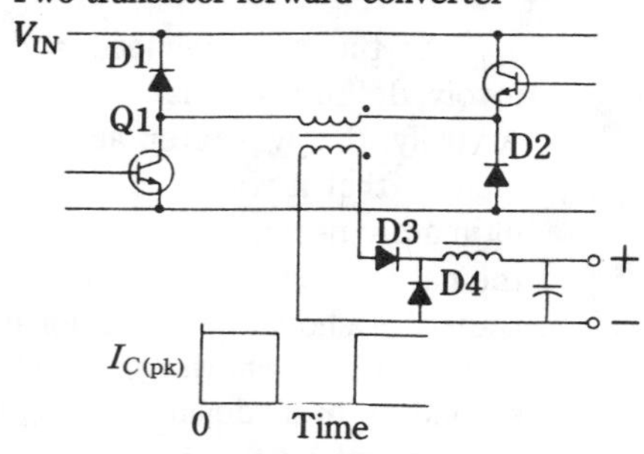

$V_{CEX} = V_{IN}$ $P_O = V_{IN}I_{C(pk)/2}$

5. Single-ended push-pull (half-wave) bridge converter

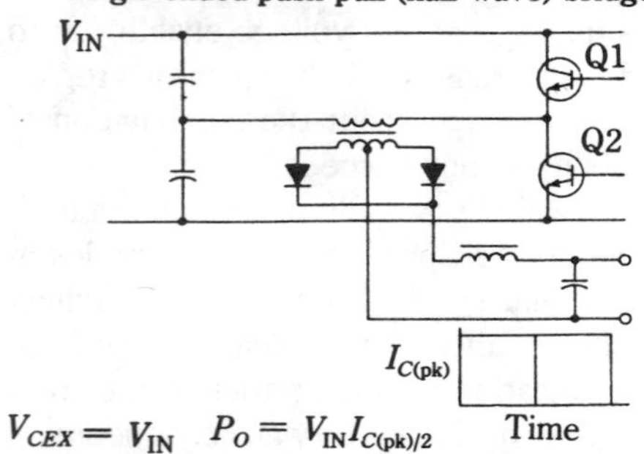

$V_{CEX} = V_{IN}$ $P_O = V_{IN}I_{C(pk)/2}$

6. Push-pull converter

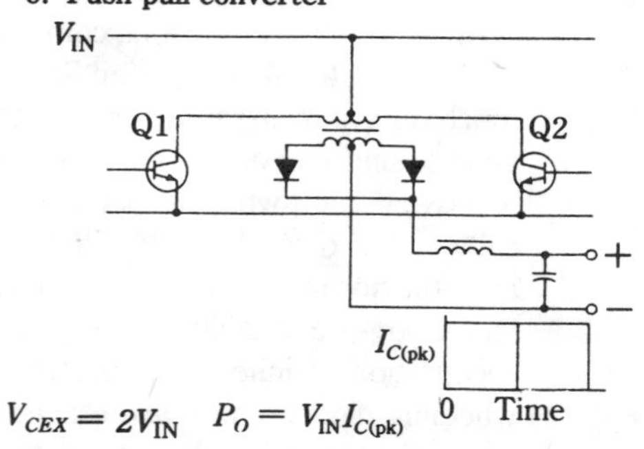

$V_{CEX} = 2V_{IN}$ $P_O = V_{IN}I_{C(pk)}$

7. Full-wave bridge converter

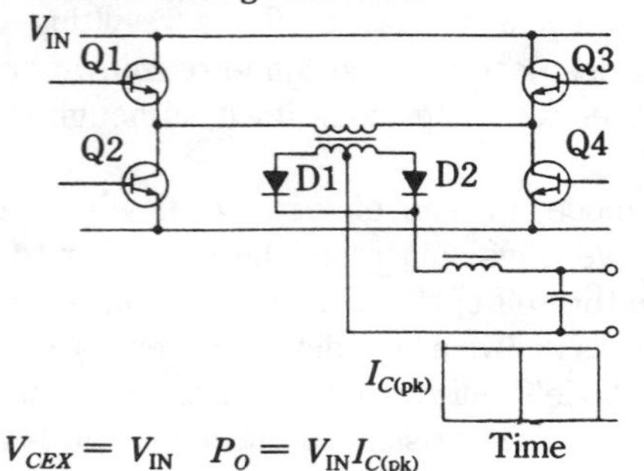

$V_{CEX} = V_{IN}$ $P_O = V_{IN}I_{C(pk)}$

15-9 Switching transistor voltage ratings for various switching circuits. Snubber networks can be added for enhanced safety. Also, notice the use of clamp windings and clamp diodes.

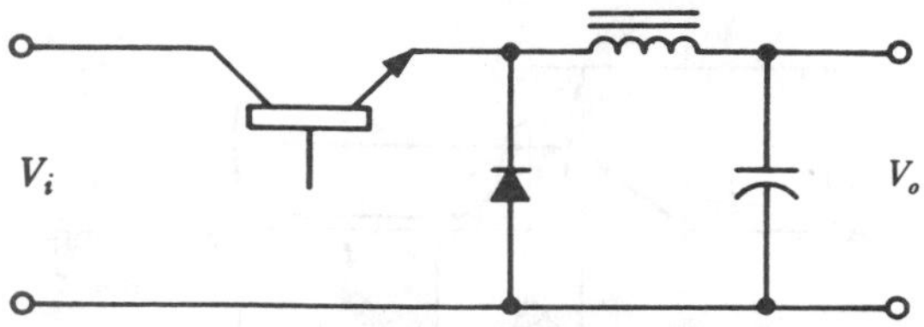

15-10 A simplified circuit that depicts a switching transistor and a free-wheeling diode. The effect of the non-ideal behavior of the diode has a pronounced effect on the safe operation of the switching transistor.

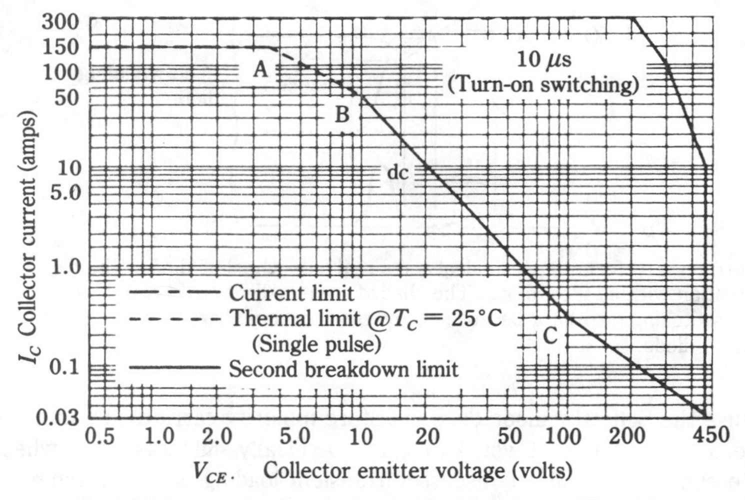

15-11 A forward-bias SOA curve for a typical switching transistor. Allowable power dissipation cannot be satisfied by just any combination of collector current and voltage. For some combinations, destruction from secondary breakdown is likely. Motorola Semiconductor Products, Inc.

select a free-wheeling diode with fast enough reverse-recovery time so as to minimize this burden on the switching transistor. The transistor turn-on waveforms of Fig. 15-12 clearly show the added power transients the transistor is subjected to because of the nonideal nature of practical free-wheeling PN-diodes.

This seems simple enough once the basic cause-and-effect mechanism is understood. In actual practice, the remedy becomes an art, as well as a science because of various interactions and trade-offs, as well as effects of stray parameters. For example, other things being equal, a switching transistor with fast current rise time is indicated. This, however, must be balanced with the fact that such transistors might trade off electrical ruggedness in other performance ratings. Then, too, the fast rise time makes it more imperative that the free-wheeling diode's reverse recovery be also fast. Moreover, the turn-on of the switching transistor might not be as rapid as its specifications suggest because of inductance in the wiring of the PC board. Sometimes, a diode's reverse recovery can prove too abrupt as manifested by increased noise production.

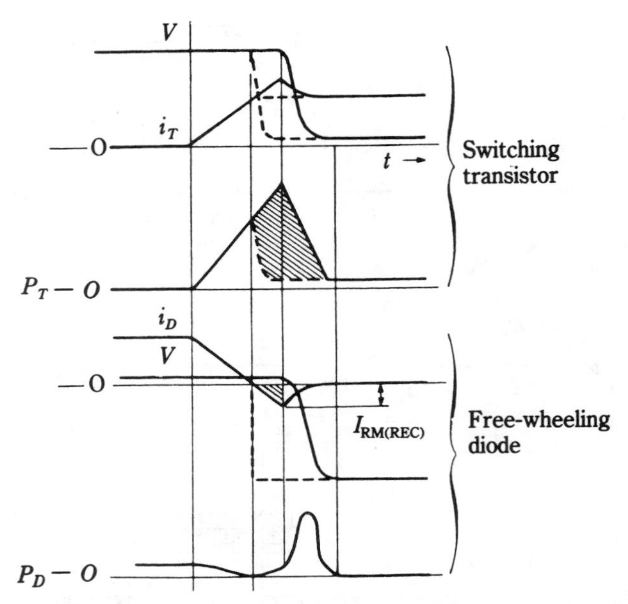

15-12 Turn-on waveforms with an ideal and a real free-wheeling diode. The dashed lines depict situation with an ideal diode. The shaded areas indicate ***additional*** power dissipation in the switching transistor because of reverse-recovery characteristic and capacitance of an actual diode. Unitrode Corp.

Because the Schottky diode does not store minority carriers and therefore has no reverse recovery behavior, it would appear to be ideally suited as a free-wheeling diode in order to eliminate the above-discussed transient loading of the switching transistor. Often, however, other factors preclude its use, such as considerations of temperature, voltage rating, and reverse leakage. Moreover, at high switching rates, the internal capacitance of the Schottky diode can cause essentially the same effect on the turn-on waveform of the switching transistor as the reverse-recovery phenomenon of the PN diode.

The repetitive overloading imposed by the free-wheeling diode is an exclusive danger, because the average temperature of the switching transistor can remain at an acceptable level. However, the cyclical penetration of the switching transistor's SOA can trigger secondary breakdown and destruction. Although this is an energy phenomenon, you can be reminded of the dielectric destruction of a capacitor by momentary over voltage. The thermal energy that destroys both the switching transistor and the capacitor is so localized that neither component need exhibit detectable temperature rise prior to their demise.

Power MOSFETs are almost immune to destruction by second-breakdown, whether in their on or off switching state. The word "almost" is appropos because a heavy slug of transient energy can certainly destroy the device. An interesting feature is the built-in internal body diode, a parasitic PN-structure that "comes along for the ride" during fabrication. It is often used as a current-return, or "free-wheeling" diode in switching circuits. However, at frequencies above about 100 kHz, this diode might be unsuitable because of its slow reverse-recovery characteristic. Although the MOSFET itself, might perform well into the several or multi-megahertz region, its parasitic diode is not so blessed.

If you want to use an external diode, the arrangement shown in Fig. 15-13 is often used. D3 is the external free-wheeling diode. Diode D2 renders the internal diode inactive by blocking its forward current path. D2 is usually a Schottky type in order to minimize the power loss, which must necessarily occur. An optional diode, D1, is also shown; this diode clamps the kickback voltage across the transformer primary, or other such inductance. It, too, should be a fast-acting type.

The power-MOSFET can withstand a much wider combination of voltage and current in its output section than the bipolar transistor. On the other hand, it exhibits its own unique vulnerability to destruction. This pertains to puncture of the thin silicon-dioxide film in its gate structure. Many power MOSFETs tend to be in dangerous territory when the gate-to-emitter voltage exceeds 20 V. Some early devices provided gate protection by means of built-in Zener diodes. This has been largely discontinued because of adverse effects on frequency. Some switching circuits use external Zener diodes or MOV elements for this purpose. Snubber networks are often used in the output circuit because switching transients can find their way to the gate via drain-gate capacitance.

Precautions are advised in the packaging and handling of power MOSFETs in order to prevent gate damage from electrostatic discharge. The relatively high capacitance of the gate reduces vulnerability somewhat; on the other hand, once a high electrostatic voltage accumulates, there is abundant energy available for gate destruction. Also, gate

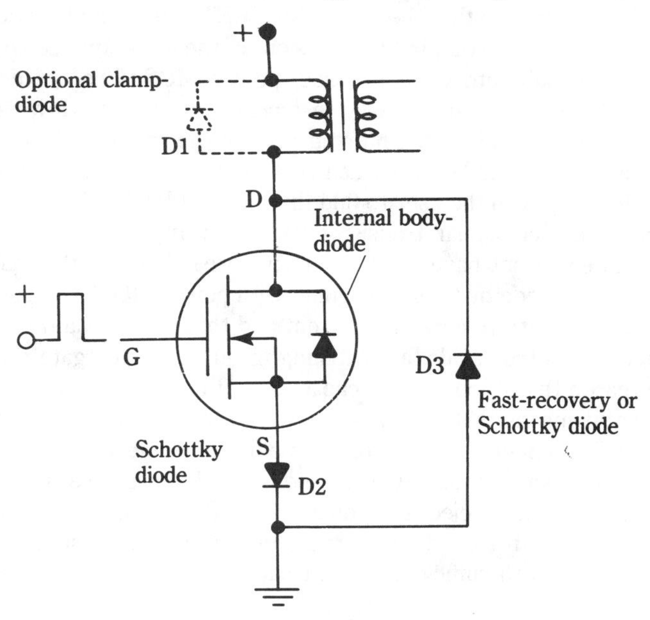

15-13 A power-MOSFET switch with external diode D3 for free-wheeling function. The diode in source-lead D2 blocks the transient current path through the internal diode of the power MOSFET.

damage is, in a sense, worse than gate destruction because it is elusive to detect. Such preliminary damage is very likely to bestow a short life-span on the MOSFET under operating conditions. Another source of danger to the gate is leakage current from a soldering iron. Suitable grounding and isolating techniques during storage, handling, and construction will pay off handsomely, for once in a proper circuit, the power MOSFET reveals itself as an electrically rugged device with a well-deserved reputation for reliability.

Because of its high transconductance and its high-frequency capability, the power MOSFET is subject to another often-destructive phenomenon. In conjunction with various stray parameters of the physical circuit, the device can easily become a self-excited oscillator. This is quite similar to the occurrence of parasitic oscillations in vacuum-tube RF amplifiers. Sometimes such oscillation will exist without much effect on the normal operation of the supply. If, however, this unplanned oscillation attains a violent level, regulation, or other aspects of performance will usually be degraded. Under such circumstances, destruction of the device is likely. The failure mechanism tends to be gate puncture from high RF voltage. Oft-used remedies for this defective performance are ferrite beads, resistors, or both in the gate lead. A 100-ohm noninductive resistor often works. It is important, of course, to follow good high-frequency practice in the layout of the PC board—leads must be as short as possible and parts placement should minimize coupling of energy from the output to the input of the circuit. A ground plane is always advisable and shielding and bypassing techniques can rarely be overdone. An incidental by-product of oscillation suppression is reduction in RFI.

The sales literature says that the power MOSFET is immune to thermal runaway. This statement is essentially true, but it comes out of the marketing department, rather than the engineering laboratory. In practice, power MOSFETs are destroyed by a phenomenon similar to thermal runaway in bipolar transistors. However, it takes a brutal overload to bring this about. Unfortunately, in the real world, heat removal and safety margins are compromised to reduce cost, and abusive operation is not uncommon. It is of little comfort to the user to find that power MOSFETs can be destroyed in similar fashion to that of bipolar transistors; it is purely academic to insist that the failure mechanism could not have been secondary breakdown or thermal runaway.

A common error made by those designing with power MOSFETs for the first time is to suppose that the gate requires no current and that drive conditions are therefore trivial. Although this is true at dc, at switching frequencies, the gate requires current for the simple reason that it looks like a capacitor to the drive source. Insufficient drive voltage and current means that the on state will not be saturated to its fullest extent. This, in turn, implies higher voltage drop, loss efficiency, and greater heat generation. As with any device, operation at higher temperature decreases reliability. If the drive current is only adequate to project the power MOSFET into its linear operating mode, speedy destruction is just around the corner. In any event, be sure to drive from a low-impedance source with sufficient current capability.

16
CHAPTER

Rectifier circuit design

BECAUSE THE SWITCHING-TYPE REGULATING SYSTEM IS MUCH MORE tolerant of the nominal voltage levels developed by its associated unregulated supply, there are misconceptions in switcher design: that it is only necessary to make certain that there is sufficient unregulated voltage under worst-case conditions; that only a primitive unregulated supply with rudimentary filtering is required; and that the switching regulator will take care of "everything else." Admittedly, there is some truth in this because an ultrasonic switching rate eliminates 60- and 120-Hz ripple in the regulated dc output. The switching process, together with the error-signal canceling action of the feedback loop, provides electronic filtering. Life becomes even simpler in some switching systems where the 60-Hz power transformer can be eliminated—a very compelling feature.

If for no other reason, the basic characteristics of unregulated supplies deserve a quick review because they are adjuncts of the switching-type regulating system. But an excessively carefree attitude with regard to unregulated supplies has led to some unfortunate consequences and less-than-optimum performance.

The important characteristics of rectifier circuits are as shown in Tables 16-1 and 16-2. Considerations of efficiency have made the full-wave circuits the most frequently encountered in single-phase power systems. Of the two configurations, the bridge rectifier circuit is usually preferred to the full-wave centertapped circuit. In the first place, the bridge provides a better transformer utilization factor; in the second, the bridge requires no tapped transformer connection. When the bridge rectifier is used in a regulating system that does not use a 60-Hz power transformer, such a system is operable from dc sources as well, and the polarity of the dc source is unimportant.

The single-phase bridge rectifier has a few shortcomings, and it might come as a surprise that the centertap arrangement (so popular during the era of tube-dominated techniques) is creeping back into sophisticated switching-type supplies. An increasing demand for regulated dc power at low voltage and high current is imposed by the

Table 16-1 Voltage relationships in rectifier circuits. Factors are normalized with respect to average dc value.

Schematic	Voltage Output Waveform	Average DC Volts Output	RMS Volts Output	Peak Volts Output	Peak Reverse Rectifier Voltage	Ripple
Half wave		1	1.57	3.14	3.14	121%
Full wave ct.		1	1.11	1.57	3.14	48%
Full wave bridge		1	1.11	1.57	1.57	48%
3ϕ Star (wye)		1	1.02	1.21	2.09	18.3%
3ϕ Bridge		1	1.00	1.05	1.05	4.2%

Table 16-2 Current relationships in rectifier circuits.

		SINGLE PHASE HALF WAVE 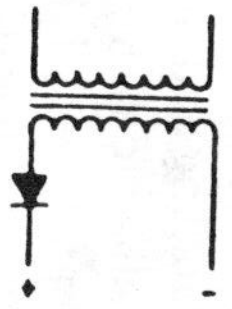	SINGLE PHASE CENTER-TAP 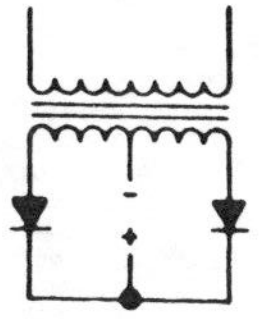	SINGLE PHASE BRIDGE 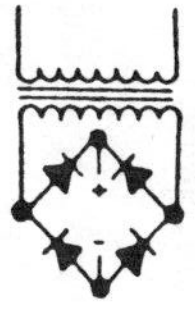	THREE PHASE STAR (WYE) 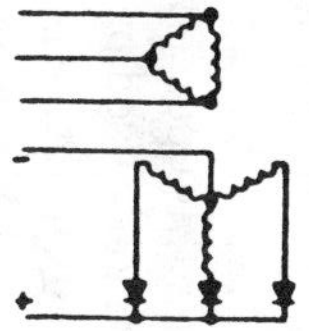	THREE PHASE BRIDGE 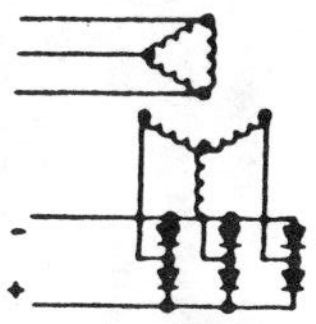
AVERAGE D.C. OUTPUT CURRENT		1.00	1.00	1.00	1.00	1.00
AVERAGE D.C. OUTPUT CURRENT PER RECTIFIER ELEMENT		1.00	0.500	0.500	0.333	0.333
RMS CURRENT PER RECTIFIER ELEMENT	RESISTIVE LOAD	1.57	0.785	0.785	0.587	0.579
	INDUCTIVE LOAD	----	0.707	0.707	0.578	0.578
PEAK CURRENT PER RECTIFIER ELEMENT	RESISTIVE LOAD	3.14	1.57	1.57	1.21	1.05
	INDUCTIVE LOAD	----	1.00	1.00	1.00	1.00
RATIO: PEAK TO AVERAGE CURRENT PER ELEMENT	RESISTIVE LOAD	3.14	3.14	3.14	3.63	3.15
	INDUCTIVE LOAD	----	2.00	2.00	3.00	3.00
		Resistive Load		Inductive Load or Large Choke Input Filter		
TRANSFORMER PRIMARY RMS AMPERES PER LEG		1.57	1.00	1.00	0.471	0.816
SECONDARY LINE CURRENT		1.57	0.707	1.00	0.578	0.816

computer industry; a common requirement is 5 V at several hundred amperes. Neither the conventional silicon junction diode nor the bridge configuration can be readily used for these applications because the voltage drop they cause greatly lowered rectification efficiency. The bridge is worse than the centertap arrangement in this respect because the forward voltage drop of two diodes is always involved. The power dissipation in a rectifying diode is proportional to the forward voltage drop, so by merely substituting Schottky diodes for conventional pn junction types in a bridge, a 200-percent improvement in rectifier dissipation can be obtained. And, if the circuit changes from a bridge to the centertap configuration, there is another 200-percent improvement in rectifier dissipation. Thus, by dispensing with the commonly used pnp junction-diode bridge, and substituting two Schottky diodes in a centertap circuit, about 400 percent less power is thrown away in the rectification process. At 30-V this might not be of great consequence, but when dealing with only 5 V together with high current, the rectifying technique assumes overwhelming significance because the "conventional" approach would in this case totally destroy the high efficiency, which a switching supply could otherwise develop.

Although this is simple circuitry, it is clear that the unregulated supply cannot always be thrown together on the premise that the switching process, feedback, or other techniques will make everything come out well. And, as might be expected, there are other pitfalls. Another problem manifests itself when conventional diodes are used to rectify frequencies above 20 kHz. This has already been covered in chapter 13, although the emphasis there was placed upon performance problems, rather than just dissipation in the rectifying diodes.

Synchronous rectification

In the early days of pn junction rectifiers, the germanium diode received great developmental impetus from the major processors of solid-state devices. These diodes, however, became nearly obsolete with the advent of silicon devices. The silicon diode proved far superior in the matters of power handling and the ability to operate at higher temperatures. As if this was not enough, the silicon devices had much less reverse-leakage current and could be made to operate at much higher voltages. However, the germanium junction diode had one feature that was superior to the best silicon device: its low forward voltage drop. Designers and experimenters have lamented the demise of the germanium diode whenever confronted with an application requiring low forward voltage drop. One such application is the rectification of low voltages at high currents. Fortunately, several of the major semiconductor firms have continued germanium technology, primarily in the form of germanium power transistors. It is well known that these transistors have an inordinately low collector-emitter saturation voltage, $V_{CE(SAT)}$, on the order of 0.3 V, even for currents as high as 50 A. It would be only natural to speculate on the feasibility of using transistors to accomplish rectification.

Motorola Semiconductor Products Company has investigated this possibility and has developed a practical circuit in the form of the synchronous rectifier, shown in Fig. 16-1. It has long been known among experimenters that transistors, even defective ones, can be used as a rectifier by using the base-collector junction. But the voltage drop across the collector-emitter terminals of a saturated transistor is even lower than that

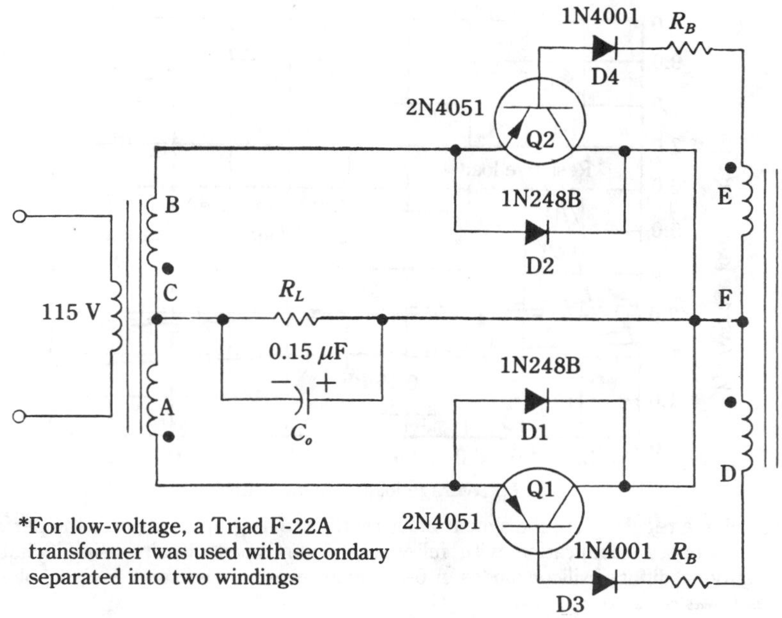

16-1 A full-wave synchronous rectifier circuit. The low collector-emitter saturation voltage of germanium power transistors is exploited to produce high rectification efficiency of low and moderate voltages at high currents. Motorola Semiconductor Products, Inc.

attainable by using the diode action of the collector-base junction. Thus, it is desirable to use the transistor action to make a synchronous rectifier. An interesting comparison of voltage regulation of a synchronous rectifier circuit with that of conventional diffused-junction and Schottky diodes is shown in Fig. 16-2. The efficiency as a function of load current for the three rectification schemes is illustrated in Fig. 16-3. Notice the exceedingly low output voltages involved. The tests were conducted with a 60-Hz square wave obtained from an inverter.

How much the frequency can be increased and still retain the high rectification efficiency of the synchronous rectifier depends somewhat upon the germanium transistors used, whether a sine or square wave was being rectified, and the nature of the load. It is probable, however, that difficulties from charge storage could be encountered above about 3 kHz. On the other hand, because germanium power transistors are used successfully at much higher frequencies in inverters, problems from charge storage could be overcome, or satisfactorily reduced, using circuitry techniques of the kind used to speed up response in logic gates.

In Fig. 16-1, diodes D1 and D2 conduct only when the transistors are forced to momentarily operate in their unsaturated modes, and this operation makes them vulnerable to destruction because the safe operating areas are then exceeded. Such operation usually results from the effect of capacitive loads. Diodes D4 and D5 also help reduce this problem by causing earlier turnoff of forward base-emitter bias than would otherwise occur.

Rectification efficiency did not change appreciably when the ambient operating

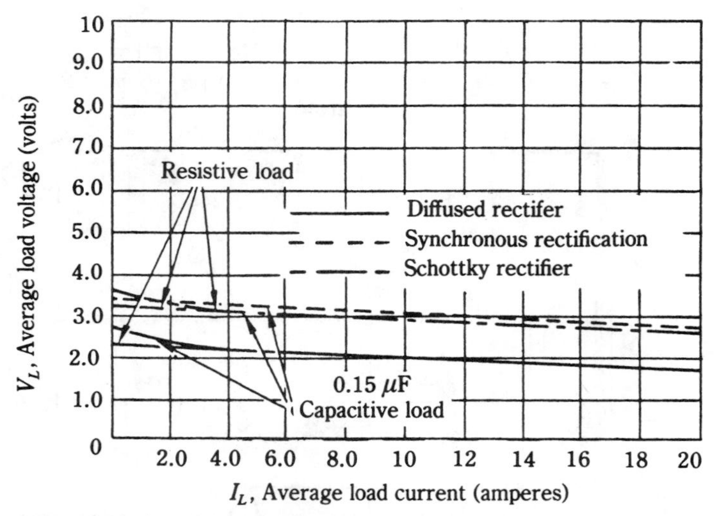

16-2 The voltage regulation of the synchronous rectifier. The graph compares the performance of synchronous rectification with full-wave centertapped circuits using Schottky and conventional diffused silicon diodes. A 6-Hz square-wave power source was employed in all cases. Motorola Semiconductor Products, Inc.

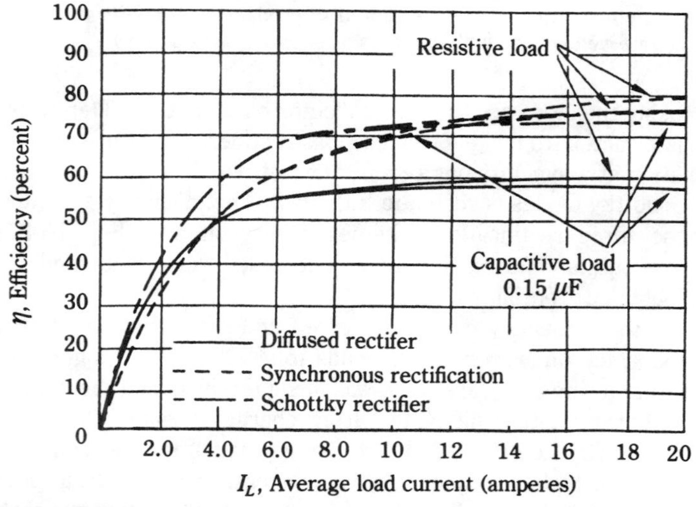

16-3 Rectification efficiency vs. load current for the three rectification systems. The graph compares the performance of synchronous rectification with full-wave centertapped circuits using Schottky and conventional diffused silicon diodes. A 60-Hz square-wave power source was employed in all cases. Motorola Semiconductor Products, Inc.

temperature was raised to 80°C. Half-wave versions of the circuit were also investigated. Within the inherent limitations of such rectification, the performance of the synchronous circuit was comparable to that of the Schottky-diode counterpart, and it was considerably superior to that of half-wave rectification using conventional diffused diodes. Thus, synchronous rectification using germanium transistors merits investigation whenever low voltages and high currents are involved. Candidates for such levels of voltage and current are the ECL, IIL, TTL, DTL, RTL, and low-voltage MOS digital systems. In other areas, low voltages or high currents are also required by thermoelectric coolers and often by electromagnetic solenoids. There are many electrochemical processes that fall into this category. A better approach could be to use power MOSFETs. Some types develop very low drain-source voltage drops, even at high frequencies and high temperatures (General Electric is developing 20-A units specifically for use in synchronous rectifiers).

Pitfalls

It is obvious that unregulated power supplies can be quite simple. It has also been pointed out that the series-transistor switching regulator is quite forgiving when it comes to input voltage variations and filtering requirements of the unregulated dc. Yet much grief has resulted from the association of the two units. It happens that the input to the switching transistor looks like a negative resistance. It can be mathematically shown that oscillation from positive feedback, shock excitation, and negative resistance are all very similar. Consequently, long leads between the switcher and the unregulated supply can often result in surprisingly strong radio-frequency oscillations. In effect, these are all of the essentials of a high-frequency, class-C oscillator — the leads function as a transmission line or tank circuit. The fact that the oscillation is inadvertent does not protect the switching transistor from destruction. Other oscillation frequencies and modes result from the use of LC filters in the unregulated supply. Generally, if inductance is used, a very large filter capacitor must be used to lower the impedance of the tuned circuit seen by the switching transistor. Sometimes it proves rewarding to connect a low resistance in series with the filter capacitor to lower the Q of the resonant "tank" circuit. Fortunately, it is often possible to dispense with the use of a filter choke; this also tends to be profitable from both a cost and a packaging standpoint.

Voltage multipliers

Voltage multipliers have been neglected until recently. Many designers associate these circuits with vacuum-tube technology, and thereby tend to overlook some good possibilities. The almost dramatic rescue made by voltage triplers and quadruplers in TV sets is quite well known. Fortunately, we don't have to solve problems relating to X-radiation in switching-type power supplies, but a voltage multiplying circuit can often be useful in further reducing package dimensions after an apparent limit is reached by the more usual techniques of high-frequency switching and the elimination of 60-Hz magnetics. In other instances, voltage multipliers can provide a neat method of providing additional output voltage using only a single-secondary transformer.

Many textbooks dwell upon the shortcomings of voltage multipliers. They are claimed to have poor regulation and increased circuit complexities. Statements of this nature involve an element of truth, but are based upon complex vacuum-tube circuits, which were always operated from a 60-Hz sine wave. The performance of voltage multipliers is greatly enhanced when driven from square, rather than sine, waves, and especially when driven at higher frequencies. At 1 kHz, and certainly at the popular 20-kHz switching rate, the voltage multiplier deserves reevaluation. Considering that the peak value and rms value are the same for a square wave, the capacitors in the multiplier circuit have much more time to accumulate charge, contrasted to the peak pulses delivered in sine-wave operation. This manifests itself as improved regulation and filtering. It is true that very good regulation can be had from sine-wave operation too, but only at the expense of larger capacitors. Some useful voltage multiplying circuits are shown in Fig. 16-4. The two topographic versions of the same hookup in (A) show how you can sometimes be misled by the draftsman's technique.

Although poor regulation is no longer a big problem with voltage multipliers, superb regulation is not necessary for use in a system where the ultimate regulation of dc output is taken care of by one or more feedback loops. In particular, certain voltage multipliers work very well with 50 percent duty-cycle inverters. Appropriate voltage multipliers are advocated for use in the unregulated power supply and will normally precede the feedback loop of the regulating circuits. Ordinarily, this application is associated with the dc-to-dc converter. For example, the 60-Hz line could be immediately rectified and doubled; this dc could next be used to power a dc-to-dc converter, which can be configured to operate as a switching regulator. Notice that this technique enables high output voltages to be attained without 60-Hz magnetics.

The voltage multiplier also makes it easier to design a good-performing inverter. The inverter transformer works best with a near unity turns ratio. Large deviations from this one-to-one ratio, particularly when stepping up voltage, often produce enough leakage inductance in the transformer windings to cause unstable operation of the inverter. As those who have experimented with inverters and converters well know, oscillation at other than the designed frequency is often the most likely malfunction in what is actually a simple circuit. And the leakage inductance can easily lead to destruction of the switching transistors. This problem can be avoided by the use of a voltage multiplier to permit the use of a near-unity turns ratio in the transformer.

When dealing with sine waves, it must be remembered that voltage multipliers operate on the peak value of the waves. Thus, a so-called voltage doubler, operating from a 100-V rms input, will show a no-load output voltage of $2 \times 1.41 \times 100 = 282$ V. Thus, if the capacitors are large and the load is relatively light, the action more closely approximates tripling of the rms input voltage. Similar reasoning prevails for the other multipliers.

Assuming equal-valued capacitors and sine-wave operation, voltage multipliers should be designed to have a minimum ΩCR product of 100, where Ω is 2π times the operating frequency in hertz, C is the capacitance value in farads, and R is the effective resistance in ohms presented by the heaviest load that will be imposed. This will result at least in 90 percent of the attainable dc voltage, and it will confine the operation to a relatively flat portion of the regulation characteristic. For square-wave operation, equivalent results can be had for an ΩCR product considerably less than 100.

In selecting a voltage-multiplying circuit, due consideration must be given the matter of grounding. In Fig. 16-4, the generator symbol usually represents the secondary of a transformer. Notice that grounding of one terminal of the transformer is permissible in half-wave circuits, but not in their full-wave counterparts if one side load

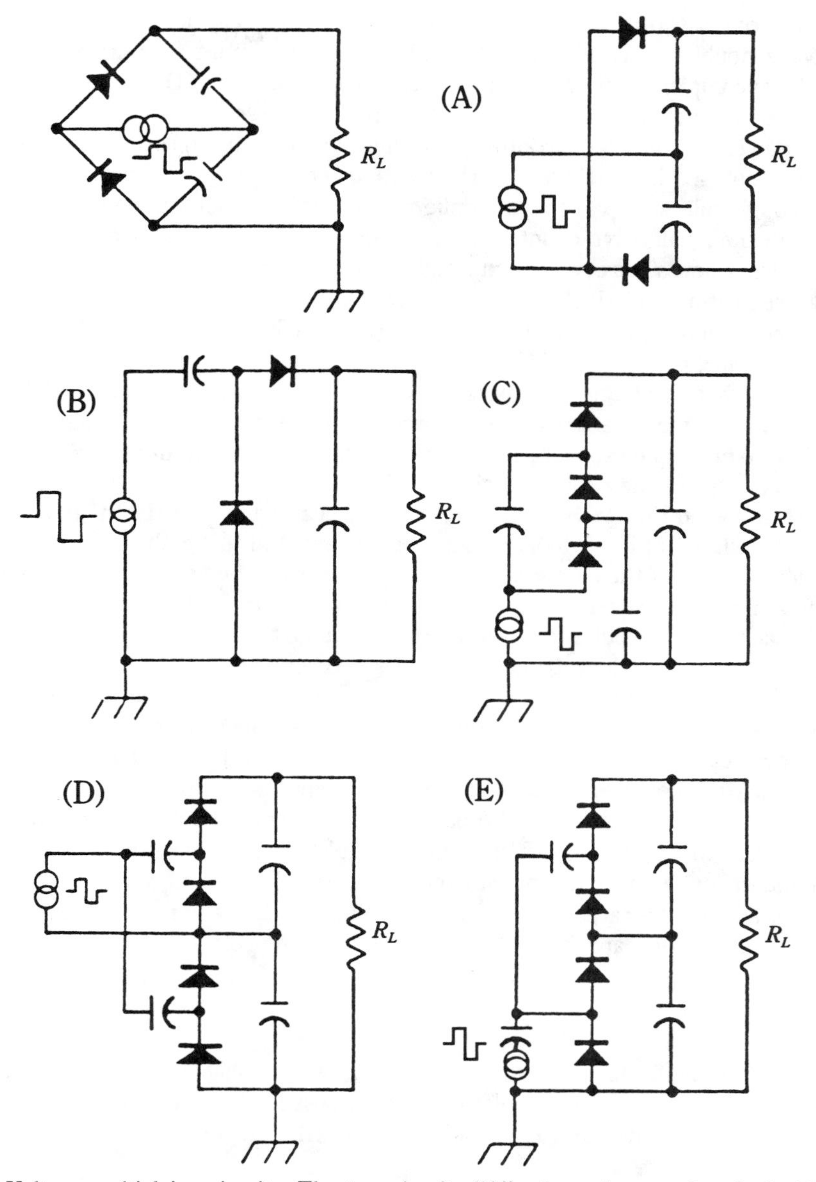

16-4 Voltage multiplying circuits. The two circuits "A" schematics are electrically identical. Notice the allowable and forbidden grounding options of the various circuits—in some cases, the generator and the load cannot use the same ground.

is to be at dc ground potential. Full-wave circuits are useful for developing dual-polarity outputs, in which one output is positive, with respect to ground, and the other output is negative, with respect to ground, and half of the total output voltage appears at each output terminal.

The circuits shown in A of Fig. 16-4 are schematically identical, being that of a full-wave doubler. Circuit B is a half-wave doubler. Circuit C operates essentially as a half-wave tripler. A full-wave quadrupler is shown as circuit D, whereas circuit E is a half-wave quadrupler. Voltage multipliers similar to these find widespread use in TV fly-back power supplies for developing the high voltages for the CRT picture tubes. They are also used in Geiger counters, lasers, electrostatic precipitators, etc.

Although full-wave voltage multipliers tend to have better regulation and less ripple than their half-wave counterparts, the differences are usually of little practical consequence when square waves and high frequencies are involved. You can always improve regulation and ripple by using larger capacitors. In general, at 20 kHz and higher, the common-ground feature of half-wave multipliers tends to influence the designers choice.

Very high dc voltages can be produced by cascading many elemental stages. Although the techniques are old, greater practicability is realized with solid-state diodes than with the erstwhile vacuum-tube rectifiers, which aggravated problems of insulation and cost because of their filament circuits. Two examples of cascaded voltage multipliers are shown in Fig. 16-5. These multiply the peak voltage of the input ac wave by eight. In the circuit of Fig. 16-5A, no capacitor is exposed to a higher voltage than $2E$. The salient feature of the circuit depicted in Fig. 16-5B is the common ground between input and output. However, capacitor-voltage ratings must be progressively higher as one approaches the output end of the cascade. Although this is a detriment in physical size and cost at 60 Hz, these penalties are less severely felt at higher frequencies. The diodes in both circuits must withstand the peak input voltage, E, but for safety should carry voltage-ratings of at least several times E. Equal-valued capacitors are commonly used in these cascades. The larger the capacitors, the better the regulation and ripple. Large capacitors, however, impose surge-current burdens on the diodes.

The cascade shown in Fig. 16-6 has found considerable favor in electronics applications. Notice that operation is from a unipolar pulse train. This is the Cockroft-Walton voltage multiplier that is found in many physics texts. Although all capacitors can be the same size, and can be rated to withstand E volts, a better engineering approach is:

First, the output capacitor, C_o, is calculated by the relationship,

$$C_o = \frac{I_o \times t}{V}$$

where I_o is the output current in amperes. Let $I_o = 40$ milliamps for sake of example, where t is the duration of the unipolar pulse in microseconds. If you assume a frequency of 20 kHz, t will be one-half the reciprocal of 20 kHz, or

$$\frac{1}{2 \times 20 \times 10^3}$$

where V is the assumed peak-peak ripple voltage. A reasonable value could be 100 millivolts then,

$$C_o = \frac{40 \times 10^3 \times 25 \times 10^{-6}}{100 \times 10^{-3}}$$

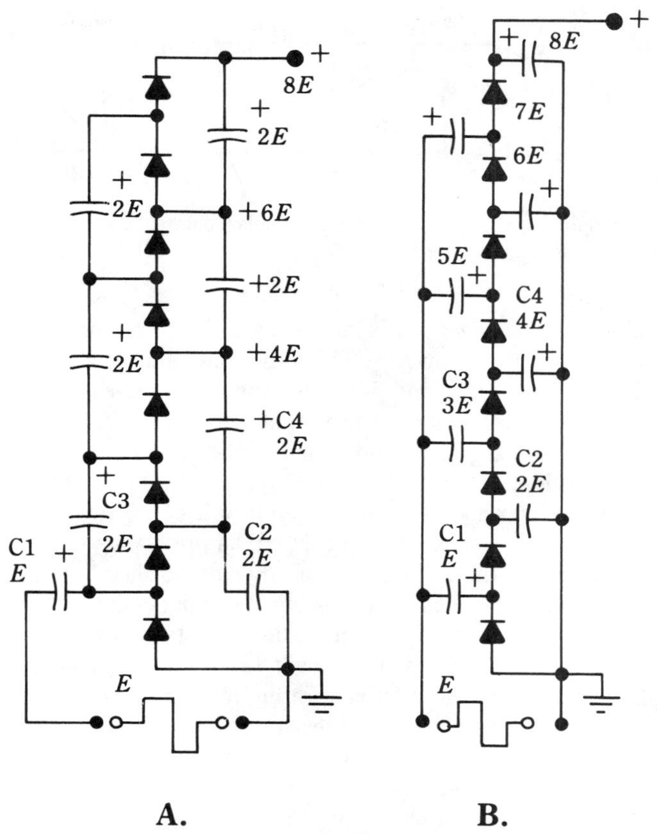

16-5 Two versions of voltage-multiplying cascades. (A) In this arrangement, no capacitor is exposed to a higher voltage than 2*E*. (B) This scheme features a common ground between input and outputs.

As you go toward the input of the cascade, the capacitors will become progressively larger by some factor times the value of the output capacitor, C_o. These calculations are straightforward, but can be tricky unless close attention is paid to the procedure. Notice the numbers alongside the capacitors in the circuit of Fig. 16-7. These are the factors C_o must be multiplied by in order to derive the actual capacities. Thus, the capacity of the capacitor designated by the number 2 is $2C$ or $2 \times 10\ \mu F$ in our example, or $20\ \mu F$. C5 has a value of $5C_o$, or $50\ \mu F$. And, the input capacitor, C11, is sized at $11C_o$, or $110\ \mu F$.

Where do these numbers come from? They represent the relative currents throughout the network. If the numbers did not appear alongside the capacitors in Fig. 16-8, you could determine their value by appropriately using the expression $(2n - 1)$. Here, n represents the number of times the input voltage is multiplied. Obviously, in the sextupler, n is 6. You commence at the input capacitor and find that $2n$-1 is 11. Then, you proceed along the bottom row of capacitors, applying successively $2n$-3, $2n$-5,

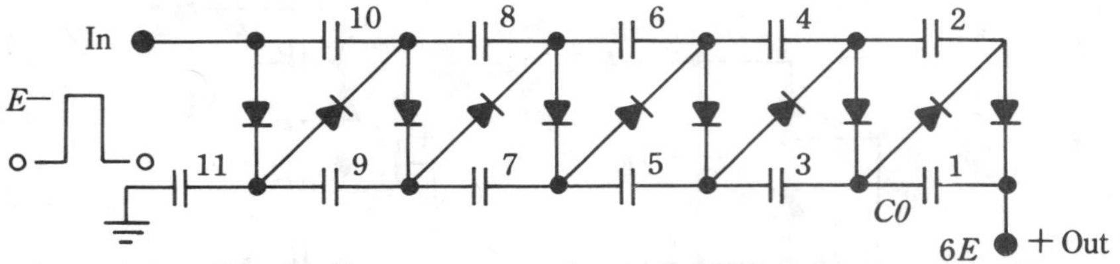

16-6 A sextupler operating from a source of unipolar pulses. The numbers alongside the capacitors are explained in the text.

$2n$-7, $2n$-9, and finally at C_o, $2n$-11. Next, follow a similar procedure, starting at the first capacitor at the left of the top row. This time, the multiplying factors of C_o are as follows: $2n$-2, $2n$-4, $2n$-6, $2n$-8, and finally for the right end-capacitor, $2n$-10.

The fact that the capacitors near the input are larger than those closer to the output is because of the charge-pumping action of the circuit, which, naturally enough commences from the input end. There are $2n$-1 transitions of charge per-cycle of operation. At each such transition, inherent energy loss occurs. This energy loss is minimal when the capacitor values are calculated, as previously outlined.

Initial testing of any of the voltage-multiplying circuits should be carried out with a variac, or by some gradual means of advancing the input voltage. Otherwise, diodes might be destroyed by the inrush current. The severity of this depends upon such factors as the size of the capacitors, the power level, the frequency, the ESR of the capacitors, and, of course, the surge-current rating of the diodes. It might be necessary to place a thermistor, or a relay-controlled resistance in the incoming line. On the other hand, many electronic applications can be implemented with no protection at all because it is so easy to obtain high surge-current ratings in diodes. Sometimes, the protection is "invisible," such as an input transformer that simply cannot supply high inrush current.

At high voltages, the forward voltage-drops of the diodes are of no consequence. At a lower voltage, the cumulative voltage-drop of the diodes can prevent attainment of the sought output voltage, and can seriously impair the efficiency of the voltage multiplier. Also, be sure that the reverse-recovery characteristics of the diodes are compatible with the frequency of operations. Otherwise, the designed voltage multiple will be "mysteriously" absent.

Special considerations to inverter and regulator rectifying systems

For the benefit of those who have successfully dealt with rectifier-filter designs in 60- and 400-Hz unregulated power supplies, a few guidelines are relevant in making the transition to high-frequency converters and switching regulators.

Where applicable, the flyback principle often proves economical because it dispenses with the need of a filter choke. Other switching circuits either incorporate a choke as part of their basic operating requirement or require a choke as part of an effective LC output filter.

More and more, high-current, low-voltage regulated supplies are used for empowerment of large mainframe computer systems. Bridge rectification is not efficient for such supplies because the forward voltage drop of two diodes becomes an appreciable fraction of the 5 V or so that is required at the output terminals. The centertap full-wave rectifying circuit is better here. And so is half-wave rectification if the trade-off in output ripple or in the cost of the filtering system can be accommodated.

As has been pointed out, another area of rectifier power dissipation lies in the charge-storage behavior of ordinary rectifying diodes, which might perform well enough below a kilohertz or so. Because modern converters and regulators operate at 20 kHz and higher, it becomes mandatory to use either fast-recovery (gold-diffused) silicon diodes, or Schottky diodes. The Schottky diodes feature very low forward voltage drop, but their reverse leakage losses tend to go up rapidly with reverse voltage and with temperature. They are fine for 5-V systems, but can be marginal for use in 15- or 20-V supplies. An attractive feature of Schottky diodes is that they can be paralleled without need for energy-dissipating ballast resistances. However, it should be ascertained that all paralleled units are essentially in the same thermal environment.

The output transformer for a high-current supply would have to be wound with "railroad track"-size conductor if low-current design practice was merely scaled up. A better approach is to use multiple secondaries of moderately sized wire. Then, particularly if Schottky diodes are used, a number of half-wave outputs are available for simple paralleling in order to provide the overall load current. This scheme is readily extended to the center-tapped full-wave configuration, further enhancing the performance of the high-current rectifier-filter system. Be sure to heed the phasing of the individual windings when this technique is used.

17
CHAPTER

Switching-type power-supply applications

ONE OF THE BEST WAYS TO BECOME FAMILIAR WITH THE TECHNIQUES USED in switching-type power supplies is to study the application notes published by the large electronics manufacturers. In order to sell their products, these firms must both develop devices to fit new applications and conceive new applications to optimally utilize devices already developed. Only a well-funded business can finance and maintain such a program. Such firms have already done much of the circuit engineering that otherwise burdens the user of their products. In the specific case of switching-type supplies, it is surprising how much money, time, and effort are needlessly invested by smaller firms in the pursuit of circuit or system objectives already worked out in the superior facilities of the large companies. The knowledgeable worker in electronics can find some interesting and diverse circuits in these engineering notes, and he can adapt portions of them to his own needs or derive techniques somewhat modified from those presented.

General guidelines

With regard to the self-oscillating type of switching regulator, the following facts are relevant:

- The input noise and ripple riding on the unregulated supply line are well rejected.
- The output ripple voltage can be determined by the designed-in hysteresis of the comparator, but its lower limit is imposed by the ESR of the output filter capacitor.
- The resonant frequency of the LC filter is commonly no greater than about one-twentieth of the switching frequency.
- The ESR of the capacitor is a basic design parameter of the switching circuit. At the same time, it must be kept in mind that a low ESR prevents excessive temperature rise in the capacitor.

- The peak-to-peak current in the inductor must cause neither current discontinuity nor core saturation. A nominal excursion for this triangular wave would be about 10 percent above and below the dc load current; this is only a ballpark figure, but it is a reasonable one for many applications.
- The driven or synchronized switching regulator has, in many respects, a similar circuit configuration to that of the self-oscillating type. Both make use of a transistor switch, an LC output filter, and a free-wheeling diode. There are, however, significant differences in both design and operation:
- Ripple and noise on the unregulated supply line result in poor output regulation, since the unregulated supply adversely affects the dc regulation of the overall system.
- Unlike the self-oscillating mode of operation, driven operation does not impose any requirement for output ripple.
- The ESR of the filter capacitor is not a basic design parameter of the driven switch circuit, but the ESR should be low in order to avoid excessive temperature rise in the capacitor from ripple current I^2R heating.
- In the driven switcher, the loop gain is governed by the amplitude of the triangular-wave drive voltage, which is superimposed upon the reference voltage. Small drive amplitudes correspond to a high loop gain, which is desirable in order to suppress ripple from the unregulated power supply; however, excessive loop gain makes the comparator vulnerable to other noise sources and can therefore degrade regulation and stability.
- The bandpass response-time characteristic tends to be less than that for a similar self-oscillating switcher operating at the same switching rate. This adversely affects response and transient recovery time.

Neither the oscillating nor the driven switching regulator can remove switching frequency ripple by the action of the feedback loop.

Regulators using shunt switching

Conventional shunt switching regulators are characterized by dc feedthrough from the unregulated power supply. This means that the output voltage is available at higher, but not lower, voltage levels than that of the unregulated input voltage. The restriction can be circumvented by using an inductor with a secondary winding. The shunt transistor, in any case, must be able to withstand relatively high peak currents, and this is why the shunt regulator will be encountered only at relatively low power levels. Although there is no hard and fast rule, the shunt regulation scheme has rarely been used beyond the 150-W output level; exceptions have occurred where factors other than cost are important. At low power levels, however, useful objectives can be accomplished, such as producing an appreciable voltage stepup. This circuit has recently appeared in electronic calculators, watches, geiger counters, and fluorescent-light power sources. Filtering problems might occur because of the waveshape of the flyback pulses generated when the switch opens. On the other hand, some workers feel that this type of switching supply tends to produce less RFI than does the conventional series switcher.

Inverters and converters

When working with inverters and converters, the "apparently" superficial capacitors, RC networks, and diodes often associated with the basic circuit should not be ignored. These are generally despiking, snubbing, or energy-absorbing provisions used to protect the active devices from transients, turn-on surges, and excessively high rates of change in either voltage or current. Without these circuits, either faulty operation or catastrophic destruction can occur.

Unless stated, or schematically indicated, the output transformer of driven inverters ordinarily does not saturate. In the simple self-oscillatory inverter, the operation is dependent upon transformer saturation, but in more sophisticated types, the requisite saturation occurs in a small base-drive transformer, rather than in the large output transformer. Look for the saturating-core symbol on the schematic.

Switcher using linear voltage regulator

Both the similarities and differences of linear and switching-type regulators have received considerable attention. Indeed, the quick change from the dissipative to the switching mode, depicted in Fig. 8-5, might be considered interesting, but not too significant. It happens, however, that the ability to convert from linear to switching-type regulation has been fortuitous in the evolution of high-performance switchers. The National Semiconductor Corporation and other manufacturers of solid-state devices, have developed an extremely useful line of linear voltage regulators; these were specialized adaptations of previously developed operational amplifiers. The linear regulators include internal temperature-stabilized voltage references and external connections enabling such operational modes as current limiting, current foldover, remote shutdown, and drive to high-current booster stages. That these functions are contained in an IC module means that the overall performance greatly exceeds that which is readily attained by using discrete devices, hybrid modules, or even monolithic op amps.

Figure 17-1A shows the circuit of the LM205. The circuitry of the IC is not meant to duplicate a circuit made with discrete elements. A discrete layout of the indicated components would be attended by formidable conflicts of interest involving thermal stability, packaging, and economic considerations. Zener diode D1 develops the reference voltage, and the current for D1 is derived from one of the collectors of Q12. The use of a constant-current source for the operation of D1 is the first step in producing a very stable voltage reference. The zener voltage across D1 is not directly utilized; rather, it is buffered by Q10. The output of Q10 is applied to the series-connected array comprising Q1, R2, R3, R4, and Q8 (diode-connected transistor structures, such as Q1 and Q8, appear commonly in monolithic circuits). In this instance, the overall effect of the series circuit is to provide a reference voltage of 2.2 V to the base of Q2—with a nearly zero coefficient of temperature. It can be seen that Q2 is part of the comparator, consisting of Q2 and Q3. Transistor Q5 boosts the voltage from the output of the comparator, and Q14 provides drive for series-pass transistor Q15. Most of the voltage amplification is developed in Q5 because its collector load is essentially the high-impedance, constant-current source from the collector of transistor Q12. Other active devices

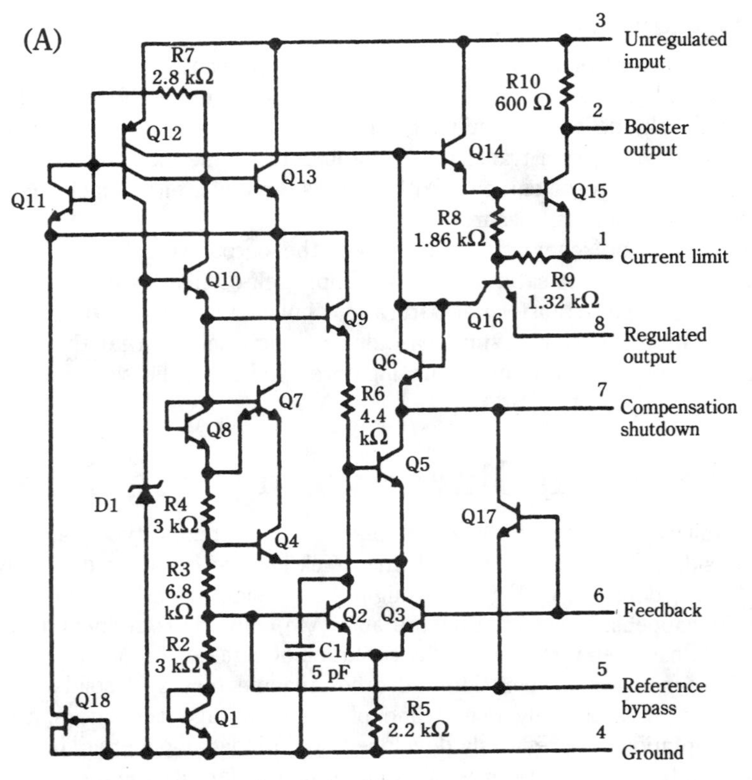

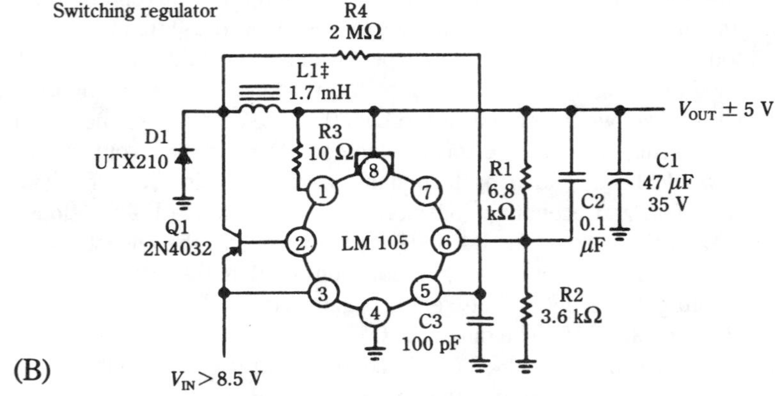

17-1 The internal circuitry of the LM105 voltage regulator and an application. (A) The semiconductor equivalent circuit of the LM105. (B) A switchmode regulator circuit using the LM105. This early IC can be considered a prototype of later dedicated control ICs, retaining much of this basic structure, but often endowed with more "bells and whistles."

National Semiconductor Corp.

are involved in auxiliary functions; for example, the field-effect structure (Q18) provides initial base bias to Q12 during startup, and thereafter Q12 receives its base bias from Q9. Transistor Q16 participates when current limiting is implemented in the external circuit; it chokes off base current to Q14, thereby allowing pass transistor Q15 to deliver no more than a predetermined current to the load. Transistor Q17 prevents paralysis of the voltage comparator, a situation arising from saturation of Q3 when overdriven.

All of the essentials of a switching-type regulator are present in the LM105. Specifically, this is an excellent voltage-reference source, a voltage comparator, error-voltage amplification, and a series-connected output transistor. This leads naturally to the switching regulator shown in Fig. 17-1B. The external transistor (Q1) extends the current output to several hundred milliamperes. This circuit constitutes a basic approach for many switching regulators. It should be appreciated that the current-boost technique provided by the external transistor can be extended by cascading progressively larger transistors. Of course, the current-carrying capacity of L1 and D1 must also be appropriately increased. Hundreds of amperes can be precisely and efficiently regulated by this cascading technique.

There are two feedback paths in the circuit of Fig. 17-1B, as well as in most self-oscillating regulators. One of these, a negative feedback path, brings about regulation of the output voltage by minimizing the error signal from the comparator. Although the action is somewhat different, the effect is very similar to that of a series-pass linear supply, and the actual circuit connections are also similar in the two types of regulators. The negative feedback is obtained through the output sampling network, R1, R2, and C2, and is applied to pin 6, the inverting input of the comparator. Capacitor C2 provides a low-attenuation feedback path for ripple and noise, thereby reducing the level of these ac components on the output line. The ripple primarily affected is that contributed by the unregulated supply, rather than that corresponding to the switching rate.

The second feedback is positive in nature and causes the self-oscillation of the system. The positive feedback stems from the connection of R4 from the input of the inductor to the noninverting input of the comparator, pin 5. Resistor R4 also establishes the peak-to-peak ripple voltage because its value governs the hysteresis of the comparator. Regulation is achieved by varying the switching duty cycle, or by changing the switching rate or the duration of on time. The voltage level of the regulated dc output can be adjusted by means of the R_1/R_2 ratio, as in a linear regulator.

The basic concepts embodied in this switching-type regulator are similar to many others. Various modules do not all coincide in pin numbers; for example, in the LM723 type, the voltage-reference source is not internally connected to the voltage comparator; rather, it appears at a pin, where it can be used for various circuit purposes.

Negative-voltage switching regulators

Although a voltage regulator, such as the LM105, is intended to be used for a positive output polarity, it can also be adapted to deliver a negative output. Still better results and easier implementation can be obtained by using ICs specifically designed for negative output voltages. An example of such a voltage regulator is the LM104, shown

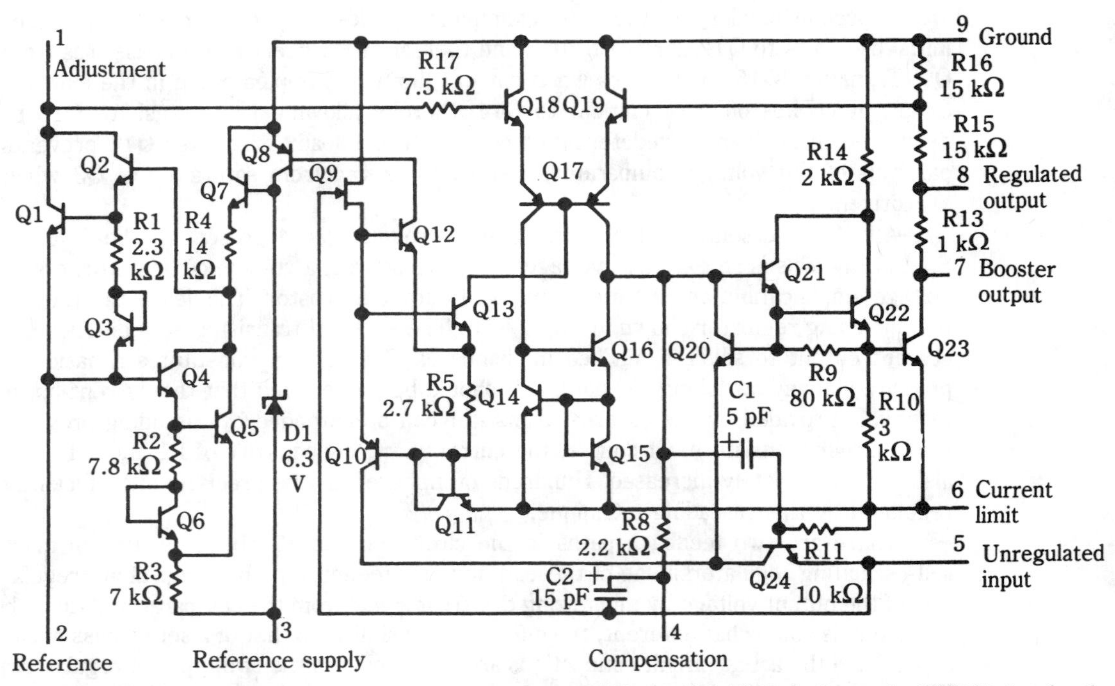

17-2 The internal circuitry of the National Semiconductor LM104. This IC is similar to the LM105, but for the convenience of the user, is simply and directly applicable to regulate a negative voltage.

in Fig. 17-2. Two negative-output switching regulators are shown in Fig. 17-3. The circuit philosophy and applications are very similar to the LM105.

In the high-current circuit of Fig. 17-3B, the extended performance is derived from several modifications. Both the inductor and the free-wheeling diode must have greater current capability. Notice that the reference supply terminal is no longer returned to the unregulated input, but is connected to the base of output transistor Q2. This stratagem is used in order to prevent pin 5 from becoming more than 2 V positive, with respect to pin 3—a condition that would adversely affect the operation of the LM104. Although these modifications accomplish its primary objective, they might produce an undesirable side effect from some applications because the line regulation is degraded by the unregulated input voltage being injected into the voltage reference source. Fortunately, this effect can be eliminated by inserting a 0.01-μF capacitor in series with positive-feedback resistor R6. The capacitor is large enough to have negligible effect upon the comparator hysteresis at the oscillation frequency, but blocks the dc component of the feedback.

As current capability is increased, the need for fast-responding semiconductor devices becomes more pressing. Not only is more heat evolved from slow elements, but at high current levels, the severity of circuit disturbances from high peak currents and chassis loop currents mounts rapidly. Low-frequency workhorse transistors and free-wheeling diodes should not be used. The junction charge storage of ordinary rectifying diodes not only results in high dissipation in the diode itself, but constitutes a short

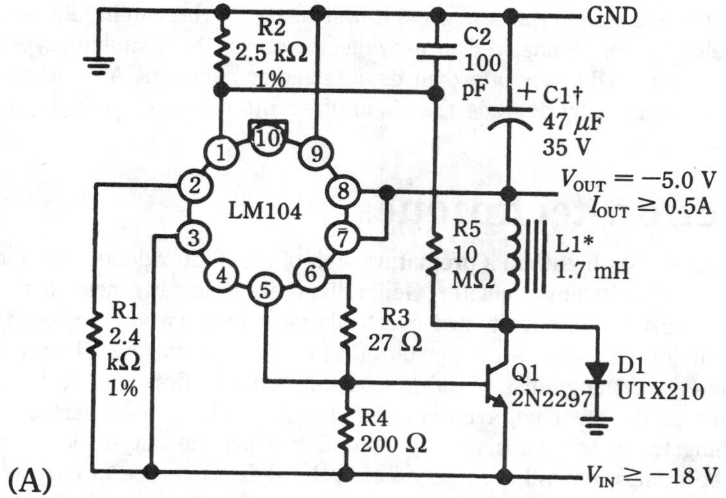

†Solid tantalum *125 turns No. 22 on Arnold Engineering A262123-2 molybdenum permalloy core

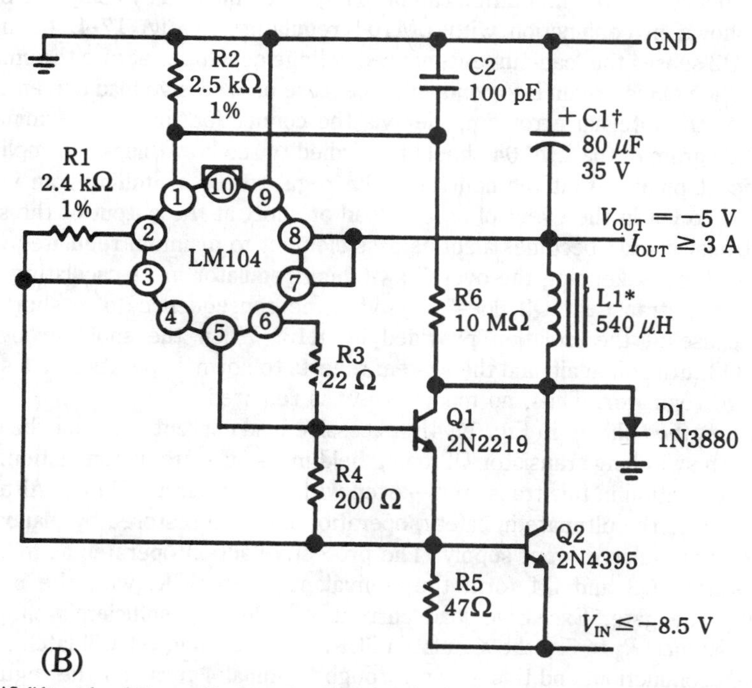

†Solid tantalum *60 turns No. 20 on Arnold Engineering A930157-2 molybdenum permalloy core

17-3 Negative voltage switching supplies. (A) A single-boost transistor design provides negative 5-V output. (B) Design with two boost transistors for increased output current. National Semiconductor Corp.

circuit for the switching transistor when it first turns on (this can hardly be classified as "free-wheeling"). As emphasized in previous chapters, the switching-type supply requires the coordinated action of proper devices and components. A single bargain-basement item can harmfully degrade the chain of events required for high efficiency and clean operation.

Protection techniques

Many a designer has breathed a premature sigh of relief after debugging and refining a newly developed switching regulator. Generally, there is an awareness that some sort of protection against short circuits and overloads must be incorporated, but this is often handled as an afterthought—a minor bit of last-minute patchwork. However, a protective function is anything but a trivial problem. A common first approach is to limit the current through the switching transistor by starving its base of forward drive. Because the switching transistor tends to stop switching when the output of the regulator is shorted, the dissipation will then be excessively high. This type of current limiting is often applied to linear regulators, but it will not provide the requisite protection here. Either the switching operation must continue or the switching transistor must be turned off.

Many hours of experimentation can probably be eliminated by using one of the two schemes shown in conjunction with LM104 regulators in Fig. 17-4. In method A, transistor Q3 senses the load-current-induced voltage developed across the small resistance R9. When Q3 is driven into conduction because of excessive load current, it exerts control over the internal error amplifier via the connection made to terminal 8. The schematic diagram of the LM104 should be studied to see how this is accomplished, but the significant point is that the control of the regulator by switching transistor Q2 is relinquished to Q3 in the event of an overload or short at the output of the supply. In essence, transistor Q3 becomes a series-pass element to maintain regulated voltage at terminal 8, thereby keeping the overall switching regulator in its oscillating mode. At the same time, transistor Q3 does not suffer the consequences of a short-circuited output because of the isolation provided by R13. When the short or overload is removed, Q3 turns off again and the system reverts to normal operation as a switching-type voltage regulator. Thus, no manual reset is required.

In the method shown in Fig. 17-4B, excessive load current turns off the switching process, with switching transistor Q2 being held in its off state of conduction. There is negligible dissipation in this transistor—even with a short-circuited load. Although this scheme provides the ultimate in safety, operation must be restored by manually interrupting the unregulated input supply. The protective action operates as follows:

Transistors Q3 and Q4 form the equivalent of an SCR, with the base of Q4 constituting the gate. Excessive load current will develop sufficient voltage across sensing resistance R_8 to fire the simulated SCR. Then, Q3 and Q4 will latch each other in saturated conduction, and this action through terminal 4 turns off the regulators, by depriving transistor Q21 of forward bias. Here is an example of versatility because terminal 4 not only is ordinarily used for frequency compensation, but it serves well for a shutdown technique. These protection circuits can be readily adapted to other regulator ICs.

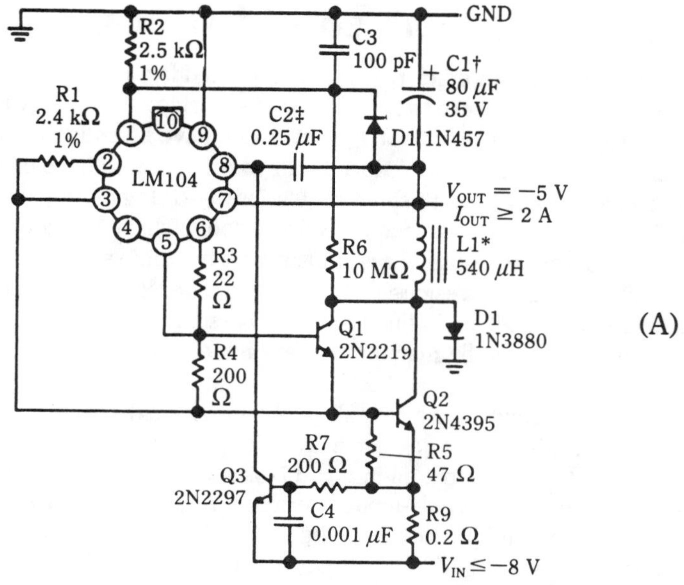

‡Ceramic or paper †Solid tantalum *60 turns No. 20 on Arnold Engineering A930157-2 Molybdenum Permalloy Core

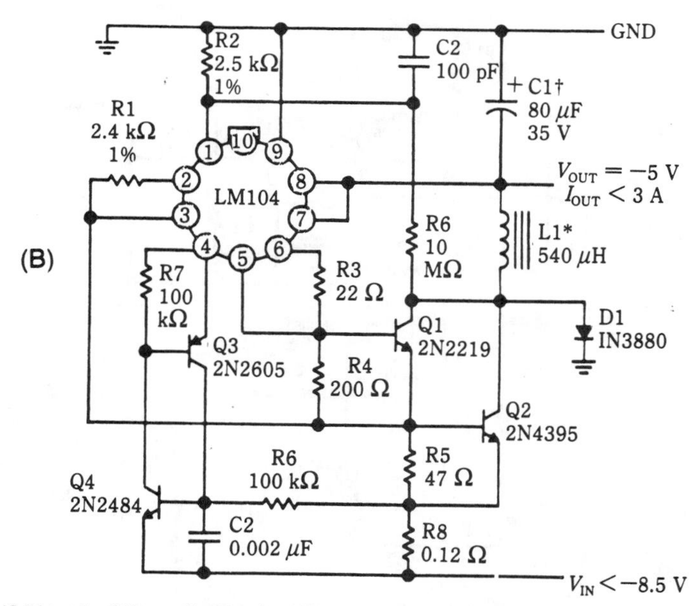

†Solid tantalum *60 turns No. 20 on Arnold Engineering A930157-2 molybdenum permalloy core

17-4 Two overload-protective techniques for the LM104 switching circuit. (A) Protection is imparted by current limiting; oscillation is maintained. (B) Protection results from a shutdown of the switching process with the switching transistor hold in its off state. National Semiconductor Corp.

A switching/linear combination

If you are not concerned about efficiency and such related factors as heat removal, cost, and packaging dimensions, there would be no dissatisfaction at all with the linear regulator. The proponents of regulation by dissipation have investigated many avenues whereby excess dissipation could be reduced. For example, there have been designs for automatic switchover from series to shunt regulation at strategic load-current values. A more popular approach is to use a second feedback loop, which functions to maintain a near constant drop across the series-pass transistor. The prime source of inefficiency in these rheostat-like regulators resides in the voltage differential between unregulated input and regulated output. The dissipation in the pass transistor will be minimal if the unregulated input voltage can be maintained just high enough to ensure proper regulator action. Further, there can be no objection to regulating the dc input voltage because such preregulation must add to the overall performance of the regulating system.

This naturally leads to the use of a switching regulator followed by a linear regulator. Such a system is illustrated in Fig. 17-5. Two LM105 modules are used in a configuration, which, despite its dual-mode operation, retains the basic circuit simplicity that these modules generally provide. Despite the 25-V differential between the nominal input voltage and the regulated 5-V output, the efficiency of this regulating scheme

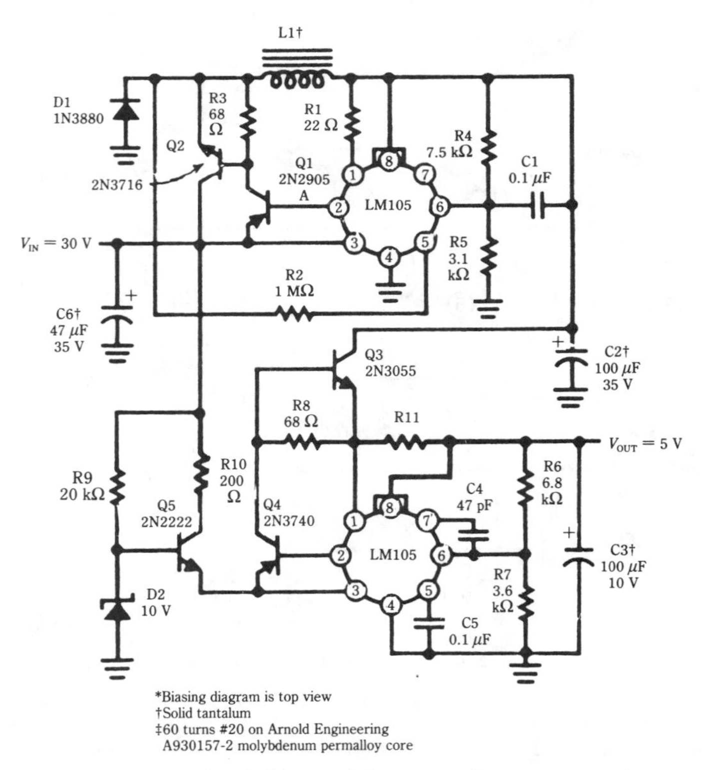

17-5 A combined switching and linear supply. National Semiconductor Corp.

can range between 60 and 70 percent. The network formed by Q5, D2, R9, and R10 comprises a constant-voltage source, which isolates the linear regulating circuitry from noise and transients on the unregulated dc line. Many commercial power supplies use such a scheme to achieve high operating efficiency, together with low output ripple and fast transient response. To scale a design for greater output current, the current capabilities of both the switching and the linear regulators must be increased. This is obvious enough, but still a point to be watched when ordering the highest current rating from any family or group of commercial power supplies.

A simple switching regulator

A natural reaction to a diagram devoid of the usual complexities is that marginal performance must ensue. Yet, the straightforward switching regulator shown in Fig. 17-6 does not involve tradeoffs with basic power-supply parameters. Inspection of the schematic diagram of the LM341 shows that despite the fact that only three connections emerge from this IC, the internal circuitry is like that of the high-performance LM104. Incorporated in the design of the LM341 is complete protection, including both current limiting and thermal shutdown.

The LM341 was initially intended for use as a point-of-load or on-card linear regulator. The use of several regulators physically situated close to the loads, deserves serious consideration in large installations. Too often, the splendid dc and ac regulating characteristics of a sophisticated and costly source of operating power is considerably degraded by the resistance and inductance of the connecting leads. It might be advantageous to relax the specifications on the main power supply and to use small point-of-load regulators where necessary. Such localized regulation has usually been accomplished by linear regulators; however, there are instances where a small, easily applied switching regulator would be very desirable. Although low efficiency tends to be ignored at low power levels, the cumulative dissipation of many small linear regulators could prove undesirable. The three-terminal switching circuit might provide an acceptable solution, but much depends upon the nature of the load, primarily its susceptibility to high-frequency ripple and RFI resulting from the switching process.

A point of caution is in order: the tab of the package is electrically identical with terminal 3. In the switching circuit shown in Fig. 17-6, terminal 3 is not directly grounded. Therefore, it would not be permissible to use a mounting method in which the tab makes electrical contact with the chassis in most situations. Similar reasoning pertains to the pnp switching transistor—its collector and case (if metallic) must not be inadvertently grounded.

Self-protected switching regulator

The switching regulator shown in Fig. 17-7 involves no significant differences from those already covered. The basic principles of operation are the same as those already described for self-oscillating switching regulators. However, the use of a new semiconductor device as the switching element gives this regulator a new dimension of performance. The LM195, shown in simplified circuitry in Fig. 17-7, has only three terminals

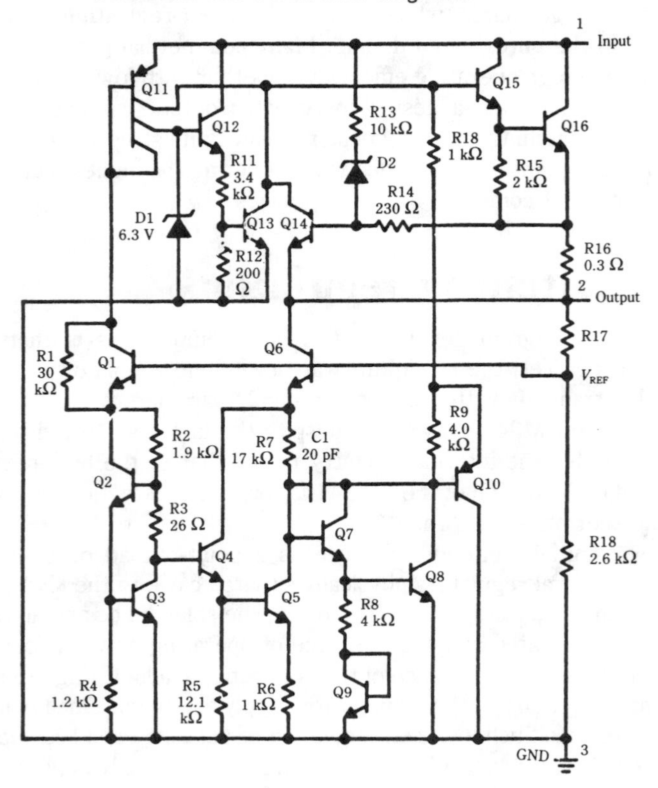

*Solid tantalum
**Needed for stability
***Heatsink Q1

$R_2 = 0.5\ \Omega$, $f \sim 45$ kHz, Ripple ~ 17 mW
$R_2 = 1\ \Omega$, $f \sim 25$ kHz, Ripple ~ 35 mW
Load = 500 mA

Switching regulator

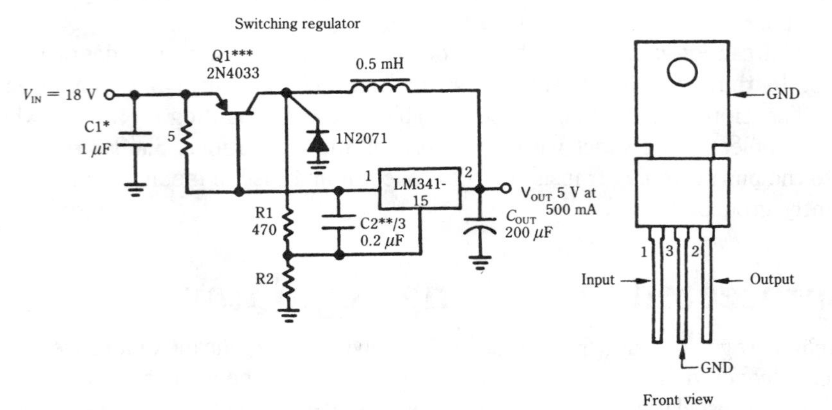

17-6 A switching regulator utilizing the LM341 three-terminal IC. National Semiconductor Corp.

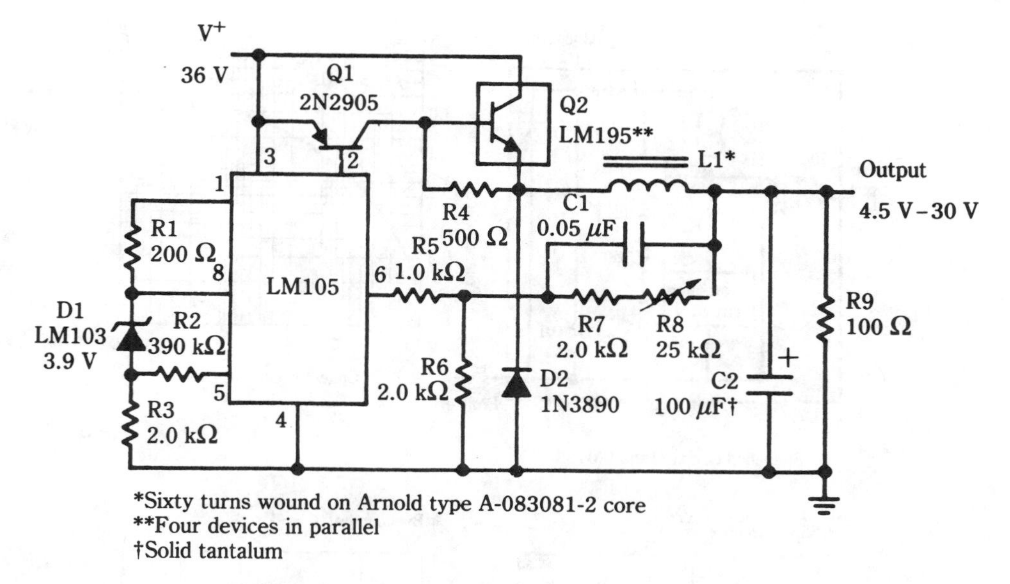

17-7 A self-protected switching regulator using the LM195 monolithic power transistor. The important function of overload protection is neatly provided through the use of this simulated power transistor. National Semiconductor Corp.

and physically resembles a conventional power transistor, and it is used in circuits very much like ordinary transistors. But there the resemblance ends, for this product is in a class by itself—it cannot be considered as just another power transistor.

This monolithic power module includes provisions that limit its current and power output and shut it off in the event of excessive temperature rise. Once properly incorporated in a circuit, the LM195 is virtually immune to destruction from any type of overload. Such behavior is particularly welcome for application to switching-type power supplies, where reliable overload protection is considerably more difficult to achieve. In conventional approaches, it is not uncommon to encounter difficult problems of interaction between the switching circuit and the added protective circuitry. This is an area where compromise is hardly desirable.

In addition to its unique self-protection features, the LM195 is a high-performance device at the currents and frequencies that are useful for switching regulators. The high transconductances shown in Fig. 17-8 were measured at 50 kHz. This contrasts favorably with many ordinary power transistors, which require high driving power. The inherent protective features of the LM195 are also depicted. Not shown is the fact that the "base" of this device can be driven up to 40 V without causing breakdown, and this too facilitates the design of a switcher. The actual monolithic circuitry is presented in Fig. 17-9. This exemplifies the clear superiority of ICs for complex functions, in which discrete devices would be expensive, bulky, and impractical.

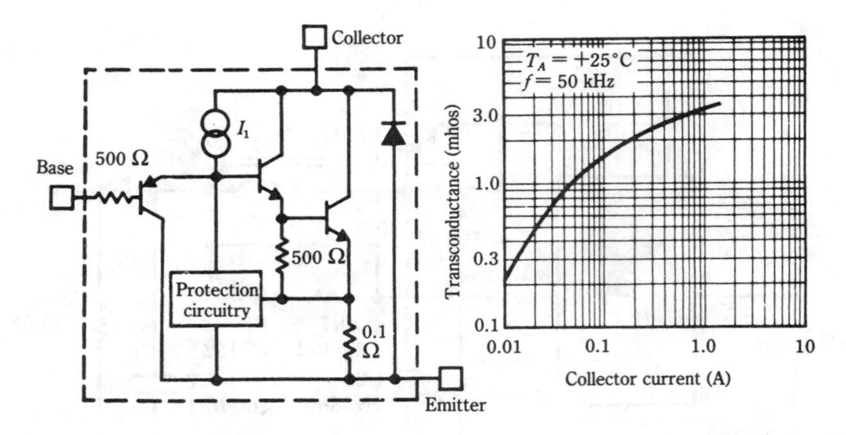

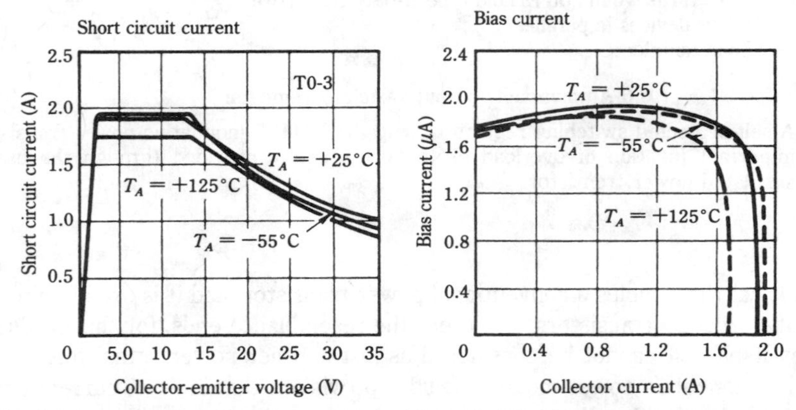

17-8 The LM195 monolithic power transistor. The simplified circuit shows the three "transistor" terminals. Curves display high transconductance range at 50 kHz and its self-protection characteristics. Power dissipation is limited to 40 W. National Semiconductor Corp.

A 100-W switching regulator with shunt chopper

Otherwise known as a flyback regulator, this circuit is most commonly seen at low power levels. Over about 150 W, such a system becomes uneconomical because of the large peak currents the switching or chopping device must withstand. However, good performance is readily achieved from this technique, and it can deliver regulated dc at a higher voltage level than that obtained from the unregulated supply. For some applications, this is a compelling feature.

The circuit shown in Fig. 17-10 is a straightforward shunt switcher. It is essentially a constant-frequency, variable-pulse width scheme. In order to accomplish this operational mode, an oscillator and a duty-cycle-controlled multivibrator are used. The

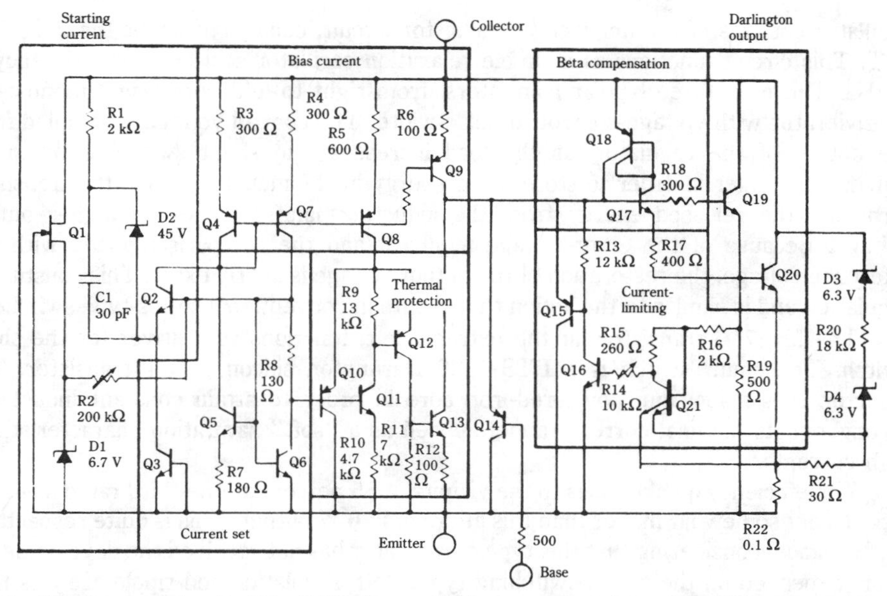

17-9 The internal circuit of the LM195 monolithic power transistor. In addition to performing the role of power transistor, this unique device also confers protection against its own destruction from overloads, and from excessive temperature rise. National Semiconductor Corp.

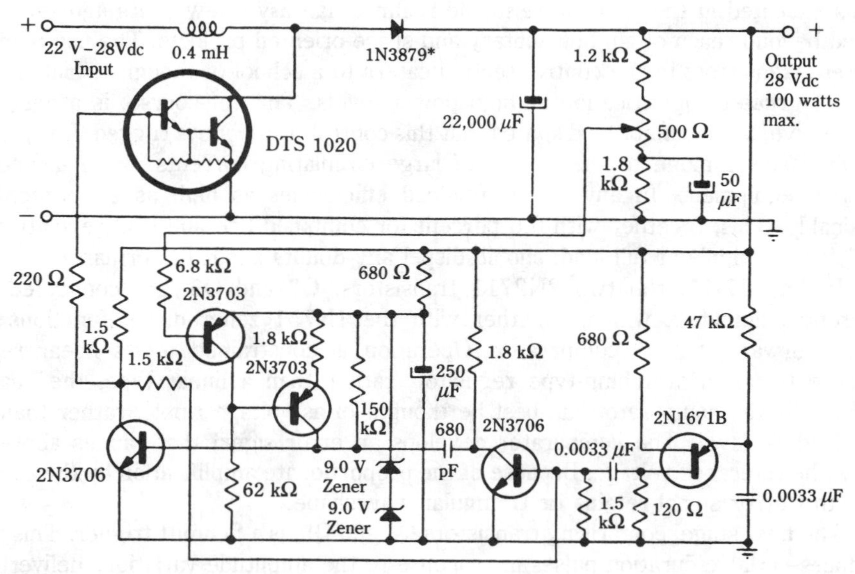

17-10 A 100-W flyback switching regulator. The 0.4-mH choke consists of 124 turns of 17 AWG wire on an Arnold B079024-3 core. Delco Electronics Division.

oscillator consists of the unijunction transistor circuit, configured about the 2N1671B UJT. This circuit functions as a simple relaxation oscillator at a nominal frequency of 9 kHz. The next three bipolar transistors, from right to left, comprise a monostable multivibrator with voltage-controlled pulse width. The control voltage is sampled from the output of the regulator; as the load increases, the shunt switching transistor remains on longer in order to store more energy in the inductor. When the transistor turns off, the released energy from the inductor manifests itself as higher output voltage. Because of the error-voltage feedback and the comparison made with the reference voltage, the restoration of the output voltage is nearly exact. This constitutes regulation and is similar to the action that occurs in conventional series-type switchers.

The 2N3706 transistor, at the extreme left, functions as a driver for the shunt switch. The shunt switch is a DTS-1020 Darlington silicon power transistor. The inductor uses a low mu, powdered-iron core in order to attain constant inductance throughout its nominal current range, as well as a "soft" saturation characteristic at high currents.

The efficiency quickly rises to the vicinity of 85 percent at one-third rated load, and it peaks out somewhat higher than this at around 80 W output. This is quite respectable performance, considering that this type of regulator has not received the developmental effort expended on the series-switching type. Both regulation and ripple are less than 1 percent at the rated 100-W output level.

High-voltage switching regulator

The relatively simple circuit of Fig. 17-11 exemplifies the performance that is now attainable with high-technology transistors and diodes. Both the voltages and the power levels indicated in the chart were in the realm of fantasy a few years ago—or were priced beyond reach of all, but military and space-oriented projects. The use of simple discrete transistors in the control section caters to a school of thought which sees no need of greater sophistication at high power levels. This, of course, is a designer's prerogative, but it cannot be disputed that this control technique is rugged. It is perhaps readily made immune to the effects of large circulating currents, often induced by external equipment. In any event, full-load efficiencies as high as 92 percent are obtainable. This, together with 0.6 percent for combined line and load regulation and 0.75 V peak ripple at full load, should dispel any doubts about performance.

In Fig. 17-11, the two 2N2711 transistors, Q2 and Q3, are connected as a differential amplifier, which, together with the 1N751 zener diode, functions as a straightforward voltage comparator. Operation is constrained to its linear region. Because this is a switching-type regulator, rather than a linear type, the quantity sampled by the comparator can best be thought of as voltage ripple, rather than as a sustained dc level. The comparator develops an error signal that ranges above and below the reference voltage. Because of the proportionate amplification by the comparator, this error signal retains its triangular waveshape.

The next stage, comprising transistors Q4 and Q5, is a Schmitt trigger. This stage produces variable-duration pulses in response to the amplitude variations delivered by the comparator; the overall control circuitry thus behaves as a pulse-width modulator.

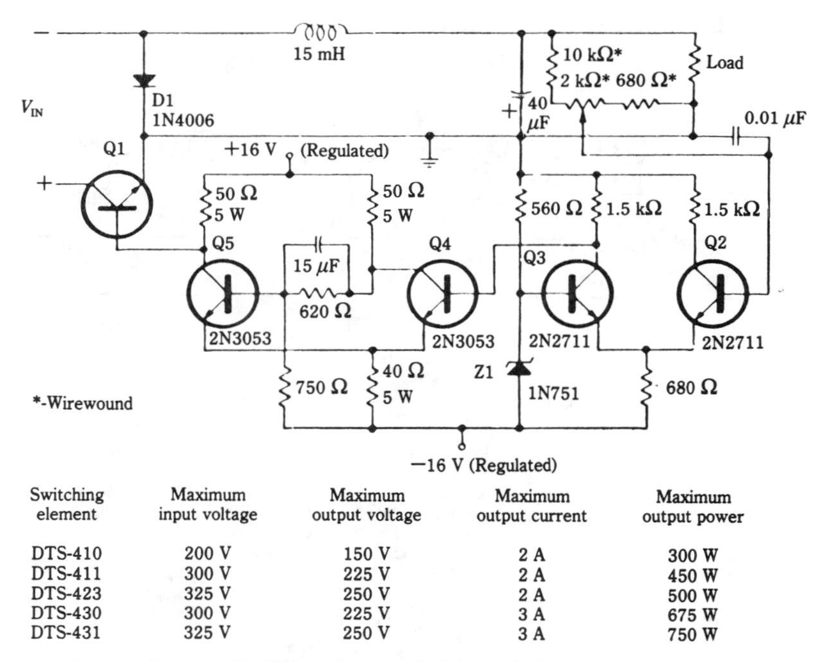

Switching element	Maximum input voltage	Maximum output voltage	Maximum output current	Maximum output power
DTS-410	200 V	150 V	2 A	300 W
DTS-411	300 V	225 V	2 A	450 W
DTS-423	325 V	250 V	2 A	500 W
DTS-430	300 V	225 V	3 A	675 W
DTS-431	325 V	250 V	3 A	750 W

17-11 A versatile high-voltage switching regulator. Delco Electronics Division.

Switching transistor Q1 is directly driven by the Schmitt trigger stage. The final act in this sequence of events is output-voltage regulation by control of the switching duty cycle. The nominal frequency of self-oscillation of this supply is in the vicinity of 7 kHz.

All-discrete 1000-W switching regulator

The circuit shown in Fig. 17-12 provides a regulated output of 500 V at 2 A. The configuration is unencumbered; its simplicity is suggestive of circuit techniques commonly used at lower power and voltage levels. The fixed switching frequency is 13 kHz, and is established by the UJT Q7. Transistor Q6 functions as a sawtooth generator, being triggered from the unijunction stage. The sawtooth waveform is modulated by dc voltage from Q8, which in turn is controlled from the output of the error-sensing differential amplifier, comprised of transistors Q9 and Q10. The consequence of the dc modulation is that the dc level of the sawtooth wave appearing at the base of transistor Q5 varies with the error signal. The error signal is generated by the input differential amplifier, Q9 and Q10, as the difference between the sampled output voltage and the reference voltage established by zener diode D4. Transistors Q5 and Q4 function as a pulse-width modulator because the varying dc level of the sawtooth wave causes Q5 to respond to varying portions of the sawtooth. Because of saturation in the pulse-width modulator, the output consists of a rectangular waveform with on and off times dependent upon the dc component of the sawtooth driving wave. Except for power level, the

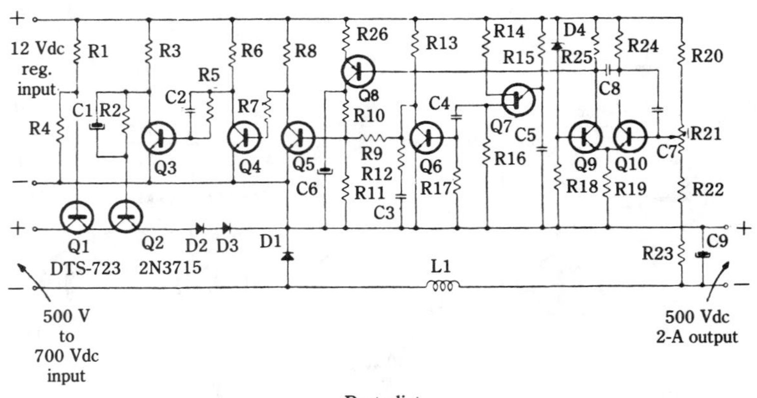

Parts list

Q1 DTS-723
Q2 2N3715
Q3, Q4, Q5, Q6, Q9, Q10 2N3706
Q7 TIS 43
Q8 2N3703
D1 See text
D2, D3 1N4001
D4 5.6 V zener
R1 6.8 Ω, 25 W (noninductive)
R2, R12 2.7 Ω
R3 47 Ω, 2 W
R4 100 Ω, 1 W
R5 560 Ω
R6, R17 1.5 kΩ
R7 8.2 kΩ
R8, R24, R25 5.6 kΩ
R9, R13, R20 2.2 kΩ
R10 6.8 kΩ
R11 1 kΩ
R14 620 Ω
R15 47 kΩ
R16 220 Ω
R18 10 kΩ
R19 15 kΩ
R21 2 kΩ potentiometer
R22 22 kΩ
R23 200 kΩ, 5 W (wire-wound)
R26 820 Ω
C1 100 μF
C2 0.082 μF
C3 0.33 μF
C4 0.033 μF
C5 1500 pF
C6 5 μF
C7, C8 0.1 μF
C9 2 μF
L1 7 mH

17-12 A 1-kW switching regulator featuring simple design. Delco Electronics Division.

pulse-width-modulated wave from Q4 is now satisfactory for controlling the duty cycle of the switching process.

Transistor Q3 is a predriver. Transistor Q2 is the driver stage for switching transistor Q1. The duty cycle of Q1 is varied in such a manner as to produce a near constant output voltage. The combination of Q2 and Q1 is of special interest—Q1 is connected in the common-base configuration with its emitter driven from Q2, which is a fall-time speedup technique. When Q1 is turned off, its emitter-base section is avalanched, quickly depleting the stored base charge.

The minimum requirements for free-wheeling diode D1 are 800 PIV, 3 A peak repetitive forward current, and 0.5 μsec minimum reverse-recovery time. The combination of a fast diode and rapid turn-off switching transistor mades possible an operational efficiency in the 90 percent range. The load regulation is on the order of ±0.5 percent with most of the change occurring at loads exceeding 1.5 A. The circuit could be scaled for operation at 20 kHz, but the output should then be restricted to 500 W.

5-kW switching regulator

Any circuit-oriented worker in electronics is bound to have his interest stimulated by the unusual topography of the 5000-W switching regulator shown in Fig. 17-13. Not only does the schematic depart from the beaten path, but the operating parameters are also uncommon. For example, despite the tremendous power capability of this switcher,

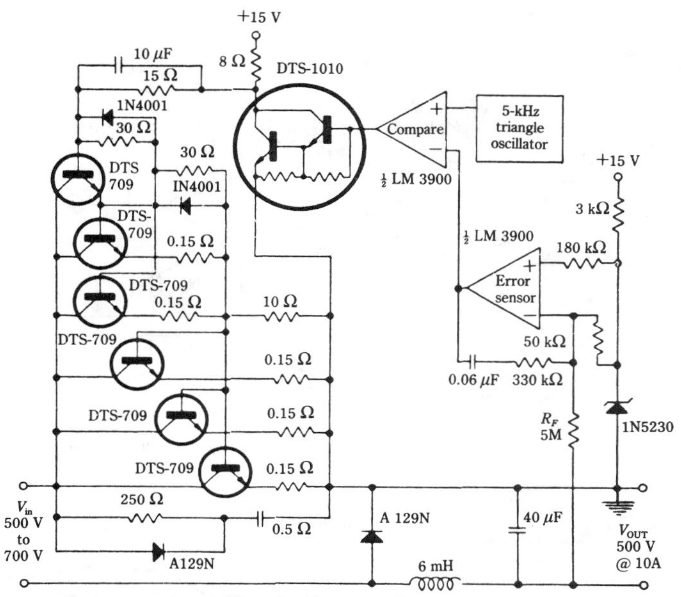

17-13 A 5-kW switching regulator. Delco Electronics Division.

the efficiency is in the vicinity of 95 percent from 3 kW to full output, and for all loads greater than 5000 W, the efficiency exceeds 90 percent. Line regulation is better than 0.1 percent over the full input-voltage swing from 500 to 700 V. And load regulation is less than 0.5 percent when the resistive load is changed from 500 W to 5 kW. This combination of outstanding performance parameters reflects the excellence of both the Delco switching transistors and the National Semiconductor quad amplifier used in the control section.

The LM3900 quad amplifier differs from conventional operational amplifiers; a more detailed look into its circuitry and its applications relevant to the regulator is provided in Fig. 17-14. The *current-mirror principle* enables simulation of a differential-amplifier input circuit; the input-biasing current, however, is inordinately low. Operation is generally restricted to ac signals, but all of the benefits of direct coupling are retained. This configuration develops high open-loop gain, with a relatively low number of active elements. Implementation is made convenient because of internal frequency compensation (for unity gain) and by virtue of the single power-supply requirement.

The regulator contains a "snubber" circuit, comprising diode AT129N, the 250-Ω resistor, and the 0.5-μF capacitor. This network safeguards the three parallel-connected DTS-709 switching transistors, maintaining their operation within the bounds of reverse bias, second breakdown. These three output transistors are driven by two parallel-connected DTS-709s, and they, in turn, receive drive from a single DTS-709. The array of DTS-709 transistors comprises a triple-Darlington arrangement. The predriver, an integral Darlington DTS-1010, provides voltage amplification of the width-modulated pulses received from the comparator. Thus, this regulator is actually quite straightforward. Its dollar-cost per kilowatt would certainly meet low-cost re-

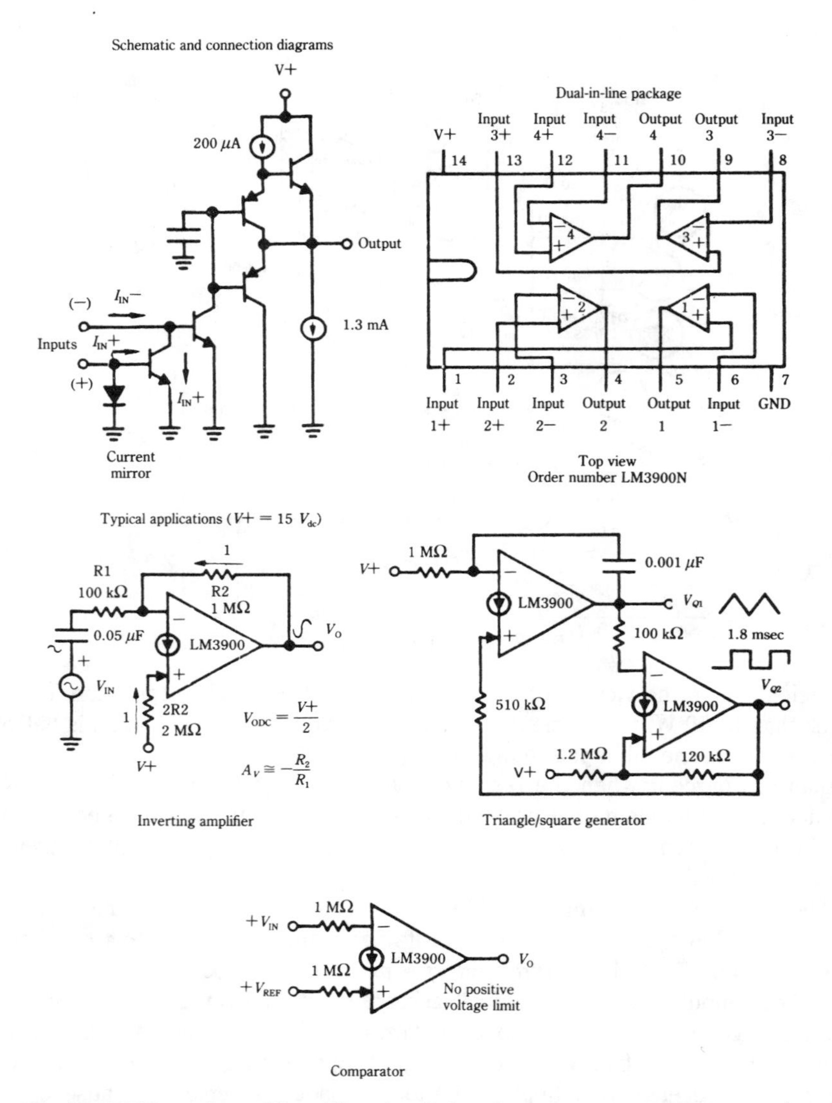

17-14 The LM3900 quad amplifier. The circuit functions shown are relevant to the control section of the 5000-W switching regulator that is depicted in Fig. 17-13. National Semiconductor Corp.

quirements. Even greater economy can be obtained by operating from a 480-V, 3-phase, full-wave-rectified line, which uses minimal filtering.

Advanced-design switching regulator

The switching regulator shown in Fig. 17-15 merits coverage because it embodies a combination of the features that have appeared singly in a number of other circuits. One

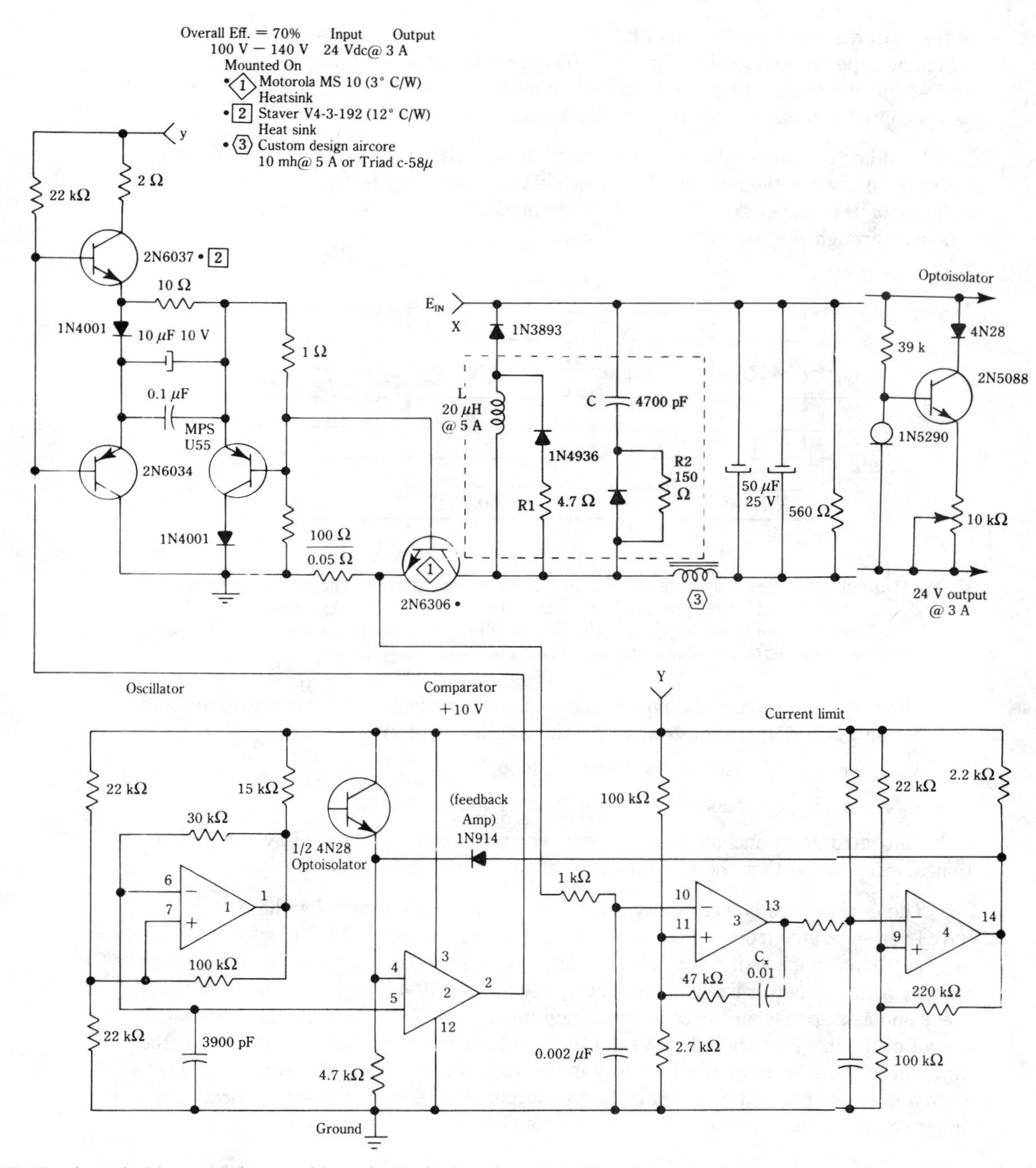

17-15 A switching regulator with unique design features. Regulator incorporates load-line shaping shaded area, optoisolator feedback, quad-function IC package, overcurrent protection, and direction operation from the ac power line. The unregulated and auxiliary power supplies are shown in Fig. 17-17. Motorola Semiconductor Products, Inc.

of the features is not only unique, but it represents a significant advance in the art of switching-type power supplies. This *load-line shaping network* comprises the elements enclosed in the shaded area. Its function will be described, but first, however, it would be appropriate to list the features of this regulator:

Load-line shaping—this is an important evolutionary step in switcher technology. Suffice it to say, for the moment, that much cleaner switching is accomplished with the addition of this network. Figure 17-16 dramatically illustrates the noise reduction attained through the use of this technique.

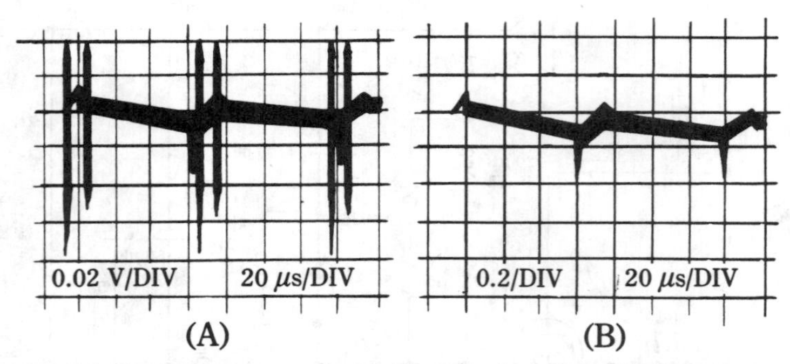

17-16 Output noise cleans up because of a load-line shaping network. The switching transistor is protected and noise generation is greatly reduced. (A) The regulator of Fig. 17-15 ***without*** a load-line shaping network. Notice "normal" switching spikes. (B) The same measurment ***with*** a load-shaping network. Motorola Semiconductor Products, Inc.

Direct operation from the power line—there is no bulky 60-Hz transformer; and despite the 20-kHz operating frequency, there is no inverter.

An optoisolator is used in the feedback loop.

Short-circuit protection and current limiting are provided.

All control logic, including the overcurrent protection, is provided by a quad-function IC package so that the circuitry is relatively unencumbered.

The unregulated power supply, together with the 10-V supply for the IC sections, have been separated from the regulator and are shown in Fig. 17-17. Transformer T1 is very small inasmuch as it has approximately a 15-VA rating. The main power is drawn directly from the power line and rectified by the MDA-980-4 bridge. The connections of these supplies appear on the schematic diagram of Fig. 17-15. Because the nominal dc output of the main supply is 160 V, and the regulator provides 24 V, it follows that the switching process occurs at a low duty cycle, with the on time being relatively short. Such a great differential between input and output voltage would be impractical with a linear regulator intended to deliver 24 V to a 3-A load.

The basic principles of load-line shaping

The objective of the load-line shaping network shown in the shaded area of Fig. 17-15 is to divert rise-time and fall-time switching losses from the switching transistor (these

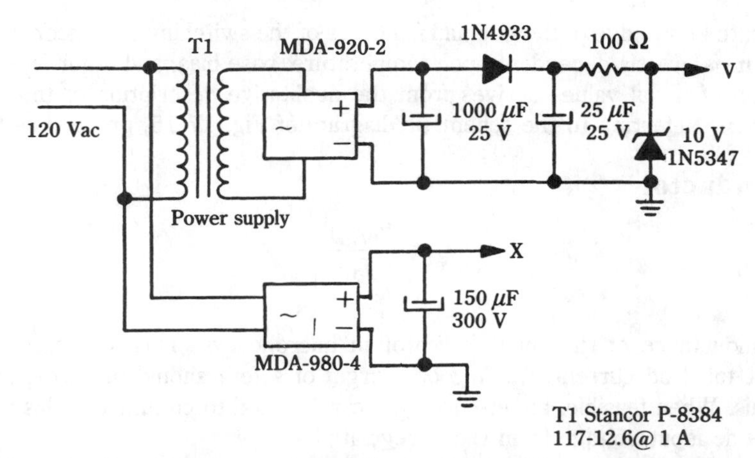

17-17 Unregulated and auxiliary power supplies for the switching regulator of Fig. 17-15. The MDA-980-4 and the MDA-920-2 are bridge rectifiers. Connections are made to the appropriate points labeled "X" and "Y" in the schematic of Fig. 17-15. Motorola Semiconductor Products, Inc.

losses are designated as areas 1 and 3 in Fig. 13-1). The energy that would otherwise be dissipated in the switching transistor is stored in the reactances of the shaping network until the full on or off conditions are attained. Then, this energy is dissipated in the resistors of the shaping network. Inasmuch as energy is dissipated in this way, this technique does not directly increase the overall efficiency of the switching regulator; it makes little difference whether energy is dissipated in the switching transistor or is diverted to resistors, but the indirect advantages are of considerable worth, the SOA requirement of the switching transistor is greatly relaxed, permitting this transistor to operate under more reliable conditions. In a sense, the load-line shaping network can be viewed as an electronic heatsink—heat that would otherwise have to be removed from the switching transistor is released by the shaping-network resistors.

The function of the small 10-μH inductor is to absorb voltage during the finite turn-on period of the switching transistor. In effect, the application of dc input voltage E_{IN} is delayed until the switching transistor is fully on. The diode/resistor combination clamps the induced voltage during turn off, but permits the inductor current to build up again well within "free-wheeling" time; that is, when the switching transistor is off. During the transition time from off to on, the switching transistor is spared the need to simultaneously operate at high current and high voltage.

The function of the small 4700-pF capacitor is to provide load current during the finite turn-off period of the switching transistor. This current does not have to be provided by the switching transistor. The diode/resistor combination used with this capacitor permits the capacitor to charge well with on time, and to act as a current source during fall-time, when the switching transistor is turning off. During the transition time from on to off, the switching transistor is also spared the need to simultaneously operate at high voltage and high current.

The design of the load-line shaping network is not at all difficult; however, empirical techniques will usually be needed for final optimization because you will not likely to have accurate knowledge of the rise and fall times of the switching transistor. Fall time, in particular, is greatly dependent upon temperature, base bias, and other factors. The computation of initial values derives from the qualitative description of the network already given. Referring to the schematic diagram of Fig. 17-15, proceed as follows:

Shunt inductor

$$L = \frac{V_{IN}(t_r)}{I_O}$$

Where:

L = inductance of the shunt indicator in microhenrys. This inductor must not saturate at full load current, I_O. A good margin of safety should be incorporated to prevent this. When feasible, an air-core type can be used to circumvent this problem.

V_{IN} = dc input voltage from the unregulated supply

t_R = rise time of the switching transistor in microseconds—best obtained by measurement

I_O = maximum load current in amperes

Resistor in the inductive arm

$$R1 = \frac{L}{t_A}$$

Where:

R_1 = resistance in ohms

L = inductance of the shunt inductor in microhenrys

t_A = time interval in microseconds (short compared to "free-wheeling" or off time.) As a start, t_A can be taken as one-tenth the minimum off time involved in the operation of the regulator.

The product $R_1 \times I_O$ yields the voltage overshoot at the collector of the switching transistor because of shunt inductor L. If this is greater than desired, the value of R1 must be reduced.

Shunt capacitor

$$C = \frac{I_o(t_F)}{V_{IN}}$$

Where:

C = capacitance in microfarads

V_{IN} = dc input voltage from the unregulated supply

I_0 = maximum load current in amperes

t_F = fall time of the switching transistor in microseconds (best obtained by measurement)

Resistor in the capacitive arm

$$R2 = \frac{t_B}{C}$$

Where:

R_2 = resistance in ohms

t_B = time interval in microseconds, which is short compared to on time. As a start, t_B can be taken as one-tenth the minimum on time involved in the operation of the regulator.

C = capacitance from previous step.

The quotient of V_{IN}/R_2 yields the current overshoot required from the collector of the switching transistor because of the capacitive arm involving C. If this value is greater than desired, the value of R2 must be increased.

The feedback circuit

The feedback circuit uses a triangular wave and an output-derived dc voltage to produce duty-cycle control of the switching transistor. In principle, this is similar to the regulation method used in some of the regulators already covered. In this regulator, the sampling of the output voltage takes place via the 4N28 optoisolator. Referring to Fig. 17-15, the sampled voltage is initially sensed at the junction of the 1N5290 constant-current diode and the 39-kΩ resistor. Because the constant-current diode dynamically simulates an extremely high resistance, the dividing effect of conventional sampling networks is avoided; that is, if the constant-current diode was replaced by a resistance, attenuation of the error signal would occur and the feedback loop would be less effective in providing tight regulation.

A voltage level proportional to the sampled output voltage appears at terminal 4 of the comparator, comprised of section 2 on the MC3320 quad-function IC module. There is complete electrical isolation between the regulator output and the comparator, for the optoisolator behaves as a transformer with both dc and ac capability. The other input of the comparator, terminal 5 of section 2, is impressed with a fixed-frequency, constant-amplitude triangular wave. This wave is obtained from the 20-kHz triangle-wave generator configured around section 1 of the quad-function module.

The output of the comparator (terminal 2 of section 2) is a rectangular wave having a duty cycle dependent upon the dc level of the error signal appearing at input terminal 4. The action of the comparator is illustrated in Fig. 17-18. Although no voltage-reference source is directly associated with the comparator, the fixed-frequency, constant-amplitude, triangular wave can be said to perform this function. A 10-V zener diode is provided in the output of the auxiliary power supply to ensure the amplitude constancy of the triangular wave.

The driving circuit for the 2N306 switching transistor encompasses the circuitry configured around the 2N6037, 2N6034, and MPS-U55 transistors. In Fig. 17-15, the driving circuit is actuated by the output of the comparator. The driving circuit is not

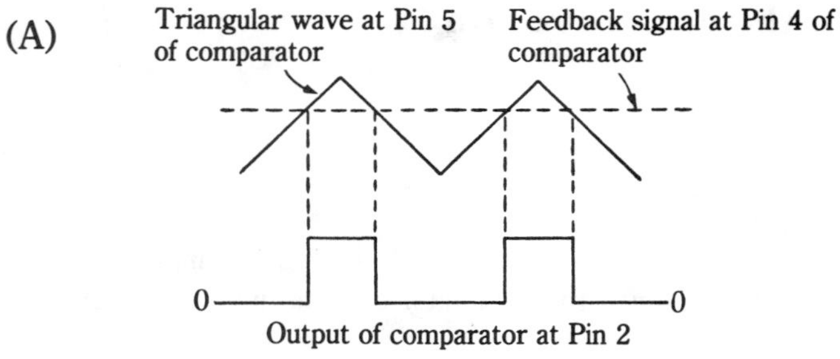

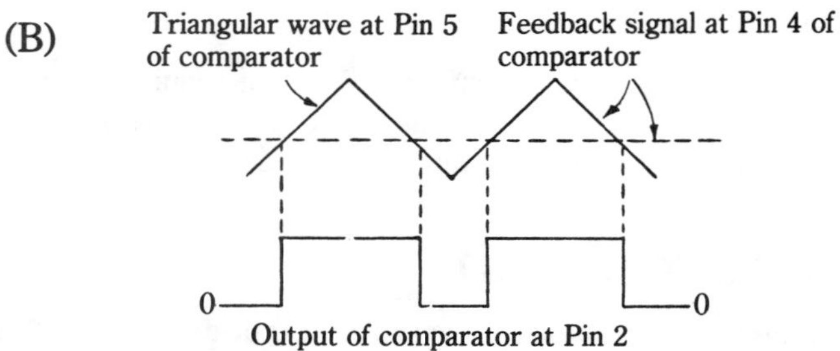

17-18 A waveform showing how the comparator responds to the sampled error signal. (A) The dc output level of the regulator is ***high.*** This results in a low duty-cycle switching wave, tending to lower the dc output-level of the regulator. (B) The dc output-level of the regulator is *low.* This results in a high duty-cycle switching wave, tending to raise the dc output level of the regulator.

merely a power booster or voltage level translator, as in conventional regulators; this driving circuit generates a negative transient at the termination of on time, which speeds up removal of stored charge in the emitter-base region of the 2N6306 switching transistor. The negative transient is derived from the 10-μF capacitor. The drive current waveform is shown in Fig. 17-19.

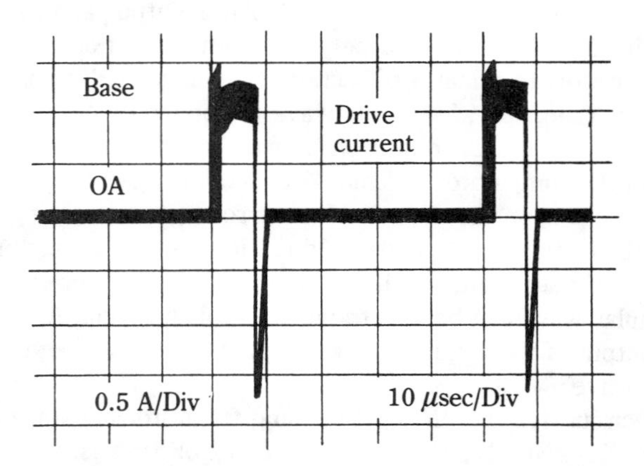

17-19 The waveform of driving current for the switching transistor. Notice the negative-going transient at turn-off. This is purposeful and speeds up reverse-recovery time of the switching transistor. Motorola Semiconductor Products, Inc.

Actually, the 2N6037 and 2N6034 "transistors" are monolithic Darlington devices. Together, they form a push-pull arrangement for the production of the "charge-cleansing" negative transient. The primary function of the third device, the MPS-U55, is to limit the forward driving current to the switching transistor. Because of this provision, the base drive to the 2N6306 switching transistor is limited to about one ampere. Excessive base drive to the switching transistor is a source of excessive dissipation in conventional regulators, and also slows fall time when the transistor is turned off.

Current limiting and short-circuit protection

When an overload or short circuit occurs at the output, the normal feedback operation is overridden and a different mode of operation takes place to protect the regulator. Although in this protective mode the regulator delivers 4 A pulses with a duration of 30 μsec and a pulse repetition rate of about 1 kHz. Thus, the regulator is kept alive in an idling state. This protective technique features the resumption of normal operation once the short or overload is removed. Such overload protection is brought about in the following way;

Referring to Fig. 17-15, section 3 of the MC3302 quad-function IC module is used as a voltage comparator to sense the voltage drop developed across the 0.05-Ω resistor connected in series with the 2N6306 switching transistor. When the output of the regulator is overloaded, this voltage becomes high enough to cause the comparator to change its conductive state, thereby discharging the 0.01-μF capacitor (C_X) in its output circuit. The capacitor requires time to recharge—in the meantime the comparator of section 4 is switched to its "high" logic state. This, in turn, is communicated through the 1N914 hold-off diode to the feedback comparator, section 2. This forms a closed loop, for comparator 2 now is forced to turn the switching transistor off. When this happens, the entire sequence of events repeats; and as long as the overload exists at the output of the regulator, this low-duty-cycle pulsing persists. The feedback comparator of section 2 is, however, ready to resume normal operation once the overload is removed. During normal operation of the regulator the feedback comparator is undisturbed by the current-limiting circuitry because of the isolating action of the 1N914 hold-off diode.

Duty-cycle controlled inverter/converter/regulator

The block diagram of Fig. 17-20 depicts the circuit function arrangement of a modern inverter or converter. As will be shown later, this arrangement is also readily made into a regulated dc power supply. Because any of the three applications are conveniently forthcoming from the basic systems approach, Fig. 17-20 appropriately serves to exemplify the basic ideas covered in the preceding pages of this book. Although the design appears complex compared to simple inverter and power supply regulators, the underlying logic is actually quite straightforward and can be quickly implemented using off-the-shelf digital ICs.

To start with, this is a driven, rather than a self-oscillatory inverter. The driver stages use digital logic ICs. The output transformer does not saturate. The driving signal is a square wave obtained from a JK flip-flop, which, in turn, is triggered from a

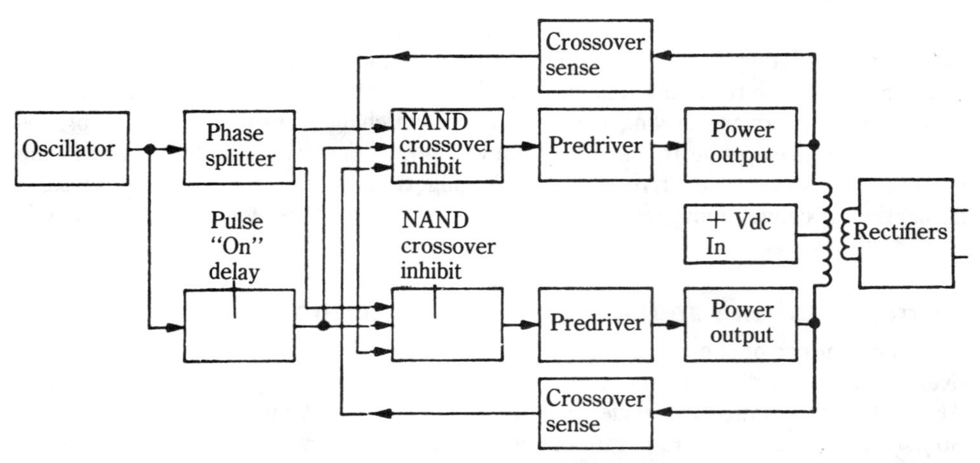

17-20 A block diagram of duty-cycle-controlled inverter or converter. A regulating feedback loop is not shown, but a suggested technique is shown in a later diagram. Motorola Semiconductor Products, Inc.

UJT oscillator. A one-shot multivibrator (pulse "on" delay) is connected with NAND gates so that the leading gate of the square-wave driving signals can be delayed. The NAND gates are inserted directly in the path of the drive signal from the JK flip-flop (phase splitter). Thus, the one-shot multivibrator can control the duty cycle of the inverter by either manual or by electrical actuation. In the latter case, a feedback loop derived from the output of the inverter could govern the time delay imposed by the one-shot to control the overall regulation of the inverter output. It is in this manner that the system becomes a regulated dc power supply.

The "crossover sense" functional blocks are peripheral to the basic operating principle outlined above. This does not diminish their importance, however, for they prevent turn on of one switching transistor before the other has turned off (the simultaneous on condition of the switching transistors is a commonly encountered failure mode of less sophisticated inverters). Even when only momentary, this operating mode is undesirable because it tends to exceed the SOA of the transistors. It is a factor that must be considered, largely because turn-off time is usually greater than turn-on time. Because of the leading-edge delay imposed by the one-shot multivibrator, it might be assumed that the crossover sensing provision is redundant. However, either manual adjustment or electrical control of the one-shot multivibrator could produce an unfavorable duty-cycle, such that a conduction overlap could result. The crossover provision protects against this. Notice in Fig. 17-20 that this protective provision can override the command of the drive circuitry in the event that one switching transistor has not turned completely off.

From the foregoing discussion, the various items shown in the schematic diagram of Fig. 17-21 should readily fall into place. The 2N2647 unijunction oscillator, together with its phase-splitting circuitry, consisting of a drive transistor and the MC663 JK flip-flop are easily discernible. The variable pulse-width one-shot circuit is comprised of the four NAND gates in the MC668 IC module. The drive-controlling NAND gates are

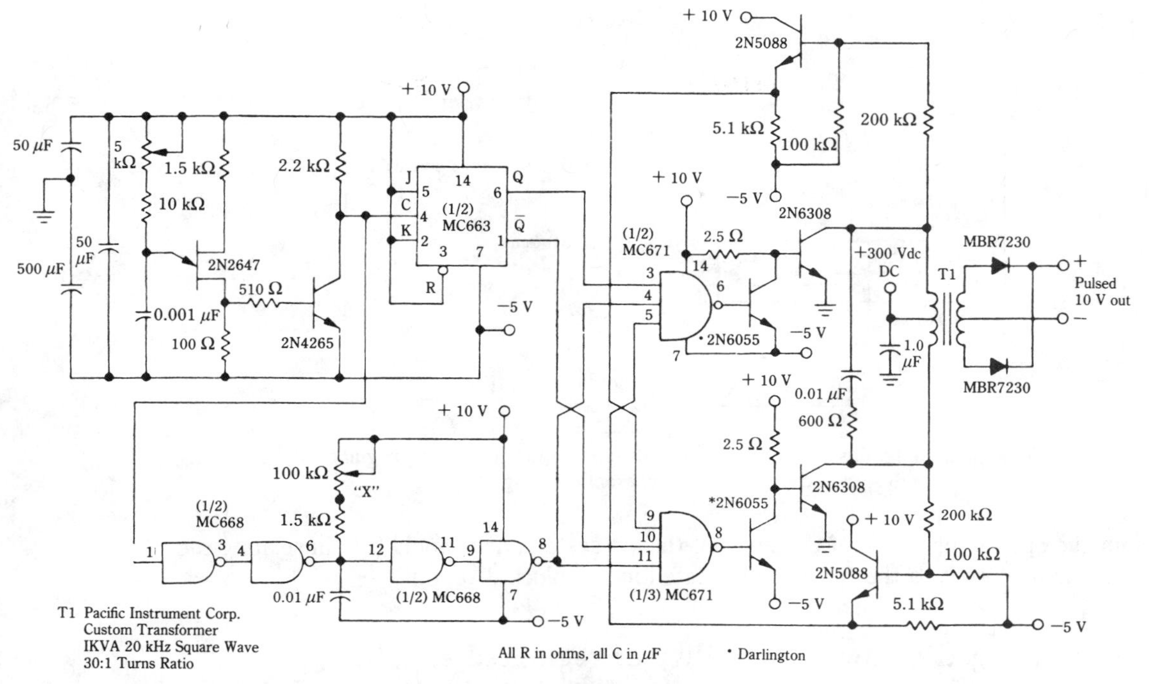

17-21 A schematic diagram of duty-cycle-controlled inverter or converter. Motorola Semiconductor Products

the two sections of the MC671. The output switching transistors are both 2N6308 and these are driven by the 2N6055 transistors. The 2N5088 transistors sense the conduction state of the output transistors and pass this information to pins 5 and 9 of the MC671 NAND gates.

As shown, the output of this inverter is manually controlled by means of the 100-kΩ variable resistance in the one-shot multivibrator circuit. However, if the circuit is severed at point X, an electrical input can be injected from a feedback path derived from the output of the converter. Filtering, preferably an LC type, is also needed to further smooth the output. Of course, our inverter then becomes a converter. A suggested feedback arrangement for regulating the converter is shown in Fig. 17-22 in block-diagram form. The MC1723 linear voltage regulator is used in place of a conventional op amp because of its self-contained voltage reference. The MOC1001 optoisolator takes care of such matters as output/input isolation and dc level translation—in essence, this device provides a dc transformer action, but without adversely affecting phase conditions in the feedback loop.

The UJT oscillates at 40 kHz, but because of the frequency division performed by the JK flip-flop, the output switching transistors are driven at a 20-kHz switching rate. One-kW output is obtainable from the circuit. The output transformer used was designed by the Pacific Instrument Corporation of Oakland, California, and provided a nominal 10-V output when used with the MBR7230 Schottky rectifiers and an L-section filter (not shown). The timing diagram of Fig. 17-23 provides additional insight into the

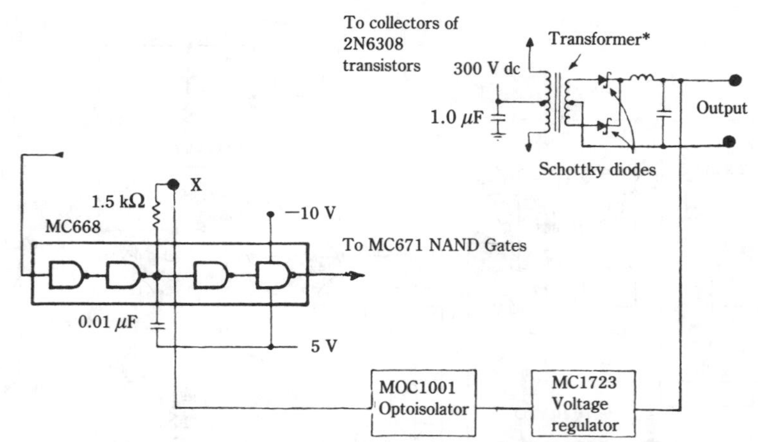

17-22 A suggested feedback technique for attaining output-voltage regulation. Point "X" refers to the designated point in the schematic of Fig. 17-21.

unique operational mode of this state-of-the-art system, and should help in relating the schematic circuit to the functions indicated in the block diagram.

Increasing the power capability of regulated supplies

One of the most common modification requirements for power supplies is the *increase* of their current or power capability. This can often be accomplished at a fraction of the cost and effort needed to design and build a new supply. Several techniques will be described for increasing the output of extant supplies.

The stratagem that generally first comes to mind is the paralleling of the power transistors. In a linear regulator, this would involve the series-pass, or in some cases, the shunt regulating transistor. In such supplies, it will generally be found that merely connecting like leads of the transistors in parallel will not work out in practice because of unequal current distribution among the transistors. As operating temperature rises, such unequality of load sharing tends to worsen in a regenerative fashion until one transistor ultimately hogs most of the load current. The scheme might be feasible, however, if the paralleled transistors are very nearly identical in characteristics and are mounted in the same thermal environment. Such a condition is generally not readily realized in practice because of the relatively wide tolerances in bipolar transistor characteristics.

If, on the other hand, power MOSFETs are used in the linear regulator, simple paralleling will work because these devices have opposite temperature coefficients from bipolar power transistors and will not undergo thermal runaway or current hogging. But MOSFETs have been used much more in switching supplies than in linear or dissipative types (our consideration of these nonswitching regulators provides preliminary insights into the problems of paralleling in switching regulators).

Figure 17-24 shows what must be done in paralleling either linear or switching power supply transistors. The small resistances inserted in the emitter leads of the bipolar transistors develops individual base-emitter biases that tend to counteract any

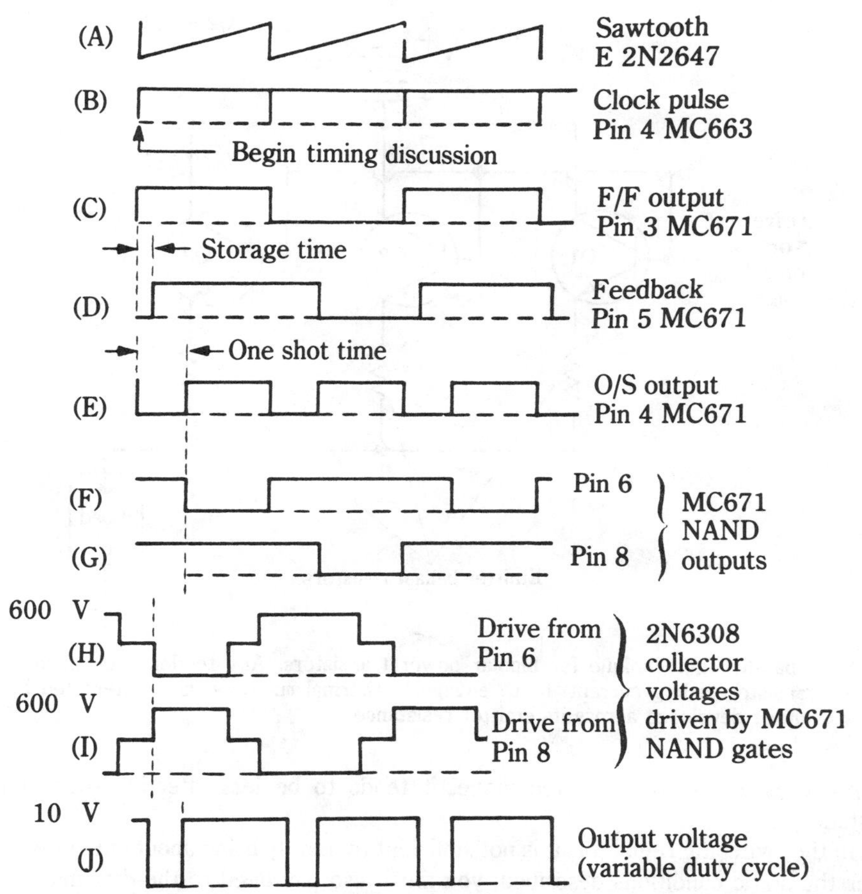

17-23 The timing diagram of duty-cycle-controlled inverter/converter. With the addition of a simple LC-filter, waveform J can be converted to smooth dc, the level of which depends upon the duty cycle of this wave. Motorola Semiconductor Products, Inc.

tendency of individual transistors to increase their proportionate share of load current. Although the use of these so-called emitter ballast resistances is very effective in discouraging current hogging or thermal runaway, you must use the smallest value of resistance that is sufficient for this objective; otherwise there will be excessive power dissipation — this could be particularly self-defeating in switching regulators where high operating efficiency is supposed to be the salient performance feature. It is not surprising, therefore, that one encounters emitter ballast resistances on the order of 0.1 Ω, 0.05 Ω, or lower, the actual value will, of course, primarily depend upon the emitter current in the particular supply. An initial value can be derived from $1/I_e$, where I_e is the maximum emitter (or collector) current.

Instead of the emitter resistances, it is sometimes possible to equalize current distribution in paralleled bipolar transistors by inserting somewhat higher resistances in the base leads. These are often seen to be in the 1- to 10-Ω range. Although overall

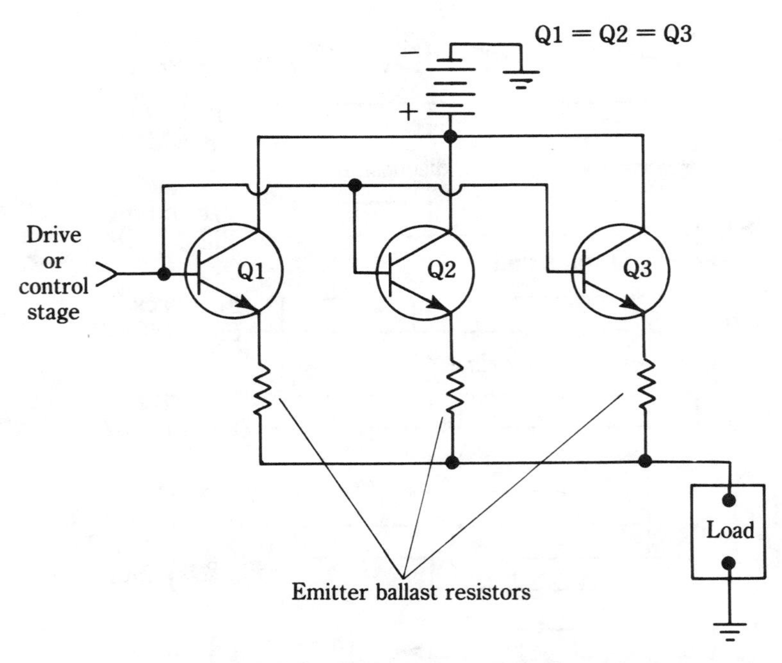

17-24 A paralleling technique for bipolar power transistors. Any tendency of an individual transistor to hog current, or to engage in thermal runaway, is counteracted by bias voltage developed across its emitter resistance.

dissipation is lower with this technique, it tends to be less effective than emitter resistances.

In the switching regulator, it is not sufficient to merely bring about current-sharing under the static conditions described; you must also pay heed to the dynamics of the switching process. This directs greater emphasis on the matching of the transistor characteristics. It is found in practice that two power transistors of the same type and designation can sometimes have appreciably different switching behavior—one might be somewhat slower than the other. Although such divergence can be rendered harmless by the emitter ballast resistances, these resistances might then have to be higher in value than would be the case with a better balance of transistor characteristics. However, even if the dynamic operation of the individual transistors in the paralleled bank are reasonably similar, the effect of unequal lead lengths or dissimilar wiring layouts can cause significant inequalities in device dissipations.

More often than not, it will be found feasible to double the power output via two paralleled bipolar transistors and it probably will not be necessary to upgrade the driver stage. Beyond this, however, it is likely that greater driving current will also be needed. Thus, with three or four, or more output transistors, paralleling will also have to be used in the driver stage. Sometimes, with the driver, it proves more expedient to substitute a transistor with a greater power rating.

Power MOSFETs can be paralleled without recourse to ballast resistances. It will frequently be found that perhaps four or more of these devices can operate from the driver stage that sufficed for one. However, the technique shown in Fig. 17-25 is recommended to prevent parasitic oscillations in the VHF and UHF range. Some experimentation might be in order with regard to the ferrite beads. Often, the lossy type with provision for two or three turns of the inserted wire proves to be effective for this application. An alternative technique is to use small film resistors in the 100- to 1000-Ω range in the gate leads. The Zener diodes shown in Fig. 17-25 are internally integrated in the structure of the particular MOSFET designated. Other MOSFET devices do not have this gate-protection feature, but the paralleling technique remains the same.

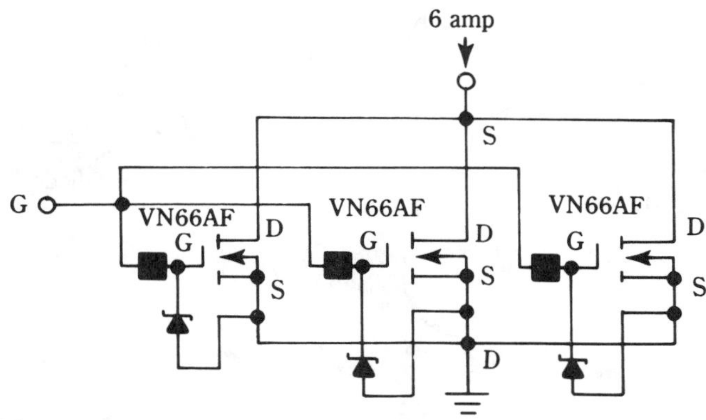

17-25 A paralleling technique for power MOSFETs. This is an easy way to increase load-current capability of both switching and linear regulators. The ferrite beads in the gate leads suppress high-frequency parasitic oscillation. The zener diodes are internal and represent a declining manufacturing process. Siliconix.

The power MOSFET switching stage can also be arranged in a series-connected format in order to accommodate higher voltage. Such an arrangement is depicted in Fig. 17-26 for two such transistors, but it can be extended to incorporate more. An interesting aspect of this technique is that only one MOSFET needs to be driven. This is because the other MOSFET has +15 V applied to its gate, with respect to ground; this MOSFET is, therefore, ready to conduct once its source circuit is completed by the turn on of the driven MOSFET. This arrangement allows twice the power to be delivered to the load that could be obtained from a single MOSFET; at the same time, each MOSFET works within its drain-source voltage rating. The RC network in the gate circuit of the upper MOSFET dynamically balances the gate waveforms of the two MOSFETs. To a first approximation, R_1C_1 should equal R_2C_2.

Because high-voltage power MOSFETs have become available, the series configuration is not used as much as when these devices first became competitive with bipolar transistors. Besides, their inherent readiness to perform well in parallel formats remains a compelling circuit design feature. The parallel configuration is easier to implement because it is more straightforward to provide the common thermal environment that is required by both connection schemes for optimum operation. The series arrangement would be selected in systems where available dc operating voltage exceeded the rating of the single MOSFET device.

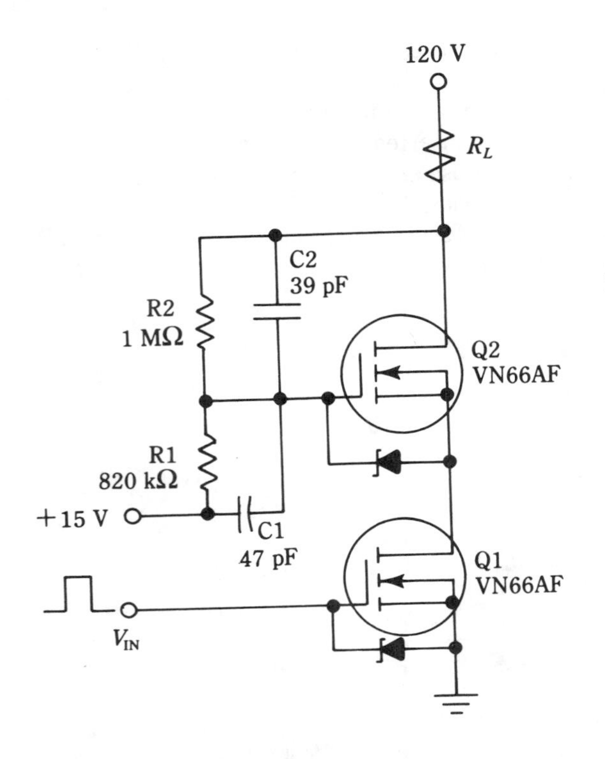

17-26 Series-connected power MOSFETs for doubled voltage operation. This basic technique can be extended for additional power MOSFETs. Notice that only one gate needs to be driven. Although the particular power MOSFET shown has an internal zoner diode, most others do not.
Siliconix.

Not only do some power MOSFET devices contain the equivalent of a Zener diode in their input circuits for gate protection, but the fabrication of these devices also can provide a "free-wheeling" diode in the output circuit. This is why many switching power supplies and motor-control circuits using power MOSFETs do not include the customary free-wheeling diode that is used in bipolar transistor circuitry. This, of course, can be an added feature in that parts count and cost is thereby reduced. When paralleling in order to attain increased power-handling capability, this can be particularly meaningful because there is no need for a high-current and an expensive "external" diode. However, the manufacturer's specifications should be studied to determine whether this is feasible for the particular application of the device being used. In some cases, an external Schottky diode, or a fast-recovery type might be needed in order to accommodate very high-speed switching rates of inductive loads.

Complementary symmetry as a means of obtaining higher power has already been alluded to, with respect to the bipolar circuits of Figs. 2-8 and 2-12. Until recently, the circuit simplification and clean performance of this technique was restricted to the use of bipolar power transistors where matched npn and pnp types were available. Now, however, several manufacturers have placed on the market p-channel MOSFET devices, which mirror the performance of certain n-channel types in reversed polarity. Therefore, it is feasible to design complementary symmetry switching circuits for power MOSFETs. Although the bipolar circuits in Fig. 2-8 and Fig. 2-12 are saturable-core oscillators, it stands to reason that only slight changes in circuitry and operating mode are needed to attain operation as driven inverters or converters. Additionally, regulated power supplies can be realized via similar feedback loops and control circuits used in other regulators.

Several semiconductor firms that now manufacture MOSFET power devices suitable for complementary symmetry applications are International Rectifier, Intersil, Supertex, and Westinghouse. The obstacles that delayed the availability of silicon pnp power transistors are not encountered to the same degree in the fabrication of p-channel MOSFETs. Accordingly, it can be anticipated that other companies will soon market paired devices for MOSFET complementary-symmetry switching applications.

Yet another power combining scheme is shown in Fig. 17-27. Here, the ac outputs of identical output stages are connected in series. This effectively combines the power capabilities of the transistors with no need for ballast resistances. It is an excellent way to circumvent the need for power transistors with higher voltage or current ratings—such devices can be either unavailable or very expensive. This arrangement is best considered during the initial design phase of an inverter or a regulated supply, for then it will be easy to prescribe appropriate input and output windings on the transformers. The phasing of the secondary windings on the output transformers must allow additive combining of the output voltages. It is relatively easy to obtain equal current sharing from the power transistors, but it is generally a good idea to ascertain that the transistors all operate in the same thermal environment. This is usually achieved via a

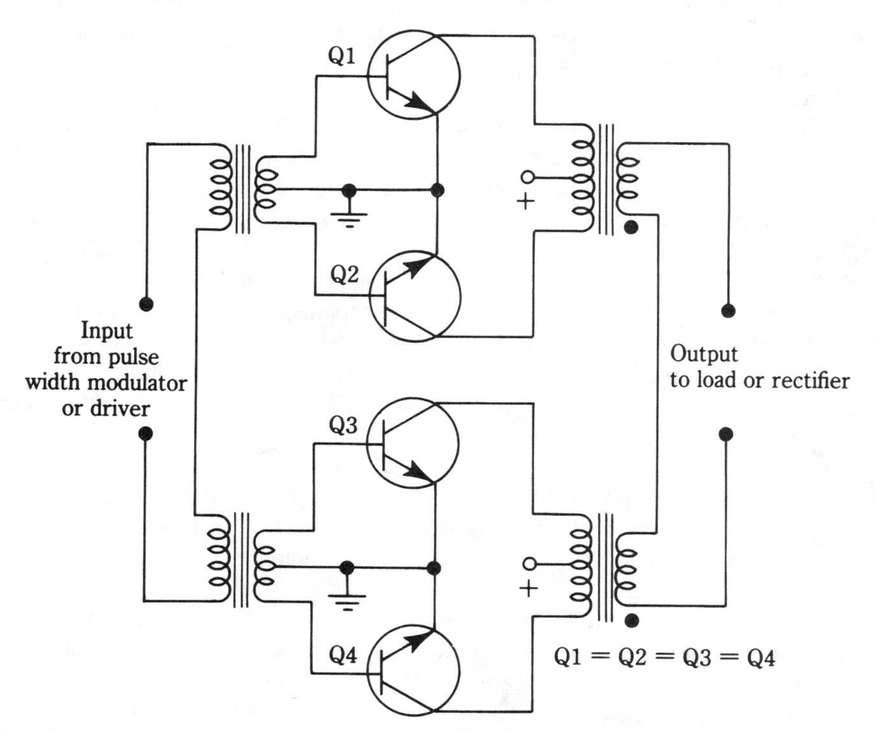

17-27 A scheme for doubling the output power from an inverter or switching regulator. This method circumvents the need for expensive or unavailable high-voltage or high-current transistors. Unlike paralleling, no power-wasting ballast resistances are required.

common heatsink. In this regard, a common-collector, rather than the common-emitter connection shown, has the desirable feature that no insulation is required between the transistor cases and the heatsink.

The disadvantages of this method involve cost, as well as size and weight. It tends to be true that two transformers are more expensive than one of double power rating. The physical dimension of two transformers will also usually exceed that of an equivalent single unit. Whether these factors are trivial or important depends, of course, on the circumstances pertinent to the particular system.

Although two output stages are shown in Fig. 17-27, the technique can accommodate a greater number of stages. But the basic approach should not be confused with the combining scheme (shown in Fig. 2-10) where a single output transformer is used, and pairs of output transistors are connected in series across the dc supply. The circuit technique of Fig. 17-27 is better suited for driven inverters and for switching power supplies, whereas that of Fig. 2-10 is more readily implemented for the saturable-core inverter. It could be argued that common-core transformers are also feasible for the combining method of Fig. 17-27. Surely, this must be so. However, the use of separate transformers, as illustrated, lend themselves particularly well for purposes of testing, evaluation, measurement, and servicing.

As an example of the flexibility of the circuit arrangement of Fig. 17-27, one pair of power transistors could conceivably be pnp type. Although this would not produce a "complementary symmetry" configuration in the usual sense, the overall dc power requirements might, in some cases, prove easier to comply with. The ac operation of the scheme would remain the same.

An interesting way to double the current capability, and, therefore, the power output of a single-transistor switching regulator is shown in Fig. 17-28. The additional switching transistor, Q2, is driven 180° out of step with the original switching transistor, Q1. This phase displacement is accomplished by means of transformer T1. Although a 1:1 ratio between the primary and secondary windings might suffice, the low impedances of the switching-transistor input circuits usually dictates a step-down transformer for optimum results—the centertapped secondary winding would then supply a lower voltage to each base-emitter circuit than is delivered to the primary. (This also tends to reduce the possibility of driving the switching transistor inputs into reverse breakdown. Although not shown, a small resistance in the base leads might also prove beneficial.)

Needed also, is inductor L2, a replica of L1. D2 is an additional "free-wheeling" diode, and is identical to D1. A doubling of the regulator's current capability is not the only way the additional switching transistor affects the performance. The ripple frequency is both doubled and reduced in peak-to-peak amplitude. Thus, with the retention of output capacitor C1, a cleaner dc level is available at the output of the regulator. Alternatively, it should be feasible to preserve the basic performance of the single-transistor switcher by reducing the size of C1, thereby saving both space and cost. If this technique is pursued during initial design, it is likely that less-costly switching transistors can be selected because each is called upon to switch at only half the rate of the output ripple frequency.

In order to use this scheme advantageously, the unregulated dc source must, of course, be able to supply twice the current demanded of it in the single-transistor

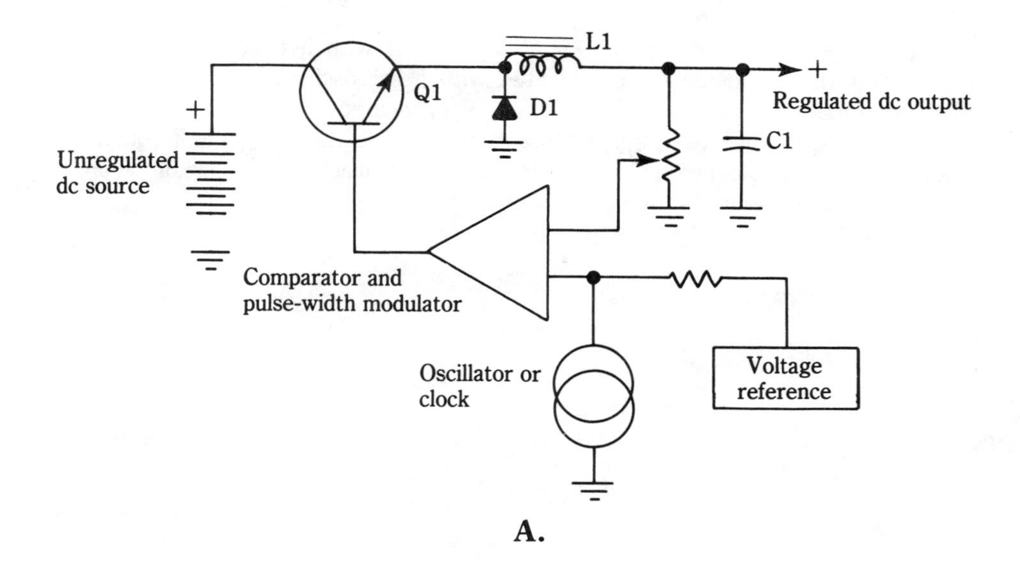

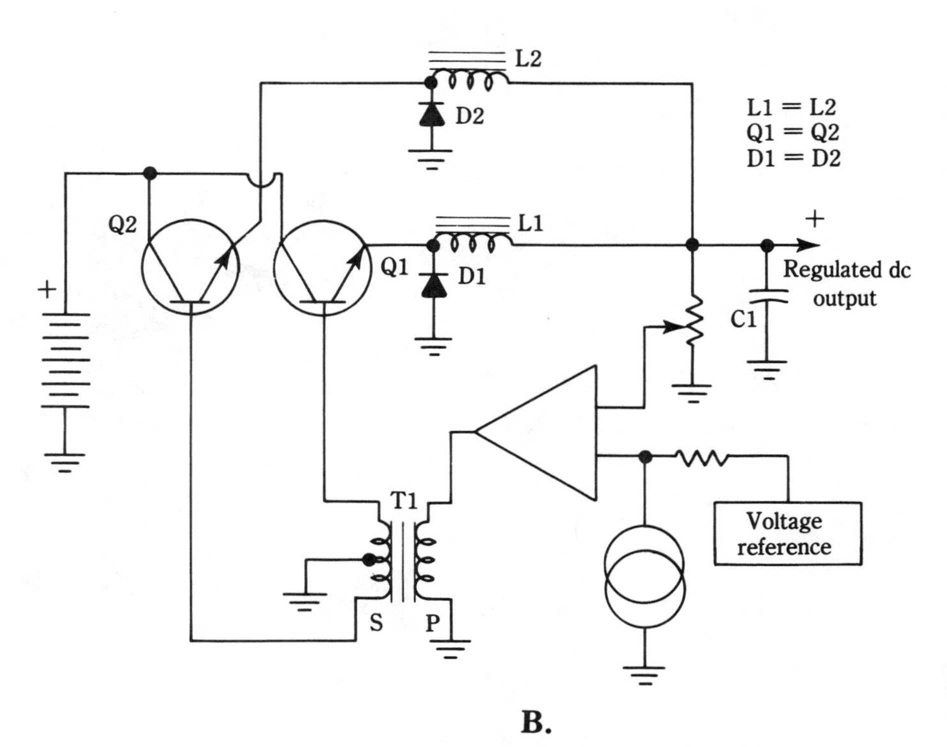

17-28 A technique for doubling the current capability of a switching regulator. This method not only provides increased load power, but reduces output ripple. (A) A simplified circuit of conventional switching regulator. (B) A modified circuit for doubling load current.

regulator. The circuits of Fig. 17-28 A and B deal with a fixed-frequency driven regulator. If the technique is implemented with a self-oscillatory regulator, certain difficulties might be encountered and some experimentation will undoubtedly be required. This is because the sampled frequency is twice the switching frequency. A possible remedy would be to sample the positive feedback from a small secondary winding placed on either L1 or L2.

18
CHAPTER

Special report on high-performance regulated supplies

BY ITS NATURE, POWER SUPPLIES, SWITCHING REGULATORS, INVERTERS AND converters were destined to undergo a long evolutionary process of development. To be sure, at any given time, there were those who declared the technology to be mature—meaning that any future progress would be limited to degrees of refinement. Although the gross error of their prognosis has been repeatedly demonstrated with the passage of years, it would, admittedly, be tempting to now make the same mistake of supposing that all major improvements have finally been made.

There is, however, a statement that can be made with certitude; state-of-the-art technology of power sources for operating electronic and electrical equipment is a quantum jump advanced from the status quo of a decade ago. It appears there has been a fortuitous convergence of progress in power devices, circuit components, circuit techniques, and in dedicated control ICs. The net result has been higher switching rates, higher operating efficiency, extended operating conveniences, expanded design flexibility, and reduced RFI and EMI. Best of all, these improvements have been gained with manageable part counts and favorable cost factors. An important factor deserving mention is that the powered equipment now performs better, too, because there is better decoupling, isolation, and regulation, as well as reduced electrical noise, transients, and heat.

The following is an investigation into some of these advanced circuits. It will be interesting to observe that some of the ideas have roots in past thinking, but that practical implementation had to await development of the appropriate hardware.

A 100-W 100-kHz PWM regulated power supply using a control IC

The regulated power supply shown in Fig. 18-1 can be viewed as a basic prototype of advanced technology. It has some notable features not readily attained in earlier

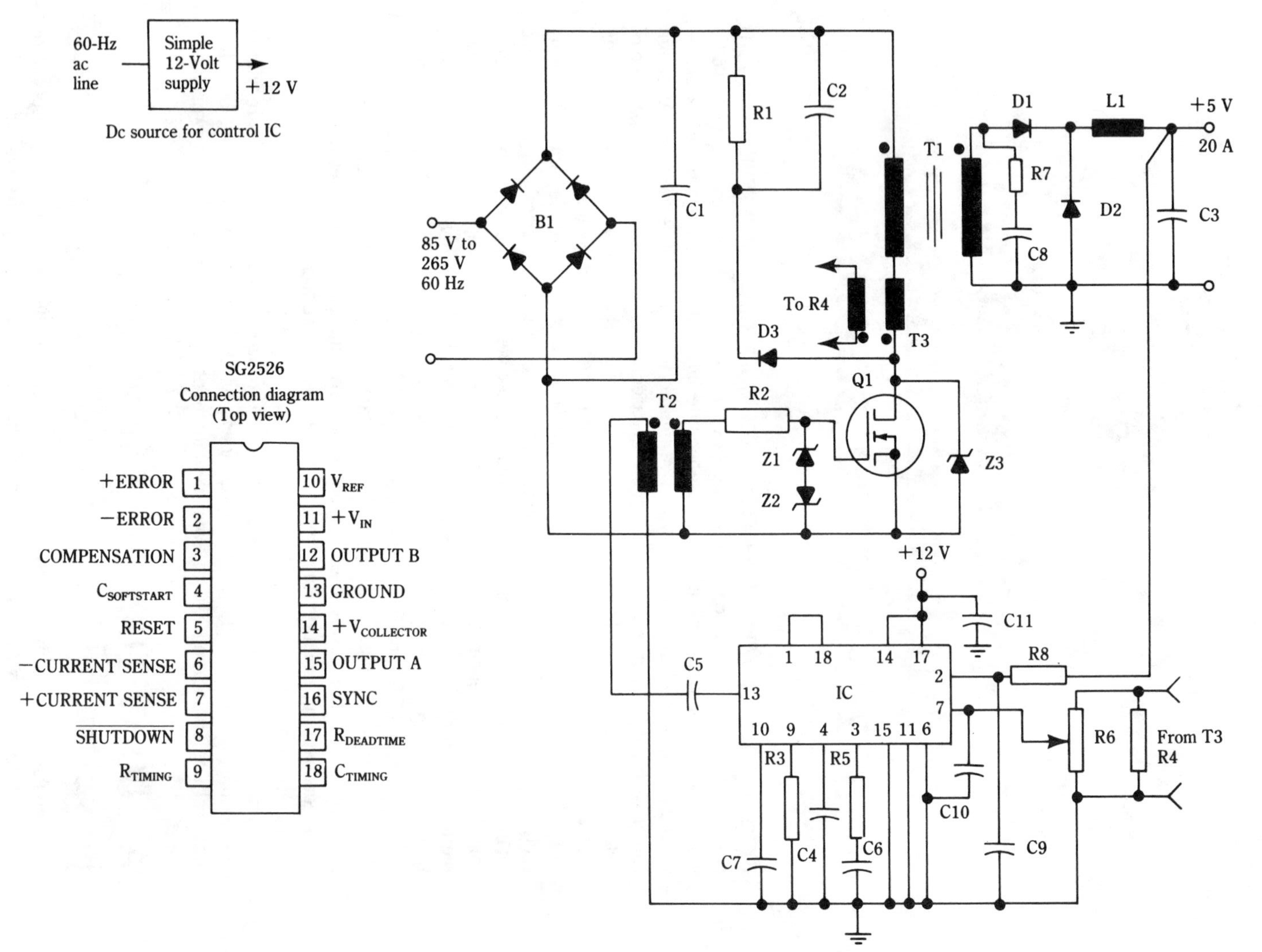

18-1 A 100-W, 100-kHz, PWM-regulated power supply using a dedicated control IC. This forward converter circuit provides about 75% full-load efficiency over an extremely wide range of ac line voltage. International Rectifier Corp.

designs. It delivers a respectable amount of power, and its switching rate is four or five times that of "the first generation" of switchmode supplies. It is controlled by a designated IC, a very significant feature because of the numerous "bells and whistles" that are for the most part, automatically implemented. It uses a power MOSFET, thereby operating with easy drive, and manifesting electrical ruggedness with high immunity from secondary breakdown and thermal runaway. Because of its efficient utilization of the PWM principle, it can operate at any ac line voltage within the 85- to 265-V range! It incorporates an adjustable current-limiting circuit. Finally, it is quite simple, involving a minimal number of parts.

In this circuit, the customary tertiary winding used for clamping the switching transient has been dispensed with; instead, the snubber network comprised of R1, R2, D3, and Z3 serves this protective function. It should be pointed out that the forward-conversion topography with transformer output (T1) provides design flexibility because design for different dc output voltages is largely a matter of selecting the appropriate transformation ratio between the primary and secondary windings of T1.

The current limiting level of the power supply is set by adjustment of potentiometer R6. This brings us to the salient feature of the recently developed family of dedicated IC controllers for power-supply regulation, the "bells and whistles" that accompany the basic regulatory function. These provide protective functions, operating conveniences, and design flexibility. To implement these "from scratch" would usually constitute an engineering project in itself. Table 18-1 lists the features specified for the Silicon General SG2526 Regulating Pulse Width Modulator used in the power supply of Fig. 18-1 (of course, it is not necessary to make use of all options designed into the IC).

You can also see that a small 12- to 15-V, 50-mA dc supply is also needed to empower the SG2526 IC. It is not imperative that this auxiliary dc source be regulated, but for isolation purposes, it should incorporate a small two-winding transformer.

Table 18-1 A list of features built into the SG2526 pulse-width controller IC. "Bells and whistles" such as these, previously had to be implemented with op amps and discrete devices—a difficult task. Silicon General.

Features

- 8- to 35-V operation
- 5-V reference trimmed to ±1%
- 1-Hz to 350-kHz oscillator range
- Dual 100-mA source/sink outputs
- Digital current limiting
- Double pulse suppression
- Programmable deadtime
- Undervoltage lockout
- Single pulse metering
- Programmable soft-start
- Wide current-limit common-mode range
- TTL/CMOS-compatible logic ports
- Symmetry correction capability
- Guaranteed 6-unit synchronization

Courtesy—Silicon General

Table 18-2 Parts list for the 100-W 100-kHz PWM supply.

Q1	IRF830 HEXFET
IC	Silicon General 3526
B1	IR KBPC 106
C1	500 μF, 450 V wkg.
C2	0.68 μF, 100 V
C3	4X 150 μF, 6 V
C4	22 μF, 16 V
C5	0.5 μF, 25 V wkg.
C6	10 nF
C7	910 pF
C8	0.0068 μF
C9	0.005 μF
C10	0.1μF
C11	22 μF, 25 V
R1	1.5 kΩ (3X500Ω, 5W)
R2	12 Ω 1/4W
R3	6.8 kΩ 1/4W
R4	10 Ω
R5	12 kΩ 1/4W
R6	100 Ω potentiometer
R7	33 Ω 1/4W
R8	560 Ω 1/4W
D1	20FQ030
D2	BYV 79-100
D3	IR 40SL6
Z1	1N4112 zener diode
Z2	1N4112 zener diode
Z3	4X 1N987B zener diodes in series
L1	Core Arnold A-930157-2, 16 turns, 2 in parallel #14
T1	Core TDK 26/20, H7C1 Primary: 20 turns, 3 in parallel #32; Secondary: 3 turns, 0.3 mm x 8 mm copper strip
T2	Core TDK H5B2T10-20-5, Primary: 60 turns #24; Secondary: 6 turns #24
T3	E2480 Core TDK H52T5-10-2.5. Primary: 1 turn; Secondary: 100 turns #32

Although this appears as an afterthought, it should not lead to any problems. The parts list, together with magnetics data for the core component, appear in Table 18-2.

Basic ideas underlying resonant-mode operation

Ideally, the pulse-width modulation (PWM) technique appears to be the answer to the quest for the near perfect regulated power supply. For, we are told, the power switch is either on or off, and control is accomplished with zero power dissipation, unlike the linear regulator, where it can be said that control is accomplished because of power dissipation in the pass element. In the real world, the PWM process does yield a reasonable approach to lossless switching at the lower switching rates, for example, in the 20- to 40-kHz region. Looking at the situation from another angle, it can be said that this is why this switching rate range enjoyed such an enduring popularity.

From the very inception of PWM regulation, designers wanted to go to higher frequencies because of the reductions that could be realized in the size, weight, and cost of magnetic core components and filter capacitors. Other advantages also appeared on the high switching-rate horizon. Reduced RFI and EMI could be anticipated by starting out with a higher frequency; one could expect relaxed problems in shielding, decoupling, isolating, and confining the circuitry. One could also expect faster response times as well as lower output impedance and ripple.

The main obstacle to higher rates resided in the practical difficulty of producing a speedy enough power switch. For, unless near instantaneous turn on and turn off of the switch could be achieved, there is the presence of both, voltage across the switch, and current through the switch during switching transitions. In other words, trapezoidal, rather than rectangular waves characterize the switching process. This, in turn, produces switching losses that tend to negate the theoretical high efficiency of the ideal switch represented by instantaneous turn on, zero resistance while on, and instantaneous turn off. Figure 18-2 compares PWM and resonant mode switching, which will be covered in more detail.

From the above, it is also evident that the ideal power switch must also give rise to no voltage drop across its terminals during on time. All things considered, it is seen that high efficiency was not an easily attainable operating condition, and certainly not at higher switching rates, until progress could be made in the solid state switching devices. It should be stated also, that accompanying progress had to be made in other devices, such as diodes, magnetics, and capacitors. It is a tribute to workers in all aspects of the technology that PWM switching rates have been extended to about 500 kHz. Nonetheless, for rates much in excess of, say 150 kHz, it tends to be better to investigate a different technique than pulse width modulation. Enter the resonant mode power supply.

The resonant-mode regulated power supply truly represents a great leap forward in the evolution of the technology. Yet, the use of resonant phenomena in inverters, converters, and power supplies pre-dates the solid-state era. For, it had been found that useful performance was often obtained through resonance effects. For example, earlier TV sets generated the needed high voltages for the picture tube by means of an RF power supply. This was an electron tube sine-wave oscillator operating in the vicinity of 150 to 300 kHz in which ac voltage step up was achieved in a self-resonant RF transformer. Essentially similar schemes are still used for producing up to at least several hundred thousand volts for various industrial and scientific research purposes. The higher voltages are often reached via the combined use of resonance and diode voltage multiplying circuits.

Also, it had been long ago found that resonating the output of an inverter greatly stabilized operation in electric motor and in arc welding applications. Generally, a large inductance was inserted in one of the leads from the dc supply to the inverter; this caused the inverter to behave as a constant current source for the load and enabled resonance to be more easily accomplished. Then, there were SCR inverters that were best described as being quasi-resonant—a resonant tank was periodically shock-

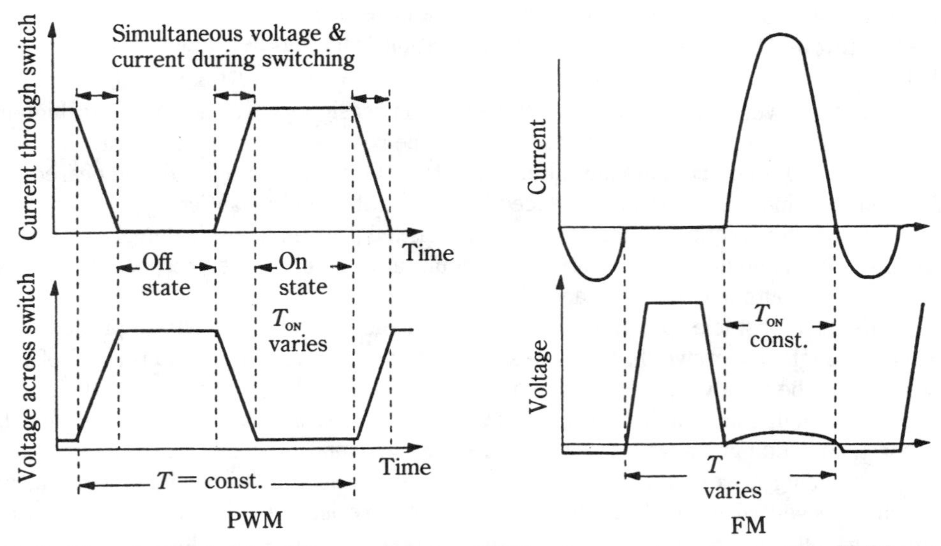

18-2 Waveshapes showing difference between PWM and resonant-mode operation. PWM operation allows switching losses because of simultaneous current *through* the switch and voltage *across* the switch. Note that this condition does *not* exist in resonant-mode operation, which makes use of FM to obtain regulation.

excited, but there was no sustained ringing. In between excitation pulses, the resonant tank would release its stored energy to the load. Examples of such alluded circuits are shown in Figs. 18-3, 18-4, and 18-5.

It is well to recall, too, that power supplies and inverters have sometimes exploited the beneficial filtering action of resonant chokes in their output circuits. Finally, not withstanding the dramatic progress from primitive power supplies, we still have with us the deleterious effects of undesired resonances. These include such detrimental performance characteristics as parasitic oscillations, RFI, EMI, "spike" generation, and related circuitry malfunctions that reduce efficiency and damage or destroy both active and passive circuit components.

With the foregoing as a prelude, it should be realized that the greatest use of resonant-mode operation has come about since the development of dedicated control ICs. These resonant-mode controllers freed designers from the various malfunctions that inevitably attended attempts to utilize resonant-mode operation in the desired several hundred to several MHz frequency range, where small components could yield worthwhile reductions in size, weight, and cost.

An interesting thing about the resonant-mode regulating supply is that it has many resemblances to the long popular pulse width modulation (PWM) scheme. Indeed, from a block-diagram concept, a variable-frequency source of constant-width pulses, together with a resonant "tank" simply substitutes for the PWM circuitry. Operationally, either the current waveform or the voltage waveform applied to the power switch(s) is a portion of a near sine wave because of the LC tank. The switching waveforms occur in such a manner that, unlike high-frequency PWM circuits, there is never simultaneous

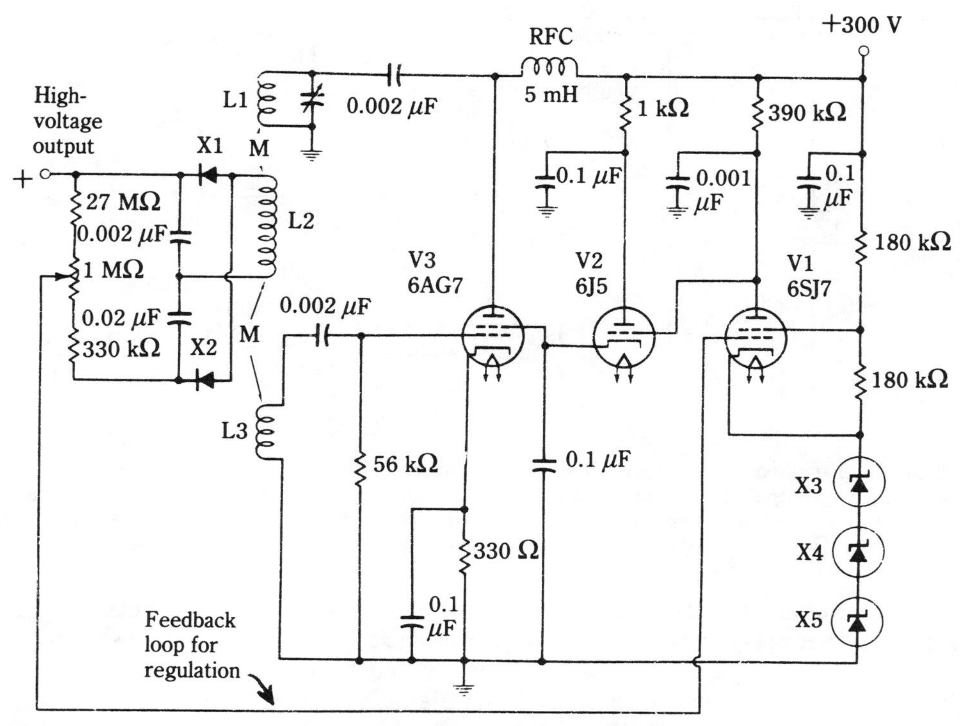

18-3 An example of resonant operation in a high-voltage RF power supply. This revamped older design uses electron-tubes in a Meisner oscillator circuit. Operating frequency is determined by step-up winding, L, in conjunction with its own distributed capacitance. No regulation is provided.

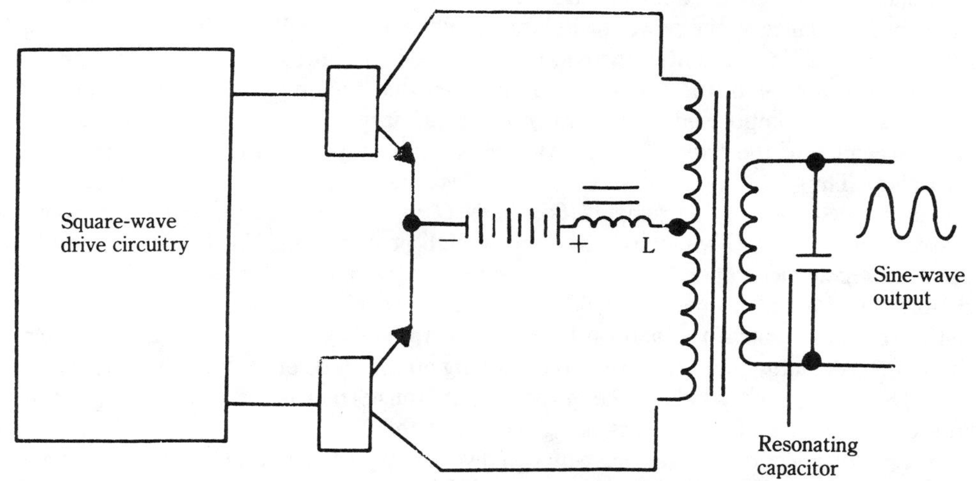

18-4 An example of a current-driven inverter with resonant output. Notice presence of large inductor, L, in dc supply lead, and the output resonating capacitor. Similar techniques apply to self-excited inverters. Circuits such as these are generally unregulated.

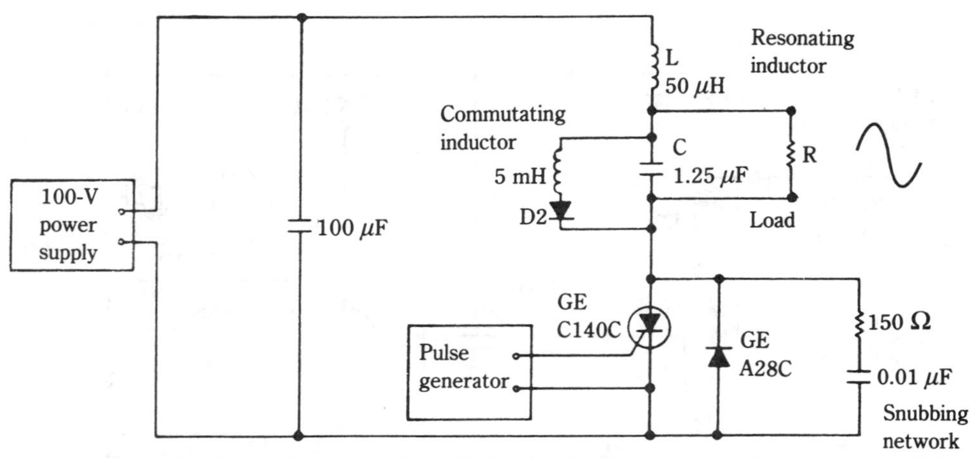

18-5 An example of a quasi-resonant single-SCR inverter. The selection of an appropriate SCR enables outputs of several kilowatts and switching rates to about 30 kHz. If the pulse rate is slightly below resonance of the series LC tank, a good sine wave will be delivered to the load. There is no regulation. General Electric Semiconductor Products Dept.

voltage across the switch and current through the switch. Therefore, switching losses tend to be negligibly small—even at high frequencies.

Figure 18-6 shows how resonant-mode operation is brought about. An error signal is derived in the same manner as in PWM supplies as a sampled and referenced portion of the dc output voltage. This error voltage is then applied to a voltage-controlled oscillator, the output of which triggers a monostable multivibrator. This modulation scheme is, in essence, a voltage-to-frequency converter. The constant width, variable repetition rate pulses from the monostable multivibrator then drives the power switch(s)—a driver amplifier is often inserted to provide high instantaneous current and low impedance to the power switch(s), usually a power MOSFET, or a pair of them.

The output of the power switch(s) is associated with a resonant LC circuit and the output transformer of the supply. It can be seen that the amplitude of the essentially sinusoidal wave impressed on the transformer primary depends upon the nearness to LC resonance of the fixed duration, variable-frequency pulses provided by the power switch(s). Thus, servo action can be produced via frequency modulation to regulate the dc output voltage of the supply. Too high Q in the LC tank will hamper power extraction; too low Q will cause excessively high peak currents in the power switch.

Resonant-mode operation can be achieved in several ways: either series or parallel resonance of the LC elements can be used. And, the nominal operating frequency can be either below or above the natural LC resonant frequency. In any of these situations, however, regulatory operation requires working on the slope of the resonance curve. In Fig. 18-6, the primary inductance of the output transformer is sufficiently high so that the LC resonance of the tank is not greatly affected.

Initial inspection of a resonant-mode power supply can sometimes prove confusing because either the resonating inductance or the capacitance, or both, will not be evident on the schematic diagram. In such cases, parasitic reactances are used, such as the leakage inductance of the output transformer, or stray or distributed capacitance.

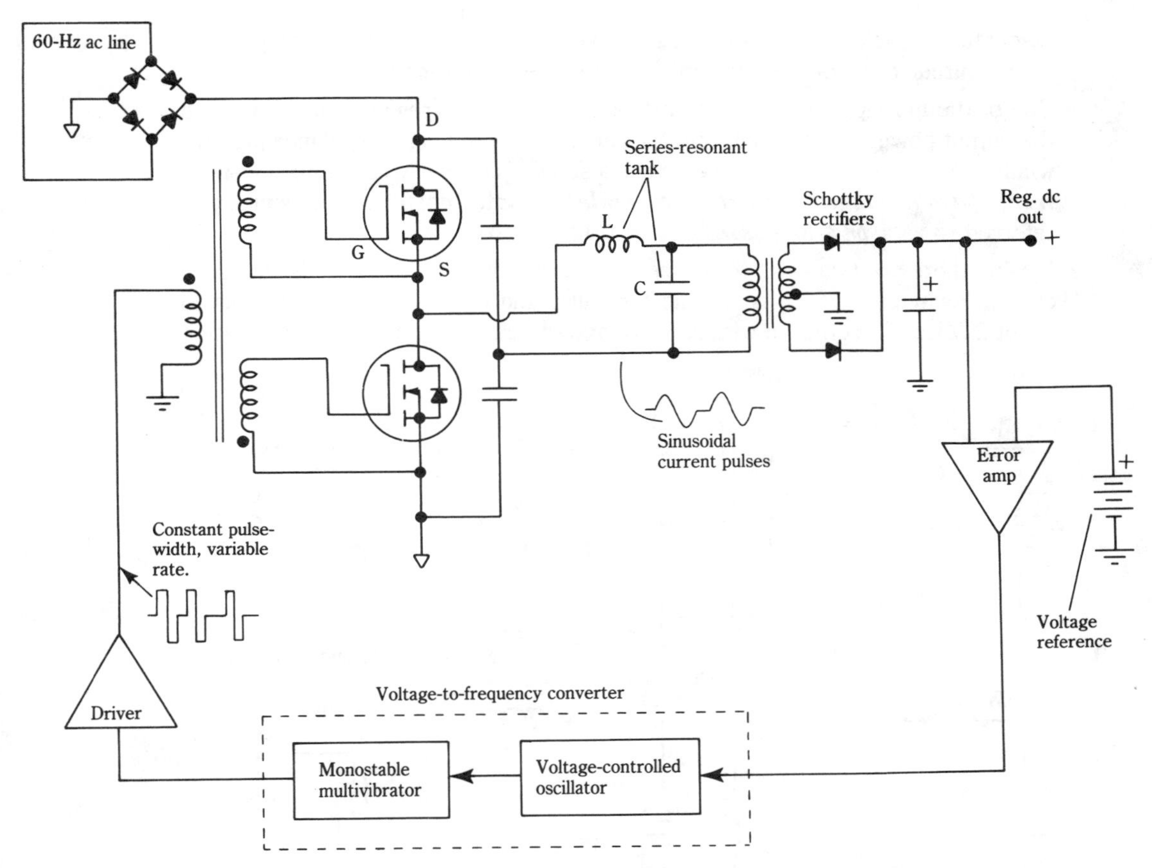

18-6 The simplified layout of a resonant-mode regulating power supply. To a first approximation, the voltage-to-frequency converter substitutes for the pulse-width modulator in the popular PWM regulator. The half-bridge topography shown conveniently lends itself to off-line operation from the ac utility source.

Sometimes the output capacitance of the power switch can be so used. Also, some circuits parallel resonate the secondary of the output transformer with physical or parasitic capacitance.

- In order to avoid confusion from sloppy semantics in the technical literature, it would be well to remember the following facts about resonant-mode regulators:
- The resonant LC tank always tends to oscillate at its self resonant frequency, regardless of the frequency of the pulses producing the shock excitation. However, in most cases, free ringing is not allowed. Rather, half sine waves are extracted and delivered to the rectifier load circuit.
- Various series and parallel resonant circuits can be used. Sometimes, the schematic diagram does not reveal any physical inductor or capacitor as part of the resonant tank. In such instances, transformer leakage inductance or stray capacitance is exploited for the purpose. There are also arrangements in which the

resonating capacitor appears in the secondary, rather than in the primary circuit of the output transformer. Several schemes are shown in Fig. 18-7.

- One of the most popular schemes makes use of a series-resonant tank in which the output power is extracted from the capacitor via the high impedance primary winding of the output transformer. Such a supply is properly called a ***series-resonant, parallel-loaded converter*** or ***regulator.*** Unfortunately, it is sometimes referred to as a ***parallel-resonant type*** (Fig. 18-7B).
- Ideally, there are two ways to realize near zero-switching loss. One is with zero current switching, which is the most popular and enables practical designs to about 2 MHz. The other in with zero voltage switching, enabling implementation

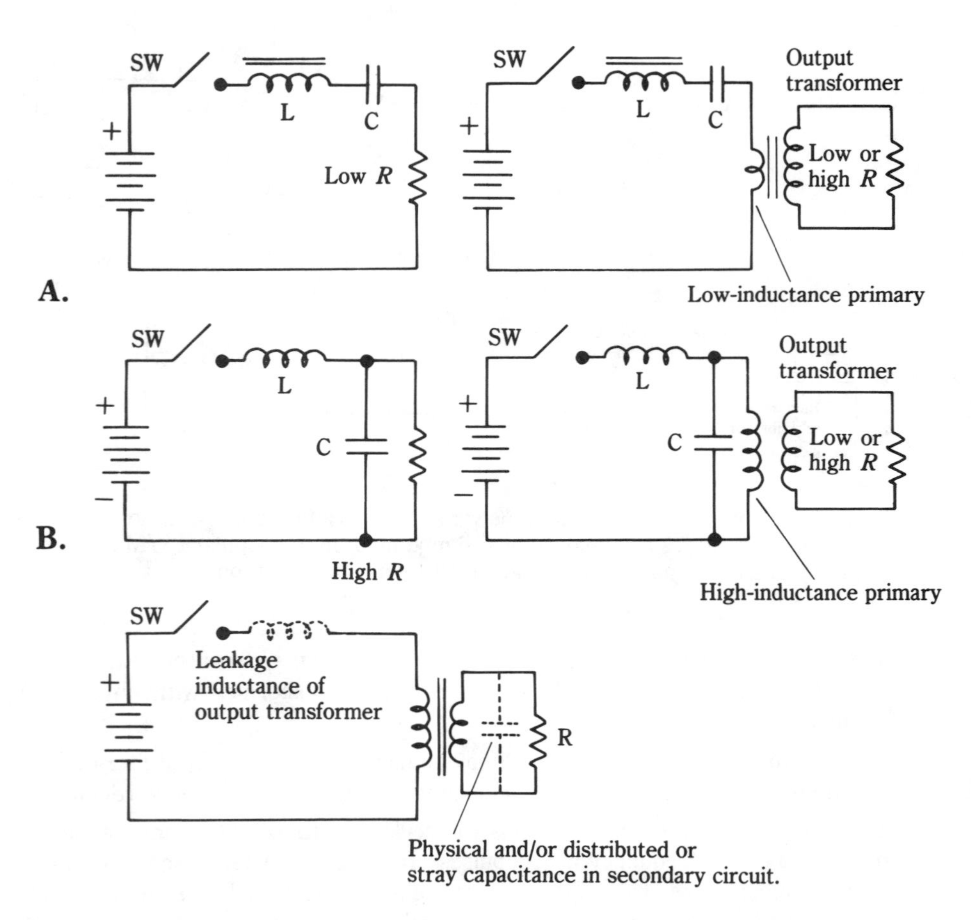

18-7 Various arrangements for extracting power from a resonant circuit. *R* is generally the combination of rectifer, filter, and load. (A) A low-impedance load with and without an output transformer. (B) A high-impedance load with and without an output transformer. (C) Resonance and power extraction via the use of stray reactances.

to the 10-MHz region. Zero-current switching utilizes constant on time pulses of variable repetition rate to shock-excite the resonant tank circuit. Constant off-time pulses are used in the zero-voltage switching mode.

- Most commonly (especially with zero current switching) the range of variable frequencies extends from a low value to about 80% of the tank resonant frequency. This assures time for the inductor current to go to zero or become negative. The pulse on time is terminated somewhere in this negative region; it isn't very critical. Negative inductor current implies that the current path transfers from the power MOSFET(s) to the free-wheeling diode(s). Pulse on time is determined by an RC network connected to the control IC. The *R* and *C* values can be conveniently read from graphs provided by the IC manufacturer. Typical data illustrating selection of RC values for determining pulse on time, as well as oscillator frequency, are shown in Fig. 18-8.
- Be certain that "switching frequency" pertains to the rate at which on pulses are applied to the resonant tank. This is not necessarily the same as the oscillator frequency in the IC controller. In some cases where push-pull operation of the power switches occurs, the oscillator frequency will be twice the switching frequency. For single-ended power switches, these frequencies will usually be the same.
- Lossless switching is approached with the supply working in the discontinuous mode. This simply means that there must be only one LC oscillatory cycle per on pulse. In practice, this requires a "dead time" between completion of one cycle of oscillation and the arrival of the next on pulse. This is why the repetition rate of the on pulses must not too closely approach the LC resonant frequency. Meeting this requirement involves a small trade off with available power output.
- Regulation is based upon the fact that the energy stored in an LC circuit is greatest when the repetition rate of shock-exciting pulses is close to the resonant frequency of the LC combination; departure of the pulse rate from this optimum condition, allows less power to be extracted. Because the resonant frequency remains constant, it is the above alluded to dead time that varies to produce regulation.
- Resonant-mode supplies often incorporate a current-sensing circuit that resembles the circuitry used in current-mode PWM supplies. Indeed, you can find reference to "current-mode" operation in the resonant supply. There is an important difference, however. In the PWM system, a current ramp is sensed, and peak current limiting of the supply occurs on a cycle per cycle basis. In the resonant-mode supply, a partial sine wave in sensed; this enables limiting of the power switch peak current, but not on an instantaneous basis. In both cases, protection is achieved, but not so quickly or precisely in the resonant-mode supply as in the PWM current-mode technique. In the PWM supply, current sensing is a feed-forward regulating technique; in the resonant-mode supply, current sensing is a shut-down technique.
- Last, but not least, the power switches in the resonant-mode supply do not experience simultaneous voltage and current during switching transitions. This

(a)

Oscillator resistance (k ohms)

Co = 680 pF

220 pF

150 pF

100 pF

47 pF

Frequency (kHz)

Maximum operating frequency (f_{max})

(b)

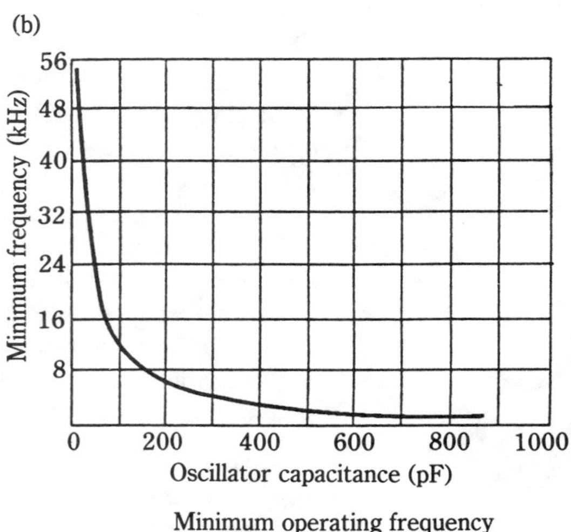

Minimum operating frequency

(c)

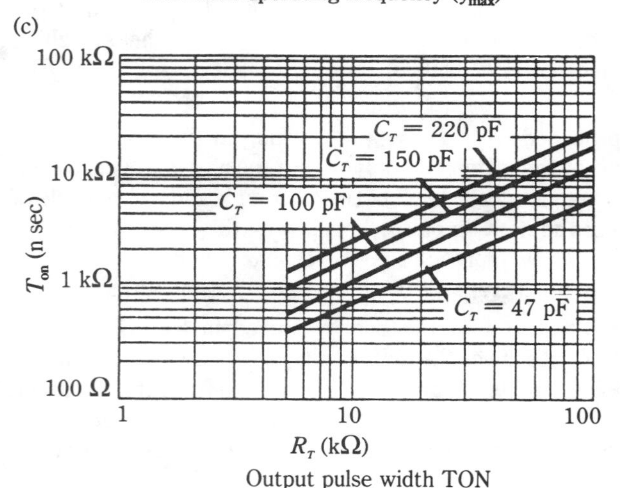

Output pulse width TON

18-8 Sample graphs for determining parameters of resonant-mode regulating supplies. These curves pertain to the GP605, but are typical of manufacturer's data. (A) Allowable resistance-capacitance combinations for any maximum oscillator frequency. (B) Allowable capacitance for any minimum oscillator frequency. (C) Resistance-capacitance combination for collected pulse-widths. This is a different RC-network from the one dealt with in A and B. Gennum Corp.

leads to high operating efficiency with significantly reduced power dissipation in the power switches. This, in turn, relaxes thermal problems—another factor that contributes to high packaging density.

A 48-/5-V 20-A high-frequency resonant converter

As previously pointed out, one of the obstacles in the path of progress of power-supply technology was the sheer complexity of designs that require numerous discrete devices to achieve performance goals. This was particularly true for the resonant-mode regulation scheme; basic principles were long understood, but practical implementation long remained awkward and uneconomic. This is no longer true, for semiconductor manufacturers have made available dedicated control ICs, which not only are inexpensive, but also incorporate almost no end of built-in housekeeping functions for protection and operating conveniences.

One of these resonant-mode control ICs is the GP605 made by the Gennum Corporation, a pioneer of resonant-mode technology. The functional block diagram of the GP605 resonant-mode controller is shown in Fig. 18-9. Notice the voltage-controlled oscillator and the monostable multivibrator. Together, these constitute a voltage to frequency converter for shock-exciting an external LC resonant tank with constant width, variable rep rate pulses.

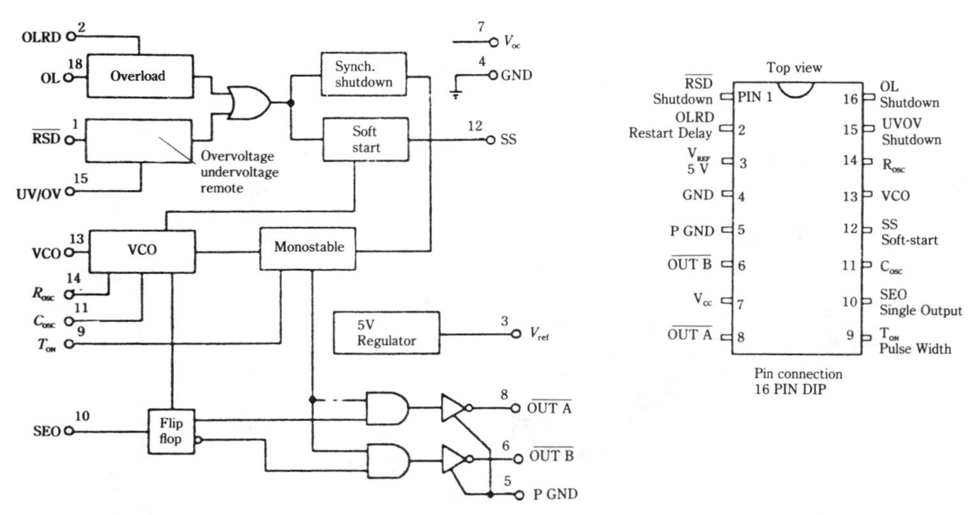

18-9 A functional block-diagram of the GP605 resonant-mode IC controller. This resonant-mode controller is particularly suitable for driving half-bridge power switches, the popular topology used with PWM supplies. Gennum Corp.

The schematic circuit of a dc-to-dc converter using the GP605 controller dc is shown in Fig. 18-10. This converter accepts 48 Vdc input and outputs 5 Vdc at 20 A. One of the salient features of resonant-mode operation is that very high switching rates can be had with minimal switching losses. In this particular design, the switching rate is in the 500-kHz region and the resonant frequency of the LC tank is about 600 kHz.

The circuit uses a pre-driver stage configured about Q3 and Q4. With PWM systems, which often switch at 100 or 200 kHz, the pre-driver can often be dispensed with. However at the higher switching rates enabled with resonant-mode operation. It is beneficial to have a means of quickly charging and discharging the rather high gate capacitance of power MOSFET switches. The drive provision must exhibit a lower impedance and deliver a greater instantaneous current than is readily obtained from many control ICs.

Schottky diodes CR1 and CR2 merit comment. These isolate and bypass the internal parasitic diode (not depicted on the schematic) of Q1, the power switch. The Schottky diodes are not slowed down by storage time effects and are better suited for the needed "free-wheeling" function.

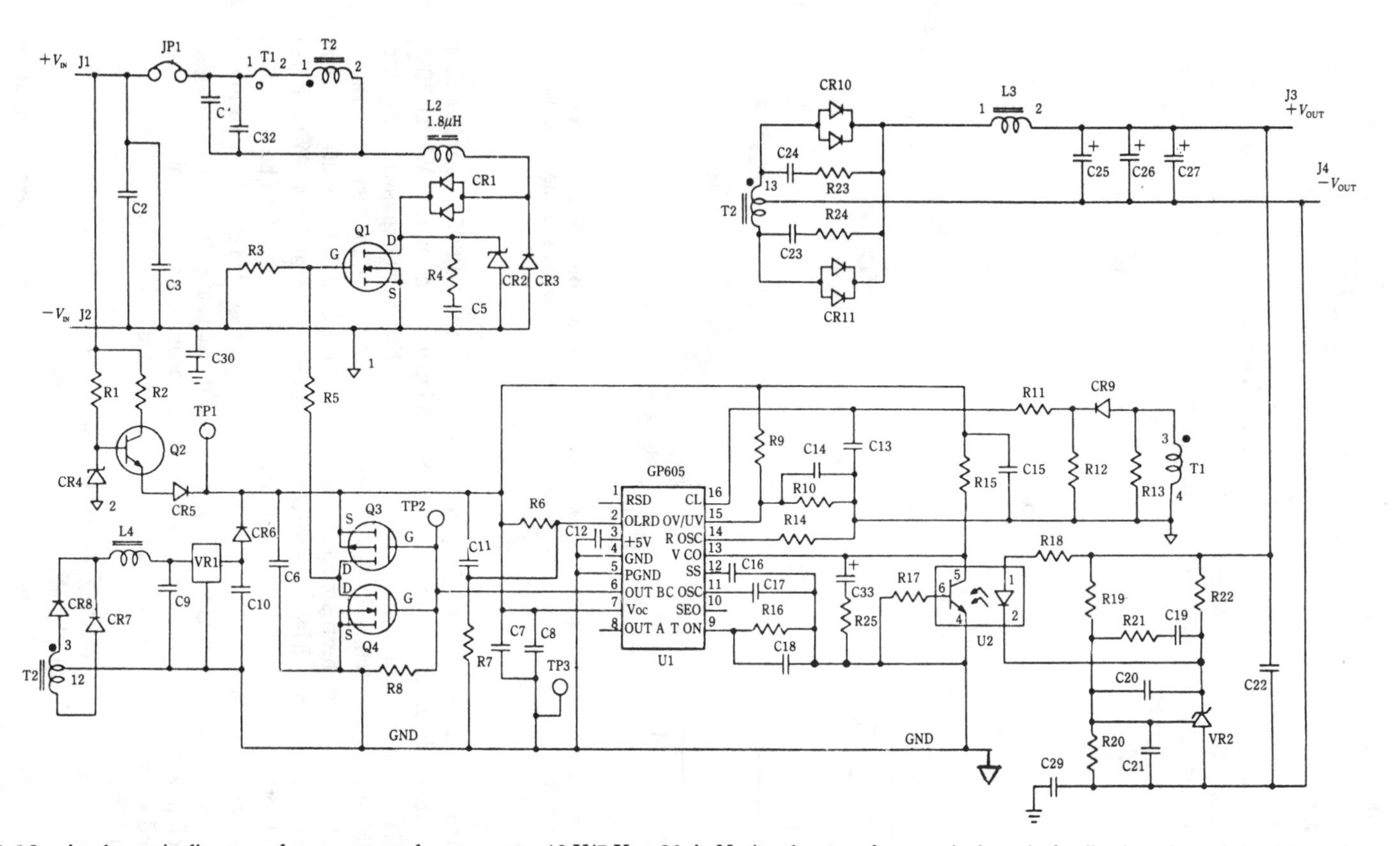

18-10 A schematic diagram of resonant-mode converter: 48 V/5 V at 20 A. Notice the use of an optoisolator in feedback path, and the driver circuit Q3/Q4. At lower frequencies and lower power levels, the driver might not be necessary. Gennum Corp.

The circuitry associated with npn transistor, Q2, comprises a simple voltage regulator, which provides an independent source of dc for the pre-driver, the under-/over-voltage circuit of the GP605, and the output transistor of the optoisolator. Table 18-3 lists the required parts.

Table 18-3 The parts list for the 48-V/5-V, 20-A resonant-mode converter. Gennum Corp.

Part No.	Description	Part No.	Description
C2,C3	6.0µF 100V ±10% Electrocube 230BIB605k	R5	2.2 Ω 1/4W 5%
C4,C32	0.022µF 200V NPO ±5% KEMET C340C223J2G5CA	R6,R7	300 kΩ 1/4W 5%
C5	680pF 200V NPO (COG) ±5% KEMET C323C681J2G5CA	R9	3.6 kΩ 1/4W 5%
C6,C8	22µF 16V tant. dipped cap. Sprague 199D226X9016DA1	R10	1 kΩ 1/4W 5%
C7,C10,C15,C19,C20,C22	0.22µF 50V Z5U ±20% Sprague 1C10Z5U224M050B	R11	220 Ω 1/4W 5%
C9	47µF 20V tant. dipped cap. Sprague 199D476X9020EE2	R12	36 Ω 1/4W 5%
C11,C16,C33	4.7µF 16V tant. dipped cap. Sprague 199D475X9016BA1	R13	22 kΩ 1/4W 5%
C14,C21,C12	0.001µF 100V ±10% Sprague 1C10X5R102K100B	R14	26.1 kΩ 1% 1/4W
C13	0.022µF 50V ±20% Sprague 1C10Z5U223M050B	R15	3.9 kΩ 5% 1/4W
C17,C18	100pF 100V ±5% Sprague 1C10C0G101J050B	R16	10 kΩ 1% 1/4W
C23,C24	0.0022µF 100V COG Sprague 1C10C0G222T050B	R17	1 MΩ 5% 1/4W
C25,C26,C27	220µF 10V 20% solid tantalum KEMET T262D227M010MS or CSR21C227KM	R18,R22	470 Ω 5% 1/4W
C29,C30	0.0022µF 500V CER. DISK Sprague 5TSD22 or 5GAD22	R19,R20	1 kΩ 1% 1/4W
CR1	Schottky diode T0-220AB DUAL Amperex BYV4335 or Motorola MBR2035CT	R21	10 kΩ 5% 1/4W
CR2	Zener diode 170V ±10% 5W IN5385A	R23,R24	51 Ω 1W 5% RCD RSF1A
CR3	UFRD 15 A 200 V T0 220 AC Single Amperex BYV29-200 or Motorola MUR1520	R25	100 Ω 5% 1/4W
CR4	Zener diode 12V ±5% I/4W IN4699	L2	2µH ±5% Core: T68-2D Micrometals 10 Turns #30 AWG x 5 (wire length = 10 inches)
CR5,CR6	IN4001 IA 50V	L3	MTI-125-02-02 GAP=.012 Multisource Technology Inductor
CR7,CR8	UFRD AXIAL 1A 200V Amperex BYV27-200 or Motorola MUR120	L4	330µH 0.6 ohm Inductor AL0410-331K Northeastern Electronics (315)455-7561
CR9	IN4148 Signal diode	T1	Current Sense Transformer: Primary: 1 turn #18 AWG Secondary: 50 turns #34 AWG Core: Ferroxcube 1041CT060/3E2A
CR10,CR11	Schottky diode T0-220, DUAL Motorola MBR2545CT or Amperex BYV 43-45	T2	MTT-125-DC-06-02-06C Multisource Technology Transformer
Q1	MOSFET IRFP250 International Rectifier T0-247AC	VR1	LM7812 12V regulator
Q2	TIP 29C TI bipolar transistor	U1	GP605 Gennum Resonant Mode Controller
Q3	MOSFET VP0104N3 Supertex or Plessy ZVP2106A	U2	CNY17-4 TRW Opto-coupler
Q4	MOSFET VN1306N3 Supertex or Plessy ZVN3306A	VR2	TL431 CLP TI Shunt Regulator
R1	180 Ω 1/2W 5%	MOUNTING HARDWARE	4 - 40 nuts and flathead bolts TO - 220 insulating bushings Bergquist insulators K4 - 90, K4 - 35 Washers for PCB spacing Baseplate Heatsink PCB Keystone #8190 terminals with screws
R2	3.9 kΩ 1/2W 5%		
R3,R8	51 kΩ 1/4W 5%		
R4	47 Ω 3W 5% RCD RSF2B or Clarostat VC-3D		

As has been mentioned, a side product of dedicated power-supply controllers is the large number of options that "come along for the ride." These provide protection to both, supply and load, various operating conveniences, and design flexibility. A few of these options available through the use of the GP605 IC controller are listed as follows:

Remote shutdown can be programmed via logic levels applied to pin 1 (RSD). Specifically, logic low will shut down the supply. Actually, logic high is not used, but if logic low is then removed, the power supply controlled by the GP605 will go into soft-start and will resume operation. The time spent in the soft-start process is determined by the size of an external capacitor connected from pin 12 (SS) to ground. In actual practice, soft-start should be slower for higher power levels. Thus, whereas 50 ms might be suitable for a 100-W supply. It would be wise to allow about 500 ms for a 500-W supply. Soft-start commences within the GP605 by initially using a low-frequency switching rate, and gradually increasing it until regulation sets in.

The maximum switching rate allowed by the VCO of the GP605 is determined by an external capacitor connected from pin 11 (C OSC) to ground and an external resistor connected from pin 14 (R OSC) to ground. Also, the fixed duration of the on pulse from the monostable multivibrator within the GP605 is set by an external RC network, connected from pin 9 (T ON) to ground.

A window comparator within the GP605, in conjunction with an external RC network associated with pin 15 (OV/UV), govern automatic shutdown levels for overvoltage and undervoltage operating conditions. Similarly, an external time constant associated with pin 2 (OLRD) determines how long the supply remains off after the initiation of an automatic shutdown. The net effect of these connections is that the power supply protects itself in a "hiccup" mode—it repetitively seeks normal operation via the soft-start process and attains it once the fault is removed from the system.

From the design standpoint, it is relevant to note that if pin 10 (SEO) is grounded, complementary output is provided for push pull power switches. If pin 10 is left open, single-ended outputs at double frequency are available from OUT A and OUT B, and these outputs can be paralleled for greater drive capability (this was not done in this application; instead, a pre-driver was used ahead of power switch Q1).

An alternate design utilizing the resonant-mode principle

Another example of resonant-mode regulation is shown in Fig. 18-11. This circuit can be used either as dc to dc converter by connecting a 300- to 350-Vdc source to input terminals E1 and E2, or as an ac-to-dc power supply by connecting input terminals E3 and E4 to the 115-V 60-Hz utility line. In either case, 28 V at 5 A of regulated dc will be available from output terminals E5 and E6. The resonant frequency of the LC tank (L1/C4) is 660 kHz. The overall efficiency at 140 W full load is about 85%. Line regulation is 0.1% and load regulation is 0.5%. The parts list for this circuit appears in Table 18-4.

Notice that, unlike the previous resonant-mode regulator, this circuit obtains its power-switch drive directly from the control IC, making an external driver unnecessary. Also, this circuit utilizes a half bridge arrangement of dual-power MOSFETs, contrasted

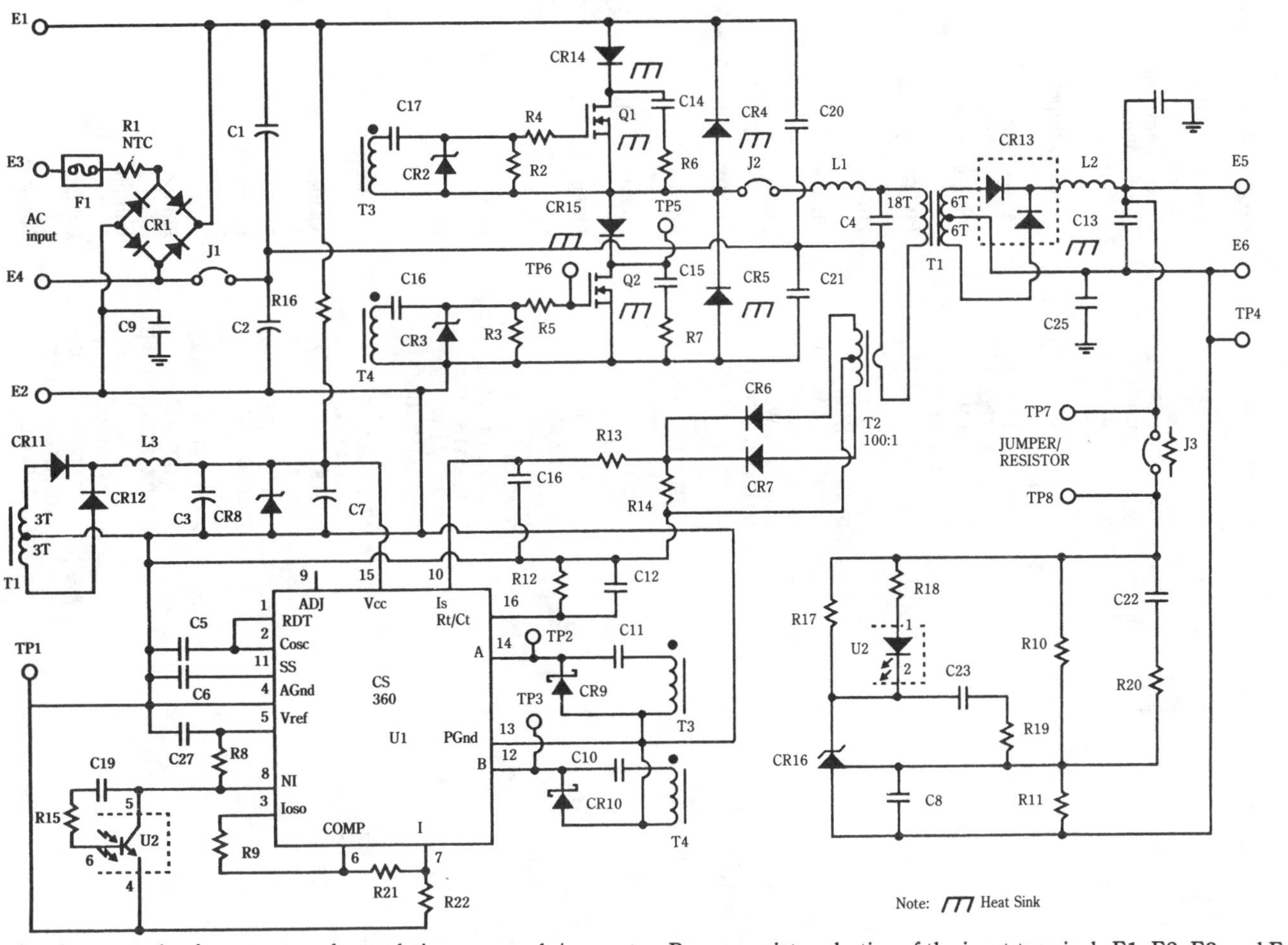

18-11 Another example of resonant-mode regulation—a supply/converter. By appropriate selection of the input terminals E1, E2, E3, and E4, this circuit can perform either as an ac-to-dc power supply or as a dc-to-dc converter. Cherry Semiconductor Corp.

Table 18-4 The parts list for the resonant-mode supply/converter.

Parts List

Reference	Qty	Description	Source/Manufacturer
R1	1	NTC WS610	Western Electronic Components
R2,R3	2	1KΩ 1/4W 5% 5043EM1K00J	Philips Components, Mineral Wells, TX
R4,R5	2	2.2Ω 1/4 W 5% 5043EM2R200J	Philips
R6,R7	2	300Ω 5W 5% SQR 5	Eurasia, Taipei
R8,R19	2	6.8KΩ 1/4 W 5% 5043EM6K800J	Philips
R9	1	5.6KΩ 1/4 W 5% 5043EM5K600J	Philips
R10	1	10.2KΩ 1/4 W 1% 5043ED10K20F	Philips
R11	1	1KΩ 1/4 W 1% 5043ED1K00F	Philips
R12	1	9.09KΩ 1/4 W 1% 5043ED9K090F	Philips
R13	1	220Ω 1/4 W 5% 5043EM220R0J	Philips
R14	1	60.4Ω 1/4 W 1% 5043ED60R40F	Philips
R15	1	1.5MΩ 1/4 W 5% 5043EM1M500J	Philips
R16	1	200KΩ 1W 1% 5073YD200K0F	Philips
R17	1	8.25KΩ 1/4 W 1% 5043ED8K250J	Philips
R18	1	7.5KΩ 1/4 W 5% 5043EM7K500J	Philips
R20	1	1.6KΩ 1/4 W 5% 5043EM1K600J	Philips
R21,R22	2	10KΩ 1/4 W 5% 5043EM10K00J	Philips
C1,C2	2	470μF 200V LLQ2D471MHSA	Nichicon
C3	1	100μF 25V USR1E101MCA	Nichicon
C4	1	5.6nF 400V Polyprop 716P56294J	SB Electronics Barre, VT
C5	1	390pF 50V NPO ±5% K391J15COGFVAWA	Philips
C6,C16	2	22nF 50V Z5U K223Z15Z5UFVCWE	Philips
C7	1	0.1μF 50V X7R ±20% K104M20X7RFVCWN	Philips
C8	1	1nF 100V X7R ±10% K102K15X7RHVAWA	Philips
C9	1	50nF Cer Disc 500V 5GAS50	Sprague
C10-11,C17-18	4	0.47μF 50V Z5U ±20% K474M30Z5UFVCWY	Philips

CSC™ 7

Table 18-4. Continued

Reference	Qty	Description	Source/Manufacturer
C12	1	330pF 50V NPO ±10% K331K15COGFVAWA	Philips
C13	1	470μF 35V Al El UPL1V471MRH6	Nichicon
C14,C15	2	390pF 500V Mica FD391G03	Cornell Dubilier Wayne, NJ
C19	1	100pF 50V NPO ±5% 1C10COG101J050B	Sprague
C20,C21	2	47nF 200V Polyprop Stk No. 89F3474	Newark Electronics
C22	1	18nF 50V X7R ±10% K183K20X7RFVBWD	Philips
C23	1	27nF 50V X7R ±10% K273K20X7RFVBWF	Philips
C27	1	10nF 50V X7R ±10% K103M15X7RFVCWA	Philips
C25,C26	2	2.2nF Cer Disc 500V 5TSD22	Sprague
CR1	1	4A/400V Br. Rect. 4PH40	Electronic Devices Inc. Yonkers, NY
CR2,CR3,CR8	3	18V Zener 1W 1N4746A	Philips
CR4,CR5	2	URFD, 400V/9A BYV29-400	Philips
CR6,CR7	2	1N914B Diode	Philips
CR9,CR10	2	Schottky, 1A/30V BYV10-30	Philips
CR11,CR12	2	VFSR, 200V/1.6A BYV36-200	Philips
CR13	1	UFRD 300V/20A BYV34-300 Diode	Philips
CR14,CR15	2	Schottky, 35V/16A PHBR1635	Philips
CR16	1	TL431LP	Motorola
Q1,Q2	2	MOSFET, 400V/2.4A BUK444-400B	Philips
U1	1	CS-360	Cherry Semiconductor
U2	1	Optocoupler CNY17-4	Siemens
L1	1	Core: T94-2D Inductor: 10.2μH (25T, Litz 20/34)	Micrometals Anaheim, Calif
L2	1	Inductor MTI-125-12A Gap: 15 mils	Multisource Technology Waltham, Mass
T1	1	Power Transformer MTT-125-DC-18C-12C-06C	Multisource Technology
L3	1	330μH Inductor AL0410-331K 0.6W	Northeastern Electronics Elbridge, NY

8 CSC™

Table 18-4. Continued

Reference	Qty	Description	Source/Manufacturer
T2	1	Toroid n:1/100 Pri: 1T 18AWG Sec: 200CT 32AWG 2x70" wire req'd Core:768XT188-3E2A	Philips Saugerties, NY
T3, T4	2	Core: 266CT125-3D3 Toroid n:1 Pri,Sec: 10T 3x32AWG	Philips
E1,E2	2	Turret Carlac 10-203-2-01	Bohemia, NY (Tel: 516 567 4200) Manuf: Concord
E3-6	4	Terminal Post	Allied Electronics Keystone No. 8190 Billerica, Mass
	2	Fuse Clips P/N: 798	Zierick, Mt. Kisko, NY (Tel: 800 882 8020)
	5	Sil Pads 3223-07FR-54	Harman & Assoc. Brookline, Mass
TP1-6	6	Solder Terminal 10-833-2-04	Concord, Carlac
F1	1	2A 250V 3AG Fuse Stk. No. 27f658	Newark Electronics
	1	Isolation Tape No.20 3M Tape No. 20 7-2/5"x 4-3/4"	Bristol Tape, Fall River, Mass
	1	Baseplate & Heatsink 1/8" 6061 Al Stock black anodized	SPS Machinery & Design Mississauga, Ontario or Harris Textiles E. Greenwich, RI
	5	Flat Head Slotted Machine Screws 91781A108 4/40-3/8"	McMaster-Carr New Brunswick, NJ
	7	4/40-1/2" Mch. Scrw 97781A110	McMaster Carr
	5	Washers 46F7485	Newark Electronics
	12	Hex Nuts 4-40 Stk No. 58F1103	Newark Electronics
	5	SPC Tech Bushing Stk. No. 28F1914	Newark Electronics
J2	1	Jumper wire, 16AWG 19x#28 tinned 2-3/4"	Belden 9980
	1	Jumper wire, 20AWG .625" long	Belden 8529
	2	.8x1 1/4 x.04" No. 4965	Bristol Tape foam
	1	42"-long 20/34 Litz MWS wire	Wire Industries Westlake Vill, Cal.
	1	Printed Circuit Board	Delta-V Richardson, TX

to the single-ended output scheme of the previous supply. Also, not readily apparent from the schematic diagrams, this resonant-mode supply gets by with PN rectifier diodes; the previous circuit uses Schottky diodes to provide the dc output. Notice the snubber networks associated with those Schottky diodes the high internal capacitance of Schottky diodes tends to resonate with transformer leakage inductance, thereby producing dangerously high ringing voltages.

In both circuits, it is desirable to control the Q of the series resonant tank by selection of an appropriate LC ratio, rather than by dissipative losses in these elements. That is, both the resonating inductor and capacitor should be as ideal as possible. In both circuits, the resonating inductors are wound with Litz wire in order to keep the effective ac resistance low. Interesting, too, both circuits use output transformers of unconventional fabrication. These are of planar magnetics construction making use of flat copper traces printed on thin dielectric substrates in place of the usual magnet wire. These transformers are pretooled, so excellent repeatability in production runs result.

These circuits, operating at about 30 times the switching rate of the "garden-variety" 20-kHz switcher, are indicative of the progress made in power-supply technology during the decade of the 1980s. Although the basic idea had been around earlier, successful implementation had to await the availability of appropriate control ICs, power switches, and passive components to enable high-frequency operation.

Designing the tank circuit for the resonant-mode regulator

The resonant frequency of the LC circuit of the resonant-mode supply can be decided upon the basin of such factors as the frequency capability of the IC controller and power switches, the size reduction strived for, the cost and availability of components, etc. Generally, the idea is to take full advantage of the resonant-mode supply to provide efficient operation at higher switching rates than are readily feasible with the pulse-width modulation regulator. Often, the designer will sense the region of diminishing returns and will select a resonant frequency that is somewhat less than the very highest that could be accommodated.

Once the resonant frequency of the LC tank has been selected, it is only natural to ponder the unique combination of L and C values that can be used. The effect of Q has already been mentioned and it is known that in a series resonant circuit, high Q corresponds to a large inductance and a small capacitance (the opposite holds true for parallel resonance). This knowledge provides some useful qualitative insights, but does not suffice to determine the sought after LC values. Further insight into the situation is provided by Fig. 18-12, showing the effect of different L and C values for a given resonant frequency.

A good way to solve this problem is to ball-park an appropriate LC combination from use of fundamental energy and power relationships. First, it is known that the energy stored in a capacitor is given by

$$E = \frac{CV^2}{2}$$

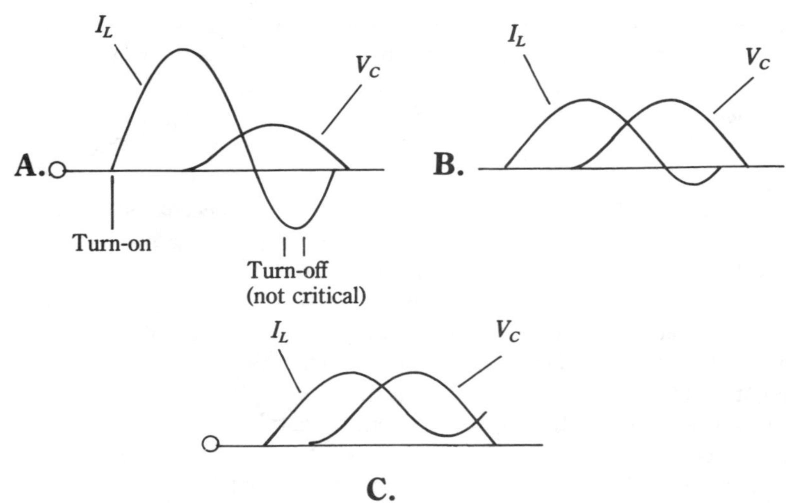

18-12 The effect of inductance and capacitance values in an LC resonant tank. Because a given resonant frequency can be attained with large inductance and small capacitance, or vice versa, it is important to know the effect of such combinations. (A) High C and low L: this LC combination results in high peak current and high losses in the free-wheeling diodes, which occur in the negative region of I_L. (B) Optimum L and C values: negative region of I_L is of small amplitude and the peak current is moderate. Low C and high L: non-return to zero of I_L. Although a further reduction in peak current is attained, this condition negates the salient feature of the resonant-mode supply—switching at zero current. Current in the MOSFET switches is zero during negative excursion of I_L.

In an oscillatory circuit this basic relationship can be expressed in terms of power by multiplying by the frequency, thus:

$$P = \frac{CV^2f}{2}$$

Next, a simple algebraic transformation is performed to express this equation in terms of C. It is

$$C = \frac{2P}{V^2f}$$

This puts you in a nice situation because the desired output power of the supply is known, as in the dc voltage impressed across the resonant tank, and the resonant frequency, itself. But, before using this derived formula for C, the following matters should receive consideration:

- Because the supply cannot be 100% efficient, a reasonable value of efficiency must be introduced as a modifier. Assume an efficiency of 80%.
- The worst conditions of operation occur at low ac line voltage and full dc output from the regulator. Under such conditions, it is necessary to estimate, or otherwise determine, the available dc voltage to shock-excite the resonant

circuit. This will be the value of V to use in calculating the minimum size of the resonating capacitor, C. Finally,

$$C = \frac{2P}{(0.80)(V)^2(f)}$$

is to be used, as stipulated above for a reasonable determination of C. Then, from the resonance relationship,

$$L = \frac{1}{4\pi^2 f^2 C}$$

Another blockbuster: the current-mode power supply

Like the resonant-mode supply, the current-mode regulating power supply adds a dramatically improved performer to the arsenal of power-supply technology. And, as with the resonant-mode technique, current-mode operation was relegated to the side lines until dedicated control ICs became commercially available. The best way to understand the current-mode supply is to view it as a basic PWM circuit with something added. The added something is an "internal" feedback path, which senses instantaneous current through the power switch. See how the performance is modified by the addition of this second feedback path.

First, consider an audio amplifier comprised of, for example, four cascaded stages (Fig. 18-13). Let there be two negative feedback paths, one from the output of the second stage to the input of the first stages, the other from the output of the final stage to the input of the first stage. This, indeed, is not an uncommon circuit format in audio practice. In such a configuration, the all-embracing feedback path—the path from the final stage to the first stage, will, if sufficient feedback is used, dominate the circuit

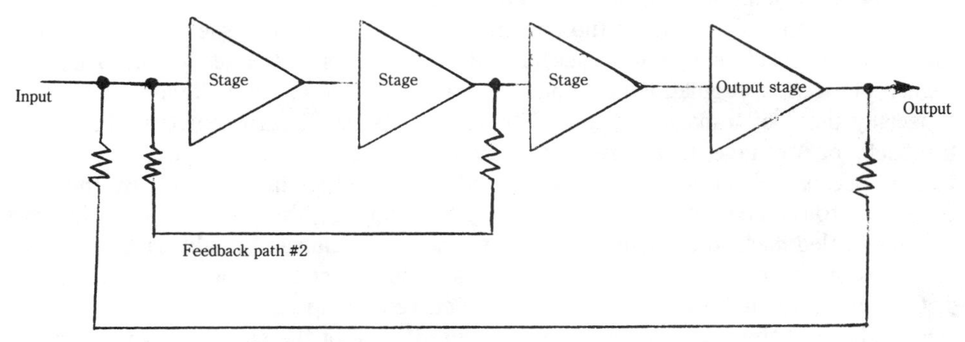

This outermost-feedback cancels any disturbance in the output that feedback-path #2 tends to produce.

18-13 An audio-amplifier format for demonstrating the effect of 2 feedback paths. The basic idea is that the all-embracing path, #1, dominates the output characteristics of the amplifier. In analogous fashion, the *outer* feedback path of the current-mode regulator continues to regulate dc output voltage, despite the introduction of a second feedback path.

behavior in modifying the input and output impedance of the amplifier, in reducing distortion and in improving linearity. To a first approximation, the amount of feedback used in the "internal" path will have relatively little effect on the gain and other performance parameters of the amplifier—the "all embracing" path will regulate out any tendency toward change, regardless of its origination.

Referring now to Fig. 18-14, the above analogy should be useful in tracing out the overall operation of the current-mode supply. The "internal" feedback path samples the current ramp through the power switch, and through a PWM modulator, which momentarily alters the width of the switching pulses. However, the output voltage of the supply is negligibly affected because the conventional PWM "voltage model" feedback path dominates load-circuit regulation. That is, the "all-embracing" feedback path stabilizes the dc output voltage of the supply, no matter what tends to change it. Having added the second feedback path, it is, however, natural to ponder what has thereby been gained. Here, "common sense" might well fail us because it might seem that the effect of the added current feedback has been altogether negated by the dominant action of the voltage-mode feedback path.

Performance benefits of the current-mode regulator

The audio amplifier analogy to the current-mode regulator serves as an aid to understanding. However, it naturally engenders a question: why add the internal feedback path if the outer feedback path dominates the output characteristics of the amplifier? The answer here is that certain performance features of the amplifier can nevertheless be improved. For example, appropriate use of the added feedback can widen the bandwidth, improve stability, extend dynamic range, and enhance linearity. This being the case, it should not come as a surprise that the performance of the PWM regulator can also benefit by addition of an inner feedback path.

One of the major benefits the current-mode regulator possesses, with regard to the "plain vanilla" (single feedback) PWM regulator is dramatically improved line regulation, that is, regulation of output voltage in the face of line-voltage variations—especially those of transient nature. Why should this be? It turns out that the single-feedback, or "voltage-mode" regulator cannot act rapidly to quick changes in line voltage because of the inherent delay caused by the output filter circuit. By the time correction to the line-voltage change is ready to be applied, the line voltage might have already settled back to its nominal level, or a new line transient might have occurred.

In contrast, the current-mode format responds almost instantaneously to any line disturbance. This is because sampling of the current ramp through the power switch occurs ahead of the output filter. Another consequence of the sampling of the current ramp is that the regulator is able to limit current on a cycle-by-cycle basis. Therefore, no additional protection is needed to prevent excessive current demand from the supply.

There is yet another benefit conferred by the current-ramp sampling process. The phase gain margin of the regulator is greatly extended because the output filter no longer imposes its 40-dB per octave roll-off characteristic. This is a mathematical subtlety, but its practical consequence is much greater stability. A corollary of this is

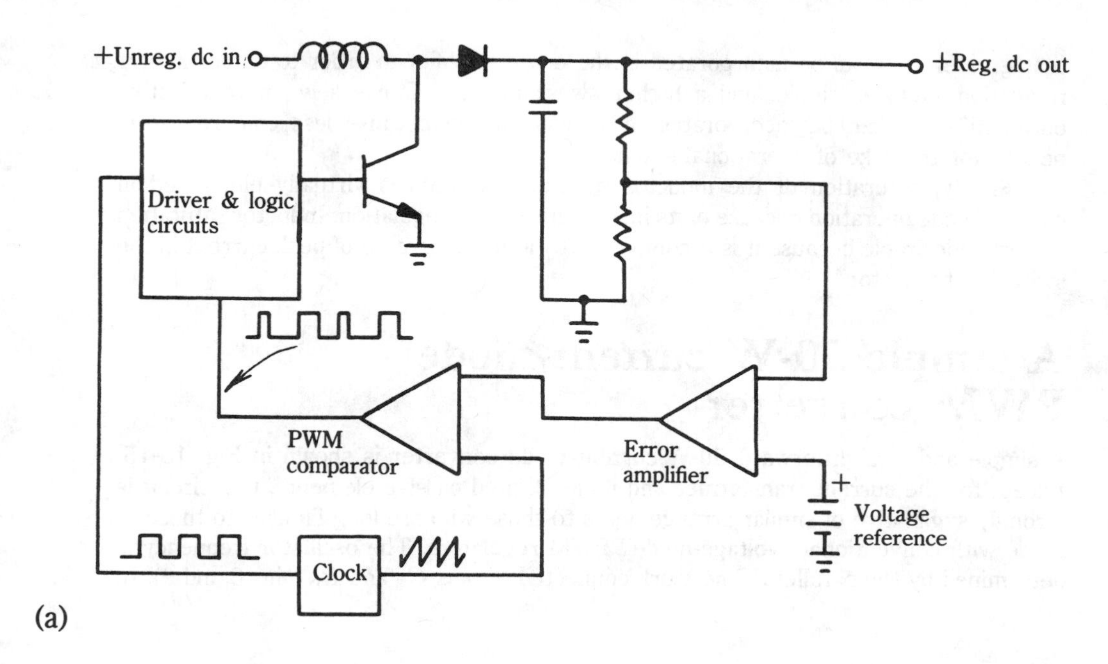

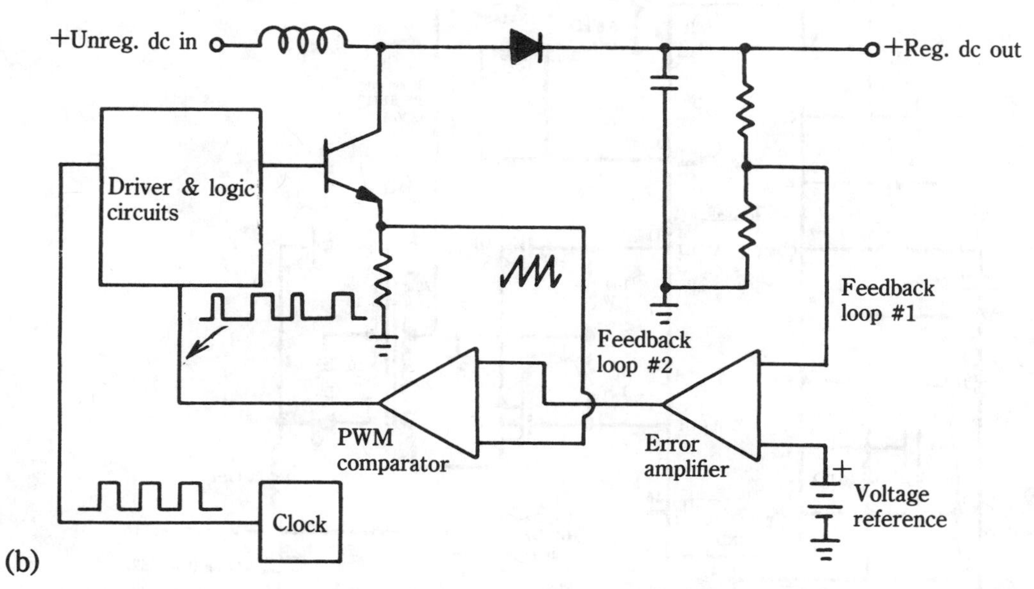

18-14 A comparison between voltage-mode and current-mode regulating supplies. This current-mode circuit contains an additional feedback path (A). A conventional voltage-mode configuration. (B) Current-mode configuration. Notice the retention of the original feedback path that is depicted in A.

that greater gain can be incorporated in the error amplifier in order to obtain tighter regulation, reduced ripple, and a higher switching rate. Conversely, more effective output filtering can be incorporated in a given design because less compromise is needed for the sake of operational stability.

Finally, saturation of the inductor by line transients is virtually eliminated in current-mode operation because of its instantaneous line regulation. Inductor saturation is very undesirable because it is accompanied by a great increase of peak current in the switching transistor.

A simple 50-W current-mode PWM converter

A simple and straightforward 50-W current-mode converter is shown in Fig. 18-15. Except for the current transformer and its associated passive elements, this circuit is strongly suggestive of similar arrangements to those who are long familiar to those at home with conventional "voltage-mode" PWM regulators. The oscillator frequency is determined by the parallel RC network connected to pins C_T and R_T (pins 8 and 9). In

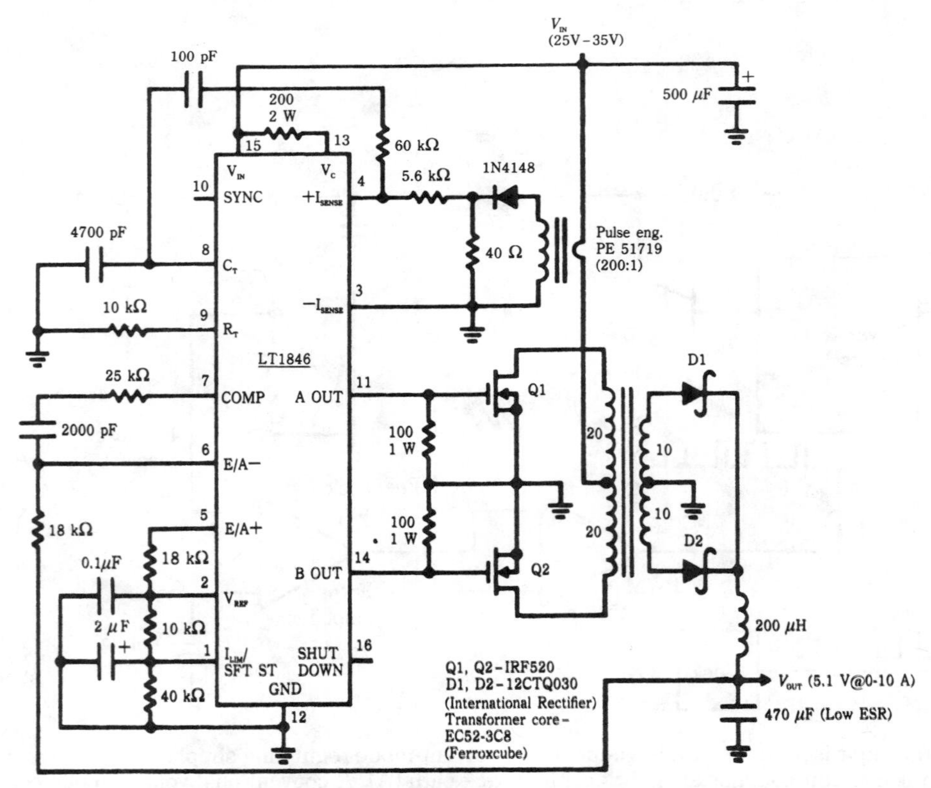

18-15 A current-mode forward converter with a 50-W push-pull output. Notice the current transformer for sampling current ramps in the Q1/Q2 switching stage. Linear Technology Corp.

this case, it is about 45 kHz. The parts count is relatively low because no pre-driver is needed, and all protective and house keeping functions are contained within the LT1846 IC controller. Although this circuit cannot be classified as a high-frequency converter, the presence of the current-mode operational principle confers worthwhile performance advantages.

In order to get the feel of the current-mode switcher, you should become familiar with the block diagram of the IC controller, in this case, the LT1846 shown in Fig. 18-16. In so doing, notice that the comparator is subjected to two error signals; one is the conventional voltage feedback from the dc output, the other is a voltage representing the current ramp in the switching stage. The "conventional" feedback (pins 5 and 6) causes stabilization of the dc output voltage, regardless of the nature of tendencies trying to change it. The current feedback (pins 3 and 4) produces near instantaneous regulation against changes in line voltage. Also, as previously mentioned, the current-mode principle tends to immunize the regulator against instability.

For the experimentally inclined, direct current sensing in a sampling resistance can sometimes be used in place of the current transformer (Fig. 18-17) for a single-ended switching stage. However, what is gained in simplicity and cost savings, has to be balanced against the power dissipation in the sensing resistance, as well as the loss in electrical isolation. Notice that the RC filter (often used with the current transformer method, as well) is only for the purpose of attenuating the switching transient. The RC time constant must not be so great as to appreciably distort the shape of the voltage representing the sampled current ramp.

A 500-W 200-kHz current-mode power supply

The current-mode regulated power supply shown in Fig. 18-18, operates directly from either a 120-V (or 240-V) 50- to 60-Hz line, and can provide up to 500 W of dc output power. Its design is predicated on a sophisticated, yet easy-to-use control IC, the CS 3842A. This is not a recommended project for the neophyte, if for no other reason that this book is intended to provide generalized guidance for the experienced electronics practitioner, not detailed "how-to-build-it" instructions. Aside from this, the combined high-frequency and high-power performance ratings are likely to tax the skills of even the specialist. Nonetheless, the parts values are indicated and the knowledgeable worker might well try for a "Chinese copy." Better still, many aspects of the circuit are amenable to various modifications without drastically changing the basic operational philosophy.

The first thing you should do is become familiar with the block diagram (Figs. 18 and 19) and pin out of the CS 3842A so that the basic reason for the circuit connections can be appreciated. Much of this supply is quite similar to ordinary circuit practice involved in conventional PWM supplies. For example, there is nothing out of the ordinary about the unregulated power supply, which converts the ac line voltage to a nominal 340 Vdc for operation of the two MOSFET power switches. When this unregulated supply is working from 120 Vac, it operates as a voltage doubler. If the line voltage selector switch is placed in its 240-V position, this dc supply then operates as a

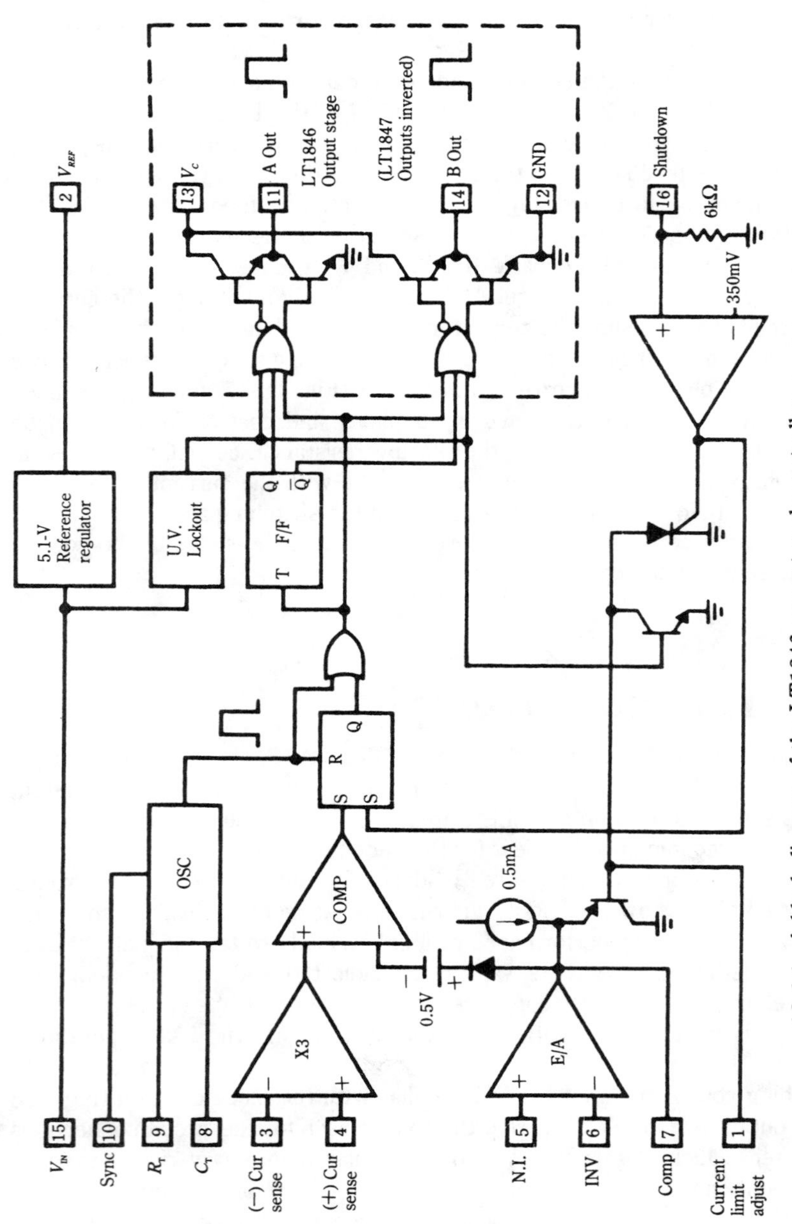

18-16 A block diagram of the LT1846 current-mode controller. Linear Technology Corp.

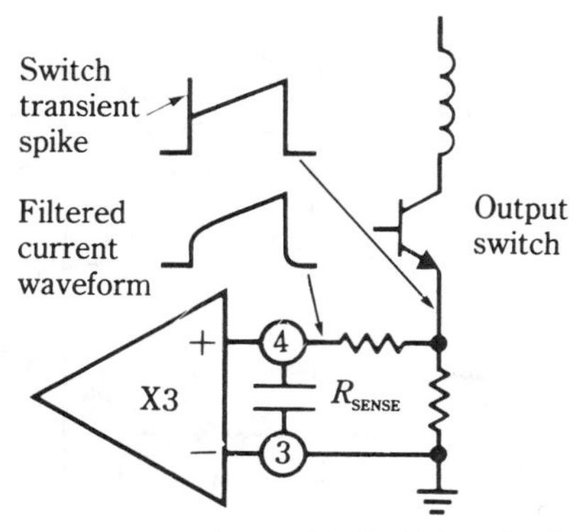

18-17 An alternative current-ramp sensing method. Although simple and cost-effective, this direct sampling method suffers from power loss in the sensing resistance. It also lacks the isolation provided by a current transformer.

full-wave bridge circuit. In either case, the dc voltage produced is the same. Caution: do not connect to a 240-Vac line with the voltage selector switch in its 120-V position.

The drive provision for the power switches is quite interesting. Drive transformer T2 has two secondaries. The phasing of these windings is very important because it is the means whereby the single-ended output of the CS 3842A IC drives the series-connected power MOSFETs. It is noteworthy that the drive capability of this control IC makes a pre-driver unnecessary. (At high frequencies, the capacitance of the gate circuit of larger power MOSFETs often makes fast switching difficult.) The power MOSFETs operate as a single-ended, not as a push-pull stage.

In this supply, an auxiliary winding on the power transformer, T1, is used as the source of power to operate the control IC after operation sets in; prior to that, the dc operating power for the control IC is derived from the main dc source through the 250-kΩ resistance. The auxiliary winding is associated with its network of rectifying and Zener diodes, as well as the 100-μF filter capacitor, as shown.

Next observe another off-the-beaten-path circuit technique. Notice the use of small transformer, T4, as an isolation element in the voltage-mode feedback path. More familiar is the use of an optoisolator for this purpose. The use of a transformer dictates the sensing of the ac component, rather than the filtered dc voltage level. Thus, the 2N2904 transistor is utilized to pass the ac component of the rectified, but unfiltered, output through the primary winding of T4. One of the secondary windings then, with the assistance of a diode and a filter capacitor, applies the "recovered" error voltage to pin 2 of the CS 3842A IC (of the two 2N914 diodes, only one performs the needed rectification; the other provides temperature compensation). Insofar as concerns pin 2, the input to the error amplifier, it does not matter whether the sampled error signal is obtained directly, from a photoisolator, or by a transformer rectifier arrangements, as used in this power supply.

The basic information for construction of T1, the power transformer follows:

- *Core* Magnetics Inc. PQ 4040, P material
- *Primary* 30 turns of six #24 wires in parallel.

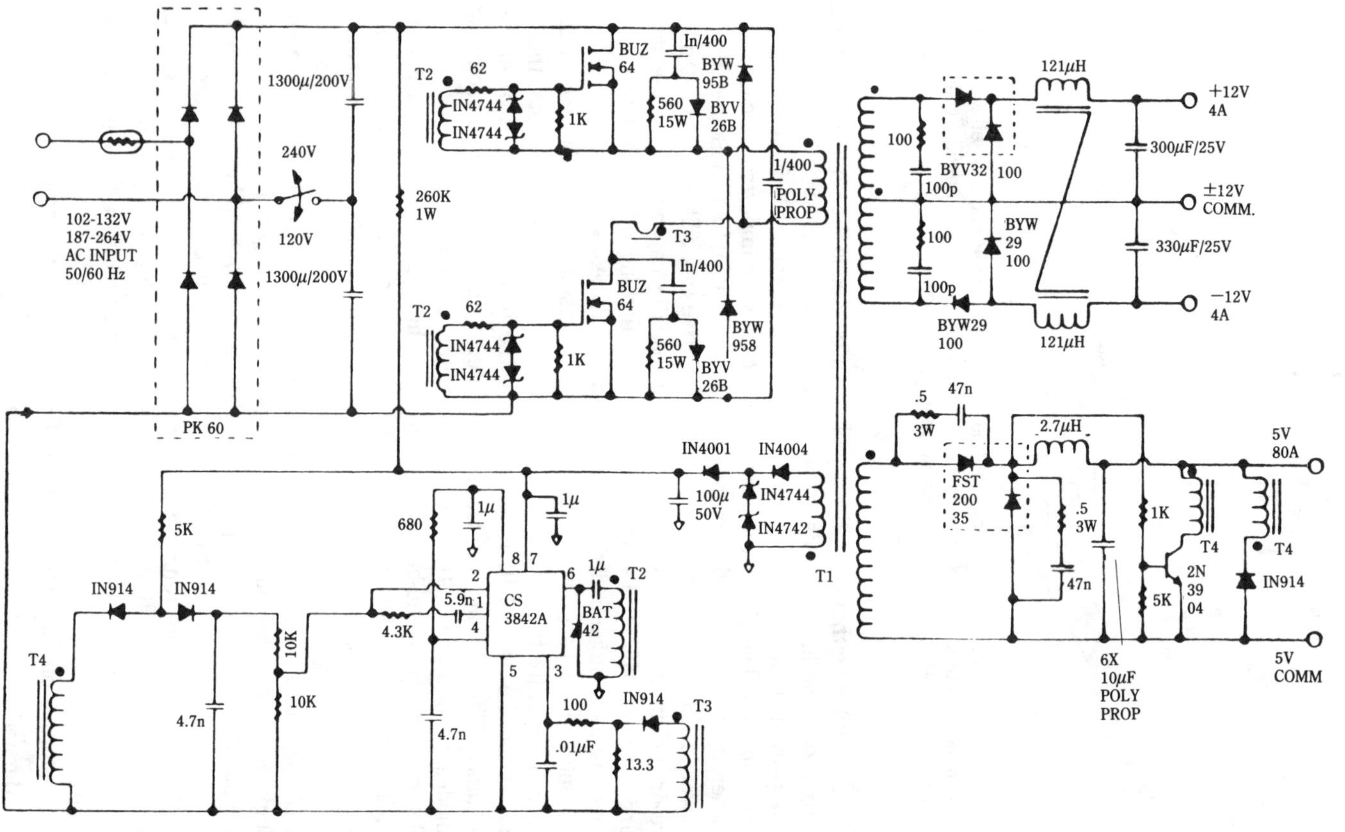

18-18 The circuit of 500-W 200-khz current-mode power supply. The combination of high power and high switching rate is noteworthy. Further circuit simplification can readily be implemented by the knowledgeable experimenter. Cherry Semiconductor Corp.

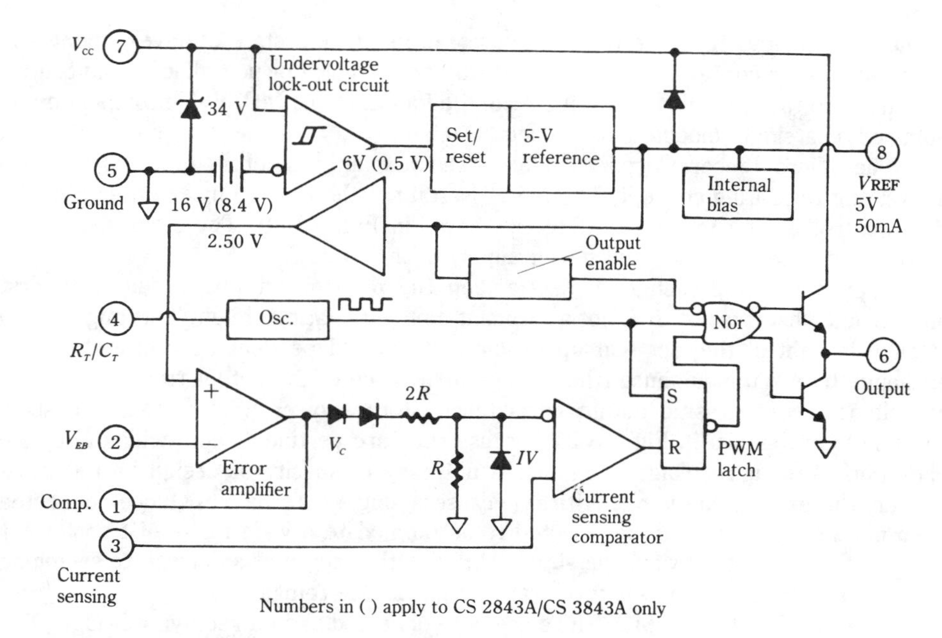

18-19 The block diagram of the CS-3842A current-mode IC controller. Cherry Semiconductor Corp.

- *5-V secondary* two turns of 16-mil copper foil, 0.9 inch in width.
- *±12-V secondary* two 5-turn sections using two #19 wires in parallel.
- *Auxiliary secondary* two physically separated windings are used and these are then connected in parallel. Use four turns of #30 wire in each of these windings.

The reason for the 6-parallel-wire primary is to reduce skin effect losses that would be higher if a single large wire was used. The sectioning of the secondary windings is to reduce leakage inductance. Indeed, the primary is also split. The basic idea is to construct a distributed winding format so that all windings are as tightly coupled as is physically possible. It is also recommended that shields be placed between the primary and the 5-V secondary windings.

The power transformer has been designed to operate well below the high-temperature saturation flux density of the core material, which is about 3000 gauss. In some cases, however, a slight air gap might prove beneficial.

The phase-modulation regulating supply

The phase-modulation supply is of more recent development than any of the regulating power supplies thus far covered. It might well be the harbinger of future designs, where a favorable mix of efficiency, cost, and easily manageable electrical noise is sought. This technique also appears well-suited to production goals because its performance is less vulnerable to the vagrancies of external components than other high-frequency designs. At first inspection, phase modulation might appear more complex than PWM or resonant-mode supplies. This might, indeed, have been perceived as a practical obstacle to economic exploitation at one time. However, in its present state of development, the user need not be concerned in this regard whether complex or not, all logic and control are built into the low-cost control IC. Also, the four external MOSFET switches can be obtained as a single module from various vendors.

The Micro Linear Corporation has been a main pioneer in practicalizing this interesting regulating concept. The block diagram of the Micro Linear Phase Modulation Controller, the model ML4818, is shown in Fig. 18-20. The simplified circuit depicting its use in a regulated power supply is shown in Fig. 18-21.

A profitable approach to understanding the principle of this circuit is to first understand what it is not. It is not a resonant-mode regulating circuit. This might seem strange in light of the fact that operation is strongly dependent upon stored energy exchange between inductance (the leakage inductance of the output transformer) and capacitance (the parasitic output capacitance of the power MOSFETs). The stray reactances are shown in Fig. 18-21 because they are pertinent, not incidental to the operation. Although leakage reactance is not easy to specify or design into a transformer, the exact value is not critical because a single external resistance connected from pin 12 to ground can be selected to accommodate a wide range of transformer designs. Stated another way, the supply is not at the mercy of any discrete switching rate for proper operation; whatever rate is chosen will remain fixed.

Also, the four power MOSFETs are not connected as in a rectifying bridge. The topography of the connection identifies it as the so-called "H" bridge, which is much used in motor and servo circuits. The H bridge is the functional opposite to the rectifying bridge; the rectifying bridge converts ac to dc, whereas the H bridge enables dc to be converted to ac. Figure 18-22 is an electromechanical analog of the H bridge. Shown are the switch states needed for one full power cycle. You can see that the H bridge is able to convert dc from a single source into true bipolar ac.

In the simplified circuit of Fig. 18-21, the four power MOSFET elements of the H bridge are logic driven in such a sequence that the output transformer receives a half power cycle when Q1 and Q2 are turned on. The remaining half of the power cycle takes place later in the logic sequence when Q3 and Q4 are turned on. The transformer experiences opposite current flow for these two conduction periods. Therefore, true bipolar ac is induced in the secondary (not shown) of the transformer. This is straightforward enough; what happens between conduction periods is, however, the phenomenon that gives rise to zero voltage switching.

A study of the waveform diagram of Fig. 18-23 reveals that conduction periods commence when A1 goes high to form the first half power cycle, then again when A2 goes high to form the second-half power cycle. Fixed delay times postpone the onset of these conductive periods sufficiently to prevent unwanted conductive states in the H bridge. The postponed time intervals are designated as T_{DELAY} and are set in by means of a resistance connected from pin 12 on the ML 4818 to ground. Notice also that the conduction periods end when B1 goes low, and then again when B2 goes low. Thus, the logic states of A1 and A2 decide the onset of the half power cycles, and the logic states of B1 and B2 signal the end of the half power cycles. Here, however, a second time delay, T_{PD1}, is encountered. This time delay, unlike the first, is variable and thereby serves to control the duration of the half power cycles. Because this variable

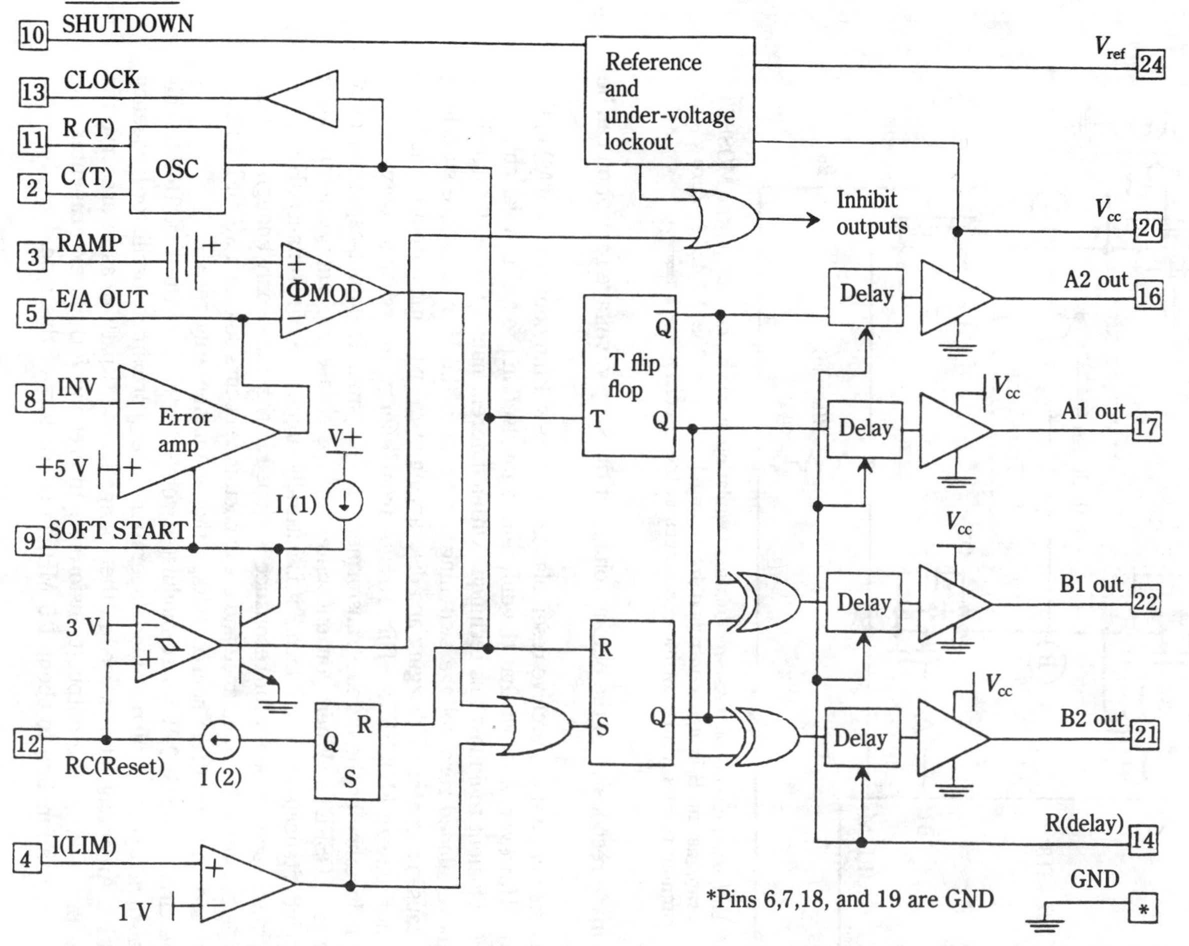

18-20 The block diagram of the ML4818 phase-modulation control IC. The four totem-pole outputs are intended to be used in conjunction with an external H-bridge output stage. Regulation can be accomplished with or without current-mode operation. A fixed switching rate is used. Micro Linear Corporation.

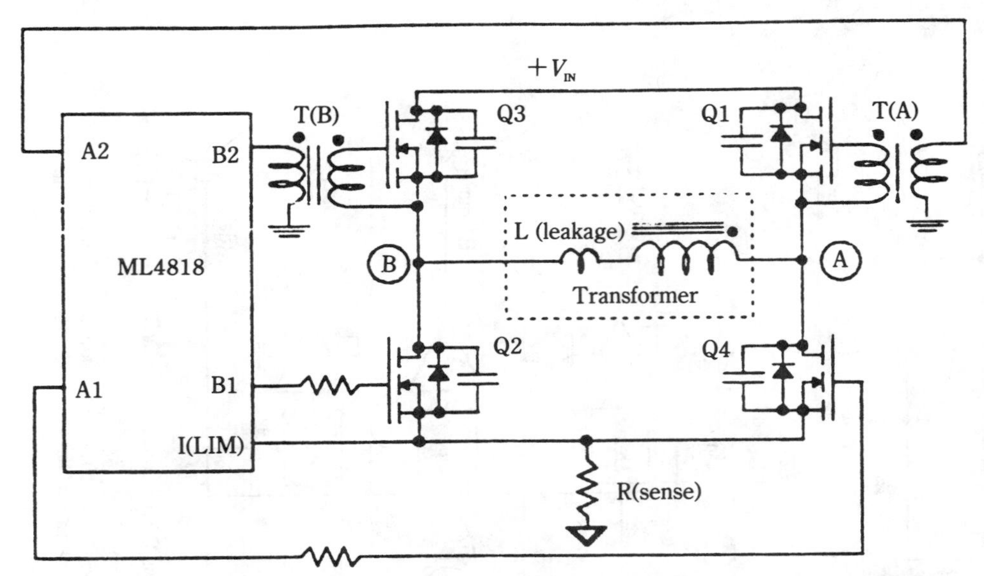

18-21 A simplified circuit of phase-modulation regulating supply. The four power MOSFETs operate as an H-bridge controlled by timing logic from the ML4818. Zero-voltage switching results from parasitic reactances in the circuit. Micro Linear Corporation.

delay is made responsive to the output voltage of the supply, voltage regulation can be accomplished.

It is only natural to perceive resemblance to pulse-width modulation (PWM) in this technique. However, it is felt that the differences involved in producing a variable time delay are sufficient to endow this technique with a different name. Especially significant is the above alluded zero voltage switching occurring in the H bridge. The switching process causes a continual charging and discharging of the parasitic output capacitance of the four power MOSFETs. This results from energy interchange between the leakage inductance of the output transformer and the parasitic output capacitances of the power MOSFETs. When you encounter such a phrase as "energy interchange between inductance and capacitance," it is naturally suggestive of resonance. However, in this circuit such tendency for resonance is nipped in the bud—only enough circulating charge between the inductance and the capacitance occurs so that a MOSFET about to be turned on has a drain source voltage that is momentarily zero.

The circuit of a 350-W phase-modulated converter is shown in Fig. 18-24. This design also incorporates current-mode operation. The approximate oscillator frequency is 175 kHz. An optoisolator is used in the voltage feedback path so as to preserve the isolation provided by the output transformer. Interesting for the experimenter, the ML4818 IC can function to about 1.5 MHz in converters of this type.

A power supply for helium neon lasers

Many inverter, converter, and power-supply topographies have been used for laser operation. This has been particularly true for the popular helium neon laser, which

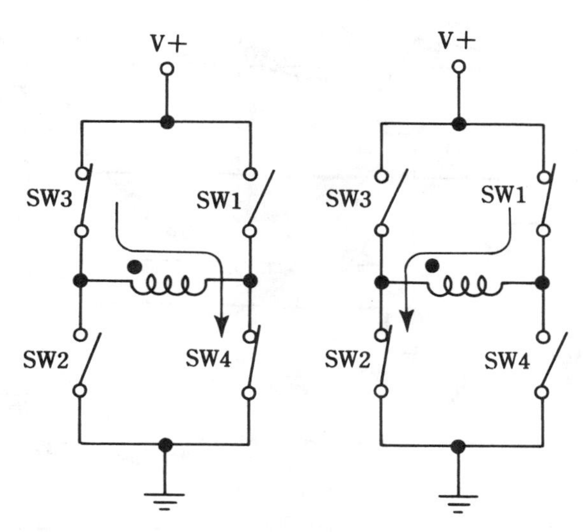

18-22 An electromechanical analog of the H-bridge. Appropriately timed action of the switches produces bipolar current through the transformer primary. This induces an ac voltage in the secondary (not shown). Notice that a single dc source suffices.

operates with a nominal 2 kV or so across its terminals, and draws 3.5 to 7 mA. It turns out that these lasers are difficult loads to power. For one thing, just like its primitive cousin, the ordinary neon lamp, the gaseous laser requires a higher than operating voltage in order to start. This can be in the 6- to 10-kV range. Much depends upon the internal gas pressure, the temperature, and the number of hours of use. There are other somewhat-elusive factors that influence the ionization process, which initiates normal operation, such as radio activity, humidity, ambient light, etc. In any event, it is desirable that start up be both reliable and automatic.

Another characteristic of gaseous conduction in the laser tube is that it exhibits negative resistance. This tends to produce instability, often of an erratic nature. This is best combated by constant-current operation, which tends to swamp out the negative resistance. In practice, a very successful approach is to use a combination of physical (positive) resistance and supply current regulation to stabilize the laser tube. It is well to point out, too, that in the interest of battery operation, a laser power supply should operate efficiently. This suggests new ways of tackling the overall problem—the laser is no longer a novelty, and it no longer suffices to energize it with simplistic- or brute-force methods.

Depicted in Fig. 18-25 is a helium neon laser power supply, which satisfies the alluded performance criteria. Additionally, if it is enclosed in a metal box, RFI can be reduced to negligible levels. The principle of operation can be inferred by identifying the several basic circuitries, which, together, produce the desired performance.

- *The LT1074 step-down switching regulator* This PWM control IC is used as a constant-current regulator. Essentially, this operational mode is achieved by applying to its FB terminal the small voltage drop developed across the 340-Ω sampling resistance inserted in series with the laser load.

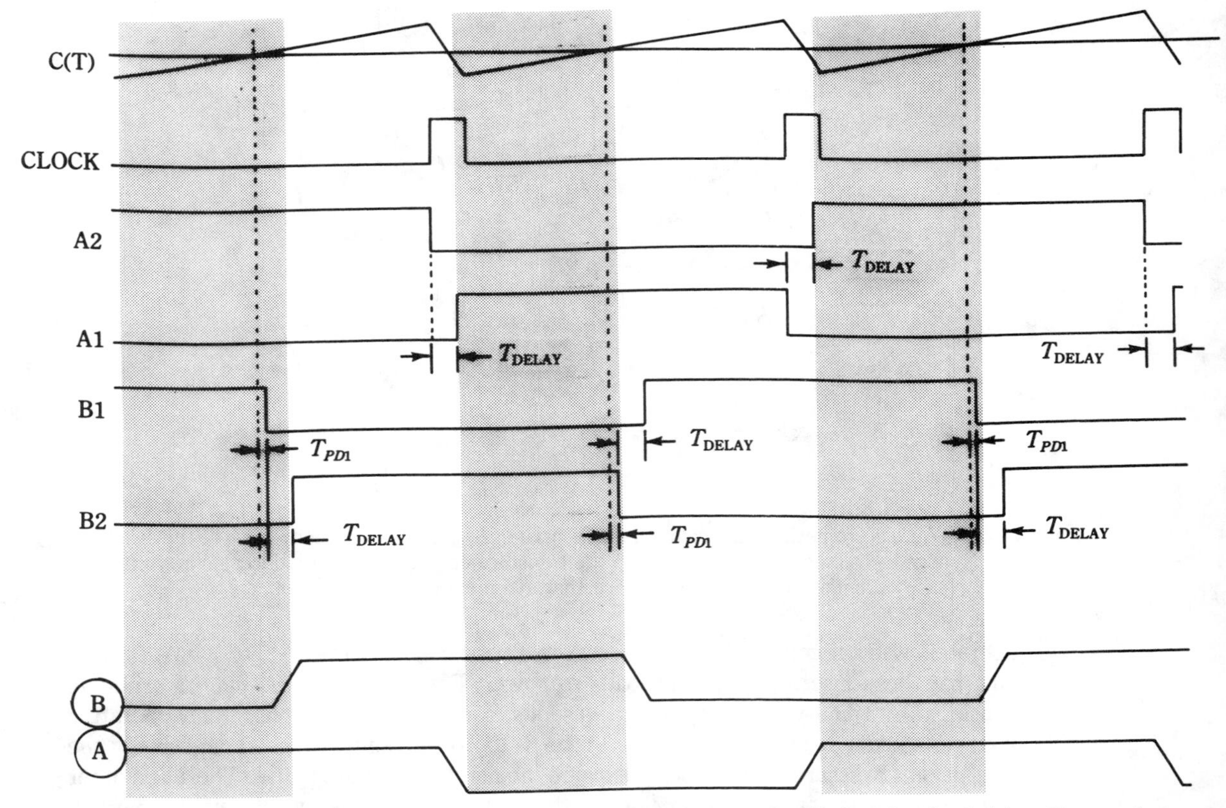

18-23 Voltage waveforms in the phase-modulation regulating supply. Notice the fixed delay T_{Delay} and the variable delay T_{PD1}. Unlike the resonant-mode regulators, the operating frequency is fixed.

- *The saturable core oscillator comprised of L2, Q1 Q2, and associated circuitry* Notice that this square-wave oscillator receives its dc supply from terminal V_{SW}, the output of the LT1074 control IC. The output is smoothed by the filter comprising L1 and C1. It can now be seen that the operating current for the laser tube can be regulated by the LT1074 via the amplitude variation of the square wave, which is generated by the saturable-core oscillator.
- *A full-wave voltage doubler made up of D1, D2, and the two associated 0.1-μF capacitors* This circuit, in conjunction with the step up in the oscillation transformer (L2), provides the high dc operating voltage for the laser tube. A good design balance must be achieved here because too much electromagnetic step up can produce problems from leakage inductance, switching transients, and unwanted self resonance. On the other hand, too much diode capacitor voltage multiplication can reduce efficiency and reliability. Also, too much of either voltage step-up technique can interfere with the stability of the feedback loop.

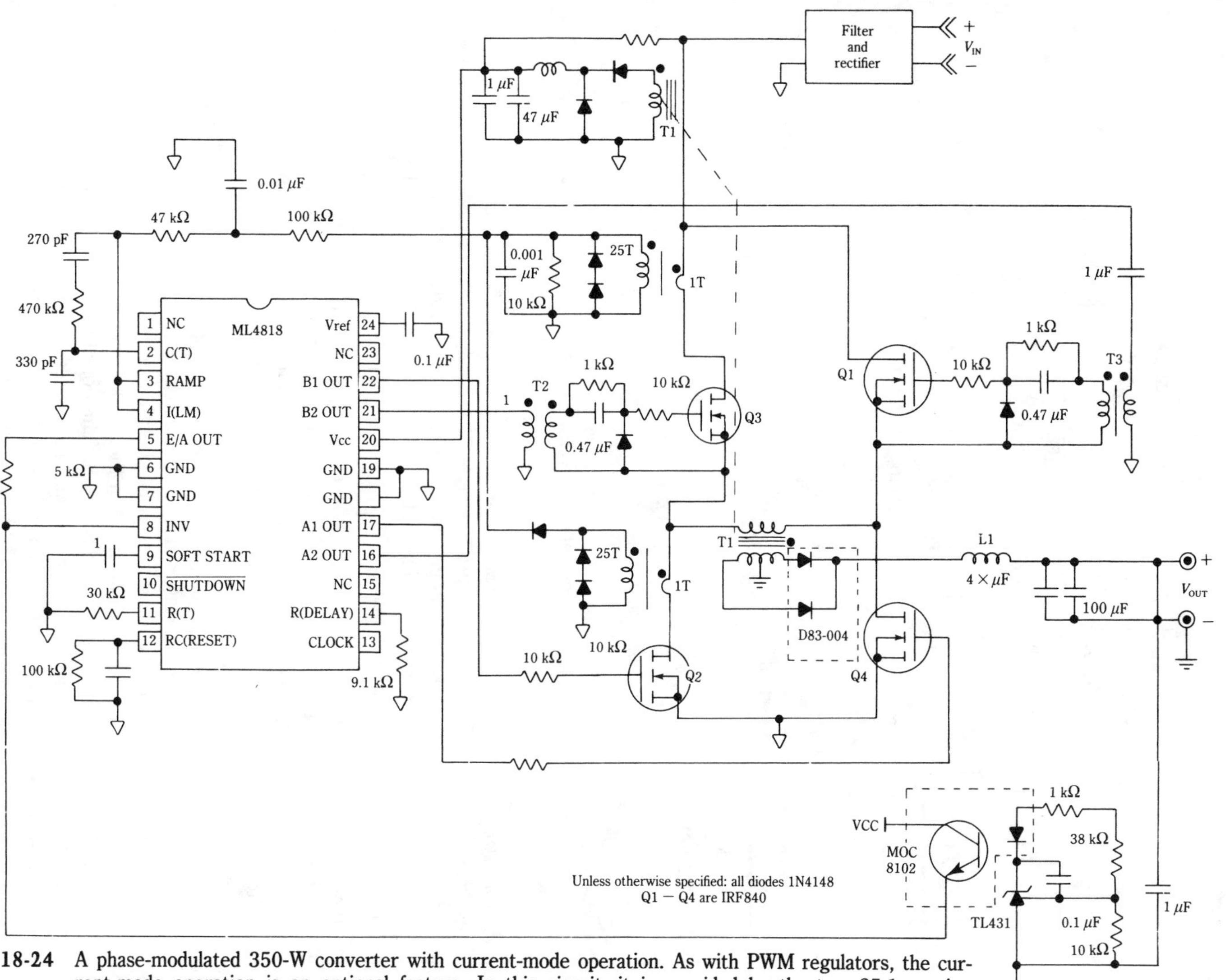

18-24 A phase-modulated 350-W converter with current-mode operation. As with PWM regulators, the current-mode operation is an optional feature. In this circuit, it is provided by the two 25:1 sensing transformers, rather than by sampling resistances. Micro Linear Corporation.

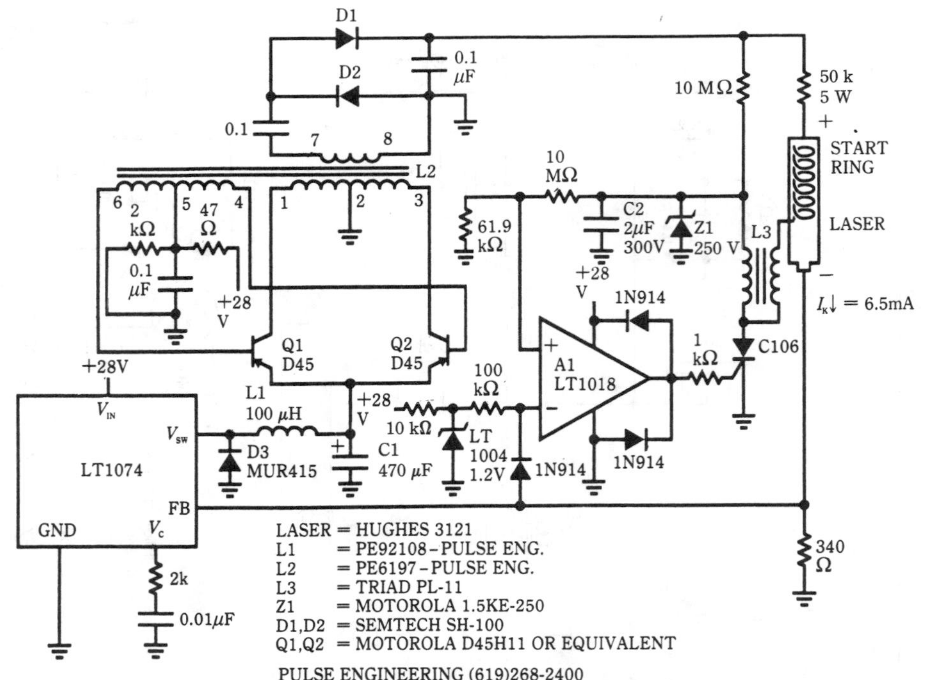

18-25 A power supply for helium-neon laser. This design makes optimum use of diverse circuitries to provide current-regulated operation, as well as automatic start. Linear Technology Corporation.

The laser tube load This must be installed with due regard to polarity. A start ring of one, or several turns of wire is used as a "third" electrode to transfer the high-voltage start pulse to the gas inside the tube. It is important that the 50-kΩ resistor be located immediately adjacent to the tube; a few inches of wire separation can give rise to parasitic oscillations and ionization instability.

The foregoing sums up the circuitries responsible for the generation and regulation of the operating power for the laser tube. The remaining circuit function is the automatic production of the high-voltage pulse, which must be delivered to the start ring; this is to initiate ionization so that the gaseous conduction needed for laser action can occur.

To see how this works, assume that the 28 Vdc has just been applied to the power supply. About 3.5 kV appears across the laser tube. As previously explained, this is not sufficient to ionize the gas within the tube and no laser action occurs. Notice that a 10-MΩ resistance allows capacitor C2 to charge from the 3.5-kV source. C2 is limited to charge to 250 V by the zener diode shunted across it. The objective is to dump the charge stored in C2 into the primary of step-up transformer L3, inducing a high-voltage transient in its secondary, and thereby trigger the laser tube through its start ring.

This is accomplished as follows: when the 28 V is first applied to the power supply, the output of comparator A1 is substantially at ground potential and nothing happens.

But as capacitor C2 charges toward its allotted 250 V, a sampled portion of this voltage appears at the noninvert terminal of the comparator, ultimately causing its output to go positive. This triggers the SCR, thereby dumping the charge stored in capacitor C2 into the primary winding of step-up transformer L3, starting the laser. As long as the laser operates in normal fashion, the comparator reverts to its initial state. This is because of the presence of a positive voltage on its inverting terminal derived from the laser current passing through the 340-Ω sampling resistance. It can be seen that should the laser become extinguished for any reason, it would be automatically restarted.

19 CHAPTER

Focus on power devices

IT IS PROBABLY SAFE TO SAY THAT THE BATTLE BETWEEN THE BIPOLAR POWER transistor (including power Darlingtons) and the power MOSFET dwindled to a draw in which each device has been recognized as appropriate in certain areas of application. Of course, there remain, considerable areas of overlapping applicability; that is why it cannot be said that the battle has ended. What has happened over and over is that anytime that one device seemed to obtain a certain advantage over a competitive device, improvements in the competitive device would very soon accrue to at least equalize the situation. For example, at one time bipolars did not perform well at switching rates much beyond the popular 20-kHz rate. Competition from the MOSFET devices, which could switch efficiently at several hundred kHz and higher, forced the development of bipolar power transistors with 100-kHz capability. Conversely, the once-inferior current and power capability of MOSFET devices has been upgraded to match the power-handling level of many bipolar products.

More recently, new or modified devices have appeared on the market that can offer compelling advantages in certain performance parameters over the aforementioned power devices. Circuitries and techniques for utilizing these newer devices have only been partially worked out. At this writing, there is considerable opportunity for interesting experimentation and creative designing.

This chapter purports to bring some of these newer devices to the reader's attention. In some instances, the best features of competing devices have been combined in one package. Other innovations provide circuit conveniences that facilitate application in more recently developed power-supply techniques. Just as switching regulators once bid for favor over the long-enduring linear regulator, now there are different types of switching techniques to consider. With the advent of resonant-mode and current-mode operation, it will be seen that off-the-beaten-path power devices have become available to best exploit the latest circuit and system concepts. Interesting new or unusual ways to use some of the older devices will be covered.

The SENSEFET tailor made for current-mode supplies

Technological evolution in electronics is often speeded up because an improved device or technique stimulates dedicated components designed to optimize the improvement. A case in point is the SENSEFET, a unique power MOSFET developed to optimize the performance, and simplify the design of current-mode regulated power supplies (see Fig. 19-1). And, after introduction of the SENSEFET, its originators, Motorola, made available a control IC that was particularly well suited for controlling the SENSEFET in current-mode supplies. The block diagram of this current-mode IC is shown in Fig. 19-2.

Consider the fact that conventional power MOSFETs actually comprise thousands of FET cells, which are electrically in parallel. All of these identical cells contribute equal increments to the drain current. In the SENSEFET, one or several of these cells are electrically isolated from the rest and connected to a separate lead, known as the *pilot* or *mirror terminal.* If this terminal is connected through a resistor to ground, a tiny current, an accurate fraction of the main drain current, can be monitored across the resistor as a voltage drop that is proportionate to the drain current. Thus, a current sampling technique is provided, a requirement for implementing current-mode supplies.

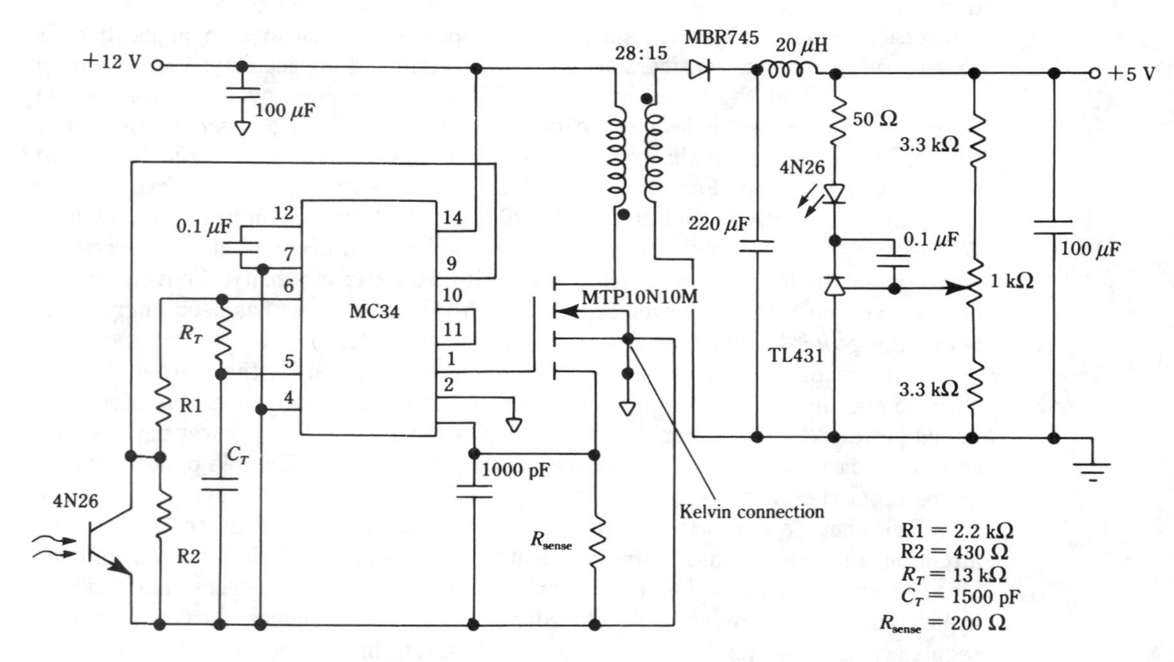

19-1 An example of the use of the SENSEFET in a dc/dc current-mode converter. The power dissipation in the current-sampling resistance, R_{sense}, is much smaller than that occurring when the power switch is a conventional power MOSFET. Motorola Semiconductor Products, Inc.

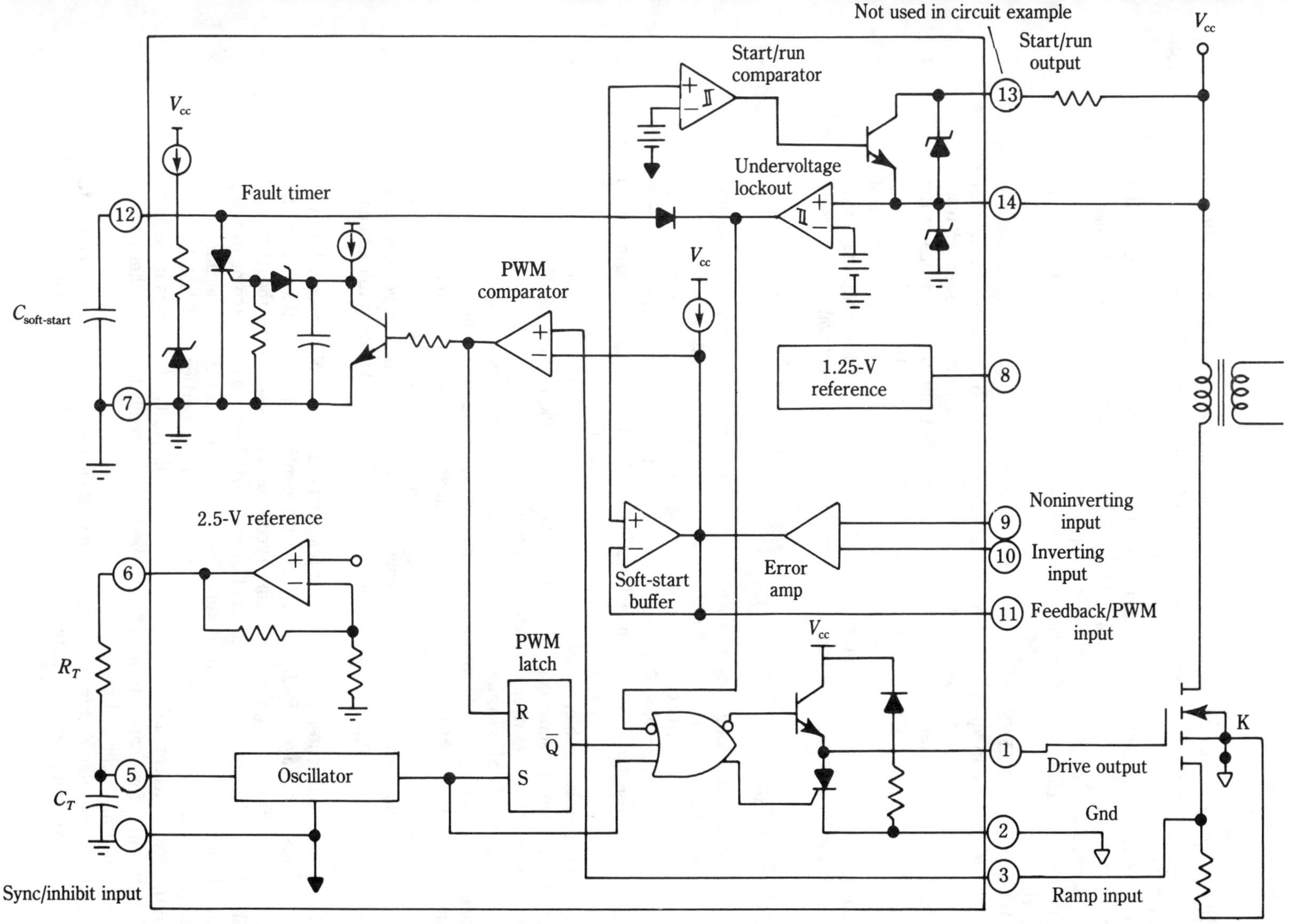

19-2 A block diagram of the MC34129 current-mode control IC. This control IC is specifically designed to work optimally from the relatively small voltage-drop in the ramp monitoring circuit of the SENSEFET. Motorola Semiconductor Products, Inc.

In conventional current-mode supplies, current sampling of the power switch is accomplished by means of a resistor in the source or drain lead of the MOSFET power switch. Superficially, the two current-sampling methods appear much the same — both monitor voltage drops related to the current ramp in the power switch. However, the sampling method used with ordinary power MOSFETs introduces a non-negligible power dissipation. On the other hand, the lost power in the pilot circuit of the SENSEFET is miniscule (the sampled current is on the order of 1/2000 part of the drain current).

Actually, the SENSEFET incorporates a further refinement in the form of another terminal, making five in all. The fifth is the Kelvin terminal. It is essentially another source terminal, but by its use the heavy drain source current is prevented from affecting the relatively feeble pilot source current via the common impedance of the source bonding wire. To take advantage of this provision, the bottom of the sampling resistor simply connects to the Kelvin terminal, rather than to the lead or terminal designated as the source. Additional insight might be gleaned by considering the Kelvin connection as closer to the true source "electrode" than the actual source lead itself.

The P-channel power MOSFET

Circuit designers have long used npn and pnp bipolar transistors in various combinations; in so doing, certain advantages have been realized. Among these are circuit simplicity, reduction of parts count, avoidance of ground conflicts, lower cost, and quite often, enhanced performance. Complementary symmetry circuits, in particular, have been popular; these tend to be push-pull or totem-pole output stages using matched npn/pnp transistors in such a way that drive requirements are simplified and direct connection output is feasible. Although these statements largely pertain to audio-amplifier practice, there are some common features between audio circuits and the circuits found in power supplies, inverters, and converters.

With power MOSFETs, both N-channel and P-channel types are now available. These correspond respectively with npn and pnp bipolar transistors. The same philosophy prevails, with respect to circuit and operational advantages that can often be had. Although P-channel power MOSFETs closely followed on the heels of npn types, their commercial development was relatively slow. Certain problems had to be overcome, particularly that of mating P- and N-channel types so that one could be the mirror image of the other.

For example, P material has a higher resistivity than does N material. When an attempt is made to make a P-channel device with identical characteristics (except for electrical polarities) to a given N-channel type, the on resistance of the P-channel MOSFET will be higher than its N-channel counterpart. This can be overcome by using a larger die for the P unit. This, however, affects the input and output capacitance, the transconductance, the threshold voltage, the thermal properties, and last (but not least), the cost. Current rating is also affected by the disparity of die sizes in "equivalent" units, but it fortunately turns out that the P device tends to have higher current capability than its N-channel mate.

All things considered, it is not possible to get all characteristics of the two types to coincide. Various compromises and trade offs are inevitable. Generally, it is desired that the on resistances of a pair of complementary MOSFETs be the same. Then, the other characteristics are tailored so that, in most practical applications, the divergences from true symmetry will not be of great consequence. Table 19-1 lists the characteristics of a pair of oppositely poled power MOSFETs. It is easy to see that in many applications, these devices will behave in essentially the same fashion. The "totem-pole" switching circuit of Fig. 19-3 exemplifies the use of P- and N-channel power MOSFETs. With a slight modification in the matter of gate biasing, this basic arrangement is excellent as a push-pull output stage for sine waves.

Table 19-1 Comparison of p- and n-channel power MOSFETs intended for complementary-symmetry applications. International Rectifier Corp.

	N-Channel	P-Channel
Device Type	IRF120	IRF9130
Drain-to-Source Voltage (Max.)	100V	−100V
Die Size	8.04mm^2	13.25mm^2
On-Resistance (Maximum)	0.3Ω	0.3Ω
On-State Drain Current @ $T_C = 90°C$	6A	−8A
Pulsed Drain Current	15A	−30A
Gate Threshold Voltage (Minimum-Maximum)	2 to 4V	−2 to −4V
Forward Transconductance (Typical)	2.5 S	3.5 S
Input Capacitance (Typical)	450pF	500pF
Output Capacitance (Typical)	200pF	300pF
Reverse Transfer Capacitance (Typical)	50pF	100pF
Maximum Thermal Resistance	3.12 deg. C/W	1.67 deg. C/W
Package	TO-3	TO-3

An exception to the statement in the preceding paragraph is of relevant interest to designers of switching regulators and power supplies operating at higher frequencies, for example, in excess of 50 kHz. Siliconix has made available the MPP500 family of complementary power MOSFETs fabricated on equal area dies. This endows selected pairs of complementary devices with equal input capacitances, rather than equal on resistances (R_D). The rationale is that this approach results in better device balance at the higher switching rates.

The selected MOSFET pairs alluded to are not offered as discrete units, but rather as arrays with the "matched" P- and N-channel devices fabricated in a single package. The drains of the two devices are internally connected together. The manufacturer points out that such a topography occurs in most practical complementary symmetry circuits so that minimal loss of application flexibility is likely to result. Power MOSFET arrays of various kinds, such as half bridges, full bridges, "smart" power packages with integrated drive and/or control logic, and bilateral topographies in which a pair of power MOSFETs are made to simulate the ac conduction behavior of triacs, have been receiving considerable developmental effort.

A nice feature of P-channel power MOSFETs is that they simplify automobile and aircraft applications, where it is desired that one load terminal be grounded. If, for example, a single-ended N-channel output stage were used, the gate driver would

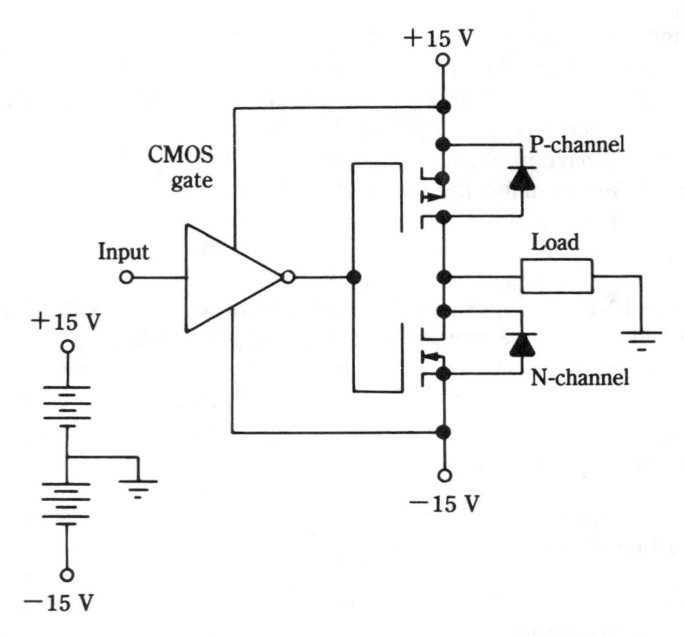

19-3 A basic complementary-symmetry "totem pole" circuit. P-channel and N-channel MOSFET pairs are much used in power supply, inverter, and regulator applications. Notice the integral diodes and the "mirror-image" aspect of the arrangement.

require an additional dc source. With a P-channel output stage, the load can be connected between the drain and ground and no extra supply is needed for the driver. Recall that it is nearly universal practice to ground the negative terminal of vehicular and small aircraft batteries. Consumer applications, where the load must be grounded for safety reasons, can also benefit from P-channel circuitry.

P-channel power MOSFETs are used in synchronous rectifier circuits in conjunction with N-channel types. Such active rectification can rival the high-frequency capability of Schottky diodes, but can exceed the voltage limitations of the Schottky diodes, and is also less likely to be adversely affected by temperature. In implementing both P- and N-channel power MOSFETs in synchronous rectification circuits, it is best to consult with the manufacturer, inasmuch as devices optimized for this application are available.

The strange operation of the MOSFET synchronous rectifying bridge

The synchronous rectifier circuit shown in Fig. 19-4 makes use of two P-channel and two N-channel power MOSFETs. Yet, it is not a complementary symmetry arrangement in the usual sense. For simplicity, the integral diodes of the MOSFETs are not shown. There is, however, another reason these "come-along-for-the-ride" diodes need not be depicted in the schematic diagram.

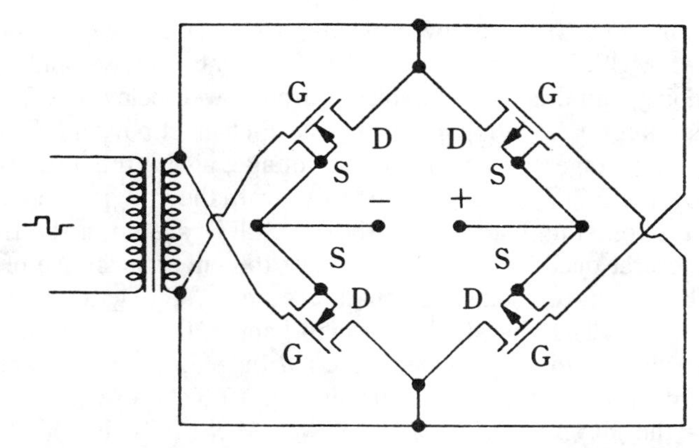

19-4 A synchronous rectifier circuit using p and n-channel power mosfets. Despite its simplicity, the operating principle of this bridge is not obvious from inspection. It can excel Schottky diodes in some applications.

The reader will, no doubt, be familiar with a number of power-MOSFET applications in which the integral diode serves some circuit function, such as a "free-wheeling" diode in power switch circuits, as a clamp diode, or as a transient snubber. Also, in regenerative braking, as seen in the vehicular application of inverters and converters, the integral diode provides the needed path for reverse current. In the synchronous rectification scheme of Fig. 19-4, however, the four integral diodes serve no electrical purpose whatsoever. Indeed, the successful operation of the bridge circuit depends upon the integral diodes being electrically inactive at all times! This is fortunate, for involvement of these "parasitic" diodes would greatly slow down the high-frequency rectification capability.

In light of the above statements, it is only natural to ponder how the integral diodes can be maintained inactive—especially in a circuit in which each power MOSFET is subjected to alternating current. The answer is in two parts. Obviously, the integral diodes cannot conduct for one polarity of the ac wave. For the opposite polarity, it is true that conduction could take place if the on resistance of the MOSFETs were high enough to permit the requisite 0.7-V drop needed to forward-bias silicon PN junctions. The trick, then, is to use MOSFETs in which the drain-source voltage drop is less than approximately 0.7 V. Toward this end, there are MOSFETs optimized for synchronous rectifier service with very low on resistances.

Keeping this in mind, another aspect, perhaps a surprising one, of MOSFET behavior is involved: power MOSFETs, unlike bipolar transistors can operate bilaterally. That is, the drain and source connections can be interchanged. For example, an N- channel device can operate with negative voltage on the drain and positive voltage on the source. This is not encountered in most situations because of the onset of conduction in the integral diode. But in the synchronous rectifier bridge circuit, special low-R_D MOSFETs are used that do not allow sufficient voltage drop to turn on the integral diodes.

From the foregoing, it is to be emphasized that all of the power MOSFETs in the rectifying bridge turn on with opposite polarized drain source voltages from conventional use. But turn on is produced by conventionally polarized gate voltage. Thus, the N-channel power MOSFETs turn on with positive gate voltage and negative drain voltage. In so doing, the drain-source voltage drop is well below the 0.7 V needed to forward-bias the integral diodes. Conversely, the P-channel power MOSFETs turn on with negative gate voltage and positive drain voltage; again, the drain-source voltage drop is well below the 0.7-V that is required to activate the integral diode. This mode of operation has an interesting feature that the MOSFETs are on in the third quadrant, but are off in the first quadrant. This is contrary to conventional use of the devices.

Suppose, by means of some attenuation scheme, the gate drive to the four MOSFETs was gradually reduced. What would happen to the performance? Surprisingly, the bridge circuit would continue to rectify! Indeed, the circuit would "degenerate" into the ordinary diode bridge circuit of Fig. 19-5. Where did the diodes come from? They are the integral diodes of the power MOSFETs that have now become active because the drain source voltage drops of the under-driven MOSFETs now exceed 0.7 V.

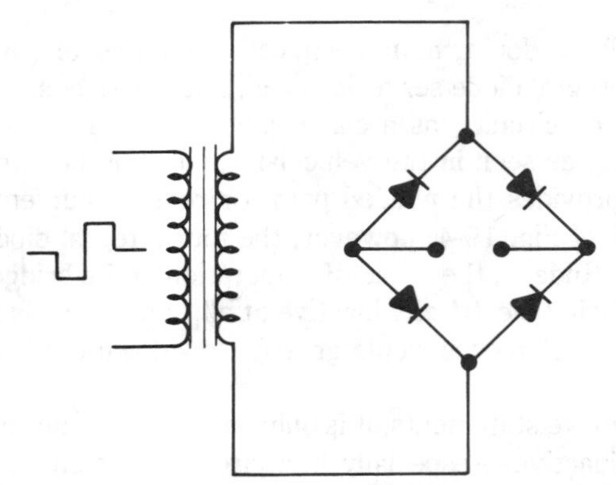

19-5 The "degenerate version" of the MOSFET bridge synchronous rectifier. This is the equivalent circuit of the MOSFET bridge synchronous rectifier with inadequate drive to the gates. The MOSFETs have thereby become inactive, but their integral diodes have become active. Thus, rectification continuous, but at poorer efficiency—especially at high frequencies.

Although the under-driven circuit would continue basic rectification, the efficiency would have suffered on two counts. First, the storage phenomenon of junction rectifiers would allow good rectification at low frequencies, for example 20 kHz, but progressively poorer rectification at the higher frequencies. Second, the 0.7+ V drop would cause more power dissipation than the lower drop produced by hard-driven MOSFETs. It can be appreciated that the overall efficiency of the MOSFET circuit most greatly exceeds that of ordinary circuits using PN junction diodes when the frequency is high, for example above 100 kHz. And the advantage of the MOSFET synchronous rectifier over Schottky diode circuits is primarily that both high frequencies and high voltages can be handled.

The simple circuit of Fig. 19-5 is at its best when the MOSFET gate-source voltages do not exceed 20 V, but are high enough so that the devices are hard-driven into saturation. Much higher rectification voltages can be accommodated if some means, such an Zener diodes or resistive networks, are used to keep the gate-source voltages within the 15- to 20-V range. This is for the sake of safety—to prevent gate puncture.

How to obtain high switching rates from the power MOSFET

The high-frequency capability of the power MOSFET is not without some qualifications. In particular, when dealing with switching rates in the several-hundred kHz to several-MHz region, attention must be directed to other factors than the theoretical wideband response of the MOSFET structure itself. Because the input section of this device is essentially a capacitor, the problem of rapid charge and discharge of the gate circuit increasingly asserts itself as higher switching rates are used. The straightforward solution is to drive the gate from a low-impedance source—the lower the better.

Toward this end, you can interpose a pre-driver stage between the control IC and the gate(s) of the power MOSFET(s) switching circuit. Such a pre-driver is usually a totem-pole configured pair of smaller power MOSFETs. Of course, in the interest of simplicity and economy, it is desirable to select a control IC capable of directly driving power MOSFETs. In practice, this capability can become marginal when driving the high capacitive inputs of large power MOSFETs at the higher switching rates. Another alternative is to use a driver IC. These are designed for this specific purpose, and can exhibit such respectable switching ratings as a 40 ns rise and fall time when working into 1000 pF of MOSFET gate capacitance.

The Cherry Semiconductor CS 2706 is a dedicated IC driver with just such switching capability. It accepts logic-level input signals and can source or sink 1.5 A from each of its two outputs. The block diagram of this driver is shown in Fig. 19-6. A nice thing about these IC drivers is that they incorporate additional "bells and whistles" in the form of control, protection, and convenience.

Another speed limitation occurs in practical circuits that utilize the intrinsic diode of the power MOSFET as a free-wheeling diode, or current-return diode. This diode usually has the same current capability as the power MOSFET itself, but it suffers from slow reverse recovery at high switching rates. The solution is to render the intrinsic diode inactive and to rely upon an external fast recovery or Schottky diode. The way in which this is done is shown in Fig. 19-7. A penalty is paid in the form of added power loss in diode D1; however, the overall efficiency of the power switch can have a worthwhile increase because of the elimination of switching losses in the slow intrinsic diode.

The depletion-mode power MOSFET

It cannot be denied that power MOSFETs have provided the power-supply designer with a multiplicity of circuit options. To begin with, wide ranges of power, current, and voltage are available. They are relatively easily driven, and can accommodate switching rates in the MHz range. Because both N- and P-channel types are readily available,

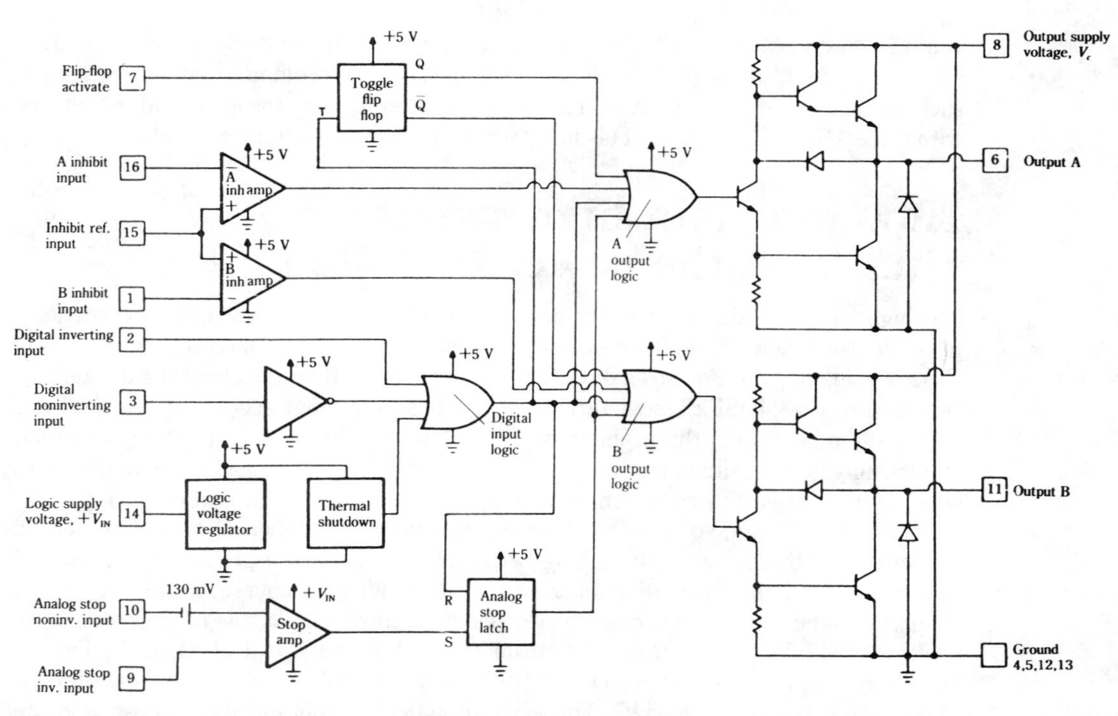

19-6 The block diagram of Cherry Semiconductor CS-2706/CS-3706 drivers. Despite the internal complexity of such dedicated ICs, practical implementation is easy. Options allow for parallel or push-pull operation of the driven power MOSFETs. Cherry Semiconductor Corp.

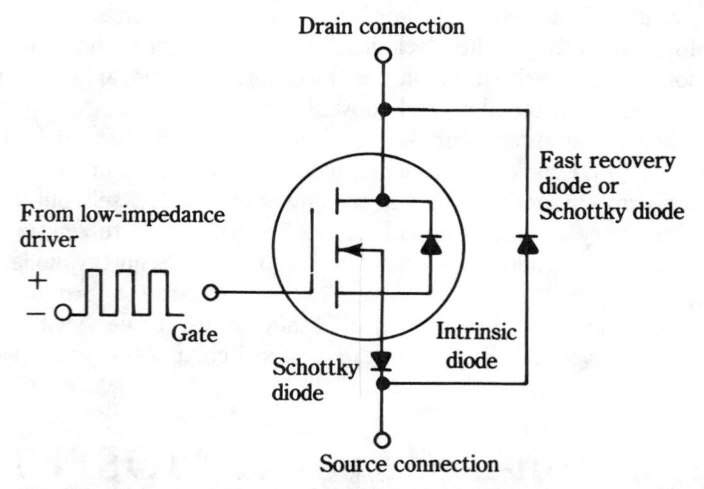

19-7 A high switching-rate circuit for the power MOSFET. This scheme renders the relatively slow intrinsic-diode inactive. In its place, the external fast-recovery diode serves as the free-wheeling, current-return, or snubber diode. Note: The self-contained diode of the power MOSFET goes by such names as internal body, intrinsic, integral, and parasitic diode.

circuitry economies and simplifications can be conveniently realized. Moreover, these devices can be operated in their third, as well as in their first quadrant, as is done in synchronous rectifier circuits. The internal diode, which is an intrinsic part of the MOSFET structure, is often useful as a free-wheeling, current-return, or transient-suppressor diode. When this diode is found to be too slow for the purpose at hand, it is easy to electrically isolate it, and use an external Schottky, or fast-recovery diode. As if all these attributes are not enough, there is the IGBT, a power MOSFET modification with very low voltage drop, resembling that of a bipolar transistor. Then, too, there is the SENSEFET—a power MOSFET with a current-sensing provision.

In light of these features, it could tax your imagination to further extend the flexibility of power MOSFETs. Consider, however, all of the power MOSFETs alluded to have one thing in common they are all enhancement-mode devices—even though some are specified to operate from logic circuits (several volts of drive turns them on hard). It is only natural to ponder the possibility of depletion-mode MOSFETs, devices that are already in their on state, and that require gate turn-off bias to shut them off. You might recall that most vacuum tubes were essentially depletion-mode devices. Also, junction FETs are depletion-mode devices (these JFETs do not have very high power ratings, but that is another story).

As might be surmised, depletion-mode power MOSFETs have become commercially available items. Because they have not received anything like the developmental effort accorded to enhancement types, they have not yet achieved much popularity. They are, however, well-suited to some specialized circuit functions, and it is likely that they are destined to become more commonly used. Whether they can be engineered to handle currents, voltages, and power levels that are comparable to those of enhancement devices remains to be seen. In the meantime, experimenters can have a field day devising interesting and useful applications centered around the depletion-mode power MOSFET.

The comparison between enhancement- and depletion-mode MOSFETs is shown in Fig. 19-8. It shows that the depletion-mode device (like many vacuum tubes) actually operates partially in both modes. However, there is no counterpart of dc grid current in the tube. Rather, the gate in the depletion-mode MOSFET continues to look like a capacitor. As such, there can be ac current, but no dc current from the driver or dc bias source.

The depletion-mode MOSFET constant-current regulator shown in Fig. 19-9 is simplicity in itself. Although inspection might not immediately display evidence of a sensing circuit, such a regulator is, nonetheless a closed loop system. For, as load current might tend to increase, the reverse gate bias thus developed across the source resistance counteracts the current increase. This, in conjunction with the fact that the device inherently maintains near constant drain current in the face of changing drain voltage, makes the arrangement a very good constant current source.

A small heatsink should be used, not so much for thermal protection, but as a means of tightening the regulation. The device has a positive temperature coefficient so that preventing excessive temperature rise is as important as attention to electrical matters. A practical implementation might involve the regulation of load current in the several-hundred mA range. At the opposite extreme, it appears feasible to regulate load currents in the vicinity of 5 mA.

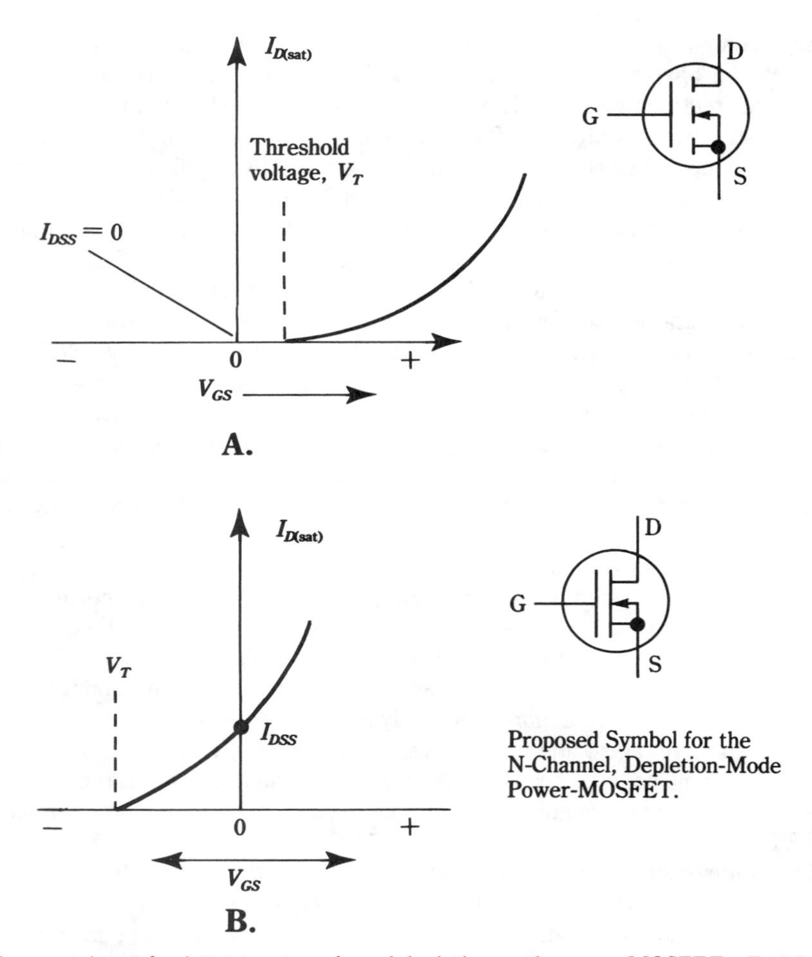

19-8 A comparison of enhancement-mode and depletion-mode power MOSFETs. For the sake of simplicity, the internal body diodes are not shown in either symbol. Their presence is, however, the same in both devices. (A) Transfer curve and commonly used symbol of the N-channel enhancement-type power MOSFET. (B) Transfer curve and proposed symbol of the N-channel depletion-type power MOSFET.

At this writing, depletion-mode power MOSFETs with multiampere ratings do not appear to be available, or at least, not readily so. This suggests the prospect of paralleling. Parallel operation of two or more of these devices should be feasible, but the matter of current sharing might require attention. By insertion of fixed resistances in the source leads of one or more of such parallel-connected devices, it should be possible to force reasonable current sharing among the devices. An interesting feature of this regulation scheme is that it provides instantaneous "line" and load regulation, despite the lack of a voltage reference.

The experimenter should remember that the depletion-mode MOSFET contains an internal body diode in the same way as enhancement-mode devices. And, similarly,

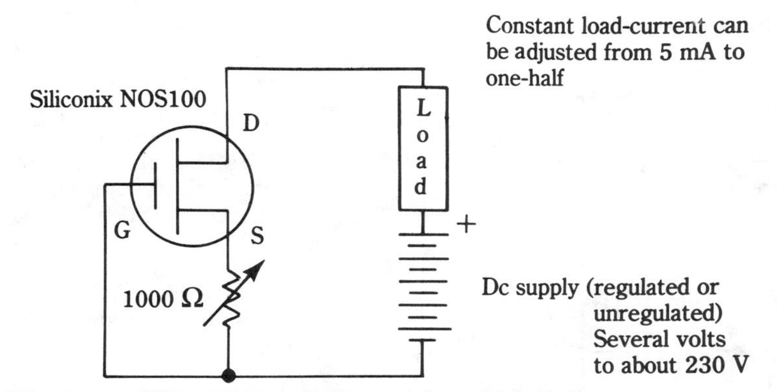

Note: Siliconix type NO2406L can also be used, but with half the current capability of the NOS100. Both devices should be capped with a small heatsink

19-9 A constant-current regulator using depletion-mode power MOSFET. A similar circuit using a conventional enhancement-mode power MOSFET would require a special biasing arrangement.

depending upon the circuit application, this diode might manifest itself either in a beneficial or a detrimental way. In any event, a nice practical feature of the circuit of Fig. 19-9 is that it is a true two-terminal regulator, contrasting to the three or more terminals that are minimally required by conventional regulating circuits. As a corollary to this feature, it will often be found that ground conflicts and other circuit inconveniences can be neatly circumvented.

The IGBT: the dream device, but not without trade offs

It has been only natural for the designers of solid-state power devices to speculate how nice it would be if the salient features of different devices could be somehow combined in a single device. There is no law against fantasizing a device that is easy to drive, has high-current and high-voltage capability, is willing to switch at MHz rates, has negligible thermal resistance for easy heat removal, etc. Indeed, there have been practical implementations that have, to some degree at least, combined performance parameters of divergent devices. BIDET op amps with FE input and bipolar transistor outputs are one example; initially these were hybrid hard-wired arrangements, but were eventually fabricated as monolithic structures. The successfully attained objective was to produce a cost-effective component with high input impedance, low output impedance, together with enhanced performance in other parameters.

With power devices, there has been the BIMOS circuit of Fig. 19-10, which again comprises the FET input and bipolar output. Here, however, the combo is made with discrete devices, a small power MOSFET input stage and a large bipolar power transistor as output. A cascode connection is used, which utilizes the bipolar transistor in its common-base configuration. It so happens that, not only is the common base connection superior to the more often used common-emitter circuit in frequency

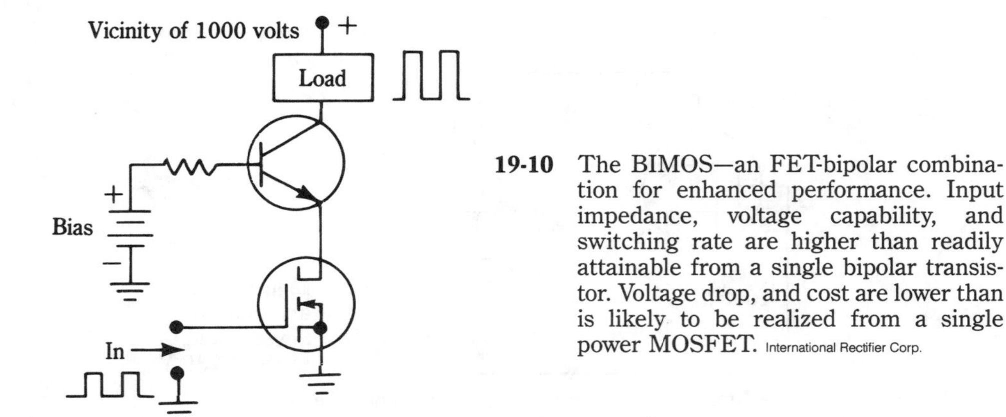

19-10 The BIMOS—an FET-bipolar combination for enhanced performance. Input impedance, voltage capability, and switching rate are higher than readily attainable from a single bipolar transistor. Voltage drop, and cost are lower than is likely to be realized from a single power MOSFET. International Rectifier Corp.

capability, but it also can withstand higher operating voltage. However, by itself, the common-base bipolar circuit is extremely hard to drive because it has very low input impedance. This is where the MOSFET input stage comes to the rescue. The overall result is an easy drive, high-voltage power switch that has respectable frequency capability, and can also handle high currents. The single high-voltage power MOSFET, which the BIMOS circuit replaces, would have high conductive losses, and would probably be more costly, as well.

Yet another FET bipolar arrangement is the parallel circuit shown in Fig. 19-11. With appropriate timing of the turn-on and turn-off waveforms applied to the gate and base of the two devices, you can approach the fast-switching transitions of the MOSFET together with the low conductive losses of the bipolar transistor. Although this scheme intrigues the imagination, it is not easy to implement in practice. For example, the timing pulses of the control circuit would be very critical in a PWM regulator attempting to make use of such a power switch.

The foregoing paragraphs have paved the way to anticipate some kind of a power device merging at least some of the attributes of field-effect and bipolar technology. This has, indeed, materialized in the form of the *insulated-gate, bipolar transistor (IGBT)*. You can correctly guess from the very nomenclature that this is a power device.

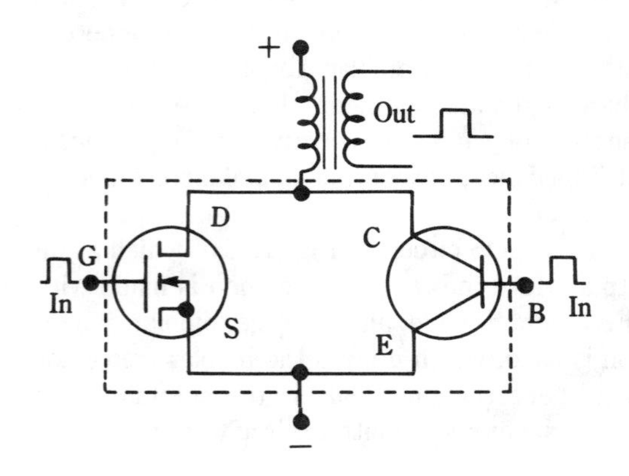

19-11 An experimental combination of FET and bipolar devices to obtain features of each. With appropriate timing and duration of input pulses, the combination circuit can yield the rapid turn-on and turn-off performance of the FET device, together with the low conductive loss of the bipolar device.

with input characteristics like a MOSFET, and output characteristics like a bipolar transistor. Keeping in mind that this has been accomplished in a single structure, it will be appreciated that a dramatic breakthrough has been achieved in the realm of solid-state power.

Surprisingly, the IGBT structure is not radically different from that of ordinary power MOSFETs. This statement implies that the IGBT is better described as a modified MOSFET, rather than a modified bipolar transistor. The modification consists of a change in the doping profile of the drain region, as can be seen in Fig. 19-12A and 19-12B. As trivial as this might appear, it greatly affects the output behavior of the device. Because minority carriers are introduced by the resultant PN junction, the new device is no longer the purely majority-carrier MOSFET prior to the doping modification. The salient performance features of the IGBT are as follows:

- The input remains capacitative, resembling that of the unmodified power MOSFET.
- The output has the low $V_{CE(sat)}$ suggestive of the bipolar transistor. This is true even at high voltages, where ordinary power MOSFETs tend to produce high voltage drops in their drain region. The on power dissipation of the IGBT is low.
- The presence of the minority carriers greatly increase the current density in the drain region. For the same size of die, the IGBT therefore has greater current-handling capability than the power MOSFET. It follows, too, that the transconductance of the IGBT is higher than that of the power MOSFET.
- The output of the IGBT looks like a forward-biased PN diode and this produces a 0.7-V offset in its drain characteristics. In this respect, it differs from the bipolar transistor. Because the IGBT is a high-voltage device, this difference is more subtle than significant.
- Unfortunately, there is a trade off for the goodies brought about by the injection of minority carriers in the drain region. The IGBT no longer has the frequency capability of the ordinary power MOSFET. The best IGBTs are limited to about a 5-kHz switching rate. However, compromised 50-kHz versions are made, which still are advantageous over both power MOSFETs and bipolars at these lower switching rates.

Both the symbol and the electrode nomenclature of the IGBT has been subject to variations in the technical literature. This investigation of the device has dealt with a gate, a source, and a drain. This has served editorial logic because the IGBT is a rather direct derivation of the power MOSFET. On the other hand, semiconductor firms understandably emphasize the bipolar transistor-like characteristics of the output of the IGBT. Accordingly, many of these companies have endowed their product with a gate, an emitter, and a collector. This appears reasonable; an understanding of the relatively new power device is, indeed, facilitated by viewing it as having the input characteristics of a MOSFET, coupled with output behavior that is suggestive of a bipolar transistor. Be prepared, however, to encounter various combinations of gate, base, source, emitter, drain, and collector. Be aware, too, that an alternate name for the IGBT is "conductivity-modulated power field-effect transistor" (bestowed by silicon "chefs," not circuit designers!).

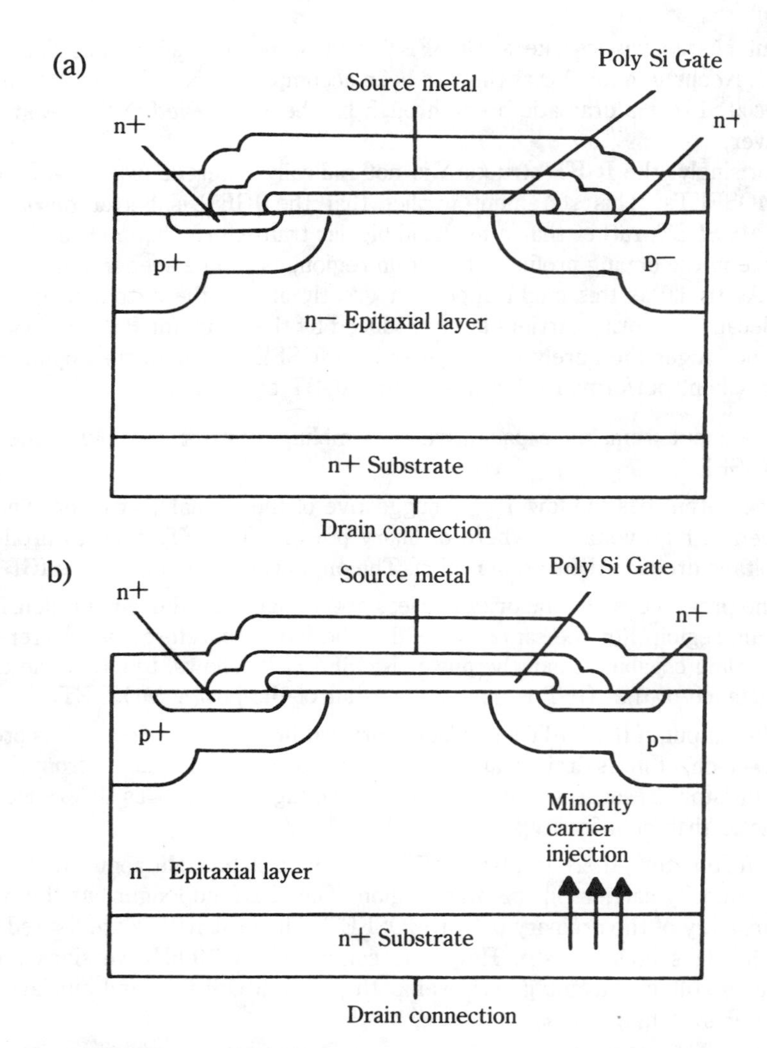

19-12 The evolution of the IGBT from the conventional power MOSFET. Modification of the doping profile in the drain region makes the difference. (A) The basic power MOSFET doping profile. (B) The IGBT doping profile showing the minority carrier injection, which simulates the behavior in bipolar devices.

In similar fashion, there has been a wide variety of symbols used to depict the IGBT. In some instances, the conventional symbols for either MOSFETs or bipolar transistors have been used. And as you might suspect, various hybrid art forms depicting both FET and bipolar symbolism have been offered by imaginative draftsmen. The symbol likely to endure and attain standardization is shown in Fig. 19-13, together with electrode nomenclature. Stylized and simplified versions of this symbol have also appeared, but still retaining the essential idea. The intrinsic diode that bridges the output circuit of the power MOSFET does not manifest itself in the IGBT, and

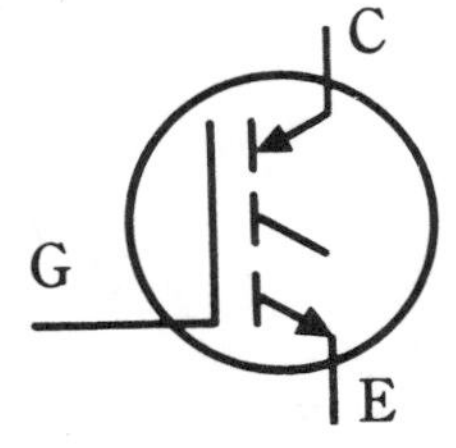

19-13 The symbol of the IGBT and accompanying nomenclature. This depiction of the IGBT appears likely to become the standard format. Notice how the idea of MOSFET input and bipolar-transistor output is conveyed. The forward-biased diode in the collector region is clearly shown.

accordingly does not accompany the basic symbol, as is sometimes the case with the power MOSFET (in applications where a free-wheeling diode, or a diode for other circuit functions is needed, an external diode must be used).

The output characteristics of the IGBT is shown in Fig. 19-14. Notice the 0.7-V offset. The reverse blocking ability of these devices is subject to considerable variation and should be investigated carefully if it is of circuit importance. The qualitative aspect of the curves in Fig. 19-14) does not reveal the all important feature of the IGBT—that its equivalent R_D is often in the vicinity of tenfold smaller than the R_D of a similar-rated power MOSFET. Keep in mind, too, that this is a high voltage (500 V and higher) and high current (15 to 80 A, typ.) device. Figures 19-15 and 19-16 depict the salient features of one manufacturer's 5- and 50-kHz switching rate, high-current IGBTs. Typical collector-to-emitter saturation voltages are 1.6 and 2.2 V, respectively. These are 25° C values, but are not substantially different at 150° C (the already-high R_D of a high-voltage power MOSFET would have doubled from 25° C to 150° C).

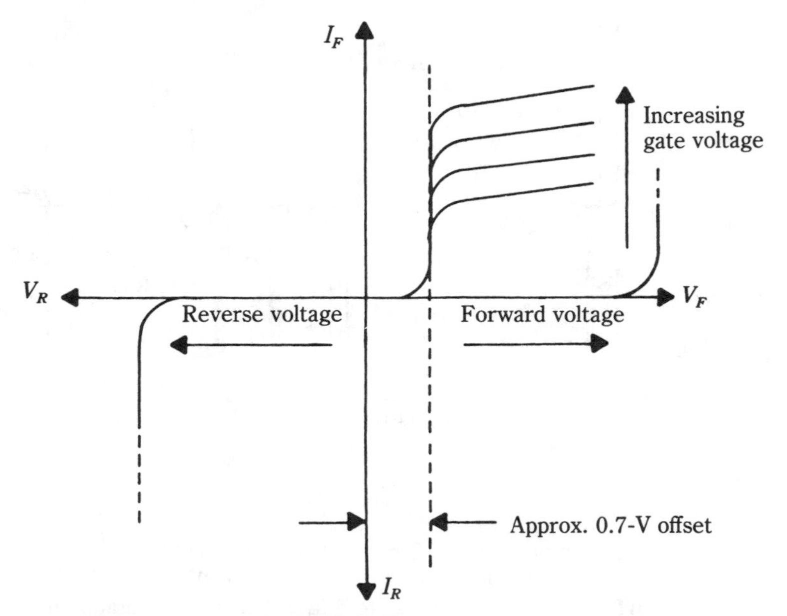

19-14 The output characteristics of the IGBT. These devices tend to have ratings of 500 V and higher. Therefore, the 0.7-V offset is of little consequence in practical circuits.

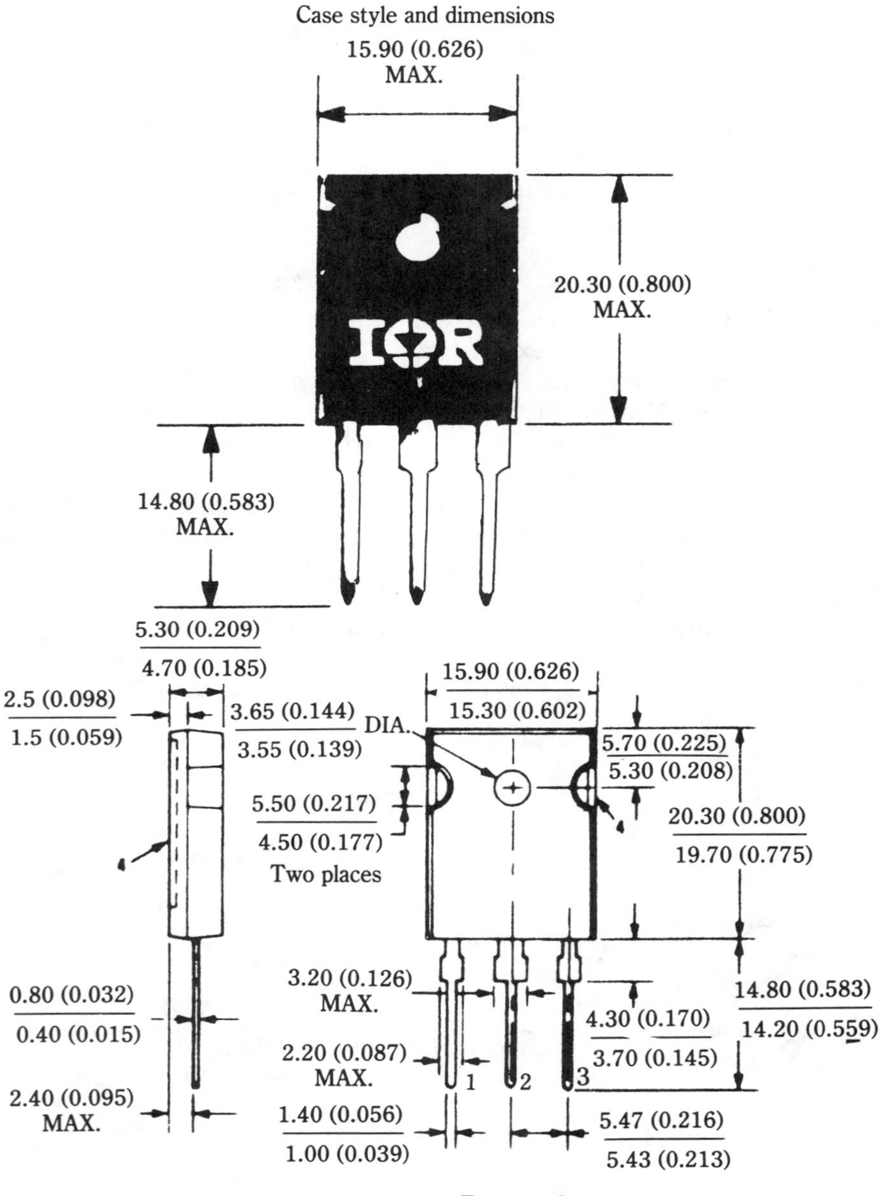

19-15 A typical IGBT for switching rates up to 5 kHz. Latch-free operation pertains to earlier IGBTs, which suffered from this performance defect. Notice the high-voltage, high-current, and the small package. International Rectifier Corp.

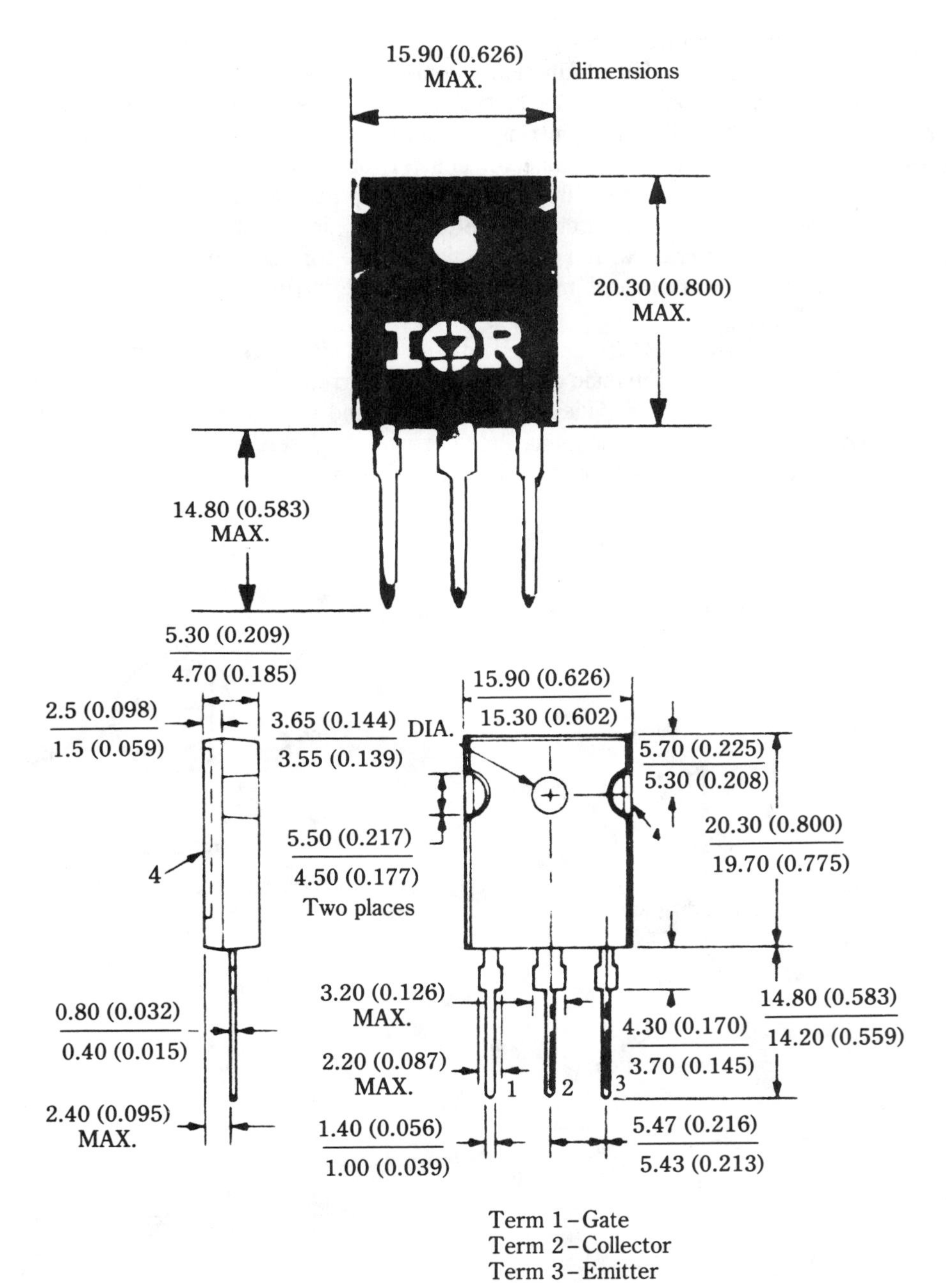

19-16 An example of IGBT for switching rates up to 50 kHz. Surprisingly, the performance ratings are nearly as good as those of the 5-kHz type. Application to the popular 20- to 40-kHz switchmode power supplies is clearly indicated. International Rectifier Corp.

At least two modified IGBTs have become commercially available for circuit designers who require ancillary functions from the devices. Examples of these are illustrated in Fig. 19-17. Figure 19-17A shows a device that is suggestive of the ordinary power MOSFET with its intrinsic diode. However, in the conventional IGBT, no such diode appears across the collector-emitter (or drain source) terminals. Sometimes this is to the detriment of circuit designers who need a diode in this position to serve as a free-wheeling diode, a current return diode, or as a transient snubber. The modified IGBT is fabricated with a self-contained diode for such purposes. Not only is this convenient, but this diode is much faster than the intrinsic diode that appears in power MOSFETs.

Another modified IGBT is depicted in Fig. 19-17B. Here, you can see a pilot element(s) for monitoring the main collector emitter current. This has direct application in current-mode regulating supplies. The principle, and the means for accomplishing such current sensing is the same as has been previously described for the SENSEFET device, a power MOSFET with just such a provision for sensing a tiny fraction of its drain source current ramp.

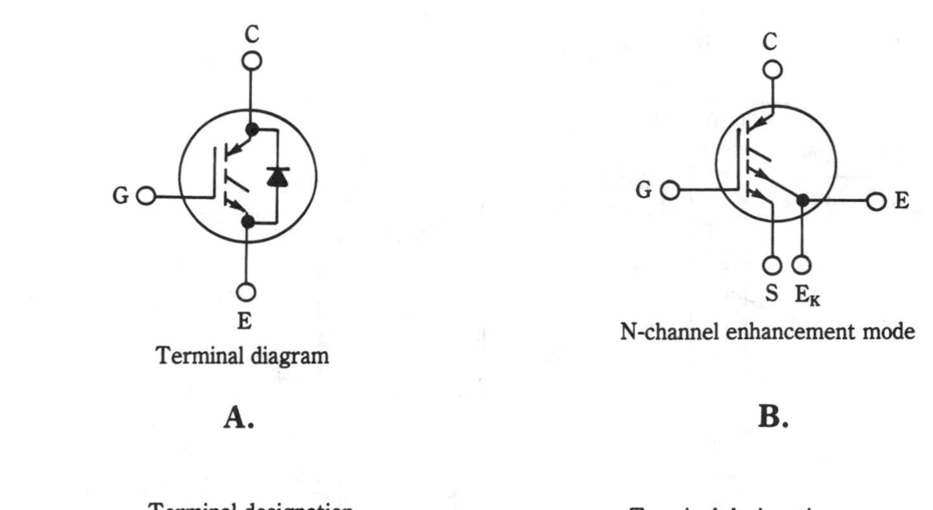

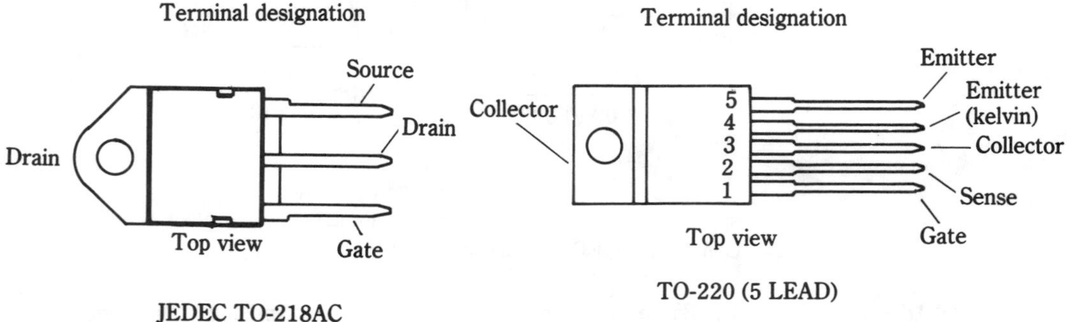

19-17 Examples of modified IGBTs. (A) An IGBT with a fast-recovery internal diode. (B) An IGBT with current-sensing "pilot" element. Similar to the SENSEFET. International Rectifier Corp.

Harris Semiconductor makes a number of these modified IGBTs with 400- and 500-V ratings, and with RMS current capability in the 10- to 25-A range (higher currents are permissible for pulsed service). The experimenter should have a field day with these 5-kHz devices. Higher switching-rate versions are almost certain to follow, for neither modification is at the expense of response time.

At this writing, the concentration of effort has gone into the development of the N-channel IGBT family of power devices. It is not clear whether P-channel versions will also appear on the market. P-channel material poses problems because of the lower mobility of its charge carriers. However, similar pessimism once slowed down the development of pnp power transistors and P-channel power MOSFETs. Perhaps problems pertaining to response, conduction losses, latch up, and cost will be solved and P-channel IGBTs will one day be readily available power devices.

The GTO again

The subheading "The GTO again" is in reference to the previous coverage of this device in chapter 5; also inferred is the periodic waxing and waning of the GTO as a popular power switching device. It seems that with every surge of popularity, other devices undergo advances that divert the attention of designers. Thus, upgraded power MOSFETs, Darlington, and discrete bipolar transistors have at one time or another damped the enthusiasm for the GTO. Also, there have been new devices, such as the IGBT and the MCT, to say nothing of ordinary SCRs. These competitive devices have featured real or perceived advantages of cost, availability, reliability, or performance capability in certain application areas. Nonetheless, interest in the GTO persists, and there are a growing number of designers who favor it as the switching power device for such applications as motor control, welders, cycloconverters, and uninterruptable power supplies.

This being the case, some extended coverage is in order. This section investigates the two-driver circuit because this aspect of using the device was not covered in chapter 5. Because the GTO requires both, turn-on and turn-off drive pulses, successful operation and control is exceptionally dependent upon the drive technique.

It is particularly relevant to again deal with the GTO inverter that was brought to your attention in chapter 5 because no detailed driving circuit was therein dealt with.

In Fig. 19-18, you can see the same 20-kHz 1200-W GTO inverter of chapter 5, but this time it is associated with a unique drive circuit. Interestingly, although both polarities of drive pulses are needed, the driver portion of Fig. 19-18 incorporates no negative dc supply. It turns out that by switching transistor Q1 on and off, both positive (turn on) and negative (turn off) pulses can be delivered to the gate of the GTO. Specifically, when Q1 is switched off, a positive turn-on pulse triggers the GTO to its conductive state. This positive pulse derives from the electromagnetic energy stored in inductors L1 and L2. Conversely, when Q1 is switched on, a negative pulse appears at the gate of the GTO and turns it off. The negative pulse derives from the electrostatic energy stored in capacitor C1. The amplitude level of the negative pulse is nearly twice the dc voltage from the auxiliary supply (70 V). However, as with conventional SCRs, several volts positive suffices to reliably turn the GTO on. Notice that PWM control of

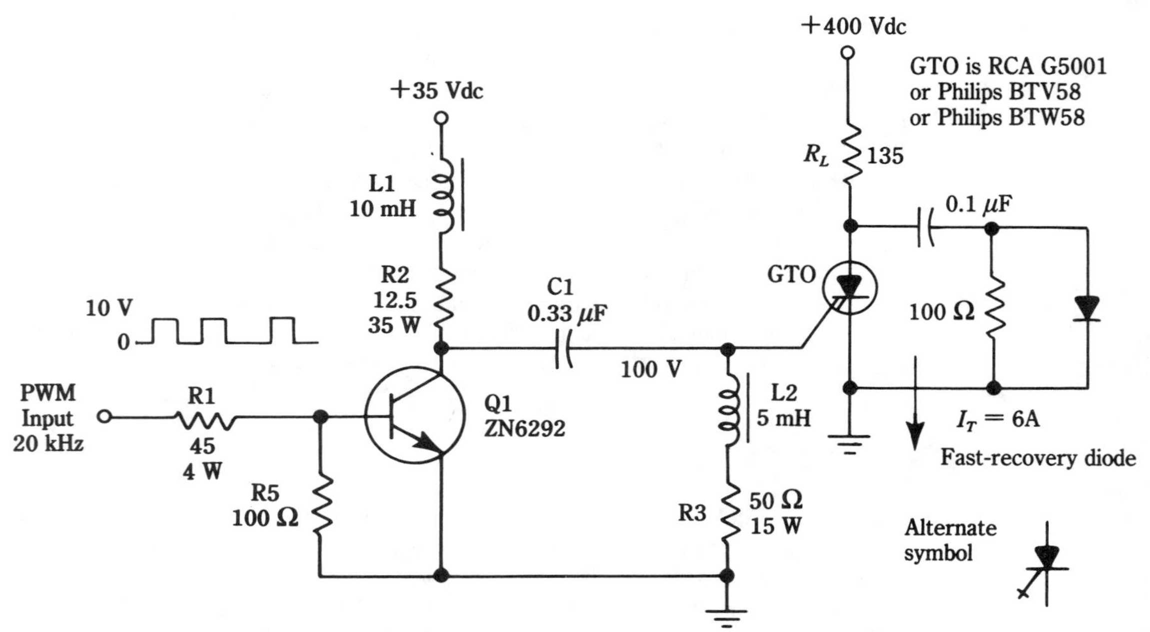

19-18 A GTO-driven converter for switching 1200 W at 20 kHz. Because of the high current density and the regenerative switching action of thyristors, this inverter can operate at 95% efficiency. Mullard and Unitrode also make GTOs. RCA.

load power is feasible by varying the spacing of the input pulses (such a control technique can also be viewed as ***pulse-position modulation***).

An alternate driver for the GTO is shown in Fig. 19-19. By appropriately selecting the complementary power MOSFETs and the auxiliary power supply voltages, the drive requirements of various GTOs can be met. The totem-pole circuit differs from the driver used in Fig. 19-18 in that no energy storage devices are used. This enables this driver to accommodate a wide range of pulse durations and pulse rates. The circuit is particularly well suited to its task because it behaves as a low-impedance source for the GTO gating pulses. This is very desirable for the negative turn-off pulse, which must be accompanied by a high peak current.

The sustained turn-on pulse that is delivered by this driver is another of its features. Under some operating conditions, such as at light load, the GTO might experience difficulty in rapidly latching into hard saturation if driven by very short-duration pulses. The GTO could suffer damage with such behavior because of excessive dissipation. This problem tends to vanish with longer turn-on pulses. Aside from electrical measurements, inadequate latching reveals itself by abnormal temperature rise—this type of self-heating unfortunately tends to be regenerative. You must, of course, know that suitable heatsinking is provided for the operating conditions.

If proper drive and heat removal are provided, it will generally be found that the GTO is an electrically rugged device. It can be successfully protected by a thermal fuse; quick-acting electronic protection is often unnecessary. This is because the likelihood of operation in its linear region tends to be slim once regenerative latching has been

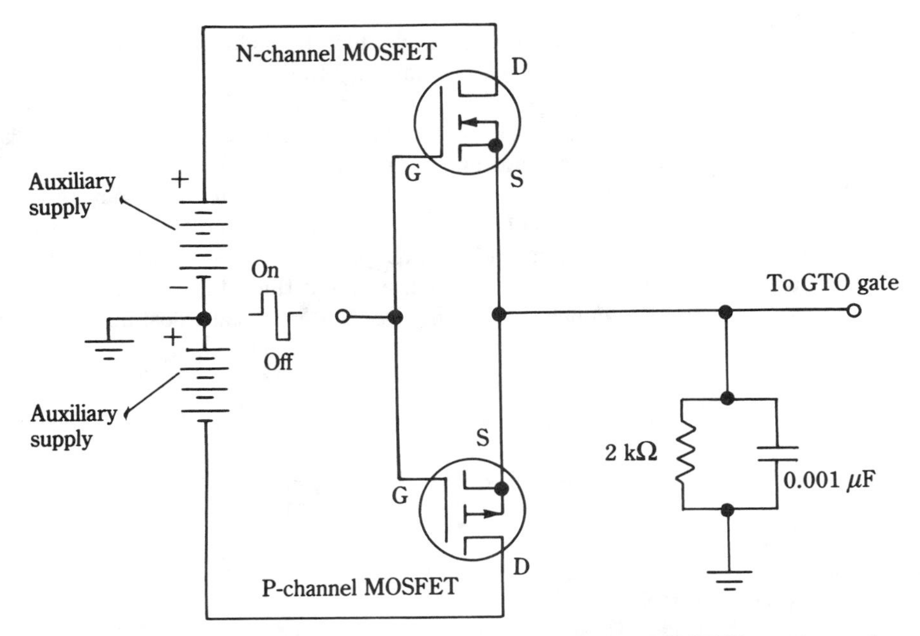

19-19 An alternate driver for GTO inverter. The complementary MOSFETs can be small types because of the high peak-current capability of these devices.

triggered. These remarks pertain to a debugged system with some history of reliable operation. During bread boarding and experimentation, caution remains a good policy.

The RC network in the output of the driver circuit of Fig. 19-19 is intended to prevent inadvertent turn on from electrical noise. This is often a consideration in the industrial environments where the high power capability of the GTO justifies its use. In this regard, care is advised before selecting certain GTOs with inordinately low turn-on voltage or turn-off current. These can be very useful in quiet environments, but it can be troublesome where energy-laden transients are the order of the day.

The MCT: still more bang for the buck

The MCT (MOS-controlled thyristor), like the GTO, is an SCR that can be turned on and off by appropriately polarized gate pulses. It is not clear whether this device has been extensively manufactured although its performance potential appears superior to that of the GTO—especially with regard to power-handling capability. As is sometimes the case with the GTO, its development might be of greater interest to overseas ventures—even though the MCT appears to have had its origin with American semiconductor firms. It might also be that the GTO has attained a state of perfection that is satisfactory for most practical purposes. A brief description of the MCT will be given because I feel that this device portends future importance to power supplies, inverters, and converters.

Also, for some purposes, it might be worthwhile to experiment with a discrete version, such as is suggested by the equivalent circuit. The ability of the MCT to turn off a megawatt in about two microseconds is not easy to dismiss as a triviality.

The equivalent circuit of the MCT is shown in Fig. 19-20. You can see that complementary MOSFETs are used to trigger the SCR (formed by the two bipolar transistors) on and off. Notice that gate pulses are supplied with respect to the SCR anode. Also, in contrast to the GTO, a negative gate pulse turns the device on; a positive gate pulse turns it off. This scheme of things stems from the objective of using a single gate terminal for both turn-on and turn-off pulses. Because of the MOSFETs, very little pulse power is needed to control the conduction state of the MCT. In compliance with the "no free lunch" law of nature, the MCT isn't destined for high-speed switching; it appears that 20 kHz, or so, can be anticipated.

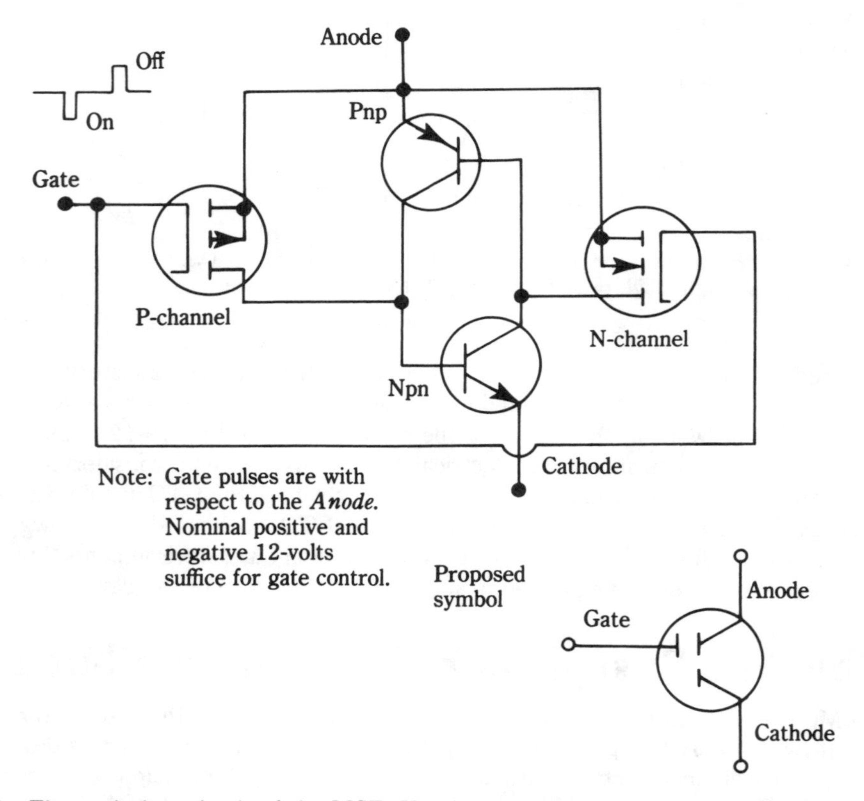

19-20 The equivalent circuit of the MCT. Via the single gate-terminal, the complementary MOSFETs direct the polarized gate pulses to the SCR in such a manner that controlled turn-on and turn-off is achieved.

Even though the MCT is controlled by short pulses, rather than by sustained input signals, it looms up as a good candidate for use in PWM regulators; as with the GTO, the input gate pulses can be duty-cycle controlled. As with the GTO and the IGBT, the application for MCTs will be for higher power systems, such as uninterrupted power supplies, welders, cycloconverters, and motor controls. Before greater popularity ensues, the MCT will have to manifest low vulnerability to latch up, reliability, thermal insensitivity, and, of course, cost effectiveness. A somewhat similar device, the MOS thyristor, has, after a several-year lifetime joined the dinosaurs in extinction. This SCR device had an integrated MOSFET in its gate circuit in order to facilitate turn on. Its shortcoming was that, like conventional SCRs, it could not be turned off via a gate signal.

The curves of Fig. 19-21 depict the generalized comparison among power devices. It is seen that the MCT stands head and shoulders above all in the matters of high current density and low voltage drop. Although instructive, graphical information of this kind must be tempered with a knowledge of other device characteristics. The behavior of each device can vary considerably as doping profiles are manipulated in quest of improvements of various parameters. Some devices are inherently capable of greater switching speed than others. For example, the 600-V power MOSFET that appears to be "low man on the totem pole" in Fig. 19-21 has, by far, the greatest frequency capability of any of the devices. Moreover, at lower voltage ratings, for example, in the 200- to 300-V region, the IGBT, Darlington, bipolar, and power MOSFET curves would tend to merge closer together. For 100-V devices, the power MOSFET could emerge supreme in the matter of low forward-voltage drop!

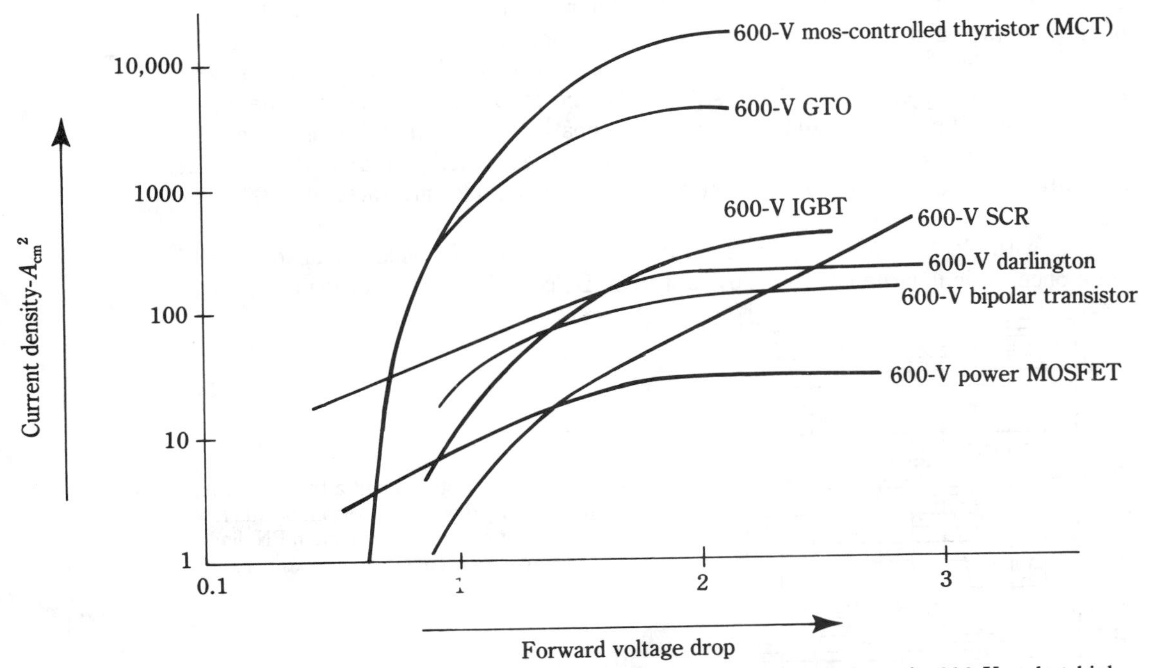

19-21 The general comparison of current density and voltage drop in 600-V devices. At 600 V and at higher voltages, IGBTs, GTOs, and MCTs, can provide higher current capability, together with lower voltage drop, than the long-used bipolar, Darlington, and MOSFET power transistors.

The material that refused to die: germanium power devices

Allusion has been made to the use of germanium power devices where fast switching speed or high frequencies are not involved. Despite the obvious fact that silicon has all but replaced germanium as the common semiconductor material, experimenters often find germanium transistors and rectifying diodes better suited than silicon devices in certain applications. As an example, 60- and 400-Hz inverters operating from vehicle batteries can develop higher operating efficiencies with germanium transistors than with silicon transistors because of the low saturation voltage of germanium transistors. As is elsewhere shown, this attribute can also merit consideration for using the germanium transistor in low-dropout regulator circuits. Remember that germanium power transistors were pnp types (there were also many germanium npn transistors for signal-level applications).

Germanium PN rectifying diodes also shine for their efficient operation in low-frequency applications. Again, this is because of their lower voltage drop, compared to silicon diodes. Forward-voltage drop curves (Fig. 19-22) might surprise the designer who has only worked with silicon PN diodes. These curves are suggestive of (or better than) the performance of silicon Schottky diodes. For 60- and 400-Hz rectification, the germanium PN diode might prove more cost effective than the silicon Schottky diode —especially if the reverse voltage is greater than 20 V or so. Although both germanium PN diodes and silicon Schottky diodes exhibit high reverse conduction near the high end of their voltage and temperature ratings, the phenomenon is not the same in the two devices. It is likely to be more severe in the silicon Schottky diode, where it can more easily be destructive. In both devices, the presence of high reverse current degrades rectification and introduces additional power dissipation. Germanium PN diodes generally will not prove troublesome in this regard if their junction temperature is not permitted to exceed about 90° C (some silicon Schottky diodes are rated at 150° C, or higher).

As with electron tubes, germanium devices are no longer widely available. They were once made in large quantities by Motorola, Delco, RCA, and other firms. Many of

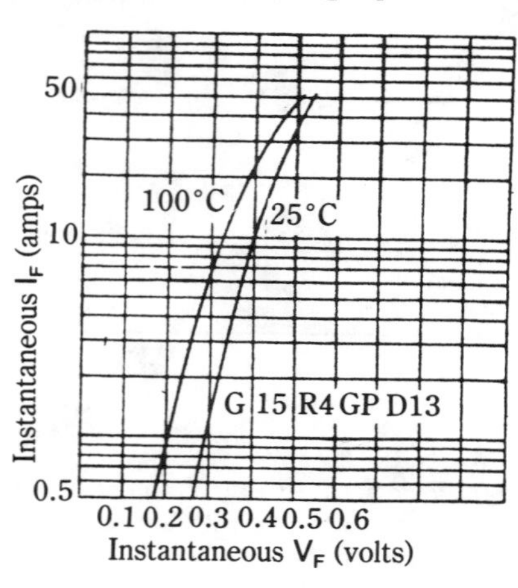

19-22 The forward voltage drop of a typical germanium rectifying diode. For certain dc and low-frequency applications, germanium PN diodes continue to merit consideration. Germanium Power Devices.

those types are now available from Germanium Power Devices Corp., Box 3065, Shawsheen Village, Station, Andover, MA 01810. Technical literature from this company should prove of particular interest to those experimenting with electric vehicles, where the use of massive heatsinks is not ordinarily considered detrimental. This firm supplies germanium pnp power transistors with at least 100-A ratings, and germanium PN diodes with at least 500-A ratings. Voltage ratings tend to fall in the 20- to 60-V range, but power transistors with V_{CBO} rating exceeding 95 V are not uncommon.

The proponents of germanium devices do not retain interest in them for sentimental or nostalgic reasons, but rather because beneficial performance can be realized in certain application areas with germanium diodes and transistors. In the foregoing paragraphs dealing with this topic, germanium diodes were always designated as PN types. Until recently, such an editorial ploy would not have been necessary. You could have correctly inferred that germanium diodes used in power circuits could not be other than PN-junction devices (point-contact germanium diodes were, and remain, signal-level devices; the same is true of tunnel diodes).

Germanium Power Devices Corp. now markets a line of germanium Schottky diodes with current ratings up to 400 A. As with silicon Schottky diodes, rectification takes place through the agency of majority carriers and there is no minority carrier storage phenomenon to slow down frequency response. Thus, the reverse recovery time is in the vicinity of 70 ns. Forward voltage drops in the 200- to 350-mV range can be attained in many cases. Thus, the germanium Schottky diode is superior to all other diodes in this respect. It can also be classified as an electrically rugged device with high immunity against damage from reverse energy, or from forward surge current. They are presently low-voltage devices with peak reverse voltage limited to 20 V. Modules of paired units are available for convenient application to centertapped (full wave) rectifier circuits.

This is a relatively new device at this writing. The firm cites successful application to 250-kHz switchmode supplies. At this switching rate, certain trade offs probably must be considered. Taking a conservative viewpoint, it is my judgment that all should be well at a 100-kHz switching rate, and that higher rates should be experimentally investigated. What counts in a specific application is the performance compared to that attainable from other devices.

20
CHAPTER

Low power, current, and voltage: small stuff

MUCH OF THE EMPHASIS IN TECHNICAL LITERATURE DEALING WITH POWER supplies, switching regulators, inverters, and converters is on transformation or delivery of high power, high current, or high voltage. The intended or inferred meaning is that you merely need to use scaling techniques to change performance ratings up or down. This tends to be essentially true, except that it is often found that special techniques usually must be used to obtain optimum results at both very high and very low power levels. In particular, in the domain of low power, for example, from fractional watt to 10 or 20 W, certain applications benefit from supplies capable of performing well at inordinately low powers, currents, and/or voltages. One manifestation of this is the relatively recent development of switch-mode supplies and converters for the 1- to 20-W range, where linear regulators had long ruled supreme.

Several examples of applications requiring high-efficiency performance at lower than conventional ratings can be cited. Battery-operated and portable systems are, of course, adversely affected by regulators or other power-conditioning circuits that dissipate power that would otherwise extend the useful life of the batteries. Distributed power systems use several or more small supplies in place of a single large one. This results in lower cost for cabling conductors, and better overall efficiency as well as better decoupling between circuits and sub-systems. Of course, these features obtain only if good performance is forthcoming from the small supplies. Another example is the integrated services digital network (ISDN), which is destined to replace ordinary telephone lines. Here, a low-power dc/dc converter will be needed at each line terminal. And, you mustn't forget computers, where substantial improvements can be had by designing for 3.3-V (or even lower) operation in place of the long-dominant 5-V technology. For solar-powered systems, the need to keep voltage drops and power losses small is self-evident. Finally, linear post-regulators for switch-mode supplies must function with minimal voltage drop to preserve overall efficiency.

Although we have learned to live with the nominally 0.7- to several-V gap or drop in semiconductor devices, these numbers can pose real obstacles in low-level supplies, regulators, and converters. Some remedies will stem from clever circuit techniques; others will best benefit from modification or substitution of semiconductor materials. It is not inconceivable that germanium devices might merit reconsideration. Help, too, will be forthcoming from batteries with higher energy densities and from multifarad (but physically small) capacitors functioning as "power supplies." All this, in any event, justifies a chapter on "small stuff."

An ultra-low dropout linear regulator with current limiting

Traditionally, bipolar power transistors have been used in most linear regulators designed around series-pass elements. This includes Darlingtons, where the higher current gain contributes to enhanced performance. The thinking on the part of designers had been that the linear regulator was the natural domain of bipolar devices, and that MOSFETs were deservedly making increased inroads as stars of switchmode supplies because of their superb switching characteristics.

There was, and there remain, much validity to this philosophy, but the steady improvement of the conductive characteristics of MOSFETs now forces a modified viewpoint in some instances. The low saturation voltage of bipolar transistors is caused by a mechanism not present (injection of minority carriers) in MOSFETs. However, continual developmental progress has produced MOSFETs with very much lower R_D than hitherto available. For practical purposes, low R_D accomplishes much the same thing as low saturation voltage ($V_{CE(sat)}$) in bipolar devices—namely, a low voltage drop across the device. This situation requires a second look at power MOSFETs as series-pass elements in linear regulators, particularly those that are designed for low drop-out voltage. Together with the low R_D, the high current gain of the MOSFET exceeds that of the Darlington. Coupled with easy drive, there are now logic-level MOSFETs, which go into hard saturation with 5 V applied to the gate. This is often more convenient than the 10 or 15 V that was previously needed.

The linear-regulator circuit of Fig. 20-1 can deliver 2.5 A to the load while incurring a drop-out voltage of about 85 mV. This stems from the exceptionally low R_D of the N-channel MTP5ONO5EL MOSFET, specified as 0.032 Ω. Other low R_D power-MOSFETs are the International Rectifier Corp. IRFZ40 (0.028 Ω), the Fuji 2SK905 (0.03 Ω), and the Harris RFG5ONO5 (0.022 Ω). These are all N-channel, enhancement devices, but are not cited as direct replacements for the MTP5ONO5EL. This regulator is designed to provide current-limiting at 3 A. The current-sense resistance is only 2 mΩ, and can consist of a 1.5-inch length of #23 copper wire. About two inches of #21 copper wire will also serve the purpose. Some experimentation will probably be needed for exactly setting the current-limiting action. In any event, the IN1006 contributes enough gain to reliably limit overload or short-circuit current to the set value. At the same time, the tiny current-sensing resistance exerts negligible degradation of efficiency.

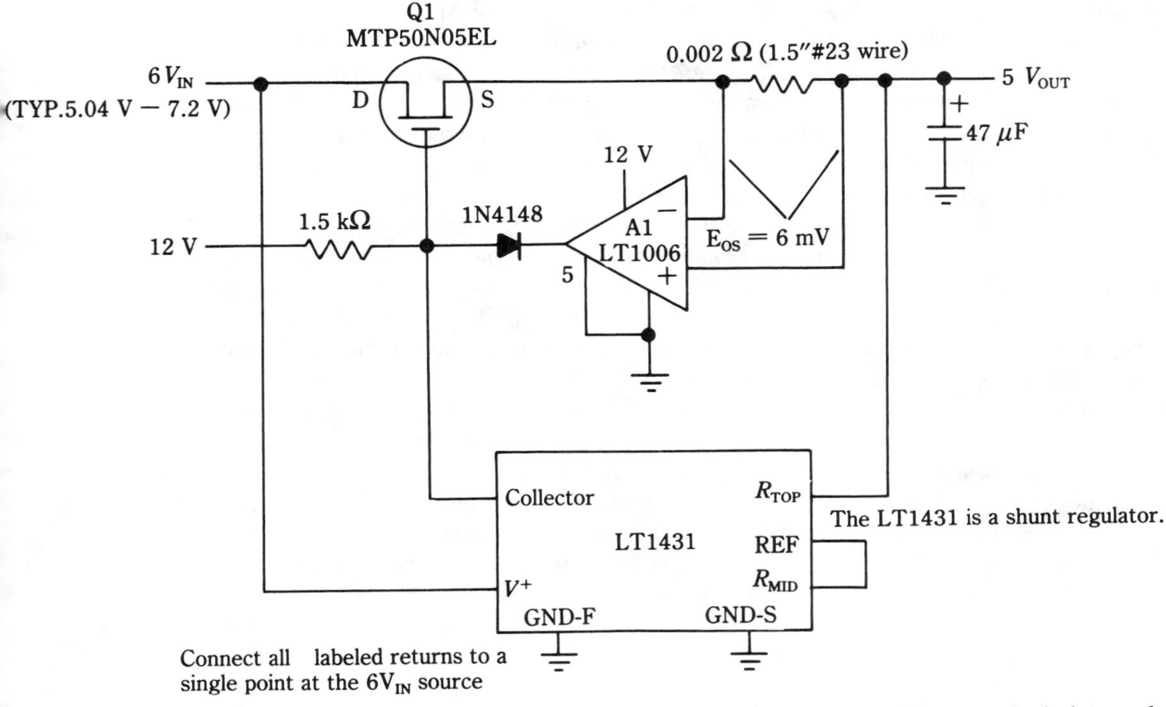

20-1 The ultra-low dropout linear regulator. Do not mistake the stylist symbol of Q1 as a depletion-mode device. Rather, it is an N-channel enhancement-type MOSFET with an inordinately low on resistance, R_D. Q1 operates as a source follower. Linear Technology Corporation.

A 12-V auxiliary dc source at a few-mA is required, but such a source is usually available in the majority of equipment because various ICs operate from a nominal 12 V.

Linear regulator with inordinately low drop-out voltage

There exists a widespread need for 5-V regulators with several-ampere capability, and with the ability to operate with as low dropout voltage as possible. *Dropout voltage* is simply the difference between dc input and output voltage with the proviso that regulation is maintained. The need for this feature can be gleaned from a practical example in which a nicad battery with a terminal voltage of about 8.2 V is regulated down to the required 5-V output. If the dropout voltage is typically 2 or 3 V, it is clear that there is little margin for long useful battery life. Increasing the battery voltage is not a good solution, for then more power from the battery must be wastefully dissipated in the series-pass transistor. If regulation could be maintained down to, for example, a half-volt drop across the pass transistor, the overall situation would be much better.

It has been learned that the series-pass transistor in monolithic IC regulators cannot readily be made to have a low saturation voltage. Although it would be satisfactory, even desirable, to drive and control the series-pass transistor by a monolithic IC, the pass-transistor must be a discrete device. This naturally suggests a hybrid arrangement, rather than a purely monolithic system. This actually is a blessing in disguise, for it permits easy optimization of saturation voltage and beta for the intended purpose. Moreover, you can even experiment with germanium transistors, which have inherently low saturation voltages. Another factor to be considered is that pnp transistors tend to have lower saturation voltages than their npn counterparts.

Using these facts naturally leads to the low-dropout regulator shown in Fig. 20-2. Its dropout voltage is only 50 mV at a load current of 1 A, and just 450 mV at 5 A of load current. The pass transistor essentially boosts the output of the LT1123 monolithic linear regulator. The MJE1123 silicon pnp transistor was specially designed for this circuit, but somewhat-similar transistors are available—the low-voltage saturation feature is the important parameter for selection, but high dc beta is also important for the sake of reliable short-circuit current limiting. It was found that a germanium-type 2N4276 provided even lower dropout-voltages, but probably at the expense of short-circuit current-limiting performance. The resistance in the base-current path of the series-pass transistor (shown as 20 Ω) should be empirically optimized, the idea being to make it as high as is consistent with acceptable drop-out voltage. Its value will depend upon the maximum input voltage expected to be used. Another feature of this regulator is the no-load low quiescent current of about 600 μA, contributing to long useful battery life.

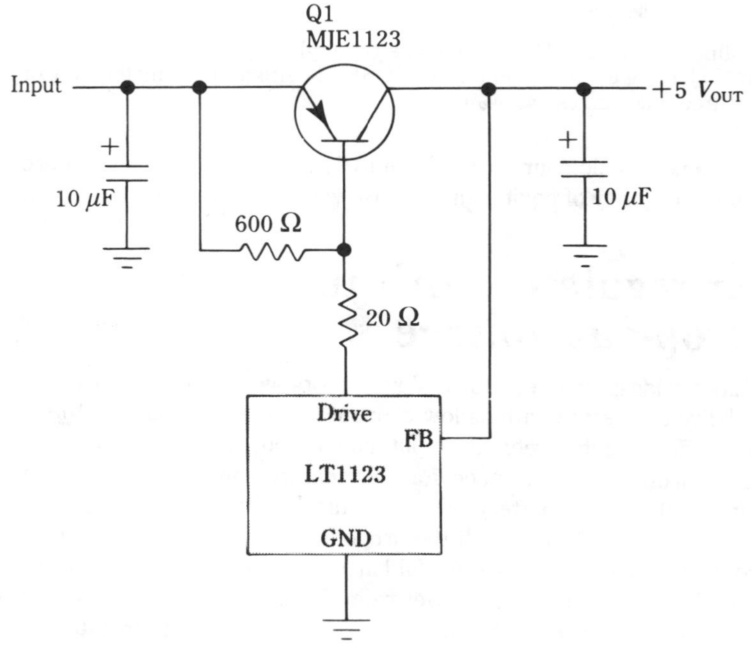

20-2 An example of linear regulator featuring low dropout voltage. The hybrid circuit topography is used because it is difficult to obtain low dropout voltage from ICs alone. Linear Technology Corporation.

Another semiconductor firm's similar low-dropout linear regulator is shown in Fig. 20-3 . The basic performance remains about the same, with 350-mV dropout cited for a 3-A load current. Again, the hybrid configuration allows a great deal of design flexibility. What differs mostly among the various low-dropout control ICs are the housekeeping (or ancillary) functions, otherwise known as "bells and whistles." These can be initially appraised with due regard to the particular application, and selection made accordingly. Most of these dedicated ICs will at least provide short-circuit and thermal overload protection. Because the pnp pass transistor is external to the IC, good heatsinking is essential. Often, a low-dropout linear regulator can be added to an already-designed switchmode supply in order to provide post-regulation. Moreover, the overall system efficiency will not be appreciatively degraded by such a modification. This cannot be said when a conventional 3-terminal voltage regulator is used for post-regulation.

At first impulse, it might appear feasible to duplicate the performance of the two low-dropout circuits just described by resorting to a conventional 3-terminal voltage

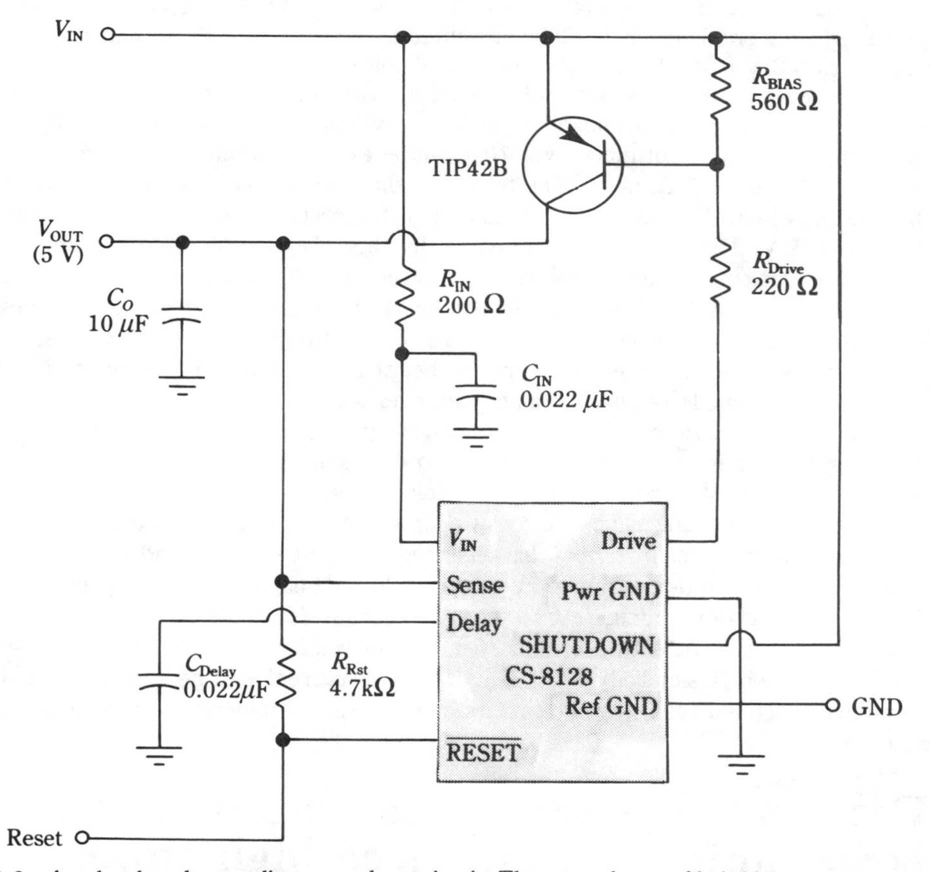

20-3 Another low-dropout linear regulator circuit. The same theme of hybrid topography with an external pnp transistor is used. Selection of the control IC is best made in terms of the desired auxiliary functions. Cherry Semiconductor Corp.

regulator and a discrete-pass transistor. However, the quiescent current (current consumed by the regulator IC that does not pass through the load) will be found to be very much higher than is the case with the special circuits. This, in itself, tends to defeat one of the objectives—not to impose additional power dissipation on the overall system.

Linear regulator for providing 3.3 V from existing 5.0-V source

It appears that the long-enduring 5-V logic technology will slowly, but steadily, be rivaled by circuitry that is designed to operate from a nominal 3.3-V supply. It has been proven that operation from the lower voltage can improve speed, packaging density, and efficiency. Although it is not clear how much 5-V logic will be retained for applications in which optimum performance parameters are not necessary, it appears certain that computer systems of the near future will contain at least some 3.3-V logic. The interesting problem which will be presented to the power-supply designer is how to convert to 3.3 V from the already built-in 5-V source.

A knee-jerk reaction would likely select a switch-mode supply for this purpose. However, both calculation and observation of available switchers indicates that an operating efficiency of not much over 70% can be expected when working from a 5-V input, and delivering 5 A, or so. The trouble is that the voltage drops associated with the power switch, the free-wheeling diode, and the rectifying diodes are too large a fraction of 5 V. The problem is aggravated by the relatively high current involved. Thus, when you consider additional factors, such as electrical noise and circuit complexity, it becomes natural to reconsider the possibility of using a linear regulator. Interestingly, the operating efficiency of a linear regulator used to convert 5.0 to 3.3 V is simply 3.3/5.0 = 66%. You can see that there can be, at best, a marginal superiority in efficiency if a switchmode regulator was chosen over a linear regulator.

Further consideration reveals that not just any linear regulator circuit will do. Rather, special design must be resorted to in order to obtain the requisite low dropout voltage under worst conditions of circuit tolerances and temperature. The Linear Technology LT1083 linear adjustable low-dropout regulator fulfills the needs of 5- to 3.3-V conversion. A nice feature of this dedicated IC is that there is no adverse behavior if operation is forced (for example, by excessive load demand) into its dropout region. Some linear regulators produce oscillations or an abrupt rise in quiescent current under such circumstances. As shown in Fig. 20-4, the application of the LT1083 to 5.0- to 3.3-V conversion is simplicity itself. 7.5 A of load current can be supplied and the protective functions against short circuits and excessive temperature rise have been designed into this IC.

Distributed power system: multiple low-power units vs. centralized high power

It is easy enough to lay out a power-distribution system on paper. Indeed, this has often been done as a last-minute job of putting the final touches on a systems design.

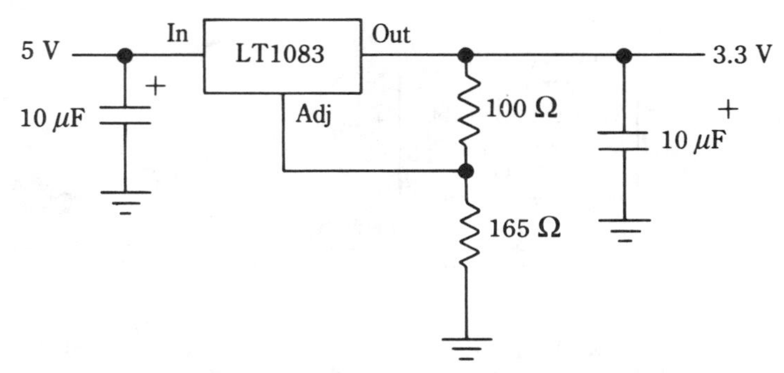

20-4 The use of dedicated IC as linear regulator in converting 5 to 3.5 V. The low-dropout requirement precludes the use of other IC regulators. Linear Technology Corp.

However, practical implementation, to say nothing of profit-gobbling costs, all too often have produced a mammoth engineering crisis at this juncture. The task of piping power to a number of subsystems at high current and low voltage is fraught with pitfalls and penalties. If the dc power originates in a large centralized regulated supply, it is likely that most benefits of regulation no longer are manifest at the various loads. Although little trouble is experienced delivering several A from a centralized source, 100 or 200 A is entirely a different ballgame. Secondary or trivial considerations attending low-current distribution become the prime technological and economic matter for high-current distribution. The weight, I^2R drops, volume, installation, and cost of the copper cabling system then jeopardize the success of the whole enterprise.

A partial solution to this dilemma was once achieved by "point of use" power distribution in which each subsystem had its own linear regulated supply—often a three-terminal voltage regulator. In this approach, a single master power supply would not be called upon to provide tight regulation; this would be the task of the individual regulators mounted on the PC boards of the subsystems. Such a scheme enabled the subsystems to be well-isolated from one another and to respond quickly to transients and to load or line changes. Some measure of reliability was realized because any small-supply failure might not disable the whole system; also, the quick replacement of a defective small supply would not involve major surgery or cost. However, the distribution voltage could not greatly exceed the dc operating voltage of the subsystems; otherwise the small point-of-use supplies would dissipate excessive power—this being the nature of linear supplies.

It was soon enough realized that what was needed was a "dc transformer" so that power distribution could be at high dc voltage and low current. Fortunately, switchmode dc/dc converters provide that very function. Not only do they behave as dc transformers, but they can be very tolerant to dc input-voltage. You could then have a simple, inexpensive centralized dc supply that didn't even need to be regulated. From this master supply, power could be distributed at a relatively high dc voltage, for example, 50 V (with current proportionally reduced in "transformer fashion"). Each point-of-use dc/dc converter then would efficiently reduce its received power to the low-voltage, high-current format required by its load (Fig. 20-5).

This elegant approach was somewhat tardy on the scene because earlier dc/dc

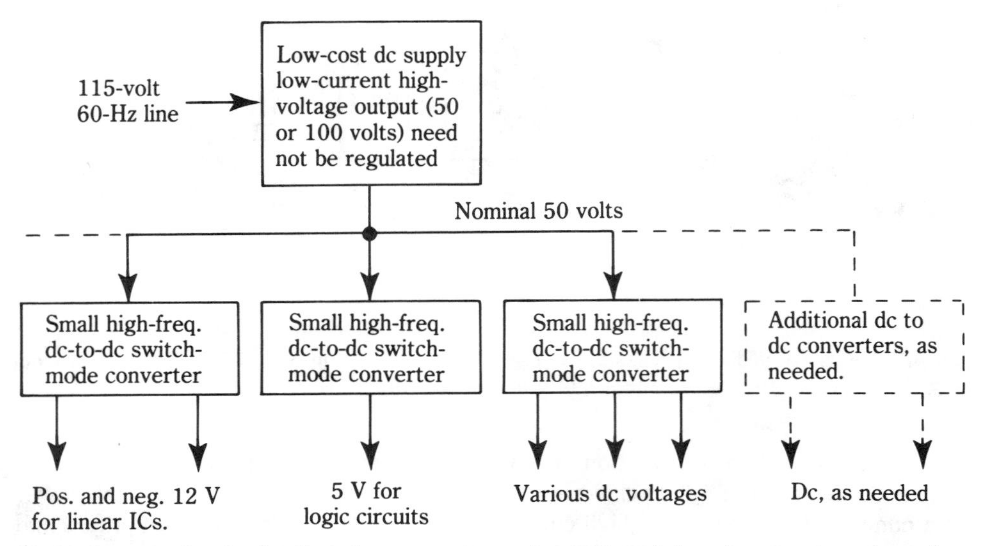

20-5 An example of a distributed power system using dc/dc switchmode converters. A number of low-power dc-to-dc converters operate from high-voltage dc feeder line and provide operating power at point-of-use. This works out better than attempting high-current, low-voltage distribution from a single regulated source.

converters, operating in the 20-35-kHz range were relatively bulky, heavy, and expensive. Also, their RFI and EMI behavior left something to be desired, and there were serious questions regarding the reliability of the switchmode technique. It must also be pointed out that these erstwhile converters were not readily available in low-power formats, this having been considered the sacred domain of the linear regulator.

The compelling features of distributed power systems comprising a master dc source and a number of low-power dc/dc converters can be summarized as follows:

Dramatic savings in cabling weight, cost, and in I^2R loss.

High efficiency performance is forthcoming from high-frequency dc/dc converters in the face of a wide range of dc input voltage.

Possibility of enhanced reliability because failure of a dc/dc converter need not necessarily disable the whole system.

Such a system lends itself to easy expandability.

Because voltage drop in the connecting cables is of little consequence, the master dc supply can be remotely located.

There is a wide selection of low-power dc/dc converters on the commercial market.

The speed of response to load and/or dc line changes is enhanced.

The dc/dc converters provide excellent isolation between subsystems.

Because of the behavior of dc/dc converters as dc transformers, such a power-distribution system allows wide design flexibility.

RFI and EMI problems are more easily handled than in a centralized distribution system conveying power at low voltage and high current.

A high-efficiency, low voltage-drop free-wheeling "diode"

A nice feature of switchmode regulators is that a wide range of input voltage can be accommodated without incurring any penalty of operating efficiency. This contrasts to the situation in linear regulators where more and more dissipation take place in the series-pass element as the input voltage is increased. However, both types of regulator share the common problem of operating from an input voltage not much greater than the desired regulated output voltage. Indeed, in both circuits, a low drop-out voltage economizes battery selection, extends useful battery life, and enhances the overall efficiency of the system. It can be appreciated that this is of utmost importance in portable equipment.

Although it is customary to assume high operating efficiency for the switchmode regulator, this no longer holds when forward voltage drops of power switches, rectifying diodes, and free-wheeling diodes are comparable to the actual output voltage. This is one of the reasons for using low R_D MOSFETs, Schottky diodes, and synchronous rectifiers. Long overlooked, however, has been improvement of the free-wheeling diode. This has been largely because the nominal 0.5 V or so drop in a Schottky diode hasn't been too objectionable in most applications.

For low-voltage operation, it turns out that a properly selected power MOSFET can outperform the Schottky diode. For sake of comparison, it is possible to produce a voltage drop on the order of 50 mV at 1 A. At the same time, the high-frequency response of the MOSFET is about comparable to that of the Schottky diode.

The application of the power MOSFET as substitute to the free-wheeling diode is straightforward, but probably not obvious. Simplified circuits showing how this is accomplished are shown in Fig. 20-6. At first glance, the electrode polarization of the substituted MOSFET appears improper for the function it is to serve. However, this MOSFET operates in the third quadrant of its characteristics as in a synchronous rectifier; it conducts only when its drain is negative and its gate is positive. Conventional conduction with positive drain never occurs because the gate then lacks turn-on bias. Moreover, the intrinsic body diode never conducts because the low R_D of the MOSFET maintains the voltage drop well below the 0.7 V required for forward bias of this diode (a small transformer could be used in place of the depicted inverter).

This scheme is best implemented with logic-level MOSFETs, which saturate with 4 or 5 V between gate and source. An even-simpler arrangement can dispense with the inverter if a p-channel MOSFET is used in place of the free-wheeling diode. However, p-channel MOSFETs tend to have a somewhat higher R_D than the n-channel devices. The drain and source connections would be interchanged if a p-channel MOSFET were used.

1-W dc/dc converter for the integrated services digital network (ISDN)

A very important need for low-power dc conversion is found in the practical and economic implementation of Integrated Services Digital Network (ISDN). Whereas the

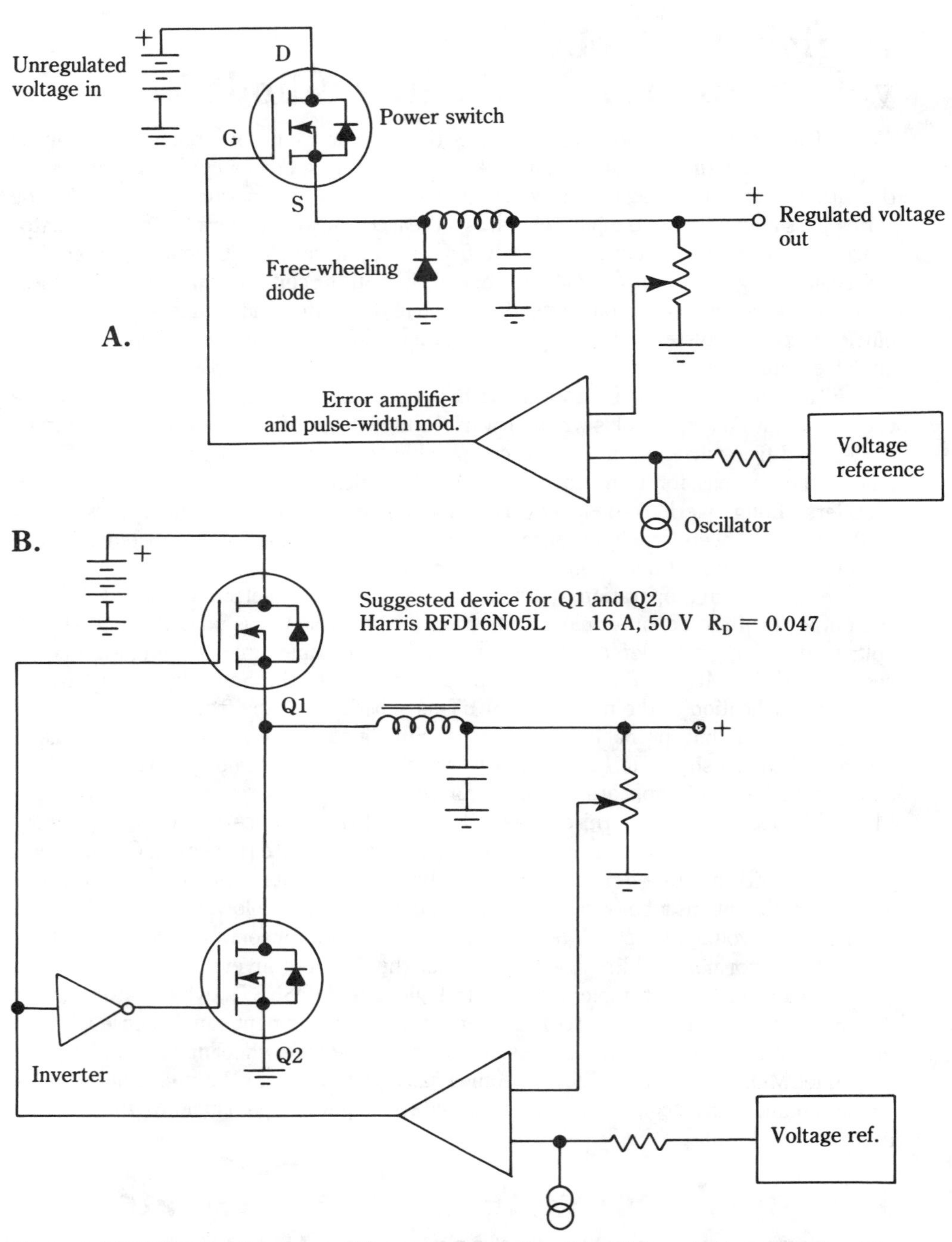

20-6 A MOSFET synchronous rectifier as free-wheeling "diode" in switching regulator. The MOSFET substitute operates in its third-quadrant and alternates its on and off states under control of the power switch. (A) A conventional circuit with a free-wheeling diode. (B) The circuit with MOSFET, Q2, substituted for the free-wheeling diode.

original telephone system was devised to accommodate analog voice transmission, the new digitized network will, in addition to voice, allow virtually any type of data, regardless of bandwidth or speed, to be handled. Just as you might plug a wide variety of appliances into the 60-Hz utility outlet, ISDN will be the common conduit of voice, satellite, TV, computer, facsimile, and other data. Moreover, this will be accomplished without modems, which impose severe restrictions on data transmission rates. Worldwide standardization will enable any type of information bit stream to be transmitted and received within a single building, or equally well between distant countries.

It has been worked out that the logic circuitry at each terminal will be based upon existent 5-V technology. Under worst conditions, the incoming dc voltage can be postulated to vary from 24 to 42 V. This calls for a switching regulator capable of converting this voltage range down to the 5-V level. It has further been decided that galvanic isolation is necessary. The called-for operating efficiency should exceed 80% with the stipulation that the minimum efficiency at extremely light load (for memory retention) should be at least 55%. Finally, the maximum power capability should be on the order of 1 W.

When the designer contends with these basic requirements, he or she finds that design philosophy is a bit different from that ordinarily used at higher power levels. Small losses, that could be dismissed as negligible at the higher power levels, tend to assume dominant influence at the 1-W level. For example, quiescent current, an unimportant factor at high power levels, can make deep inroads into operating efficiency at low power levels (quiescent current is the sum of all currents consumed by the supply or converter, but which don't continue to load current). Also, the notion of "the higher the switching rate, the better" is no longer valid because the "tiny" losses from switching, eddy currents, and hysteresis exert appreciable effect on low power efficiency. Rather, it is found that it is best to adopt as low a switching rate as possible—this turns out to be about 18 kHz.

Siliconix has developed a family of dc/dc converters and converter control ICs specifically targeted to meet the electrical and reliability specifications of ISDN terminals. One of these, the Si9105, has its own monolithically integrated power MOSFET, which develops the requisite 1-W output. It is used in a flyback circuit with an external inductor, rectifier circuits, and an associated handful of passive components. The block diagram of the Si9105 is shown in Fig. 20-7.

Although the functional blocks bear close resemblance to other control ICs, several features are worthy of special mention. CMOS technology is used in the monolithic architecture. This helps keep the quiescent current at low levels. The unique feature is, of course, the self-contained power MOSFET output stage. A second power MOSFET is used as a preregulator; notice that this is a depletion-mode type. The oscillator frequency is twice the switching rate. The internal connections are such, that current-mode operation takes place; sensing of the external inductor current ramp occurs as the result of the voltage drop developed across an external resistance connected from pin 4 to ground.

The schematic diagram of the overall dc/dc converter using the Si9105 is shown in Fig. 20-8. It is basically a flyback circuit in which extra winding N2 provides both bootstrapped operating power, and galvanic-isolated feedback to pins 6 and 14, respectively. Incidentally, it is common to refer to the core component in flyback supplies as a

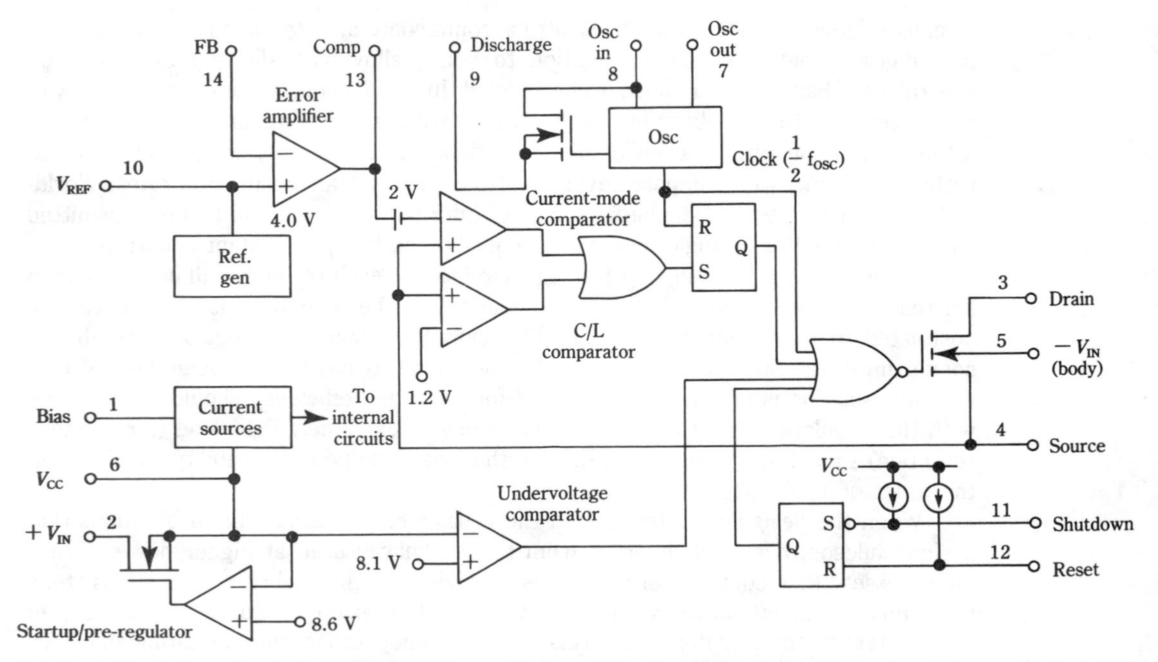

20-7 A block diagram of the SI9105 IC. By combining with a few external passive components, an efficient 1-W dc/dc converter can be formed to serve the needs of ISDN terminals. Siliconix, Inc.

"coupled inductor," when secondary windings give it the appearance of a transformer. It's largely a matter of technical semantics—conventional transformers are designed to minimize stored energy; inductors in flyback circuits are designed so that the primary inductance stores electromagnetic energy during current ramp-up, and then suddenly releases this energy to the secondary or rectifier circuits. To obtain this kind of operation, the designer tries to ensure that there will be no magnetic saturation during the current ramp and must, at the same time, hold-down resistance in the primary winding. The secondary windings are, accordingly, often dealt with as an "after-thought" in this type of core component.

The efficiency vs. output-power curve of Fig. 20-9 reveals that attention given to a number of ordinarily trivial factors has paid off. The operating efficiency over most of the load range is actually in the vicinity of 85%. Considering the modest parts count and the economy of mass production, it is likely that a dc/dc converter, such as this will prove a cost-effective way to power ISDN terminals. Other uses will no doubt suggest themselves to the experimenter because of the inherently easy way to provide output windings suited to the application at hand. The load regulation is compromised a bit as a trade-off for galvanic isolation, but is tight enough for most purposes. Line regulation, derived from current-mode operation is, however, very responsive and provides high immunity to transients on the dc input line.

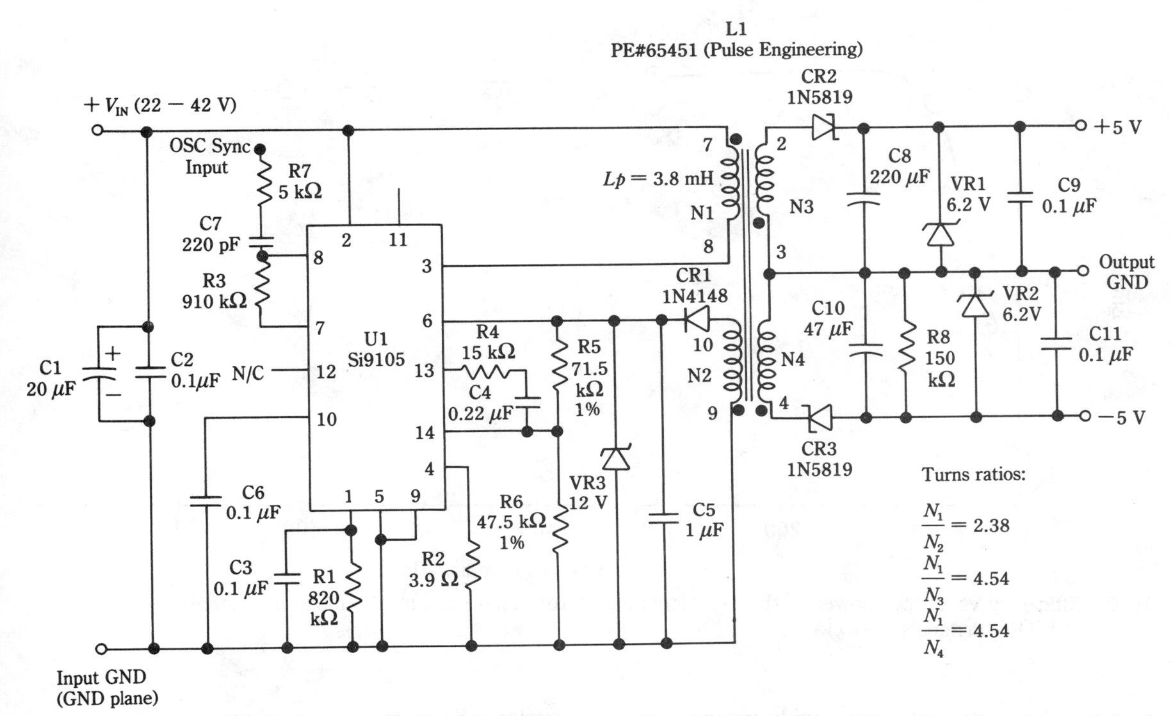

20-8 The schematic diagram of the 1-W dc/dc converter. This is a current-mode regulator based on a flyback circuit. Siliconix, Inc.

"Look ma, no inductors:" switching capacitor supplies

The need for inductors and transformers in power supplies has long been taken for granted. After all, these magnetic-core elements perform the vital role of energy storage, from which you can also begat energy transformation, filtering, and current limiting, to name a few subfunctions. At the same time, the practical implementation of magnetic-core components is all too often plagued by such negatives as saturation, core and copper loss, nonlinearity, weight, bulk, and cost, and unreliable availability. The departure from ideal characteristics can damage the supply, as well as the load, and can cause problems with RFI, EMI, unwanted resonances, insulation breakdown, etc. Because a capacitor is also an energy-storage device, it is only natural to ponder the possibility of substituting capacitors for inductors.

This can, indeed, be done. While direct substitution is not feasible, it turns out that special circuits can be devised to exploit the basic energy-storage nature of capacitors. When this is done, it is possible to design for most of the performance features ordinarily associated with circuits using inductors and transformers. For example, such

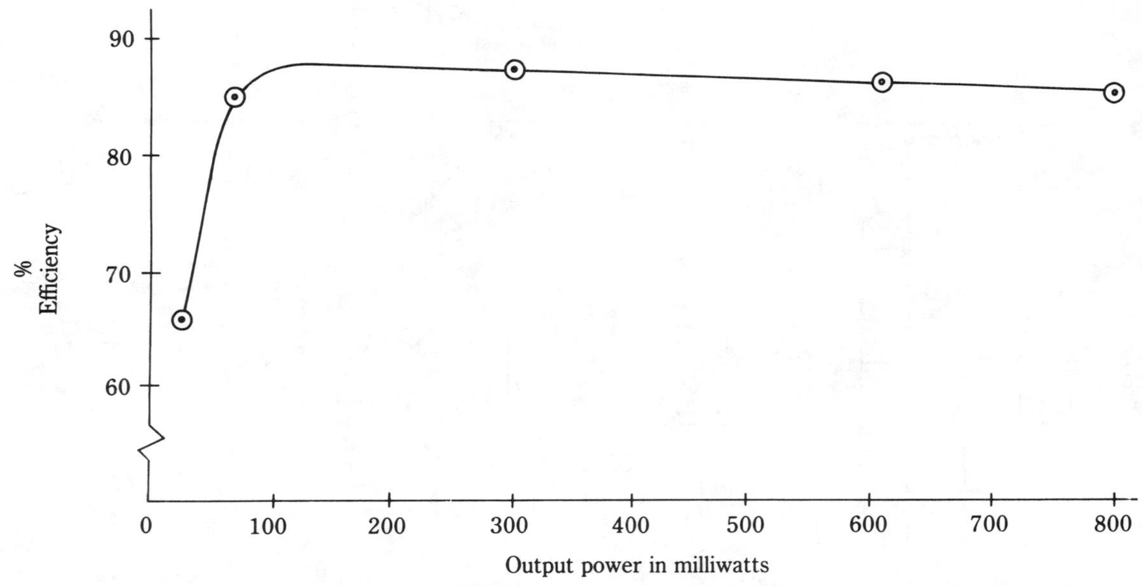

20-9 Efficiency vs. output power of the 1-W dc/dc converter. The performance more than meets requirements of ISDN terminals. Siliconix, Inc.

"inductorless" designs can provide polarity inversion, voltage step-up and step-down, galvanic isolation, voltage multiplication, and regulation. These are all accomplished by switching techniques in which accumulated charge is conveyed from one capacitor to another. The general process can be classified as the controlled use of charge-pumping circuitries.

Although there are many ways of switching charges between capacitors, a certain law of nature prevents complete recovery of the initial energy level! This can be demonstrated by a single example. Refer to Fig. 20-10. Suppose a 10-μF capacitor has been charged to 20 V. Its energy content is given by 1/2 CV^2, or 1/2 $(10 \times 10_{-6})(20)^2 =$

$$\frac{(5 \times 10^{-6})(400)}{2}$$

= 1000 microjoules.

Now, suppose this charged 10-μF capacitor is switched in parallel with an uncharged 10-μF capacitor. In the new situation, it is clear that you have 20 μF of capacitance charged to 10 V. Common sense would likely assume that the net energy in the system has remained constant, or very nearly so for high-quality capacitors—your reflex thoughts might well suppose that only the manner of storing the initial energy has changed. However, when the new numbers are plugged into the energy equation, we find that 1/2 CV^2 = 1/2 $(20 \times 10^{-6}(10)^2 =$

$$\frac{(10 \times 10^{-6})(100)}{2}$$

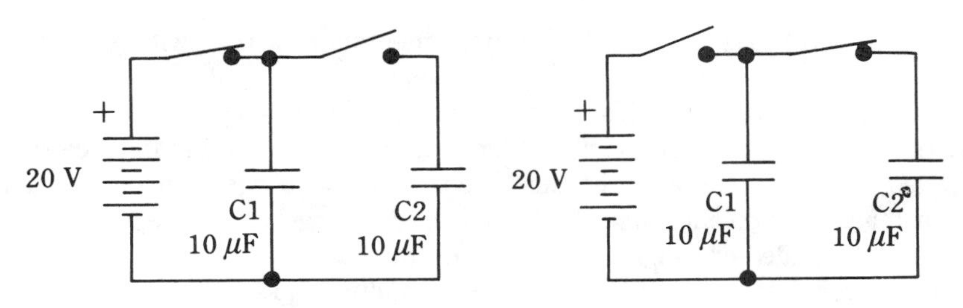

Stored energy in $C_1 = \frac{CV^2}{2} = \underline{1000}$ microjoules

where $C = 10\ \mu F$
where $V = 20$ V

A.

Stored energy in C_1 and $C_2 = \underline{500}$ microjoules
New $C = C_1 + C_2 = 20\ \mu F$
New $V = 20/2 = 10$ V

B.

20-10 A demonstration of the inherent energy-loss in capacitor switching. Despite assumed high quality of the capacitors, 50% of the energy initially stored in C1 will be lost when the energy is switched to the parallel combination of C1 and C2. This type of loss can be largely overcome by the selection of capacitor sizes and by appropriate charge/discharge timing. (A) C1 is charged to 20 V and thereby stores 1000 microjoules of energy. (B) Charge in C1 is shared with C2. Calculation shows that the net energy stored in the parallel combination of C1 and C2 is only 500 microjoules.

= 500 microjoules. Thus, only half of the initial energy is stored in the parallel capacitors! What has happened to the missing 500 microjoules?

You must, of course, attribute the energy deficit to "losses." Interestingly, the energy equation doesn't seem to be concerned whether the capacitors are charged and discharged through a mΩ or a MΩ. Indeed, it turns out that the presence of series resistance only affects the time it requires to charge and discharge the capacitors. In every situation involving equal-valued capacitors, 50% of the initial energy will be lost when the capacitors are switched in parallel. Nature, as reflected by the energy equation seems to know that half the initial energy will be lost via I^2R heat, sound, light, and RF radiation. Then, too, you can postulate dielectric leakage and dielectric hysteresis. Admittedly, this is not an easy concept to grasp; it might be easier to accept the phenomenon on a simple mathematical basis—things that happen when numbers are halved, doubled, and squared. The vital point is that charging and discharging capacitors incurs energy loss. The example given was chosen for illustration, being a particularly aggravated situation. In practice, the energy loss can be greatly reduced by appropriate selection of charging and discharging cycles and by the relative sizes of the capacitors involved.

Unlike conventional switch-mode circuits that use inductors, the efficiency of a switched capacitor supply tends to be optimized over a band of relatively low frequencies. Still lower switching rates result in lower efficiency because the capacitors do not receive enough charge replenishment to meet the power demand (*energy* $\times$ *frequency* = *power*). However, the power dissipated in the inherent losses of the switching circuit also increase with switching rate. It, therefore, turns out that a balance or compromise must be made with a too-low and a too-high switching rate. This would be true even with ideal capacitors and switches. Accordingly, switching rates are on the order of 10 to 30 kHz in the practical implementation of the switched-capacitor technique.

Now, putting it all together, this is what you find: it is not necessary to lose a large portion of the energy in a capacitor when dumping its charge into another capacitor. This happened in the example given because the transfer of charge occurred as a single event. If, instead, the replenishment and transfer of charge occurs repetitively, the energy loss can be made very slight. At the same time, this action must not be done at too great a rate, for then the tiny switching losses multiply up to serious levels. Thus, efficient operation is obtained when there is sufficient charge-replenishment—higher switching rates thereafter serve only to again degrade the efficiency of the process. Dimensionally, you can see how power can be supplied to a load because: *power* = (*energy*)(*rate of energy transfer*). A basic repetitive-switching scheme is shown in Fig. 20-11. When S1 and S3 are closed, S2 and S4 are open, and vice versa. Polarity inversion is inherent in the process. Also, as indicated by the dotted line, voltage-doubling is readily accomplished.

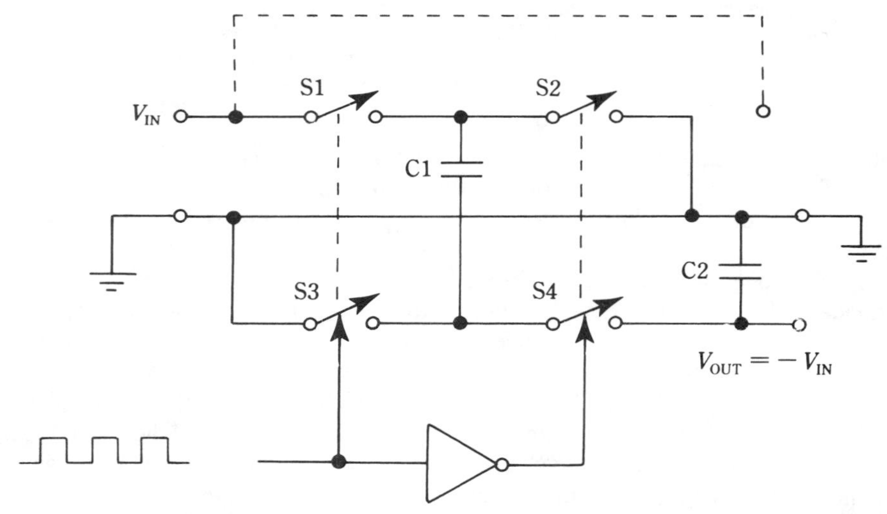

20-11 The basic switching arrangement used in many charge-pump inverters and converters. This scheme produces polarity inversion, and can also be used as a voltage doubler (dotted-line terminal).

Examples of switched-capacitor applications are shown in Figs. 20-12 and 20-13. In order to avoid confusion from nomenclature differences used by various companies, keep in mind that the switched-capacitor converter and the charge pump refer to the same circuit technique. Also, the use of the word *inverter* traditionally pertains to circuits that change dc to ac. These can be either self-oscillating, or driven. Unfortunately, circuits that reverse the dc polarity of the source are loosely referred to as *inverters* also. It would be better to call these circuits *polarity inverters.*

The LT1054 is a combined switched-capacitor converter and voltage regulator. Its block diagram is shown in Fig. 20-14. Pins are brought out to provide considerable

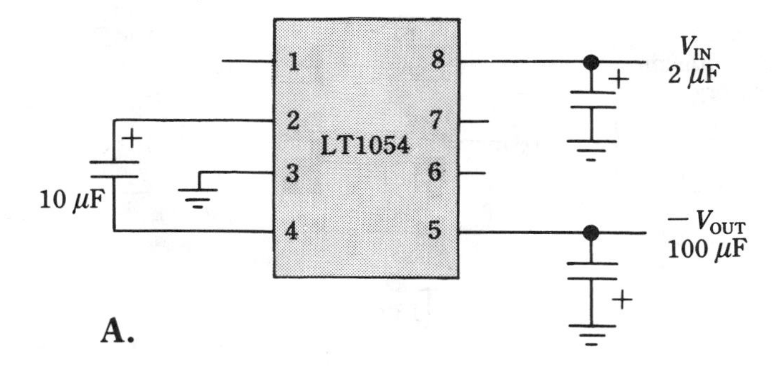

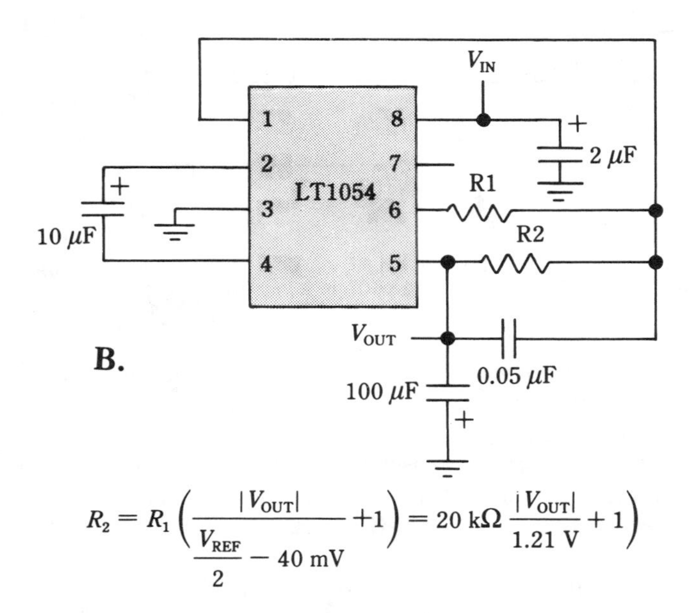

$$R_2 = R_1\left(\frac{|V_{OUT}|}{\frac{V_{REF}}{2} - 40\text{ mV}} + 1\right) = 20\text{ k}\Omega\left(\frac{|V_{OUT}|}{1.21\text{ V}} + 1\right)$$

20-12 Typical applications of the LT1054 switched capacitor IC. (A) A voltage inverter with unregulated output. (B) A voltage inverter with regulated output. Linear Technology Corporation.

flexibility in implementation. For example, the use of the regulator portion of the IC is optional. Many arrangements are possible for various polarity inversion, voltage doubling, and dc/dc conversion functions. The simplest application is the voltage inverter of Fig. 20-12A. Good results can be obtained with tantalum solid-state capacitors. The dc output voltage of Fig. 20-12A is unregulated. However, the voltage loss at 100-mA load current is on the order of 1 V; this is a "stiff" enough supply in many instances. The input voltage source can encompass the 3.5- to 15-V range. The nominal switching frequency is about 25 kHz, where operating efficiency is optimized. However, the internal oscillator frequency can be lowered by connecting a capacitor from pin 7 to ground; it can be raised by connecting a capacitor from pin 7 to pin 2. Despite the

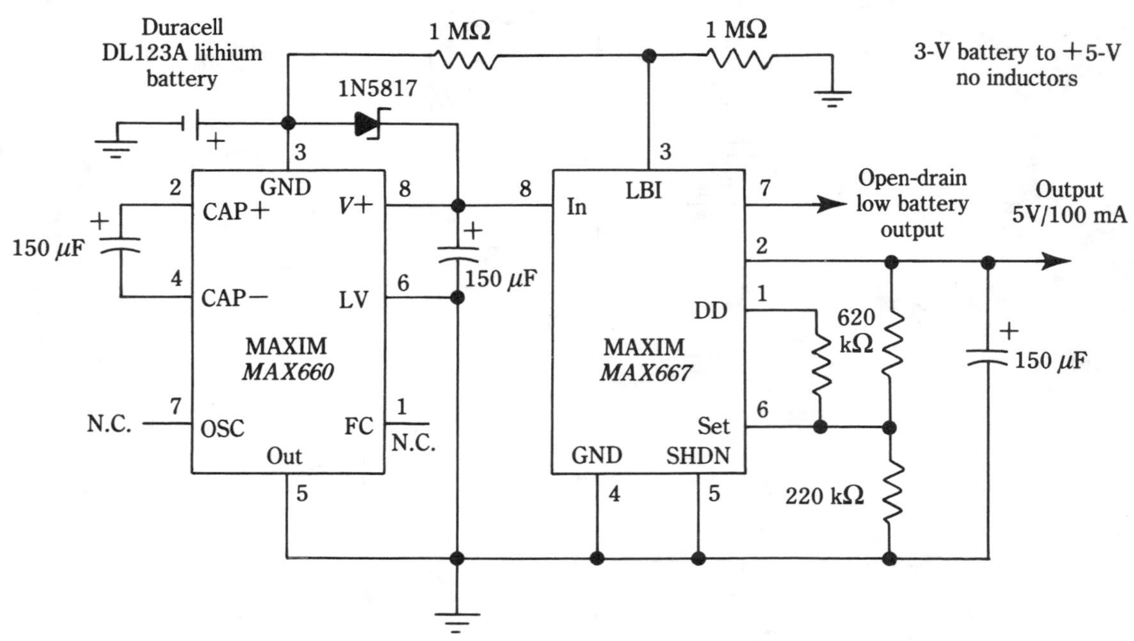

20-13 An inductorless conversion of single-cell source to a regulated 5-V output. The MAX660 is a switched-capacitor voltage-doubler. The MAX667 is a low dropout-voltage linear regulator. MAXIM Integrated Products, Inc.

relatively low frequencies involved, these capacitors will probably only be some tens of picofarads in size. Oscillator synchronization can be achieved by injecting a slightly higher signal frequency at pin 7.

The manufacturer cautions the user not to allow pin 5 to become positive, with respect to any of the other pins. This is because pin 5, in addition to being the output pin, is also tied to the substrate of the IC. No such danger exists in the two example applications.

By adding several passive components, the output voltage can be regulated against both variations in dc input voltage, and variations in load current. This results in the circuit of Fig. 20-12B. Regulation is achieved by servo-controlling the effective resistance of the pnp switch shown in the block diagram. Although this pnp transistor is part of the charge-pump switching circuit, for regulation purposes, it is controlled similarly to the series-pass element in a linear regulator. For this reason, you must not allow too much voltage differential between the input and output voltages in order to stay within the dissipation ratings of the LT1054.

A useful application of the charge pump or switched-capacitor technique is shown in Fig. 20-13. The arrangement comprised of the two ICs delivers regulated 5 V at 100 mA from a single 3-V lithium cell. 16 hours of continuous operation can be had with a 40-mA load.

The MAX660 IC is a charge pump connected to perform as a voltage doubler. Thus, a nominal 6 V is applied as V_{IN} to the MAX667 low dropout voltage regulator. The MAX660 operates at 95% efficiency. The dropout voltage of the MAX667 is less

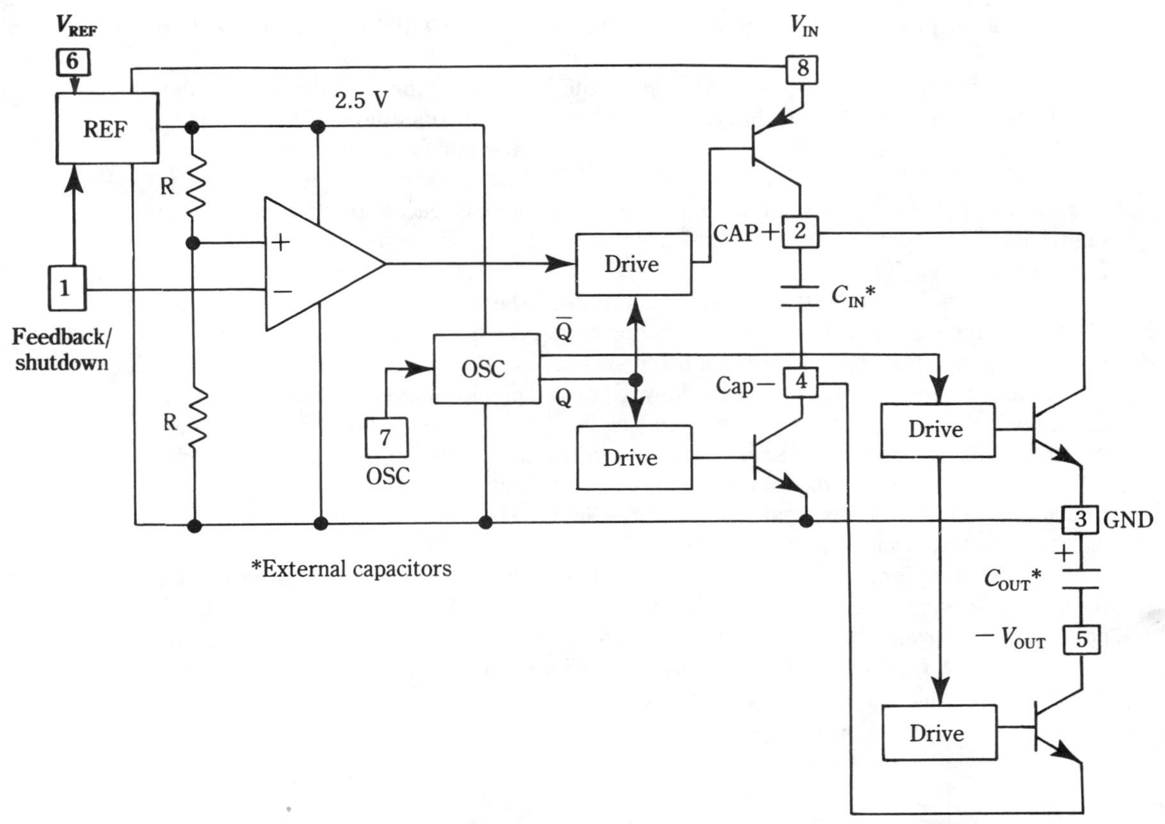

20-14 The block diagram of the LT1054 switched-capacitor voltage converter and regulator. Being essentially a charge-pump circuit, no inductor is needed. Linear Technology Corporation.

than 100 mV at 100 mA. It can be readily seen that the quoted performance from a lithium cell could not be attained via the use of conventional IC regulators—the efficiency would be too low, the dropout voltage would be too high.

Almost-free power for analog circuits in digital systems

A common laboratory experience is to be evaluating a logic system with the aid of a 5-V regulated power supply, suitable in all respects, except for the absence of low-current 12-V outputs for the requirements of analog ICs and circuitry. Generally, the need for close regulation is not paramount and usually the power demand is a small fraction of that needed for the logic system itself. Yet another generalized statement is in order when this situation is encountered—the 5-V supply is likely to be a buck-type switch-mode circuit. Assuming there is a convergence of these generalities, a simple means might exist for providing one or more 12-V auxiliary outputs with minimal modification of the supply.

A secondary winding on the inductor of the back regulator provides the needed supply voltage(s). Fine wire can be used and a little experimentation is in order to determine the volts per turn. A simple half-wave rectifier and filter capacitor suffice for the required dc. This arrangement is best-suited to those situations where the 5-V load is both heavy and fairly constant; these are the commonly encountered conditions in digital logic systems ("constant load" here refers to the average current demanded from the 5-V supply).

As long as the new load is a small fraction of the 5-V load, there should be no electrical difficulties resulting from this minor surgery. This modification is particularly easy to implement if the inductor is a toroidal type. Various arrangements of this basic theme may be used as circumstances allow. For example, full-wave rectification may be used. Or, as shown in the example circuit of Fig. 20-15, a 7-V output can be implemented, which then can be series-connected to the 5-V output to produce the sought 12-V output. Also, Zener-diode regulation of the 12-V output is a worthwhile consideration if the current of the analog circuitry is light. The basic idea is not to try for excessive power drain—this is basically a low-power technique.

It is only natural to ponder the use of this technique with boost, flyback, and polarity-inverting switchmode supplies. A little reflection shows that the technique is not applicable to such supplies because they do not normally incorporate a load-current carrying inductor. However, the filter choke in a linear power supply can be used in this manner to provide a tiny bit of auxiliary dc power.

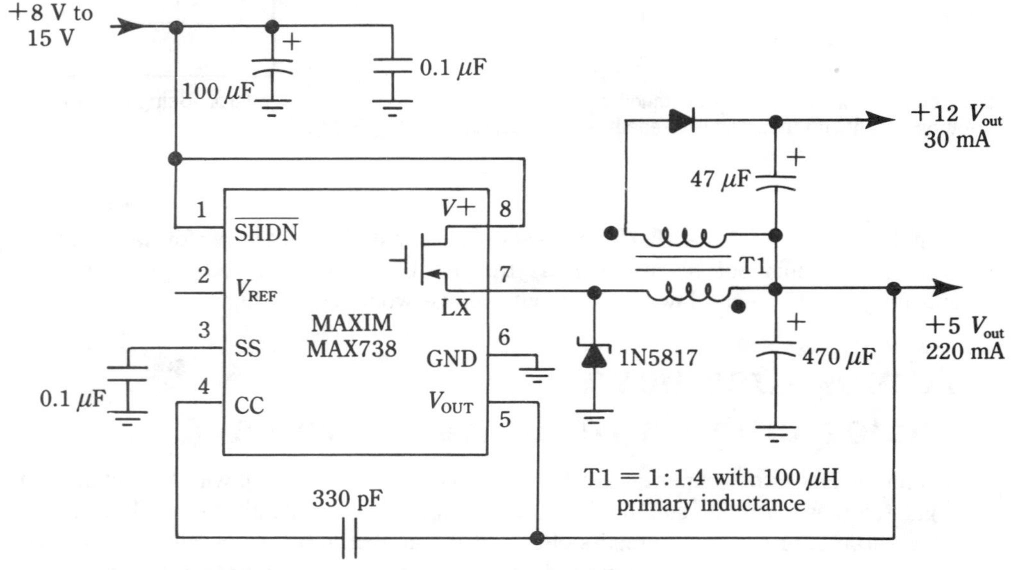

20-15 An example of an auxiliary low-power source derived from a buck regulator. Many 5-V supplies lack low-power 12-V outputs for operating analog circuits that are associated with the 5-V logic. The extra winding on the inductor of the buck regulator provides an easy solution to this problem. MAXIM Integrated Products, Inc.

Extending battery life in portable instruments using LCD readouts

Portable instruments, lap-top computers, and the like, usually make use of liquid-crystal display (LCD) readouts. In order to be user-friendly, such readouts should be clearly visible under ordinary room-ambient lighting levels. Backlighting of the display has enabled this objective to be attained, but too often the backlighting illumination system has been a major culprit in limiting battery life. The most efficient optical device for producing LCD backlighting has been found to be the *cold-cathode fluorescent lamp (CCFL)*. In order to make the best use of the CCFL, special attention must be focused on the design of its power supply, where much of the wasted battery current has hitherto been consumed.

There are a number of CCFLs with various ratings. In general, best overall operation of these lamps occurs with 300 to 400 Vac at a frequency ranging from 20 to 100 kHz. It might be argued that a square wave would be most suitable, but the RFI implications make sine-wave operation the better choice in actual practice (the lamp behaves as a good antenna for the high-frequency harmonics in square waves). At the indicated range, there is no flicker or appreciable degradation of light quality from the use of sinusoidal excitation. It should be mentioned that RFI is not the only waveshape consideration—dc or any dc component in the waveshape supplied to these lamps can seriously reduce their lifespan—even though a quick assessment might suggest that they perform well with dc or with any waveshape.

Previously designed power source for the CCFL backlighting lamp have had current capabilities of 5 mA or more. This is fine from the standpoint of light intensity, but it turns out that good results will generally be obtained with a maximum current of about 1 mA. At this lower current level, less demand is made on the battery. Another shortcoming of erstwhile CCFL power supplies is the high quiescent current—current consumed by the supply for its own internal operation that never gets to the lamp.

It appears clear that our objective to extend battery life by optimizing the efficiency of the backlighting system should be quite straightforward. It must be remembered, however, that these lamps require an initial starting-voltage in the vicinity of 600 V. Also, it is desirable that their light intensity be smoothly controllable from maximum down to near extinction (anyone who has had experience with home light dimmers can appreciate this design specification).

The schematic circuit of an efficient, low-current CCFL supply is shown in Fig. 20-16. Actually, this system is in the nature of an inverter because it operates from a 2- to 6-V battery source and delivers near-sinusoidal ac to the fluorescent lamp. It is well to mention, too, that from the standpoint of the lamp, the inverter acts as a constant-current source. This has two desirable attributes. First, such operation enables the lamp's intensity to be smoothly varied without danger of de-ionization and the annoying hysteresis that would thereby result. Second, constant-current operation swamps out the negative-resistance characteristic of the lamp, avoiding the feedback instability that could otherwise take place.

The circuitry associated with Q1 and Q2 forms a saturable-core oscillator. Notice the presence of inductor L2 in the dc feedpoint of this oscillator. This inductor causes

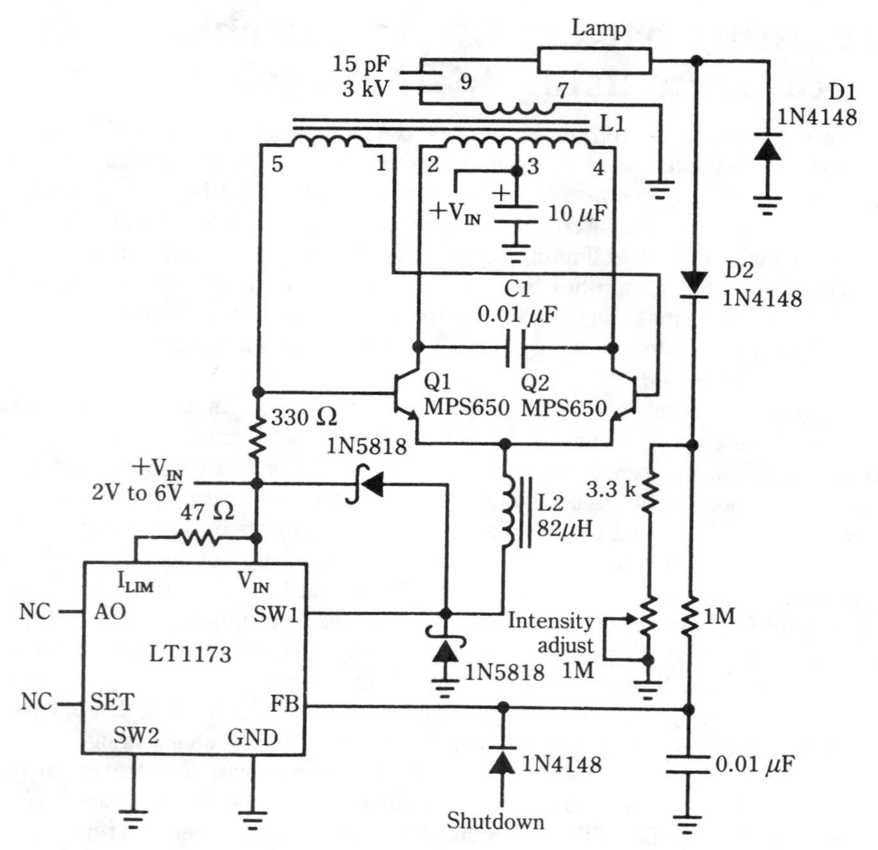

C1 = Must be a low loss capacitor.
Metalized polycarb
WIMA FKP2 (German) recommended.
L1 = SUMIDA-6345-020 or coiltronics-CTX110092-1.
Pin numbers shown for coiltronics units
L2 = TOKO 262LYF-0091K
DO NOT SUBSTITUTE COMPONENTS

20-16 Extending battery-life in portable instruments using LCD readouts. The features of this unqiue design are high efficiency, low RFI, low quiscent current, and smooth control of light intensity. Linear Technology Corporation.

the oscillator to behave as though it was current driven, rather than voltage driven; it also permits resonating the transformer with tank capacitor C1 to convert the normally generated square wave into a sine wave. Also, it will be seen that a small capacitor, about 15 pF, is inserted in series with the lamp; this provides additional ballast action to stabilize the lamp current.

The remainder of the circuit has to do with the control and regulation of the direct current allowed to pass through L1 to operate the saturable-core oscillator. As you would now suspect, this is primarily the task of the control IC, the LT1173. A lamp-current control range of 1 mA down to 1 μA is feasible with this overall arrangement. Thus, the light intensity can be manually adjusted from a maximum to virtually zero.

The LT1173 is not the conventional PWM controller. Rather, it operates on the burst-modulation principle. In this mode, its operation is intermittent—periods of near shut-down are interspaced with active periods in which pulses of dc are delivered to the load. Regulation is achieved by varying the duty cycle of these inactive and active periods. The salient feature of this operational mode is that the quiescent current of the IC is very low, compared to the situation in a conventional PWM controller. Notice that there is no relationship between the switching rate of the burst-modulation IC and the oscillation frequency of the saturable-core oscillator. The internal oscillator to the LT1173 IC is set at about 25 kHz. As explained, this oscillator is gated on or off for intervals governed by the sampled output voltage in the regulation feedback loop.

Linear regulator with automatic adaptation to ac line voltage

Inability to handle a wide range of input voltages is a notable shortcoming of the linear voltage regulator. Thus, ordinarily, if it was desired to operate a linear regulator from different ac line voltages, some kind of strapping or switching arrangement would be necessary in order to maintain the dc input voltage to the regulator at about the same level. However, the linear-regulator system about to be described can derive its input power from either 110- or a 220-Vac line without incurring dissipation problems.

Referring to Fig. 20-17, it is clear that if the dc voltage developed across the 5000-μF capacitor could be kept the same for either a 110- or 220-Vac line, the operation of the LT317A linear regulator would remain unaltered. This, indeed, is what is automatically done by the additional circuitry associated with the LT317A. To see how this is accomplished, it is best to first focus on transformer T1, and what might initially appear to be a full-wave rectifier connected to it. Such a perception would be misleading—the association of the SCR and the lN4002 diode with the transformer is not that of a full-wave rectifying circuit. Rather, it is two half-wave circuits—one using the SCR, the other using the diode. Notice the interesting logic between these two rectifiers; if the SCR is in its off state, the diode provides rectification of half of the total secondary voltage of the transformer. If the SCR is gated to its on state, the diode rectifier will be back-biased and thereby rendered inactive. Under this condition, the SCR rectifies the total secondary voltage.

Once this interaction is grasped, it is easy to see how the system operates. The LT1011 comparator monitors the ac voltage at the secondary of power transformer T1. When this voltage is relatively low because of a 110-Vac line, the comparator's output is high. This allows positive bias to reach the gate of the SCR, turning it on. The SCR then acts as a rectifier for the full secondary voltage of T1. This action, at the same time, back-biases the lN4002 diode, and effectively removes it from the circuit. Conversely, when a 220-Vac line is used, the comparator is driven to its low state. This shunts down the positive-bias supply for the SCR, which now remains in its off state. Rectification is, under this condition, performed by the lN4002 diode, which delivers half the voltage

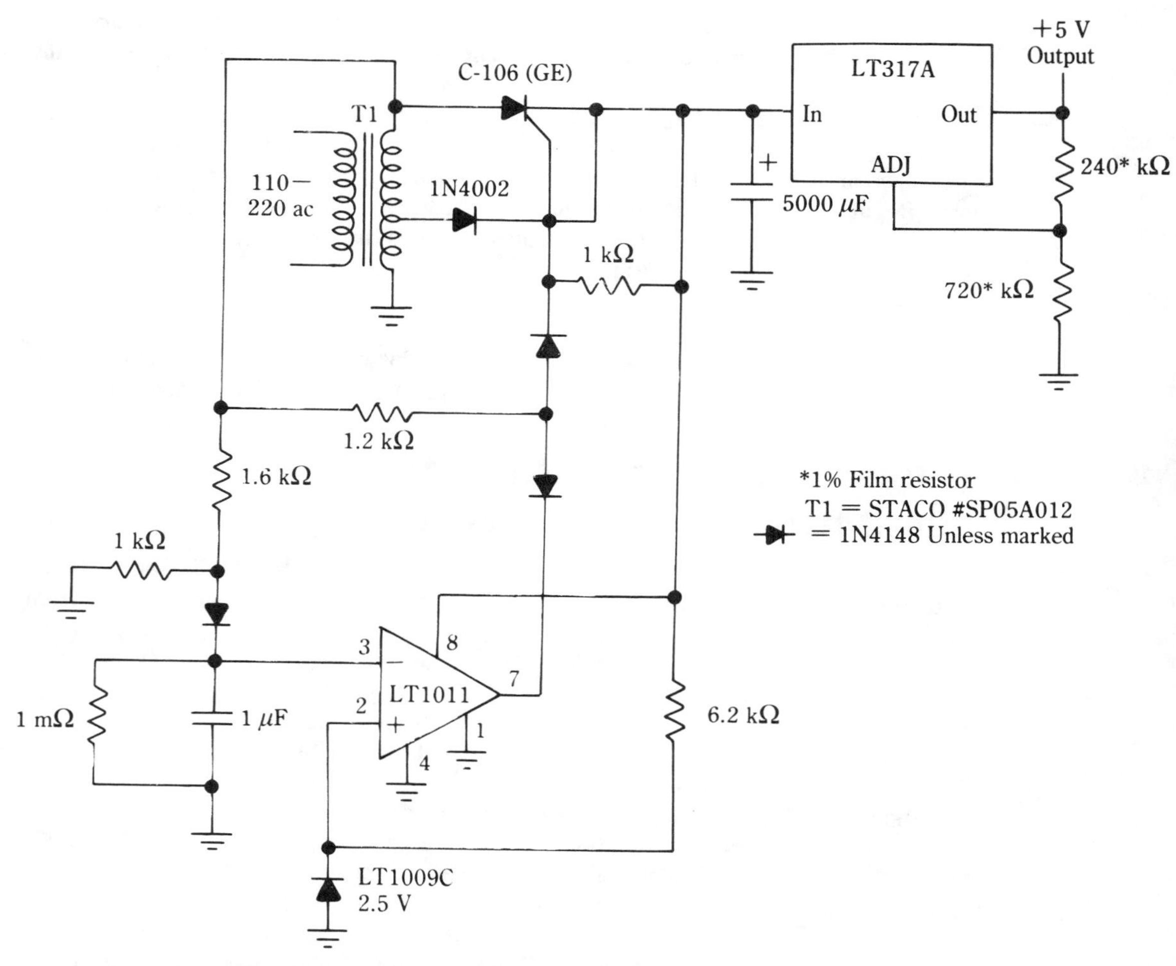

20-17 A linear regulator that operates equally well form 110- or 220-V ac line. No tap changing or other circuit alteration is needed—the accommodation to either line voltage is automatic. The linear regulator sees the same dc input-voltage range for either type of ac line. Linear Technology Corporation.

induced in the secondary of T1. Therefore, whether 110 or 220 Vac is applied to the primary of T1, the 5000-μF capacitor and the input of the linear regulator receive the same dc voltage.

The prevention of excessive dissipation in linear regulators

Although the switching-type regulator has gradually assumed domination over linear regulators, many applications benefit from the low-noise and fast response readily attainable only when the linear technique of regulation is used. Also, wherever feasible, many designers have "bent over backwards" to use linear regulation, as much as its poor efficiency could be justified. In situations where the input voltage can be reliably held at a near-constant level and the pass transistor can be operated on a low-dropout basis, the efficiency of such a linear regulator can often be acceptable. The best of both worlds' approach has already been covered; this involves the use of both switching and linear regulation. Usually, the switching regulator serves as a preregulator so that the output linear regulator can operate with very small voltage drop and thereby be spared the burden of wasting power it would otherwise have forced on it. Notice that in such an arrangement, the load can be quite well isolated from the noise-generating preregulator.

It is instructive to investigate a couple of examples of this basic idea of keeping the input voltage to the linear regulator low—not too much higher than the voltage-regulated output that is required by the load. Most recently, the wide-spread application of electronics to consumer products appears to have imparted renewed interest in such combination regulator systems.

The circuit shown in Fig. 20-18 exemplifies the objective of preventing high-dissipation in a linear voltage regulator, while, at the same time, benefiting from its desirable performance features. Initial inspection might suggest that this circuit is simply a cascaded arrangement of an input switching regulator followed by an output linear regulator. This, however, is not the case. Rather, a unique operational mode prevails—one that is likely to find expanded use in forthcoming power-supply technology. The basic idea here is to servo-control the voltage-drop, V_Z, across the linear regulator. Specifically, operation is such that V_Z is forced to remain constant at 3.7 V over a wide range of dc input voltage (3.7 V is a "comfortable" drop for the LT350A linear regulator, so it can operate properly while incurring minimal power dissipation).

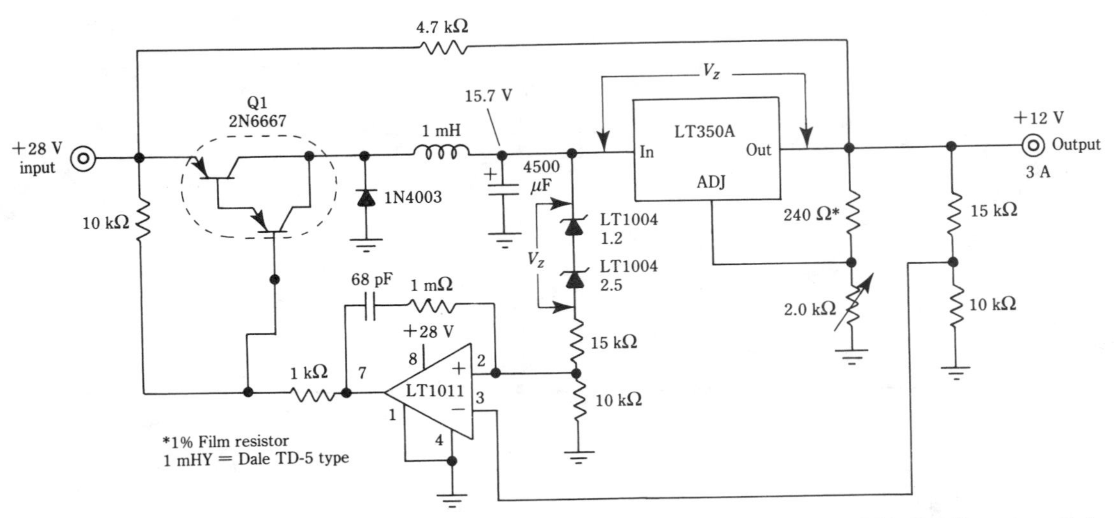

20-18 The circuit for holding dissipation of a linear regulator at a minimal level. The Darlington switching-stage operates in such a manner that charge is delivered to the 4500-μF capacitor at a rate that maintains V_Z constant over a wide range of dc input voltage. Linear Technology Corporation.

Focus now on the two 15 to 10-kΩ voltage-sampling networks, the LT1004 voltage references, the LT1011 voltage comparator, the Darlington switching transistor (Q1), and finally, on the large 4500-μF capacitor at the input of the linear voltage regulator. These components are involved in a feedback loop, which pours charge into the 4500-μF capacitor at a rate just sufficient to maintain V_Z at 3.7 V.

The comparator, because of its position in the feedback loop, tries to zero the voltage appearing across its input terminals. This occurs when both input terminals sample 3.7 V, with respect to ground. This, in turn, is true when the dc input voltage to the linear regulator is 15.7 V (the 12-V output plus the 3.7 V of the LT1004 level-shifting references). The important aspect of this balance is that V_Z is automatically maintained at the minimal level of 3.7 V.

Suppose that this balance exists, but that the 28-Vdc input increases to a higher value. Q1 has been in its on state, but now, because of the higher voltage appearing at the noninverting terminal (2) of the comparator, its output goes high, thereby turning Q1 off. This allows the voltage across the 4500-μF capacitor to commence a discharge cycle. It will discharge to the extent that the comparator goes low, in which case Q1 is again turned on, and the 4500-μF capacitor undergoes a charge cycle. The switching of Q1 continues and the dc input voltage to the linear regulator is thereby servo-controlled at a level very close to 15.7 V. Notice that Q1 is neither a driven nor a self-oscillatory switching stage in the conventional sense. Rather, it changes its conductive state only when deviation from 15.7-V input to the linear regulator is sensed.

In this operation, the linear regulator stabilizes its own dc output voltage in a normal manner, except that V_Z is servo controlled. This, of course, makes the linear regulator "happy"—its internal dissipation is limited to a safe and efficient level, and its already-tight regulation is enhanced because of the constant "line" voltage it sees. As

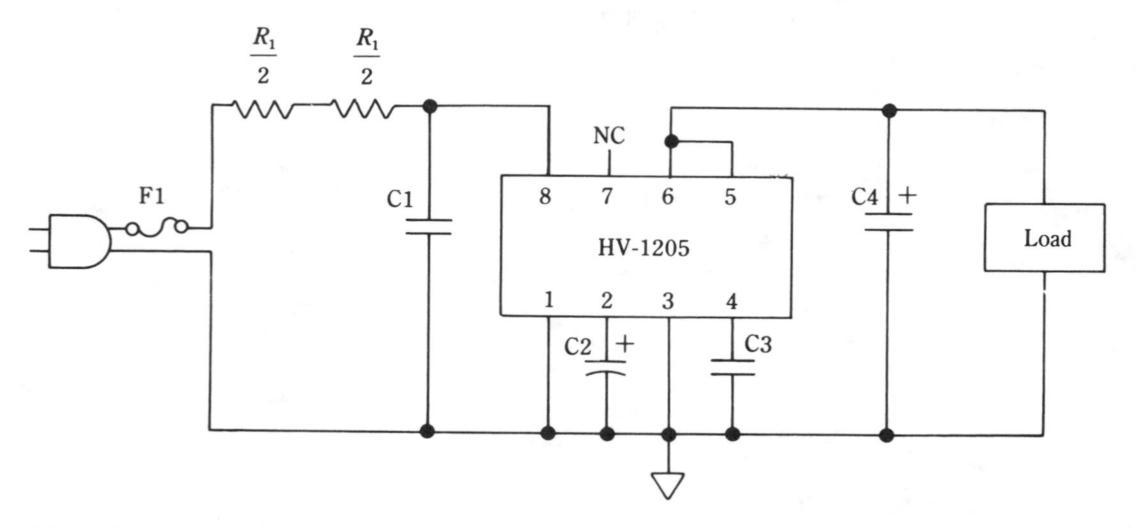

20-19 Direct operation from ac line using a dedicated IC linear regulator. Low-dissipation accommodation to the line voltage takes place because of a switching regulator also incorporated in the IC. The circuit conditions are: Input: 120 V, 60 Hz, Output: regulated 5 V, at 50 mA, R1: 150 Ω, C1: 0.05 μF, C2: 470 μF, C3: 150 pF, C4:I-μF. Harris Commercial Products Group.

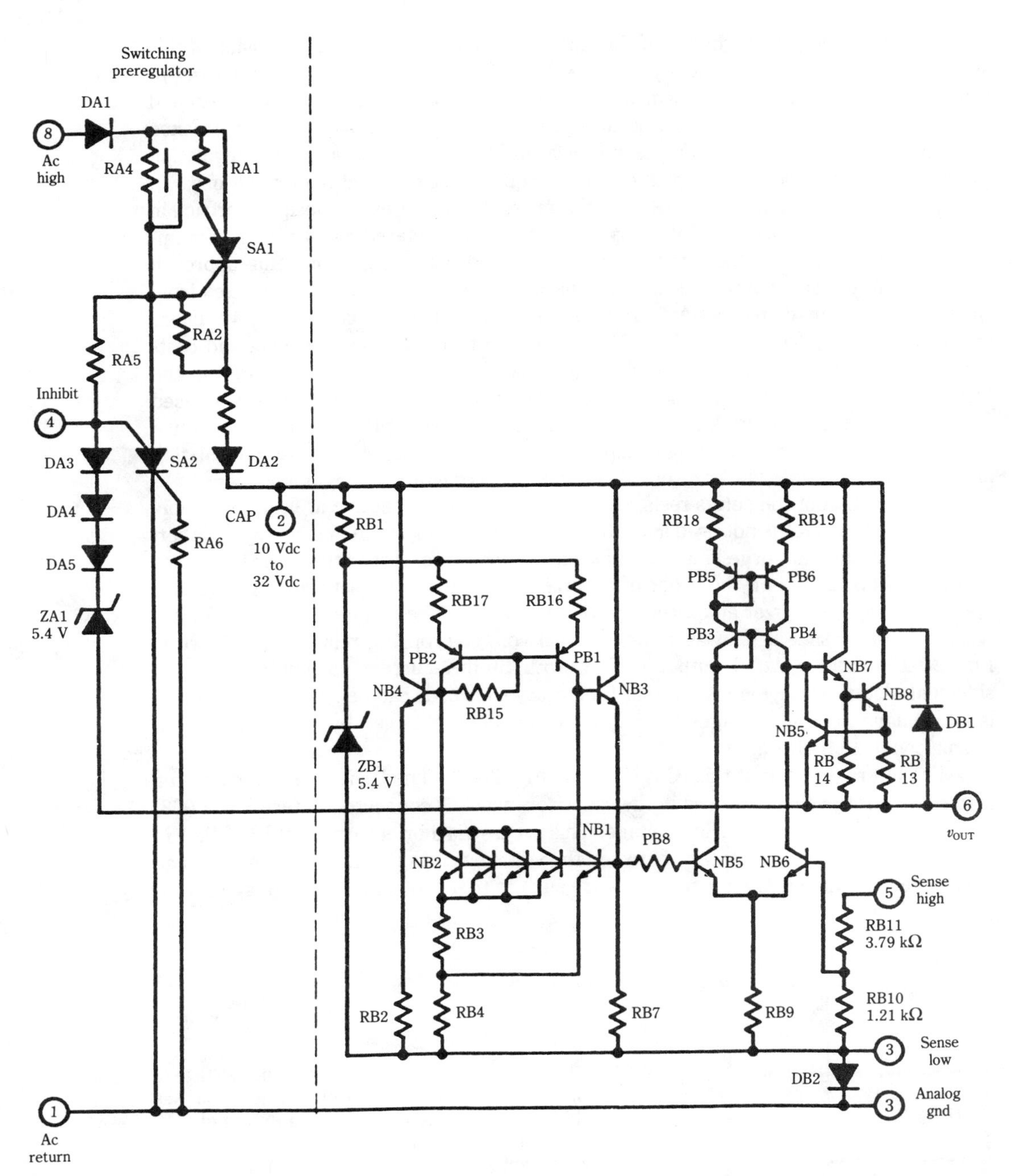

20-20 Internal circuitry of the HV-205 versatile voltage regulator. Operation is premissed n the once-per-cycle replenishment of charge in the external capacitor connected to CAP pin 2. Servo-action maintains the voltage in this capacitor about 6 V above the regulator output voltage. Harris Commercial Products Group.

would be suspected, the high gain of the comparator contributes to the precision of this scheme. In the interest of stability, an 80-mV hysteresis characteristic is imparted to the comparator's switching performance by the RC network, bridging its non-invert and output terminals. This regulation system supplies up to 3-A to 12-V loads, in accordance with the ratings of the LT350A linear regulator. It is easy to see that this regulation scheme can readily be scaled up or down as required for different load situations.

The setup shown in Fig. 20-19, which represents an interesting design approach in that circuit and performance philosophy, is primarily dictated by practical circumstances, rather than by the designer's notions of idealized behavior. This approach allows short-cuts and compromises that generally invoke only modest departures from rigorous operating demands. At the same time, the user is happy to have linear regulation directly from the ac line. Moreover, the method is space-saving and cost-effective. This is a linear voltage regulator preceded by a switching stage to present the excessive power dissipation, which would occur if a brute-force technique was used (that is, if the total required drop in line voltage was absorbed by resistance). Notice also that no transformer or bridge rectifier is needed. Of course, there is no isolation provided between line and load, but this is not a detriment in many applications.

The function of the series resistance, R_1, tends to be deceptive at first inspection. As inferred above, R1 is not used in brute-force fashion to drop the line voltage. Rather, it serves as a current-surge limiter. To be sure, some power dissipation does take place in R1, but this is only on the order of several watts. Moreover, satisfactory regulator operation can be had over an extremely wide range of ac line voltage: 28 to 132 V; this would not be feasible if R1 was a conventional voltage-dropping resistor. As shown in Fig. 20-19, the regulated output is 5 V with 50-mA current capability. There is sufficient electrical ruggedness so that temporary overloads can be readily endured. As is later shown, output voltages (at 50 mA) can also be obtained by means of a slight circuit connection alteration.

The internal circuit of this IC is shown in Fig. 20-20. The basic idea of the scheme is to replenish the charge stored in the capacitor, which is connected to terminal 2 (C2) once for each cycle of the ac line voltage. This occurs via a brief turn-on of PUT SA1 in response to the sensed difference between the voltage across C2 and the output voltage of the linear regulator. Circuit constants are such as to servo-control the voltage across

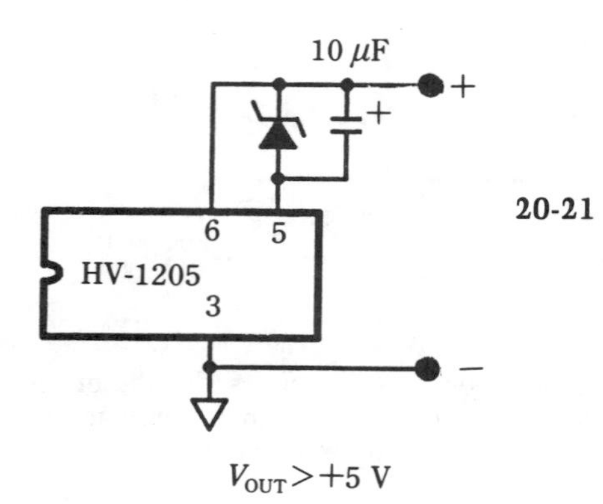

20-21 The technique for increasing the output voltage of the HV-1205 IC. Insertion of a Zener diode, as shown, enables output voltages in excess of 5 V. The value of output voltage so obtained is the zener voltage plus 5 V. Harris Commercial Products Group.

C2 so that it tends to be about 6 V greater than the regulator output voltage. The apparent complexity of the circuit is in good part caused by the number of series and parallel-connected elements, a practice common in IC architecture.

The regulated output voltage of the HV-1205 IC can be increased beyond 5-V to an allowable 24 V. Although this can be accomplished by inserting resistance in the pin 5 lead, a better way is to insert a Zener diode, as depicted in Fig. 20-21. The output voltage will then be the zener voltage plus 5-V. For practical purposes, the precision of the output voltage obtained will be governed by the Zener diode. The 10-μF capacitor prevents too-rapid ramp-up of the storage capacitor connected to pin 2, and is essentially a protective measure for the IC. The 50-mA current rating prevails for all output voltages.

Index